Advancement in the Cancer Treatment

Advancement in the Cancer Treatment

Edited by:

Preethi Kartan

Delve
PUBLISHING

www.delvepublishing.com

Advancement in the Cancer Treatment

Preethi Kartan

Delve Publishing
224 Shoreacres Road
Burlington, ON L7L 2H2
Canada
www.delvepublishing.com
Email: orders@arclereducation.com

ISBN: 978-1-77407-786-3 (Hardcover)

DECLARATION

Some content or chapters in this book are open access copyright free published research work, which is published under Creative Commons License and are indicated with the citation. We are thankful to the publishers and authors of the content and chapters as without them this book wouldn't have been possible.

ABOUT THE EDITOR

Preethi is a postgraduate in Biotechnology from University of Leeds, UK. She is currently working as a Scientific Associate at one of the CRO's in India and her interest lie in life sciences related writing.

TABLE OF CONTENTS

LIST OF CONTRIBUTORS

Stephan Kruger
Department of Medicine III, University Hospital Munich, LMU Munich, Marchioninistr. 15, D-81377 Munich, Germany
Center of Integrated Protein Science Munich (CIPS-M) and Division of Clinical Pharmacology, Department of Medicine IV, University Hospital, LMU Munich, Munich, Germany

Matthias Ilmer
Department of General, Visceral, and Transplantation Surgery, University Hospital, LMU Munich, Munich, Germany
German Cancer Consortium (DKTK), Partner Site Munich; and German Cancer Research Center (DKFZ), Heidelberg, Germany

Sebastian Kobold
Center of Integrated Protein Science Munich (CIPS-M) and Division of Clinical Pharmacology, Department of Medicine IV, University Hospital, LMU Munich, Munich, Germany

Bruno L. Cadilha
Center of Integrated Protein Science Munich (CIPS-M) and Division of Clinical Pharmacology, Department of Medicine IV, University Hospital, LMU Munich, Munich, Germany

Stefan Endres
Center of Integrated Protein Science Munich (CIPS-M) and Division of Clinical Pharmacology, Department of Medicine IV, University Hospital, LMU Munich, Munich, Germany

Steffen Ormanns
Institute of Pathology, LMU Munich, Munich, Germany

Gesa Schuebbe
Department of Medicine III, University Hospital Munich, LMU Munich, Marchioninistr. 15, D-81377 Munich, Germany

Bernhard W. Renz
Department of General, Visceral, and Transplantation Surgery, University Hospital,

LMU Munich, Munich, Germany
German Cancer Consortium (DKTK), Partner Site Munich; and German Cancer Research Center (DKFZ), Heidelberg, Germany

Jan G. D'Haese
Department of General, Visceral, and Transplantation Surgery, University Hospital, LMU Munich, Munich, Germany

Hans Schloesser
University Hospital of Cologne, Cologne, Germany

Volker Heinemann
Department of Medicine III, University Hospital Munich, LMU Munich, Marchioninistr. 15, D-81377 Munich, Germany
German Cancer Consortium (DKTK), Partner Site Munich; and German Cancer Research Center (DKFZ), Heidelberg, Germany

Marion Subklewe
Department of Medicine III, University Hospital Munich, LMU Munich, Marchioninistr. 15, D-81377 Munich, Germany
German Cancer Consortium (DKTK), Partner Site Munich; and German Cancer Research Center (DKFZ), Heidelberg, Germany
Gene Center LMU, Munich, Germany.

Stefan Boeck
Department of Medicine III, University Hospital Munich, LMU Munich, Marchioninistr. 15, D-81377 Munich, Germany
German Cancer Consortium (DKTK), Partner Site Munich; and German Cancer Research Center (DKFZ), Heidelberg, Germany

Jens Werner
Department of General, Visceral, and Transplantation Surgery, University Hospital, LMU Munich, Munich, Germany

Michael von Bergwelt-Baildon
Department of Medicine III, University Hospital Munich, LMU Munich, Marchioninistr. 15, D-81377 Munich, Germany
German Cancer Consortium (DKTK), Partner Site Munich; and German Cancer Research Center (DKFZ), Heidelberg, Germany
Center for Molecular Medicine Cologne (CMMC), Cologne, Germany
Gene Center LMU, Munich, Germany.

Zihai Li
Hollings Cancer Center, Medical University of South Carolina, Charleston, SC 29425,

USA
Chinese American Hematologist and Oncologist Network, New York, NY, USA

Wenru Song
Chinese American Hematologist and Oncologist Network, New York, NY, USA

Mark Rubinstein
Hollings Cancer Center, Medical University of South Carolina, Charleston, SC 29425, USA

Delong Liu
Chinese American Hematologist and Oncologist Network, New York, NY, USA
New York Medical College, New York, NY, USA

Chongxian Pan
Chinese American Hematologist and Oncologist Network, New York, NY, USA
University of California, Davis, CA, USA

Hongtao Liu
Chinese American Hematologist and Oncologist Network, New York, NY, USA
University of Chicago, Chicago, IL, USA

Elizabeth Robins
Pelotonia Institute for Immuno-Oncology, The Ohio State University, Columbus, OH, USA

Wenru Song
Chinese American Hematologist and Oncologist Network, New York, NY, USA
Kira Pharmaceuticals, Cambridge, MA, USA

Delong Liu
Chinese American Hematologist and Oncologist Network, New York, NY, USA
New York Medical College, Valhalla, NY, USA

Zihai Li
Chinese American Hematologist and Oncologist Network, New York, NY, USA
Pelotonia Institute for Immuno-Oncology, The Ohio State University, Columbus, OH, USA

Lei Zheng
Chinese American Hematologist and Oncologist Network, New York, NY, USA
Johns Hopkins University, Baltimore, MD, USA

Julian A. Marin-Acevedo
Department of Internal Medicine, Mayo Clinic, Jacksonville, FL, USA

Aixa E. Soyano
Department of Hematology and Oncology, Mayo Clinic, 4500 San Pablo Road, Jacksonville, FL 32224, USA

Bhagirathbhai Dholaria
Department of Hematology and Oncology, Mayo Clinic, 4500 San Pablo Road, Jacksonville, FL 32224, USA
Current address: Department of Blood and Marrow Transplant and Cellular Immunotherapy, Moffitt Cancer Center, Tampa, FL, USA

Keith L. Knutson
Department of Immunology, Mayo Clinic, Jacksonville, FL, USA

Yanyan Lou
Department of Hematology and Oncology, Mayo Clinic, 4500 San Pablo Road, Jacksonville, FL 32224, USA

Weijing Sun
University of Pittsburgh Cancer Institute, 5150 Centre Avenue, Fifth Floor, Pittsburgh, PA 15232, USA

Yanyu Pang
Department of Dermatology, Affiliated Hospital of Xuzhou Medical University, Xuzhou 221002, China

Xiaoyang Hou
Department of Dermatology, Affiliated Hospital of Xuzhou Medical University, Xuzhou 221002, China

Chunsheng Yang
Department of Dermatology, Affiliated Huai'an Hospital of Xuzhou Medical University, the Second People's Hospital of Huai'an, Huai'an 223002, China

Yanqun Liu
Department of Dermatology, Affiliated Hospital of Xuzhou Medical University, Xuzhou 221002, China

Guan Jiang
Department of Dermatology, Affiliated Hospital of Xuzhou Medical University, Xuzhou 221002, China

Felix S. Lichtenegger
Department of Medicine III, University Hospital, LMU Munich, Germany
Laboratory of Translational Cancer Immunology, Gene Center, Munich, Germany

Christina Krupka
Department of Medicine III, University Hospital, LMU Munich, Germany
Laboratory of Translational Cancer Immunology, Gene Center, Munich, Germany

Sascha Haubner
Department of Medicine III, University Hospital, LMU Munich, Germany
Laboratory of Translational Cancer Immunology, Gene Center, Munich, Germany

Thomas Köhnke
Department of Medicine III, University Hospital, LMU Munich, Germany
Laboratory of Translational Cancer Immunology, Gene Center, Munich, Germany

Marion Subklewe
Department of Medicine III, University Hospital, LMU Munich, Germany
Laboratory of Translational Cancer Immunology, Gene Center, Munich, Germany
German Cancer Consortium (DKTK), Partner Site, Munich, Germany
German Cancer Research Center (DKFZ), Heidelberg, Germany

Masanobu Chinami
BFSR Institute, Fukuoka, Japan

Kaoru Iwabuchi
BFSR Institute, Fukuoka, Japan

Yoshiteru Muto
The Research Institute of Health Rehabilitation of Tokyo, Tokyo, Japan

Yasuhiko Uchida
The Research Institute of Health Rehabilitation of Tokyo, Tokyo, Japan

Ryu Arita
Fukuoka MSC Medical Clinics, Fukuoka, Japan

Rana A. Shuraim
BFSR Institute, Fukuoka, Japan

Chaker N. Adra
BFSR Institute, Fukuoka, Japan
The Adra Institute, Boston, MA, USA

Jinsha Koroth
Institute of Bioinformatics and Applied Biotechnology, Electronic City Phase 1, Bangalore, Karnataka 560100, India
JK, SN, and VG are graduate students registered under Manipal Academy of Higher Education, Manipal 576104, India

Snehal Nirgude
Institute of Bioinformatics and Applied Biotechnology, Electronic City Phase 1, Bangalore, Karnataka 560100, India
JK, SN, and VG are graduate students registered under Manipal Academy of Higher Education, Manipal 576104, India

Shweta Tiwari
Department of Pharmaceutical Chemistry, KLE Academy of Higher Education and Research, KLE College of Pharmacy, Rajajinagar, Bangalore, KN, India.

Vidya Gopalakrishnan
Institute of Bioinformatics and Applied Biotechnology, Electronic City Phase 1, Bangalore, Karnataka 560100, India
JK, SN, and VG are graduate students registered under Manipal Academy of Higher Education, Manipal 576104, India
Department of Biochemistry, Indian Institute of Science, Bangalore 560012, India

Raghunandan Mahadeva
Institute of Bioinformatics and Applied Biotechnology, Electronic City Phase 1, Bangalore, Karnataka 560100, India

Sujeet Kumar
Department of Pharmaceutical Chemistry, KLE Academy of Higher Education and Research, KLE College of Pharmacy, Rajajinagar, Bangalore, KN, India.

Subhas S. Karki
Department of Pharmaceutical Chemistry, KLE Academy of Higher Education and Research, KLE College of Pharmacy, Rajajinagar, Bangalore, KN, India.

Bibha Choudhary
Institute of Bioinformatics and Applied Biotechnology, Electronic City Phase 1, Bangalore, Karnataka 560100, India

Ellas Spyratou
2nd Department of Radiology, Medical School, National and Kapodistrian University of Athens, 12462 Athens, Greece

Mersini Makropoulou
Department of Physics, School of Applied Mathematical and Physical Sciences, National Technical University of Athens, 15780 Athens, Greece

Efstathios P. Efstathopoulos
2nd Department of Radiology, Medical School, National and Kapodistrian University of Athens, 12462 Athens, Greece

Alexandros G. Georgakilas
Department of Physics, School of Applied Mathematical and Physical Sciences, National Technical University of Athens, 15780 Athens, Greece

Lembit Sihver
Atominstitut, Technische Universität Wien, Stadionallee 2, 1020 Vienna, Austria

Young Kwang Chae
Developmental Therapeutics Program of the Division of Hematology Oncology, Early Phase Clinical Trials Unit, 645 N. Michigan Avenue, Suite 1006, Chicago, IL 60611, USA
Robert H. Lurie Comprehensive Cancer Center of Northwestern University, 645 N. Michigan Avenue, Suite 1006, Chicago, IL 60611, USA
Northwestern University Feinberg School of Medicine, 645 N. Michigan Avenue, Suite 1006, Chicago, IL 60611, USA.

Ayush Arya
Developmental Therapeutics Program of the Division of Hematology Oncology, Early Phase Clinical Trials Unit, 645 N. Michigan Avenue, Suite 1006, Chicago, IL 60611, USA

Wade Iams
Northwestern University Feinberg School of Medicine, 645 N. Michigan Avenue, Suite 1006, Chicago, IL 60611, USA.

Marcelo R. Cruz
Developmental Therapeutics Program of the Division of Hematology Oncology, Early Phase Clinical Trials Unit, 645 N. Michigan Avenue, Suite 1006, Chicago, IL 60611, USA

Sunandana Chandra
Developmental Therapeutics Program of the Division of Hematology Oncology, Early Phase Clinical Trials Unit, 645 N. Michigan Avenue, Suite 1006, Chicago, IL 60611, USA
Robert H. Lurie Comprehensive Cancer Center of Northwestern University, 645 N. Michigan Avenue, Suite 1006, Chicago, IL 60611, USA

Northwestern University Feinberg School of Medicine, 645 N. Michigan Avenue, Suite 1006, Chicago, IL 60611, USA.

Jaehyuk Choi
Robert H. Lurie Comprehensive Cancer Center of Northwestern University, 645 N. Michigan Avenue, Suite 1006, Chicago, IL 60611, USA
Northwestern University Feinberg School of Medicine, 645 N. Michigan Avenue, Suite 1006, Chicago, IL 60611, USA.

Francis Giles
Developmental Therapeutics Program of the Division of Hematology Oncology, Early Phase Clinical Trials Unit, 645 N. Michigan Avenue, Suite 1006, Chicago, IL 60611, USA
Robert H. Lurie Comprehensive Cancer Center of Northwestern University, 645 N. Michigan Avenue, Suite 1006, Chicago, IL 60611, USA
Northwestern University Feinberg School of Medicine, 645 N. Michigan Avenue, Suite 1006, Chicago, IL 60611, USA.

Yasumasa Ishida
Division of Biological Science, Nara Institute of Science and Technology, 8916-5 Takayama-cho, Ikoma-shi, Nara 630-0192, Japan

Gabriele D'Errico
Department of Biochemistry, School of Medicine, Autónoma University of Madrid, Calle del Arzobispo Morcillo 4, 28029 Madrid, Spain

Heather L. Machado
Department of Biochemistry and Molecular Biology, Tulane University School of Medicine, 1430 Tulane Ave, #8543, New Orleans, LA 70112, USA

Bruno Sainz Jr.
Department of Biochemistry, School of Medicine, Autónoma University of Madrid, Calle del Arzobispo Morcillo 4, 28029 Madrid, Spain
Department of Cancer Biology, Instituto de Investigaciones Biomédicas "Alberto Sols" CSIC-UAM, Madrid, Spain
Enfermedades Crónicas y Cáncer Area, Instituto Ramón y Cajal de Investigación Sanitaria (IRYCIS), Madrid, Spain.

Kirill Peskov
M&S Decisions, Moscow, Russia
Computational Oncology Group, I.M. Sechenov First Moscow State Medical University of the Russian Ministry of Health, Moscow, Russia

Ivan Azarov
M&S Decisions, Moscow, Russia

Lulu Chu
Quantitative Clinical Pharmacology, Early Clinical Development, IMED Biotech Unit, AstraZeneca Pharmaceuticals, Boston, MA, United States

Veronika Voronova
M&S Decisions, Moscow, Russia

Yuri Kosinsky
M&S Decisions, Moscow, Russia

Gabriel Helmlinger
Quantitative Clinical Pharmacology, Early Clinical Development, IMED Biotech Unit, AstraZeneca Pharmaceuticals, Boston, MA, United States

Imerio Capone
Department of Oncology and Molecular Medicine, Istituto Superiore di Sanità, Rome, Italy

Paolo Marchetti
Department of Oncology, Sapienza University of Rome, Sant'Andrea Hospital, Rome, Italy

Paolo Antonio Ascierto
Unit of Melanoma, Cancer Immunotherapy and Innovative Therapy, Istituto Nazionale Tumori Fondazione G. Pascale (IRCCS), Naples, Italy

Walter Malorni
Center for Gender-Specific Medicine, Istituto Superiore di Sanità, Rome, Italy

Lucia Gabriele
Department of Oncology and Molecular Medicine, Istituto Superiore di Sanità, Rome, Italy

Yuji Miyamoto
Department of Gastroenterological Surgery, Graduate School of Medical Sciences, Kumamoto University, 1-1-1 Honjo, Kumamoto 860-8556, Japan

Koichi Suyama
Cancer Center, Kumamoto University Hospital, Kumamoto 860-8556, Japan

Hideo Baba
Department of Gastroenterological Surgery, Graduate School of Medical Sciences, Kumamoto University, 1-1-1 Honjo, Kumamoto 860-8556, Japan

LIST OF ABBREVIATIONS

ALL	acute lymphoblastic leukemia
AML	acute myeloid leukemia
ACT	adoptive cell therapy
ABMs	agent-based models
AARC	American Association of Cancer Research
ASTCT	American Society for Transplantation and Cellular Therapy
ASCO	American Society of Clinical Oncology
ASH	American Society of Hematology
ADCs	Antibody drug conjugates
APC	antigen presenting cell
BCMA	B cell maturation antigen
BiTE	bispecific T cell engager
CDDF	Cancer Drug Development Forum
CEA	carcino-embryonic antigen
CPIs	checkpoint inhibitors
CAR	Chimeric antigen receptor
CAHON	Chinese American Hematologist and Oncologist Network
CSCO	Chinese Society of Clinical Oncology
CCA	Cholangiocarcinoma
CLL	chronic lymphocytic leukemia
CMML	chronic myelomonocytic leukemia
CRC	colorectal cancer
DAMPs	damage-associated molecular patterns
DC-CIK	dendritic cell and cytokine-induced killer cell
DCR	disease control rate
DART	dual-affinity re-targeting
EMT	Epithelial to mesenchymal transition
EPR	Enhanced Permeability and Retention

EGFR	epidermal growth factor receptor
EOC	Epithelial ovarian cancer
EREs	estrogen response elements
EHA	European Hematology Association
EMA	European Medicines Agency
ESMO	European Society for Medical Oncology
ESMO	European Society of Medical Oncology
ECD	extracellular domain
FcγRs	Fc-gamma receptors
FDA	Food and Drug Administration
GARP	Glycoprotein A Repetitions Predominant
GM-CSF	granulocyte-macrophage colonystimulating factor
HNSCC	head and neck squamous cell carcinoma
HSCT	hematopoietic stem cell transplantation
HCC	hepatocellular carcinoma
HIV	human immunodeficiency virus infection
HLA	human leukocyte antigen
ICB	immune checkpoint blockade
ICIs	immune checkpoint inhibitors
IFT	interstitial fluid pressure
LDH	Lactate dehydrogenase
LET	linear energy transfer
LEM	Local Effect Model
MHC	Major Histocompatibility Complex
MM	malignant melanoma
MPM	malignant pleural mesothelioma
NIH	National Institutes of Health
NLME	nonlinear mixed-effects
NSCLC	non-small cell lung cancer
ORR	objective response rate
ODE	ordinary differential equation
PDAC	pancreatic ductal adenocarcinoma
PTT	Photothermal therapy
PFS	progression-free survival

PREFACE

In human body, under normal conditions reproduction of cells is carefully monitored and controlled, but due to certain changes in the genome of cells, cell reproduce abnormally resulting in abnormal cell growth in the form of tumour or mass or Lump. Cancer can occur in any part of the body. Cancer is a general word to described abnormal growth of cells, specific cancer name is used based on the area cancer developed, hence cancer is not single disease it is considered as different cancer diseases, eg: Breast cancer, Lung cancer.

Treatment for cancer depends on type of cancer, locality of tumour, and its stage of progression. There are different cancer treatments available. Widely used treatment methods include Surgery, radiation-based surgical knives, chemotherapy, and radiotherapy. Few latest treatments include hormone-based therapy, anti-angiogenic modalities, stem cell therapies, and dendritic cell-based immunotherapy. Based on various parameters such as patient's age, type and location of the cancer, whether the disease has spread etc patient can be treated with single method or with a combination of treatments.

While there is lot of development seen in area of cancer treatment, still cancer continues to be a major health concern, therefore continuous efforts and extensive research are being carried out for the development of new treatment approaches.

This book gives idea of some of the new breakthroughs and discoveries made in the field of cancer treatment.

1

Advances in Cancer Immunotherapy 2019 – Latest Trends

Stephan Kruger[1,3] , Matthias Ilmer[2,6] , Sebastian Kobold[3] , Bruno L. Cadilha[3] , Stefan Endres[3] , Steffen Ormanns4 , Gesa Schuebbe[1] , Bernhard W. Renz[2,6], Jan G. D'Haese[2] , Hans Schloesser[5] , Volker Heinemann[1,6], Marion Subklewe[1,6,8], Stefan Boeck[1,6], Jens Werner[2] and Michael von Bergwelt-Baildon[1,6,7,8]

[1] Department of Medicine III, University Hospital Munich, LMU Munich, Marchioninistr. 15, D-81377 Munich, Germany

[2] Department of General, Visceral, and Transplantation Surgery, University Hospital, LMU Munich, Munich, Germany

[3] Center of Integrated Protein Science Munich (CIPS-M) and Division of Clinical Pharmacology, Department of Medicine IV, University Hospital, LMU Munich, Munich, Germany

[4] Institute of Pathology, LMU Munich, Munich, Germany

[5] University Hospital of Cologne, Cologne, Germany

[6] German Cancer Consortium (DKTK), Partner Site Munich; and German Cancer Research Center (DKFZ), Heidelberg, Germany

[7] Center for Molecular Medicine Cologne (CMMC), Cologne, Germany

[8] Gene Center LMU, Munich, Germany.

Citation: Kruger, S., Ilmer, M., Kobold, S. et al. Advances in cancer immunotherapy 2019 – latest trends. J Exp Clin Cancer Res 38, 268 (2019). https://doi.org/10.1186/s13046-019-1266-0.

ABSTRACT

Immunotherapy has become an established pillar of cancer treatment improving the prognosis of many patients with a broad variety of hematological and solid malignancies. The two main drivers behind this success are checkpoint inhibitors (CPIs) and chimeric antigen receptor (CAR) T cells. This review summarizes seminal findings from clinical and translational studies recently presented or published at important meetings or in top-tier journals, respectively. For checkpoint blockade, current studies focus on combinational approaches, perioperative use, new tumor entities, response prediction, toxicity management and use in special patient populations. Regarding cellular immunotherapy, recent studies confirmed safety and efficacy of CAR T cells in larger cohorts of patients with acute lymphoblastic leukemia or diffuse large B cell lymphoma. Different strategies to translate the striking success of CAR T cells in B cell malignancies to other hematological and solid cancer types are currently under clinical investigation. Regarding the regional distribution of registered clinical immunotherapy trials a shift from PD-1 / PD-L1 trials (mainly performed in the US and Europe) to CAR T cell trials (majority of trials performed in the US and China) can be noted.

BACKGROUND

The importance of immunotherapy has been acknowledged by the *Nobel prize for physiology or medicine* 2018 awarded for the discovery of *cytotoxic T-lymphocyte-associated protein* (CTLA-4) to James P. Allison and *programmed cell death protein 1 / programmed cell death protein ligand 1* (PD-1 / PD-L1) to Tasuku Honjo [1]. Malignant tumors take advantage of the inhibitory PD-1 / PD-L1 or CTLA-4 pathways to evade the immune system [2]. Disruption of this axis by blocking monoclonal antibodies can induce durable remissions in different cancer types and has led to numerous FDA and EMA approvals, among others, for the treatment of melanoma, lung cancer, urothelial cancer, head and neck squamous cell carcinoma (HNSCC), renal cell cancer (RCC) and Hodgkin's disease [3]. Up-to-date reviews providing a comprehensive overview of approved indications for different CPIs have been published previously [3, 4].

This review focuses on clinical and pre-clinical findings that might guide future clinical application of CPIs in general. We identified potentially trendsetting studies on CPIs for combinational approaches, perioperative use, new tumor entities, response prediction, toxicity management and use in special patient populations. Further, we identified studies focusing on efficacy and toxicity of anti- CD19 CAR T cells in larger patient cohorts as well as seminal findings on adoptive T cell therapy in other hematological and solid malignancies.

CHECKPOINT INHIBITORS

Combinational Therapy

Combination with Chemotherapy

Traditionally, chemotherapy and radiotherapy were believed to mediate their anti-cancer effect by direct killing of cancer cells. This concept was challenged over a decade ago by Zitvogel and co-workers who discovered that the antineoplastic effect of chemotherapy, in part, depends on the immunogenic cell death of cancer cells. This leads to immune stimulatory signals via activation of the innate immune system through pattern recognition receptors such as toll-like receptor 4 (TLR4) [5]. Different studies confirmed the immunological effects of chemotherapeutic drugs, in particular, platinum-based agents, and paved the way for the development of combinational regimens using PD-1 / PD-L1-blockade together with established chemotherapeutic drugs [6,7,8,9,10,11]. Last year saw the completion of several practice-changing phase III trials showing the efficacy of combining PD-1 / PD-L1-blockade with chemotherapy in small cell lung cancer (SCLC), non-small cell lung cancer (NSCLC), HNSCC and breast cancer [12,13,14,15]. Currently, more than 170 studies are investigating the promising combination of PD-1/PD-L1 blockade plus chemotherapy in different cancer entities [4].

Combination with Radiotherapy

Anecdotal reports on systemic anti-tumor response after irradiation of a single tumor lesion date back more than one century [16]. Regression of non-irradiated lesions after localized radiotherapy of a single lesion was first termed 'abscopal effect' in 1958 [17]. The underlying mechanism remained unexplained for a long period and it took almost another 50 years, before Demaria et al. concluded that "*Ionizing radiation inhibition of distant untreated tumors (abscopal effect) is immune mediated"* [18]. Nowadays, the causative link between local radiation, immunogenic cell death and systemic tumor response is well-established [19]. While the abscopal effect remains a sporadic event, numerous strategies are now under investigation to harness the immunogenic effect of radiotherapy [19].

Given the clinical success of checkpoint blockade, combining radiotherapy with PD-1 / PD-L1 blockade is of special interest. Pre-clinical evidence highlights the synergistic potential of this combination [20]. Translational results from an ongoing phase I/II trial (NCT01976585) investigating local radiotherapy in combination with local application of immunostimulatory agents in patients with indolent lymphoma further support the combination of radiotherapy and PD-1 / PD-L1 blockade [21]. In this trial, patients received 2 Gy of local radiotherapy as part of a so-called "in situ vaccination" (ISV: radiotherapy plus intratumoral application of *Fms*-related tyrosine kinase 3 ligand [Flt3L] and a Toll-like receptor 3 [TLR3] ligand). ISV was able to induce systemic ("abscopal") tumor regression in three out of eleven treated patients. Importantly, in non-responding patients, the induction of tumor infiltrating PD-1^+ CD8^+ T cells was observed, prompting a follow-up trial, which is now recruiting patients for ISV in combination with PD-1 blockade (NCT03789097).

Despite these encouraging findings, negative results for the combination of radiotherapy and checkpoint-blockade have also been recently reported. In a phase II trial in metastatic HNSCC, the addition of local radiotherapy to systemic PD-1 blockade was not able to boost the effect of PD-1-blockade. Here, patients were randomized to receive either nivolumab monotherapy or nivolumab plus stereotactic body radiation therapy (SBRT) of a single tumor lesion. The primary study endpoint - response rate in none-irradiated tumor lesions – was not met. Response rate in patients receiving nivolumab plus SBRT was 22.2% (95% confidence interval [CI]: 10.6–40.8%) versus 26.9% (95% CI: 13.7–46.1%) for single agent nivolumab [22].

The placebo-controlled, randomized phase III PACIFIC trial investigated the addition of durvalumab (anti-PD-L1) to platinum-based chemoradiotherapy in locally advanced (stage III) NSCLC. The addition of durvalumab resulted in an impressive increase in progression-free (PFS) and overall survival (OS) (17.2 versus 5.2 (PFS) and 28.7 months versus "not reached" (OS), respectively) [23, 24]. In this context, the timely administration of PD-1 blockade appeared to be important: patients receiving durvalumab within 14 days after completion of chemoradiotherapy had a better overall survival than patients starting durvalumab-treatment at a later time point [25].

While recent results encourage further in-depth investigation of checkpoint blockade plus radiotherapy, successful concepts might depend on additional combination partners like the above-mentioned in situ-vaccination or chemotherapy. Additional well-designed clinical trials are necessary to identify optimal strategies for combinations and treatment sequences.

Combination with Immunomodulatory Drugs

The first CPI approved for clinical use was ipilimumab, targeting CTLA-4. Given the success of ipilimumab and the even greater success of PD-1-blockade, it is not surprising, that - with more than 250 clinical trials - the combination of PD-1 and CTLA-4 blockade is the most vigorously investigated combinational approach of two immunomodulatory drugs [4].

Due to the large number of clinically approved immunomodulatory agents (currently more than 25) and many more in pre-clinical and clinical development, there is an almost infinite number of combinatorial regimens for further clinical evaluation. In this regard, it is important to note, that the combination of two immunomodulatory drugs can also have antagonistic instead of synergistic effects [26]. Wise selection strategies based on pre-clinical data to select combinatorial approaches for clinical testing are important [26]. In light of this, Tauriello et al. provided an example for an elaborate pre-clinical model system. By using a quadruple mutant colorectal mouse model, they were able to recapitulate important immunological hallmarks of microsatellite stable colorectal cancer (MSS CRC) [27]. While PD-1 / PD-L1 blockade showed only marginal efficacy in this setting paralleling results of clinical trials with PD-1/PD-L1 blockade in MSS CRC, impressive effects were achieved when PD-1/PD-L1 blockade was combined with inhibition of transforming growth factor beta (TGF-β) [27].

Building on pre-clinical and early clinical data for simultaneous targeting of CD40 and PD-1 / PD-L1 in pancreatic cancer (a disease for which all immunotherapeutic efforts have failed so far), a phase I trial investigating the combination of CD40, durvalumab and chemotherapy was initiated. The promising results were recently presented at the annual meeting of the AACR (2019), making this combinational strategy one to keep track of in the years to come [28,29,30].

Peri-operative Use

Up to now, the clinical use of CPIs has been mainly restricted to advanced tumor stages. Yet, efficacy of checkpoint blockade has been reported to be dependent on baseline tumor burden (with better efficacy observed in patients with low tumor burden), making peri-operative usage of checkpoint blockade an attractive treatment option from a theoretical point of view [31, 32].

Although ipilimumab was approved for the adjuvant treatment of melanoma patients by the FDA (but not by the EMA) based on a placebo-controlled phase III trial reporting superior recurrence-free and overall rates, its use was internationally disputed given the relatively high frequency of serious immune-related adverse events in patients receiving treatment with ipilimumab [33,34,35]. In Europe, nivolumab was the first checkpoint inhibitor approved for adjuvant treatment of melanoma patients, based on results of the *CheckMate 238* study reported in 2017 [36]. In this study, nivolumab was compared to ipilimumab as adjuvant therapy for patients after resection of stage III-IV melanoma. Recurrence-free survival was reported to be superior while severe adverse events were significantly lower in patients treated with nivolumab (12-month recurrence-free survival: 70.5% vs 60.5%; grade 3 or 4 adverse events: 14.4% versus 45.9% for patients receiving nivolumab or ipilimumab, respectively).

A logical next step to consider would be neoadjuvant use of CPIs. Theoretically, neoadjuvant immunotherapy might be able to prime systemic immunity for tumor surveillance after complete resection – at a time point when tumor antigens are still abundantly present [37]. This concept is supported by recent translational findings from an early clinical study in patients with resectable melanoma: in a randomized phase Ib study, neoadjuvant treatment with nivolumab and ipilimumab induced a higher number of tumor specific T cell clones than adjuvant treatment [38]. Early clinical findings reported from patients with NSCLC, HNSCC and microsatellite unstable (MSI) CRC further emphasize the high potential of neoadjuvant treatment [39,40,41]. In the latter study, seven out of seven patients with MSI CRC (100%) responded to neoadjuvant treatment with complete remissions observed in 4/7 (57%) patients [41].

A large number of clinical trials is currently investigating neoadjuvant immunotherapy for different disease entities (for example, we identified nine clinical trials for neoadjuvant anti- PD-1 / PD-L1 treatment in NSCLC: NCT03197467, NCT02938624, NCT02259621, NCT03694236, NCT03732664, NCT02994576, NCT03030131, NCT02716038, NCT02818920). Given the considerable side effects of checkpoint blockade – particularly, if administered as combinational therapy - wise selection of patients that might benefit from neoadjuvant or adjuvant treatment is mandatory. One possibility for adjuvant treatment stratification might be detection of minimal residual disease (MRD) by circulating tumor DNA (ctDNA), a strategy, that is currently investigated by a clinical trial in triple-negative breast cancer (TNBC) (NCT03145961) [42].

New Tumor Entities

Current studies show the efficacy of CPIs in patients with malignant melanoma (MM), NSCLC or neoplasms with mutational defects in DNA mismatch repair proteins (micro satellite instability or MSI) independent of the actual tumor entity. Intriguingly, all of these tumors share a relatively high mutational load when their genetic characteristics are comparatively analyzed [43]. This common characteristic leads to increased expression of neo antigens in the tumor, stimulating an increased infiltration of the tumor by immune cells, which in turn can be "*activated*" by CPI administration. This fact can also be used to explain why CPI studies in certain tumor entities (among others pancreatic ductal adenocarcinoma (PDAC) or colorectal carcinoma (CRC) without DNA mismatch repair protein defects) haven't been successful as of yet.

On average, breast cancer and AML are also characterized by a low mutational load [43]. With that background, two remarkable studies from 2018 should be mentioned here in more detail. On the one hand, the phase III trial *IMpassion130* tested the combination of atezolizumab (anti-PD-L1) plus nab-paclitaxel versus nab-paclitaxel monotherapy in treatment-naïve patients with metastatic, triple-negative breast cancer (TNBC). The addition of atezolizumab not only improved the patients' PFS (PFS), but also their overall survival (OS) [14]. For patients with TNBC, this was the first phase III study that showed a strong benefit of targeted (immune) therapy. A total of 144 studies on PD-1 / PD-L1 blockade in TNBC are currently registered on *clinicaltrials.gov* (Fig. 1a).

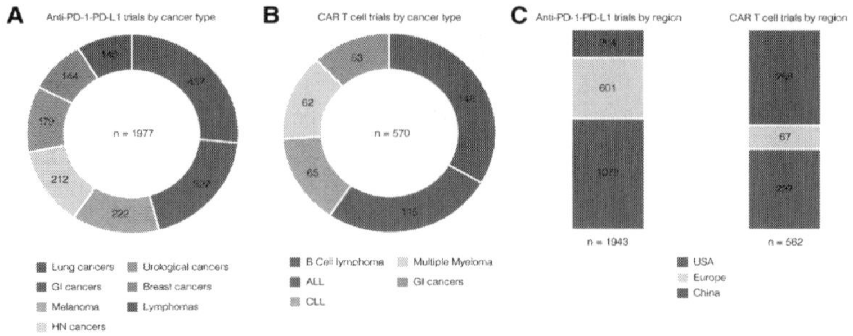

Figure 1: Included tumor types (a, b) and regional distribution (c) of clinical PD-1 / PD-L1 and CAR T cell trials in 2019. ClinicalTrials.gov was searched for "pd-l1" OR "pd-1" OR "programmed death ligand" OR "car t cell" OR "chimeric antigen receptor". All registered trials were sorted for tumor type and country/region. Search was performed on 2019-05-06. Most frequent tumor types (a, b) and regions (c) are shown as indicated. Several clinical trials included multiple tumor types or were performed in more than one country/region. Abbreviations: GI: gastrointestinal, HN: head and neck.

On the other hand, for AML, data on nivolumab maintenance therapy in high-risk AML patients was presented at the annual meeting of the American Society of Clinical Oncology (ASCO) in 2018. This study investigated whether the administration of nivolumab might prolong the time of complete remission (CR) in patients that do not qualify for an allogenic stem cell transplantation. In 14 patients that were followed-up for a median of 19.3 months, the median duration of CR averaged 8.3 months, whereas the median OS had not been reached at the time of presentation of the data. Despite the very limited number of patients, this study shows an exciting treatment concept for this specific treatment group [44].

In conclusion, both studies exemplify that successful CPI concepts might also be feasible for tumor entities with a low mutational load. Numerous clinical trials are currently investigating the use of CPIs in different cancer entities (Fig. 1a). It will be interesting to see whether further positive results for tumor entities with low mutational burden will follow in the future.

Biomarkers for Response Prediction of Checkpoint Blockade

Determination of PD-L1 expression by immunohistochemistry is an FDA-approved diagnostic test and a prerequisite for treatment with anti-PD-1 / PD-L1 therapy in various indications (e.g. monotherapy treatment of urothelial cancer with atezolizumab or pembrolizumab). However, determining PD-L1 expression does not identify all patients that profit from anti-PD-1 / PD-L1 therapy, highlighting the need for additional and better biomarkers [45].

Tissue Biomarkers

Microsatellite instability and tumor mutational burden

Another approved biomarker test (for pembrolizumab) is the determination of microsatellite instability (MSI) or deficient mismatch repair (dMMR). Pembrolizumab was the first drug that was FDA-approved with a "tumor-agnostic" indication based on findings from five different clinical trials including 15 tumor entities with MSI/dMMR tumors (KEYNOTE -012, −016, −028, −158 and−164). MSI/dMMR results in increased tumor mutational burden (TMB) with subsequent increase in neoantigens and immune cell infiltration, rendering tumors susceptible to PD-1 /PD-L1 blockade [46]. In different studies, the direct determination of TMB was also established as predictive biomarker for immunotherapy [47,48,49]. However, recently presented data suggests that not all patients with MSI/dMMR tumors also have a high TMB [50]. Furthermore, TMB^{high} is also observed in the absence of MSI/dMMR [46]. More studies are therefore necessary to inform strategies on selection of MSI/dMMR or TMB as biomarker for response to checkpoint blockade.

Tumor mutational burden and PD-L1 expression

It was previously described that TMB does not correlate to PD-L1 expression [51]. This finding was confirmed and put into therapeutic context by the ChekMate227 trial [52]. In this trial, NSCLC patients were stratified according to tumoral PD-L1 expression ($\geq$ 1% vs < 1%). Patients were then randomized (1:1:1) between either chemotherapy, nivolumab (nivolumab plus chemotherapy for patients with < 1% PD-L1 expression, respectively) or nivolumab plus ipilimumab. One predefined endpoint was response rate in patients with a TMB^{high} (defined as > 10 mutations per megabase). Independent of PD-L1 expression, nivolumab plus ipilimumab was superior to chemotherapy in patients with high TMB [52].

Inflammatory gene signatures

Apart from the biomarkers mentioned above, different inflammatory TMB-signatures determined in tumor tissues can serve as biomarkers for checkpoint blockade. These signatures indicate infiltration by a specific immune cell subset (e.g. effector T cells) or activation of a specific signaling pathway (e.g. interferon-γ signaling). Recently published data from the *IMmotion150* trial suggests that these signatures could even be superior to TMB in patients with metastatic renal cell carcinoma: patients were randomized between the combination of atezolizumab (anti-PD-L1) +/− bevacizumab versus sunitinib. T-effector, interferon-γ and myeloid inflammatory gene expression signatures were superior to TMB in predicting response to atezolizumab [53]. It should be noted, that these analyses were exploratory.

Further research is necessary to integrate the aforementioned tissue biomarkers into one clinical applicable diagnostic algorithm. Well-designed translational studies might also be able to identify completely new tissue biomarkers to predict clinical response to CPI treatment. One example are gene fusions producing immunogenic neoantigens. Such gene fusions were recently shown to predict response to checkpoint blockade in HNSCC patients with low TMB and minimal immune cell infiltrate [54].

Soluble Biomarkers

Identifying soluble biomarkers for response prediction in peripheral blood would have several advantages over tissue biomarkers. For instance, they are easily and noninvasively accessible and can be sampled repetitively for continuous response prediction. The soluble forms of PD-1 and PD-L1 (sPD1 and sPD-L1) are also present in the peripheral blood [55, 56]. Only few studies have investigated sPD-1 and sPDL-1 as biomarkers for response to checkpoint blockade. One small study in NSCLC patients suggested that high sPD-L1 levels predict poor response to nivolumab [57], a finding that is somewhat contrary to tissue PD-L1, because high PD-L1 tissue expression indicates higher likelihood of response to checkpoint blockade. Findings from patients with pancreatic cancer suggest that sPD-1 and sPD-L1 are rather indicators of systemic inflammation and independent from tumoral PD-L1 expression [56]. Together these findings question the aptitude of sPD-1 and sPD-L1 as biomarkers for checkpoint blockade.

An emerging soluble biomarker for checkpoint blockade is ctDNA in peripheral blood. It can be used for different applications. First, ctDNA can be used to determine tumor mutational burden (TMB) [58]. TMB measured in peripheral blood has been shown to predict response to checkpoint blockade in NSCLC patients [58, 59]. In patients receiving conventional chemotherapy, repeated ctDNA measurement can be used for early response prediction [60]. Recently published studies suggest that changes in ctDNA levels can also be early predictors for response to immunotherapy [61, 62]. Importantly, it might also aid to distinguish pseudo-progression from truly progressive disease in patients treated with immunotherapy [63].

Immune related Adverse Events as Biomarker for Tumor Response

Different studies suggested that immune related adverse events (IrAEs) indicate response to checkpoint blockade [64, 65]. These studies, however, were not controlled for *lead time bias* [66] and it is therefore not clear, whether IrAEs are truly independent predictors for response or merely reflect a longer time under treatment. Recent studies controlled for *lead-time bias* reported conflicting data: a large monocentric study including different cancer types presented at ESMO 2018 did not find a correlation between IrAEs and response to checkpoint blockade after controlling for lead-time bias [67]. Yet, another recent study in renal cell carcinoma reported better efficacy of nivolumab in patients with IrAEs after controlling for lead-time bias [68].

Toxicity Management

Use of Steroids

The occurrence of immune-mediated side effects (e.g. colitis, autoimmune hepatitis, endocrine or neurological side effects) requires treatment with glucocorticoids (e.g. prednisolone) as early as possible depending on the severity [69]. Whether the use of glucocorticoids has a negative effect on the success of CPI treatment remains controversial. A study presented at the annual meeting of the ASCO in 2018 retrospectively investigated NSCLC patients who received glucocorticoids at the beginning of CPI therapy. The reasons for glucocorticoid administration included the treatment of symptoms caused by brain metastases as well as respiratory distress or fatigue. In a multivariate analysis which included performance status and presence of brain metastases, patients who received glucocorticoids at the start of treatment responded significantly worse to CPI administration [67]. On the other hand, as mentioned in the biomarker section, it is often postulated that patients who develop immune-mediated side effects (and receive glucocorticoids) benefit from CPI therapy over a longer period of time (or at least not shorter) than patients without immune-mediated side effects.

As a practice-based approach, immune-mediated side effects (depending on the severity and type of side effects) should be treated early with glucocorticoids to prevent permanent damage [69]. On the other hand, the need for symptomatic and sustained administration of steroids for other reasons (e.g. brain metastases or respiratory distress) during CPI therapy should be critically scrutinized in everyday clinical practice.

Special populations: patients with pre-existing autoimmune disease or HIV

Most clinical trials on CPI therapy have excluded patients with pre-existing autoimmune diseases or human immunodeficiency virus infection (HIV). In this regard, it remained unclear whether a CPI therapy is also conceivable in these patients.

The safety and efficacy of CPIs in patients with pre-existing autoimmune diseases has been recently studied in a French registry study including different tumor entities [70]. Patients with and without pre-existing autoimmune diseases were included (patients with pre-existing autoimmune disease: $n=45$, patients without pre-existing autoimmune disease: $n=352$). Although the incidence of immune-mediated side effects was significantly increased in the group of patients with pre-existing autoimmune diseases (44% versus 23%), there was no difference in overall survival between the two groups.

For the use of CPIs in patients with HIV, data from a small HIV-positive cohort of patients ($n=20$) with NSCLC or multiple myeloma was presented at the annual meeting of the European Society of Medical Oncology (ESMO) in 2018. Overall, the therapy with CPIs was well tolerated in patients with HIV and no immune-mediated side effects were observed. An increase in HIV viral load was observed only in one patient who had paused his antiretroviral therapy. A response to therapy (PR or CR) was observed in 24% of patients [71].

Overall, both studies suggest that CPI therapy might be feasible and effective in patients with pre-existing autoimmune disease or HIV. Due to limited data on these special patient groups, a careful assessment of potential benefit versus potential harm is mandatory before starting CPI therapy in these patients.

CELLULAR IMMUNOTHERAPY

Chimeric Antigen Receptor T Cells

Tisagenlecleucel and *axicabtagen-ciloleucel* were the first two cellular cancer immunotherapies receiving FDA and EMA approval in 2017 and 2018, respectively. They are approved to treat patients with acute lymphoblastic leukemia (ALL, *tisagenlecleucel*) and diffuse-large B cell lymphoma (DLBCL, *tisagenlecleucel* and *axicabtagen-ciloleucel*). Approval was based on impressive response rates observed in the ELIANA trial (relapsed or refractory [r/r] ALL in pediatric patients or young adults treated with *tisagenlecleucel*), JULIETH trial (r/r DLBCL, *tisagenlecleucel*) and ZUMA-1 trial (r/r DLBCL, *axicabtagen-ciloleucel)* [72,73,74].

Tisagenlecleucel and *axicabtagen-ciloleucel* are autologous T cell products. After leukapheresis, T cells are genetically engineered to express an anti-CD19 chimeric antigen receptor (anti-CD19 CAR T cells). Re-infusion of CAR T cells is preceded by a lympho-depleting chemotherapy to allow for subsequent in vivo expansion of CAR T cells (Fig. 2).

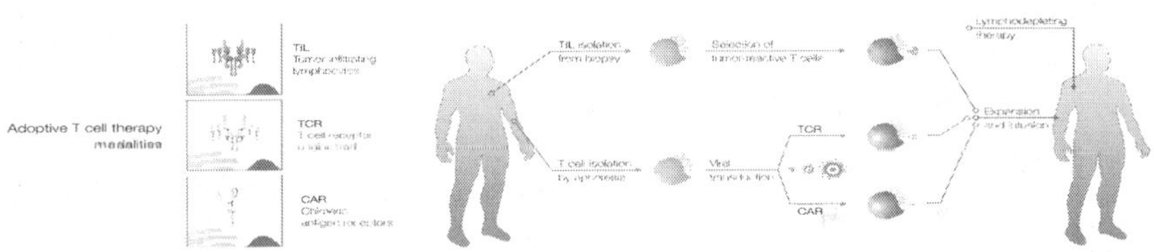

Figure 2: Different strategies for adoptive T cell therapy. Abbreviations: CAR: chimeric antigen receptor, TCR: T cell receptor, TIL: tumor infiltrating lymphocytes.

Numerous clinical trials (as of May 2019 more than 550, Fig. 1b) are investigating CAR T cell therapies for different hematological and solid cancer types [75]. Of interest and in harsh contrast to trials on PD-1 / PD-L1 blockade is the regional distribution of clinical trials on CAR T cell therapy (Fig. 1c). The USA and China by far outcompete the EU in terms of registered CAR T cell trials. This regional imbalance has been described and discussed previously and should be addressed by researches and health care policy makers in the European Union [76].

Recently reported studies on cellular therapy mainly addressed two important questions: (I) Long term and "real world" experience regarding toxicity and efficacy of CAR T cells (II) Can the striking success of CAR T cells in ALL and DLBCL be translated to other hematological and – more importantly - solid malignancies?

Updated Results from CD19 CAR T Cells Clinical Trials

Follow-up results for efficacy and toxicity from the ELIANA, JULIETH and ZUMA-1 trial were recently presented at the annual meetings of the European Hematology Association (EHA) and the American Society of Hematology (ASH).

Efficacy

As of 2018, 97 patients aged ≤21 years with r/r ALL were enrolled in the ELIANA trial, 79 patients were infused with CD19 CAR T cells and a complete remission was achieved in 65 patients. After a median follow-up of 24 months, response was ongoing in 29 patients (45%), with a maximum (ongoing) duration of response of 29 months [77]. For r/r DLBCL patients treated with *tisagenlecleucel,* the updated analysis presented at EHA 2018 included 111 infused patients. Overall response rate (ORR) was 52% (40% CR, 12% PR) [78]. After a median follow-up time of 14 months, median duration of response was not reached. Median overall survival for all infused patients was 11.7 months [79]. For *axicabtagen-ciloleucel,* the 2-year follow-up data was presented at ASH 2018. A total of 108 r/r DLBCL patients had at least one year of follow-up. ORR in this cohort was 82% (58% CR). An ongoing response was observed in 42% of all patients after a median follow-up of 15.4 months, no updated overall survival data was reported [80].

For *axicabtagen-ciloleucel, "real world"* efficacy was confirmed by data from seventeen US academic centers who evaluated *axicabtagen-ciloleucel* outside of clinical trials, independent of the manufacturer after commercialization. The authors reported an ORR of 79% (50% CR), confirming the results reported in the clinical trials mentioned above [81].

While these results support the high therapeutic potential of CAR T cell therapy, a cohort of patients does not respond to – or relapses after – CAR T cell therapy. Considering the latter group (relapse after an initial complete response), it is important to explore further treatment options for these patients. One possibility might be allogeneic stem cell transplantation, which has recently been reported to improve prognosis after anti-CD19 CAR T cell therapy for ALL patients who had not received a previous stem cell transplantation [82].

Toxicity

The updated data for ELIANA, JULIETH and ZUMA-1 confirm the previously described safety profile with cytokine release syndrome (CRS, incidence of CRS grade$\geq$3: 7 to 48%) and neurologic events (NE, incidence of NE grade$\geq$3: 11 to 31%) as most significant adverse events [78,79,80,81].

In the pivotal trials for anti-CD19 CAR T cells, treatment-related deaths have been reported [77]. No treatment-related deaths were observed in a US multi-center cohort of 165 patients who received *axicabtagen-ciloleucel* for r/r DLBCL after commercialization outside of clinical trials [81]. Recently, safety of *axicabtagen-ciloleucel* was also confirmed in patients $\geq$65 years [83]. Further it has been reported that neurotoxicity is fully reversible in most patients [84].

While the mentioned results are reassuring regarding saftey of CAR T cell therapy, different strategies are currently under investigation to further improve the safety profile of CAR T cells. These strategies include: (I) modification of the chimeric antigen receptor cell itself [85, 86]; (II) identification of predictive biomarkers for CAR T cell toxicity [84]; (III) "safety switches" such as inducible suicide genes [87]; and (IV) novel drugs to mitigate CRS and NE [88].

Adoptive T Cell Therapy in other Hematological and Solid Malignancies

Chimeric antigen receptor T cells for hematological and solid malignancies

The success of CAR T cells in ALL and B cell lymphoma led to the initiation of numerous follow-up trials in these disease entities (Fig. 1b). Regarding other cancer types, chronic lymphocytic leukemia, multiple myeloma and gastrointestinal cancers are the ones with most clinical CAR T cell trials underway (Fig. 1b).

Additionally, a large variety of strategies to improve efficacy of CAR T cells in solid malignancies are under pre-clinical investigation [89,90,91,92,93,94]. Yet, the direct translation of the CAR T cell approach to solid malignancies is often impeded by the lack of a suitable cancer specific antigen resulting in either disappointing efficacy or substantial off target toxicity in early clinical trials [95]. Another important consideration is the tumor environment which is substantially different to the one seen in the above referenced hematological cancers and impedes CAR T cell efficacy [96].

Alternative approaches are genetic modification of the T cell receptor (TCR) itself or the adoptive transfer of "naturally" occurring tumor reactive T cells (also termed tumor infiltrating lymphocytes or TILs) isolated from autologous tumor tissue or tumor draining lymph nodes (Fig. 2). The manufacturing of TCR-modified T cells is complex, dependent on a specific human leukocyte antigen (HLA)-haplotype and can lead to unexpected off-target toxicity [97, 98]. On the other hand, the use of tumor reactive (TCR-native) T cells has been investigated in numerous clinical studies (mainly in melanoma patients) with promising results [99, 100]. Recent studies suggest that this approach could also be successfully translated to other solid malignancies.

Ex vivo expansion and reinfusion of autologous tumor reactive T cells

In contrast to CAR T cells, tumor reactive T cells recognize tumor cells via their native (unmodified) T cell receptor (Fig. 2). Tumor reactive T cells can be isolated from tumor tissue or tumor draining lymph nodes [101,102,103,104,105,106]. After a potential selection step followed by ex vivo expansion, tumor reactive T cells are re-infused after lymphodepleting chemotherapy – typically with parallel intravenous administration of interleukin 2 [101]. The high potential of this approach was recently confirmed in melanoma patients after failure of PD-1 / PD-L1 blockade [107] and is currently investigated in a phase III trial as first-line treatment for advanced melanoma patients (NCT02278887). In other solid tumor entities an ongoing early clinical trial (NCT01174121) is currently investigating immunotherapy with tumor reactive T cells in patients with metastatic gastrointestinal, urothelial, breast, ovarian or endometrial cancer. Case reports from three individual patients described striking responses for this treatment approach for cholangiocarcinoma, colorectal cancer and breast cancer, respectively [104,105,106]. Further studies are necessary to evaluate the expansion of this promising treatment approach to larger patient populations.

CONCLUSION

Immunotherapy of cancer is a rapidly evolving field. Results of currently ongoing studies on checkpoint blockade will most likely expand the use of CPIs to additional patient populations (e.g. new tumor entities, perioperative use, use in special patient populations) and might identify new combination partners for CPI.

The major challenge for adoptive T cell therapy in years to come is the translation of this treatment modality to solid malignancies. A successful strategy has yet to be defined and might include more advanced genetic engineering of CAR T cells as well as the development of more advanced protocols for the use of tumor reactive (TCR-native) T cells.

Regarding the regional distribution of clinical trials on immunotherapy a shift from the European region (for PD-1 / PD-L1-trials) towards China (leading in terms of number of available CAR T cell trials) is evident and should be met by intensified research efforts on cellular immunotherapy in Europe.

REFERENCES

1. Altmann DM. A Nobel prize-worthy pursuit: cancer immunology and harnessing immunity to tumour neoantigens. Immunology. 2018;155(3):283–4.
2. Gajewski TF, Schreiber H, Fu YX. Innate and adaptive immune cells in the tumor microenvironment. Nat Immunol. 2013;14(10):1014–22.
3. Gong J, Chehrazi-Raffle A, Reddi S, Salgia R. Development of PD-1 and PD-L1 inhibitors as a form of cancer immunotherapy: a comprehensive review of registration trials and future considerations. J Immunother Cancer. 2018;6(1):8.
4. Tang J, Shalabi A, Hubbard-Lucey VM. Comprehensive analysis of the clinical immuno-oncology landscape. Ann Oncol. 2018;29(1):84–91.
5. Apetoh L, Ghiringhelli F, Tesniere A, Obeid M, Ortiz C, Criollo A, et al. Toll-like receptor 4-dependent contribution of the immune system to anticancer chemotherapy and radiotherapy. Nat Med. 2007;13(9):1050–9.
6. Liu WM, Fowler DW, Smith P, Dalgleish AG. Pre-treatment with chemotherapy can enhance the antigenicity and immunogenicity of tumours by promoting adaptive immune responses. Br J Cancer. 2010;102(1):115–23.
7. Hato SV, Khong A, de Vries IJ, Lesterhuis WJ. Molecular pathways: the immunogenic effects of platinum-based chemotherapeutics. Clin Cancer Res. 2014;20(11):2831–7.
8. Galluzzi L, Buque A, Kepp O, Zitvogel L, Kroemer G. Immunological effects of conventional chemotherapy and targeted anticancer agents. Cancer Cell. 2015;28(6):690–714.
9. Tesniere A, Schlemmer F, Boige V, Kepp O, Martins I, Ghiringhelli F, et al. Immunogenic death of colon cancer cells treated with oxaliplatin. Oncogene. 2010;29(4):482–91.
10. Shalapour S, Font-Burgada J, Di Caro G, Zhong Z, Sanchez-Lopez E, Dhar D, et al. Immunosuppressive plasma cells impede T-cell-dependent immunogenic chemotherapy. Nature. 2015;521(7550):94–8.
11. Vincent J, Mignot G, Chalmin F, Ladoire S, Bruchard M, Chevriaux A, et al. 5-fluorouracil selectively kills tumor-associated myeloid-derived suppressor cells resulting in enhanced T cell-dependent antitumor immunity. Cancer Res. 2010;70(8):3052–61.

12. Paz-Ares L, Luft A, Vicente D, Tafreshi A, Gumus M, Mazieres J, et al. Pembrolizumab plus chemotherapy for squamous non-small-cell lung Cancer. N Engl J Med. 2018.
13. Gandhi L, Rodríguez-Abreu D, Gadgeel S, Esteban E, Felip E, De Angelis F, et al. Pembrolizumab plus chemotherapy in metastatic non–small-cell lung Cancer. N Engl J Med. 2018;378(22):2078–92.
14. Schmid P, Adams S, Rugo HS, Schneeweiss A, Barrios CH, Iwata H, et al. Atezolizumab and nab-paclitaxel in advanced triple-negative breast Cancer. N Engl J Med. 2018;379(22):2108–21.
15. Burtness B, Harrington KJ, Greil R, Soulières D, Tahara M, De Castro JG, et al. LBA8_PRKEYNOTE-048: phase III study of first-line pembrolizumab (P) for recurrent/metastatic head and neck squamous cell carcinoma (R/M HNSCC). Ann Oncol. 2018;29(suppl_8):mdy424.045-mdy424.045.
16. McCulloch HD. On the analogy between spontaneous recoveries from Cancer and the specific immunity induced by X ray irradiations of the lymphatic glands involved. Br Med J. 1908;2(2494):1146–8.
17. Mole RH. Whole body irradiation; radiobiology or medicine? Br J Radiol. 1953;26(305):234–41.
18. Demaria S, Ng B, Devitt ML, Babb JS, Kawashima N, Liebes L, et al. Ionizing radiation inhibition of distant untreated tumors (abscopal effect) is immune mediated. Int J Radiat Oncol Biol Phys. 2004;58(3):862–70.
19. Brix N, Tiefenthaller A, Anders H, Belka C, Lauber K. Abscopal, immunological effects of radiotherapy: narrowing the gap between clinical and preclinical experiences. Immunol Rev. 2017;280(1):249–79.
20. Park SS, Dong H, Liu X, Harrington SM, Krco CJ, Grams MP, et al. PD-1 restrains radiotherapy-induced Abscopal effect. Cancer Immunol Res. 2015;3(6):610–9.
21. Hammerich L, Marron TU, Upadhyay R, Svensson-Arvelund J, Dhainaut M, Hussein S, et al. Systemic clinical tumor regressions and potentiation of PD1 blockade with in situ vaccination. Nat Med. 2019;25(5):814–24.
22. McBride SM, Sherman EJ, Tsai CJ, Baxi SS, Aghalar J, Eng J, et al. A phase II randomized trial of nivolumab with stereotactic body radiotherapy (SBRT) versus nivolumab alone in metastatic (M1)

head and neck squamous cell carcinoma (HNSCC). J Clin Oncol. 2018;36(15_suppl):6009.

23. Antonia SJ, Villegas A, Daniel D, Vicente D, Murakami S, Hui R, et al. Durvalumab after Chemoradiotherapy in stage III non–small-cell lung Cancer. N Engl J Med. 2017;377(20):1919–29.
24. Antonia SJ, Villegas A, Daniel D, Vicente D, Murakami S, Hui R, et al. Overall survival with Durvalumab after Chemoradiotherapy in stage III NSCLC. N Engl J Med. 2018.
25. Faivre-Finn C, Spigel DR, Senan S, Langer CJ, Raben D, Perez B, et al. 1363OEfficacy and safety evaluation based on time from completion of radiotherapy to randomization with durvalumab or placebo in pts from PACIFIC. Ann Oncol. 2018;29(suppl_8):mdy291–mdy.
26. Shrimali RK, Ahmad S, Verma V, Zeng P, Ananth S, Gaur P, et al. Concurrent PD-1 blockade negates the effects of OX40 agonist antibody in combination immunotherapy through inducing T-cell apoptosis. Cancer Immunol Res. 2017;5(9):755–66.
27. Tauriello DVF, Palomo-Ponce S, Stork D, Berenguer-Llergo A, Badia-Ramentol J, Iglesias M, et al. TGFbeta drives immune evasion in genetically reconstituted colon cancer metastasis. Nature. 2018;554(7693):538–43.
28. Beatty GL, Chiorean EG, Fishman MP, Saboury B, Teitelbaum UR, Sun W, et al. CD40 agonists alter tumor stroma and show efficacy against pancreatic carcinoma in mice and humans. Science (New York, NY). 2011;331(6024):1612–6.
29. Luheshi NM, Coates-Ulrichsen J, Harper J, Mullins S, Sulikowski MG, Martin P, et al. Transformation of the tumour microenvironment by a CD40 agonist antibody correlates with improved responses to PD-L1 blockade in a mouse orthotopic pancreatic tumour model. Oncotarget. 2016;7(14):18508–20.
30. O'Hara MH. CT004 - a phase Ib study of CD40 agonistic monoclonal antibody APX005M together with gemcitabine (gem) and nab-paclitaxel (NP) with or without nivolumab (Nivo) in untreated metastatic ductal pancreatic adenocarcinoma (PDAC) patients. In: Annual Meeting of the American Association for Cancer Research; Atlanta; 2019.
31. Huang AC, Postow MA, Orlowski RJ, Mick R, Bengsch B, Manne S, et al. T-cell invigoration to tumour burden ratio associated with anti-PD-1 response. Nature. 2017;545(7652):60–5.

32. Joseph RW, Elassaiss-Schaap J, Kefford R, Hwu WJ, Wolchok JD, Joshua AM, et al. Baseline tumor size is an independent prognostic factor for overall survival in patients with melanoma treated with Pembrolizumab. Clin Cancer Res. 2018;24(20):4960–7.

33. Garcia CA, El-Ali A, Rath TJ, Contis LC, Gorantla V, Drappatz J, et al. Neurologic immune-related adverse events associated with adjuvant ipilimumab: report of two cases. J Immunother Cancer. 2018;6(1):83.

34. Eggermont AM, Chiarion-Sileni V, Grob JJ, Dummer R, Wolchok JD, Schmidt H, et al. Adjuvant ipilimumab versus placebo after complete resection of high-risk stage III melanoma (EORTC 18071): a randomised, double-blind, phase 3 trial. Lancet Oncol. 2015;16(5):522–30.

35. Eggermont AM, Chiarion-Sileni V, Grob JJ, Dummer R, Wolchok JD, Schmidt H, et al. Prolonged survival in stage III melanoma with Ipilimumab adjuvant therapy. N Engl J Med. 2016;375(19):1845–55.

36. Weber J, Mandala M, Del Vecchio M, Gogas HJ, Arance AM, Cowey CL, et al. Adjuvant Nivolumab versus Ipilimumab in resected stage III or IV melanoma. N Engl J Med. 2017;377(19):1824–35.

37. Liu J, Blake SJ, Yong MC, Harjunpaa H, Ngiow SF, Takeda K, et al. Improved efficacy of neoadjuvant compared to adjuvant immunotherapy to eradicate metastatic disease. Cancer Discovery. 2016;6(12):1382–99.

38. Blank CU, Rozeman EA, Fanchi LF, Sikorska K, van de Wiel B, Kvistborg P, et al. Neoadjuvant versus adjuvant ipilimumab plus nivolumab in macroscopic stage III melanoma. Nat Med. 2018.

39. Forde PM, Chaft JE, Smith KN, Anagnostou V, Cottrell TR, Hellmann MD, et al. Neoadjuvant PD-1 blockade in Resectable lung Cancer. N Engl J Med. 2018;378(21):1976–86.

40. Bell RB, Duhen R, Leidner RS, Curti BD, Ballesteros-Merino C, Piening B, et al. Neoadjuvant anti-OX40 (MEDI6469) prior to surgery in head and neck squamous cell carcinoma. J Clin Oncol. 2018;36(15_suppl):6011.

41. Chalabi M, Fanchi LF, Van den Berg JG, Beets GL, Lopez-Yurda M, Aalbers AG, et al. LBA37_PRNeoadjuvant ipilimumab plus nivolumab in early stage colon cancer. Ann Oncol. 2018;29(suppl_8):mdy424.047-mdy424.047.

42. Garcia-Murillas I, Schiavon G, Weigelt B, Ng C, Hrebien S, Cutts RJ,

et al. Mutation tracking in circulating tumor DNA predicts relapse in early breast cancer. Sci Transl Med. 2015;7(302):302ra133.

43. Alexandrov LB, Nik-Zainal S, Wedge DC, Aparicio SAJR, Behjati S, Biankin AV, et al. Signatures of mutational processes in human cancer. Nature. 2013;500:415.
44. Kadia TM, Cortes JE, Ghorab A, Ravandi F, Jabbour E, Daver NG, et al. Nivolumab (Nivo) maintenance (maint) in high-risk (HR) acute myeloid leukemia (AML) patients. J Clin Oncol. 2018;36(15_suppl):7014.
45. Colwell J. Is PD-L1 expression a biomarker of response? Cancer discovery. 2015;5(12):1232.
46. Fabrizio DA, George TJ Jr, Dunne RF, Frampton G, Sun J, Gowen K, et al. Beyond microsatellite testing: assessment of tumor mutational burden identifies subsets of colorectal cancer who may respond to immune checkpoint inhibition. J Gastrointest Oncol. 2018;9(4):610–7.
47. Endris V, Buchhalter I, Allgauer M, Rempel E, Lier A, Volckmar AL, et al. Measurement of tumor mutational burden (TMB) in routine molecular diagnostics: in silico and real-life analysis of three larger gene panels. Int J Cancer. 2019;144(9):2303–12.
48. Rizvi NA, Hellmann MD, Snyder A, Kvistborg P, Makarov V, Havel JJ, et al. Cancer immunologyMutational landscape determines sensitivity to PD-1 blockade in non-small cell lung cancer. Science (New York, NY). 2015;348(6230):124–8.
49. Chan TA, Yarchoan M, Jaffee E, Swanton C, Quezada SA, Stenzinger A. Development of tumor mutation burden as an immunotherapy biomarker: utility for the oncology clinic. Ann Oncol. 2019;30(1):44–56.
50. Salem ME, Puccini A, Grothey A, Xiu J, Goldberg R. Kim ES, et al. 1835PDComparative molecular analysis between microsatellite instability-high (MSI-H) tumors with high tumor mutational burden (TMB-H) versus MSI-H tumors with TMB-intermediate/low. Ann Oncol. 2018;29(suppl_8):mdy303.005-mdy303.005.
51. Carbone DP, Reck M, Paz-Ares L, Creelan B, Horn L, Steins M, et al. First-line Nivolumab in stage IV or recurrent non-small-cell lung Cancer. N Engl J Med. 2017;376(25):2415–26.
52. Hellmann MD, Ciuleanu T-E, Pluzanski A, Lee JS, Otterson GA, Audigier-Valette C, et al. Nivolumab plus Ipilimumab in lung Cancer

with a high tumor mutational burden. N Engl J Med. 2018;378(22):2093–104.

53. McDermott DF, Huseni MA, Atkins MB, Motzer RJ, Rini BI, Escudier B, et al. Clinical activity and molecular correlates of response to atezolizumab alone or in combination with bevacizumab versus sunitinib in renal cell carcinoma. Nat Med. 2018;24(6):749–57.
54. Yang W, Lee KW, Srivastava RM, Kuo F, Krishna C, Chowell D, et al. Immunogenic neoantigens derived from gene fusions stimulate T cell responses. Nature medicine. 2019.
55. Frigola X, Inman BA, Lohse CM, Krco CJ, Cheville JC, Thompson RH, et al. Identification of a soluble form of B7-H1 that retains immunosuppressive activity and is associated with aggressive renal cell carcinoma. Clin Cancer Res. 2011;17(7):1915–23.
56. Kruger S, Legenstein ML, Rosgen V, Haas M, Modest DP, Westphalen CB, et al. Serum levels of soluble programmed death protein 1 (sPD-1) and soluble programmed death ligand 1 (sPD-L1) in advanced pancreatic cancer. Oncoimmunology. 2017;6(5):e1310358.
57. Costantini A, Julie C, Dumenil C, Helias-Rodzewicz Z, Tisserand J, Dumoulin J, et al. Predictive role of plasmatic biomarkers in advanced non-small cell lung cancer treated by nivolumab. Oncoimmunology. 2018;7(8):e1452581.
58. Gandara DR, Paul SM, Kowanetz M, Schleifman E, Zou W, Li Y, et al. Blood-based tumor mutational burden as a predictor of clinical benefit in non-small-cell lung cancer patients treated with atezolizumab. Nat Med. 2018;24(9):1441–8.
59. Kim ES, Velcheti V, Mekhail T, Leal TA, Dowell JE, Tsai ML, et al. LBA55Primary efficacy results from B-F1RST, a prospective phase II trial evaluating blood-based tumour mutational burden (bTMB) as a predictive biomarker for atezolizumab (atezo) in 1L non-small cell lung cancer (NSCLC). Ann Oncol. 2018;29(suppl_8):mdy424.067-mdy424.067.
60. Kruger S, Heinemann V, Ross C, Diehl F, Nagel D, Ormanns S, et al. Repeated mutKRAS ctDNA measurements represent a novel and promising tool for early response prediction and therapy monitoring in advanced pancreatic cancer. Ann Oncol. 2018;29(12):2348–55.
61. Lee JH, Long GV, Boyd S, Lo S, Menzies AM, Tembe V, et al. Circulating tumour DNA predicts response to anti-PD1 antibodies in

metastatic melanoma. Ann Oncol. 2017;28(5):1130–6.

62. Goldberg SB, Narayan A, Kole AJ, Decker RH, Teysir J, Carriero NJ, et al. Early assessment of lung Cancer immunotherapy response via circulating tumor DNA. Clin Cancer Res. 2018;24(8):1872–80.
63. Guibert N, Mazieres J, Delaunay M, Casanova A, Farella M, Keller L, et al. Monitoring of KRAS-mutated ctDNA to discriminate pseudo-progression from true progression during anti-PD-1 treatment of lung adenocarcinoma. Oncotarget. 2017;8(23):38056–60.
64. Ricciuti B, Giglio AD, Brambilla M, Bassanelli M, Metro G, Ludovini V, et al. Immune-related adverse events to predict survival in patients with advanced non-small cell lung cancer treated with nivolumab: a multicenter analysis. J Clin Oncol. 2018;36(15_suppl):9084.
65. Haratani K, Hayashi H, Chiba Y, et al. Association of immune-related adverse events with nivolumab efficacy in non–small-cell lung cancer. JAMA Oncol. 2018;4(3):374–8.
66. Kfoury M, Voisin AL, Najean M, Champiat S, Laghouati S, Michot JM, et al. 1141PDAssociation between immune-related adverse events and efficacy in patients treated with anti-PD-(L)1. Ann Oncol. 2018;29(suppl_8):mdy288.014-mdy288.014.
67. Arbour KC, Mezquita L, Long N, Rizvi H, Auclin E, Ni A, et al. Deleterious effect of baseline steroids on efficacy of PD-(L)1 blockade in patients with NSCLC. J Clin Oncol. 2018;36(15_suppl):9003.
68. Verzoni E, Carteni G, Cortesi E, Giannarelli D, De Giglio A, Sabbatini R, et al. Real-world efficacy and safety of nivolumab in previously-treated metastatic renal cell carcinoma, and association between immune-related adverse events and survival: the Italian expanded access program. J Immunother Cancer. 2019;7(1):99.
69. Haanen JBAG, Carbonnel F, Robert C, Kerr KM, Peters S, Larkin J, et al. Management of toxicities from immunotherapy: ESMO clinical practice guidelines for diagnosis, treatment and follow-up†. Ann Oncol. 2017;28(suppl_4):iv119–iv42.
70. Danlos F-X, Voisin A-L, Dyevre V, Michot J-M, Routier E, Taillade L, et al. Safety and efficacy of anti-programmed death 1 antibodies in patients with cancer and pre-existing autoimmune or inflammatory disease. Eur J Cancer. 2018;91:21–9.
71. Gobert A, Veyri M, Lavolé A, Montaudié H, Cloarec N, Doucet L, et al. 1213P_PRTolerance and efficacy of immune-checkpoint

inhibitors for cancer in people living with HIV (PWHIV). Ann Oncol. 2018;29(suppl_8):mdy424.051-mdy424.051.

72. Maude SL, Laetsch TW, Buechner J, Rives S, Boyer M, Bittencourt H, et al. Tisagenlecleucel in children and young adults with B-cell lymphoblastic leukemia. N Engl J Med. 2018;378(5):439–48.

73. Schuster SJ, Bishop MR, Tam CS, Waller EK, Borchmann P, McGuirk JP, et al. Tisagenlecleucel in adult relapsed or refractory diffuse large B-cell lymphoma. N Engl J Med. 2019;380(1):45–56.

74. Neelapu SS, Locke FL, Bartlett NL, Lekakis LJ, Miklos DB, Jacobson CA, et al. Axicabtagene Ciloleucel CAR T-cell therapy in refractory large B-cell lymphoma. N Engl J Med. 2017;377(26):2531–44.

75. Results retrieved from ClinicalTrials.gov on 2019-05-03, search: "other terms" = "car t cells" [internet]. 2019 [cited 2019-05-03].

76. Hartmann J, Schussler-Lenz M, Bondanza A, Buchholz CJ. Clinical development of CAR T cells-challenges and opportunities in translating innovative treatment concepts. EMBO Mol Med. 2017;9(9):1183–97.

77. Grupp SA, Maude SL, Rives S, Baruchel A, Boyer MW, Bittencourt H, et al. Updated analysis of the efficacy and safety of Tisagenlecleucel in pediatric and young adult patients with relapsed/refractory (r/r) acute lymphoblastic leukemia. Blood. 2018;132(Suppl 1):895.

78. Borchmann P, Tam CS, Jäger U, McGuirk JP, Holte H, Waller EK, et al. An updated analysis of JULIET, a global pivotal phase 2 trial of Tisagenlecleucel in adult patients with relapsed or refractory diffuse large B-cell lymphoma. Abstract S799, presented at the annual congress of the European Hematology Association (EHA) 2018.

79. Schuster SJ, Bishop MR, Tam C, Borchmann P, Jaeger U, Waller EK, et al. Sustained disease control for adult patients with relapsed or refractory diffuse large B-cell lymphoma: an updated analysis of Juliet, a global pivotal phase 2 trial of Tisagenlecleucel. Blood. 2018;132(Suppl 1):1684.

80. Neelapu SS, Ghobadi A, Jacobson CA, Miklos DB, Lekakis LJ, Oluwole OO, et al. 2-year follow-up and high-risk subset analysis of Zuma-1, the pivotal study of Axicabtagene Ciloleucel (Axi-Cel) in patients with refractory large B cell lymphoma. Blood. 2018;132(Suppl 1):2967.

81. Nastoupil LJ, Jain MD, Spiegel JY, Ghobadi A, Lin Y, Dahiya S, et al. Axicabtagene Ciloleucel (Axi-cel) CD19 chimeric antigen receptor (CAR) T-cell therapy for relapsed/refractory large B-cell lymphoma:

real world experience. Blood. 2018;132(Suppl 1):91.

82. Summers C, Annesley C, Bleakley M, Dahlberg A, Jensen MC, Gardner R. Long term follow-up after SCRI-CAR19v1 reveals late recurrences as well as a survival advantage to consolidation with HCT after CAR T cell induced remission. Blood. 2018;132(Suppl 1):967.
83. Sano D, Nastoupil LJ, Fowler NH, Fayad L, Hagemeister FB, Lee HJ, et al. Safety of Axicabtagene Ciloleucel CD19 CAR T-cell therapy in elderly patients with relapsed or refractory large B-cell lymphoma. Blood. 2018;132(Suppl 1):96.
84. Faramand R, Kotani H, Morrissey D, Yu B, Locke FL, Jain MD, et al. Prediction of CAR T-related toxicities in R/R DLBCL patients treated with Axicabtagene Ciloleucel using point of care cytokine measurements. Blood. 2018;132(Suppl 1):95.
85. Ying Z, Huang XF, Xiang X, Liu Y, Kang X, Song Y, et al. A safe and potent anti-CD19 CAR T cell therapy. Nature medicine. 2019.
86. Du H, Hirabayashi K, Ahn S, Kren NP, Montgomery SA, Wang X, et al. Antitumor responses in the absence of toxicity in solid tumors by targeting B7-H3 via chimeric antigen receptor T cells. Cancer Cell. 2019;35(2):221–37 e8.
87. Gargett T, Brown MP. The inducible caspase-9 suicide gene system as a "safety switch" to limit on-target, off-tumor toxicities of chimeric antigen receptor T cells. Front Pharmacol. 2014;5:235.
88. Mestermann K, Julian R, Silke F, Hermann E, Hudecek M. Abstract A037: A novel pharmacologic "ON/OFF" switch to modulate CAR-T-cell function in vitro and in vivo. Cancer Immunol Res. 2019;7(2 Supplement):A037-A.
89. Geyer MB, Riviere I, Senechal B, Wang X, Wang Y, Purdon TJ, et al. Safety and tolerability of conditioning chemotherapy followed by CD19-targeted CAR T cells for relapsed/refractory CLL. JCI Insight. 2019;5.
90. Kim MY, Yu KR, Kenderian SS, Ruella M, Chen S, Shin TH, et al. Genetic inactivation of CD33 in hematopoietic stem cells to enable CAR T cell immunotherapy for acute myeloid leukemia. Cell. 2018;173(6):1439–53 e19.
91. Rafiq S, Yeku OO, Jackson HJ, Purdon TJ, van Leeuwen DG, Drakes DJ, et al. Targeted delivery of a PD-1-blocking scFv by CAR-T cells enhances anti-tumor efficacy in vivo. Nat Biotechnol. 2018;36(9):847–

56.

92. Chmielewski M, Abken H. CAR T cells releasing IL-18 convert to T-bet(high) FoxO1(low) effectors that exhibit augmented activity against advanced solid tumors. Cell Rep. 2017;21(11):3205–19.
93. Chmielewski M, Hombach AA, Abken H. Of CARs and TRUCKs: chimeric antigen receptor (CAR) T cells engineered with an inducible cytokine to modulate the tumor stroma. Immunol Rev. 2014;257(1):83–90.
94. Kloess S, Kretschmer A, Stahl L, Fricke S, Koehl U. CAR-expressing natural killer cells for Cancer retargeting. Transfus Med Hemother. 2019;46(1):4–13.
95. Akce M, Zaidi MY, Waller EK, El-Rayes BF, Lesinski GB. The potential of CAR T cell therapy in pancreatic Cancer. Front Immunol. 2018;9:2166.
96. Tokarew N, Ogonek J, Endres S, von Bergwelt-Baildon M, Kobold S. Teaching an old dog new tricks: next-generation CAR T cells. Br J Cancer. 2019;120(1):26–37.
97. Legut M, Dolton G, Mian AA, Ottmann OG, Sewell AK. CRISPR-mediated TCR replacement generates superior anticancer transgenic T cells. Blood. 2018;131(3):311–22.
98. Raman MC, Rizkallah PJ, Simmons R, Donnellan Z, Dukes J, Bossi G, et al. Direct molecular mimicry enables off-target cardiovascular toxicity by an enhanced affinity TCR designed for cancer immunotherapy. Sci Rep. 2016;6:18851.
99. Rosenberg SA, Restifo NP. Adoptive cell transfer as personalized immunotherapy for human cancer. Science (New York, NY). 2015;348(6230):62–8.
100. Leung W, Heslop HE. Adoptive immunotherapy with antigen-specific T cells expressing a native TCR. Cancer Immunol Res. 2019;7(4):528–33.
101. Dahl K, Karlsson M, Marits P, Hoffstedt A, Winqvist O, Thörn M. Metinel node—the first lymph node draining a metastasis—contains tumor-reactive lymphocytes. Ann Surg Oncol. 2008;15(5):1454–63.
102. Karlsson M, Marits P, Dahl K, Dagöö T, Enerbäck S, Thörn M, et al. Pilot study of sentinel-node-based adoptive immunotherapy in advanced colorectal cancer. Ann Surg Oncol. 2010;17(7):1747–57.

103. Marits P, Karlsson M, Dahl K, Larsson P, Wanders A, Thorn M, et al. Sentinel node lymphocytes: tumour reactive lymphocytes identified intraoperatively for the use in immunotherapy of colon cancer. Br J Cancer. 2006;94(10):1478–84.
104. Tran E, Turcotte S, Gros A, Robbins PF, Lu YC, Dudley ME, et al. Cancer immunotherapy based on mutation-specific CD4+ T cells in a patient with epithelial cancer. Science (New York, NY). 2014;344(6184):641–5.
105. Tran E, Robbins PF, Lu YC, Prickett TD, Gartner JJ, Jia L, et al. T-cell transfer therapy targeting mutant KRAS in Cancer. N Engl J Med. 2016;375(23):2255–62.
106. Zacharakis N, Chinnasamy H, Black M, Xu H, Lu YC, Zheng Z, et al. Immune recognition of somatic mutations leading to complete durable regression in metastatic breast cancer. Nat Med. 2018;24(6):724–30.
107. Sarnaik AA TS, Davar D,, editor Safety and efficacy of cryopreserved autologous tumor infiltrating lymphocyte therapy (LN-144, lifileucel) in advanced metastatic melanoma patients following progression on checkpoint inhibitors. 33rd annual meeting & pre-conference programs of the Society for Immunotherapy of Cancer (SITC 2018); 2018 November 06.

2

Recent Updates in Cancer Immunotherapy: a Comprehensive Review and Perspective of the 2018 China Cancer Immunotherapy Workshop in Beijing

Zihai Li[1,2], Wenru Song[2] , Mark Rubinstein[1] and Delong Liu[2,3]

[1] Hollings Cancer Center, Medical University of South Carolina, Charleston, SC 29425, USA
[2] Chinese American Hematologist and Oncologist Network, New York, NY, USA
[3] New York Medical College, New York, NY, USA

ABSTRACT

The immune system is the hard-wired host defense mechanism against pathogens as well as cancer. Five years ago, we pondered the question if the

Citation: Li Z, Song W, Rubinstein M, Liu D. Recent updates in cancer immunotherapy: a comprehensive review and perspective of the 2018 China Cancer Immunotherapy Workshop in Beijing. J Hematol Oncol. 2018;11(1):142. Published 2018 Dec 21. doi:10.1186/s13045-018-0684-3.

era of cancer immunotherapy was upon us (Li et al., Exp Hem Oncol 2013). Exciting progresses have been made at all fronts since then, including (1) sweeping approval of six agents by the US Food and Drug Administration (FDA) to block the PD-1/PD-L1 pathway for treatment of 13 cancer types; (2) a paradigm shifting indication of PD-1 and CTLA4 blockers for the management of a broad class of cancers with DNA mismatch repair defect, the first-ever tissue agnostic approval of cancer drugs; (3) real world practice of adoptive T cell therapy with two CD19-directed chimeric antigen receptor T cell products (CAR-T) for relapsed and/or refractory B cell malignancies including acute lymphoid leukemia and diffuse large B cell lymphoma, signaling the birth of a field now known as synthetic immunology; (4) the award of 2018 Nobel Prize in Physiology and Medicine from the Nobel Committee to Tasuku Honjo and James Allison "for their discovery of cancer medicine by inhibition of negative immune regulation" (www.nobelprize.org/prizes/medicine/2018); and (5) the emerging new concept of normalizing rather than amplifying anti-tumor immunity for guiding the next wave of revolution in the field of immuno-oncology (IO) (Sanmamed and Chen, Cell 2018).

This article will highlight the significant developments of immune-oncology as of October 2018. The US FDA approved indications of all seven immune checkpoint blockers, and two CD19-directed CAR-T products are tabulated for easy references. We organized our discussion into the following sections: introduction, cell therapy, emerging immunotherapeutic strategies, expediting oncology drug development in an era of breakthrough therapies, new concepts in cancer immunology and immunotherapy, and concluding remarks. Many of these topics were covered by the 2018 China Cancer Immunotherapy Workshop in Beijing, the fourth annual conference co-organized by the Chinese American Hematologist and Oncologist Network (CAHON), China FDA (CFDA; now known as China National Medical Product Administration (NMPA)), and the Tsinghua University. We significantly expanded our discussion of important IO developments beyond what were covered in the conference, and proposed a new Three Rs conceptual framework for cancer immunotherapy, which is to reverse tolerance, rejuvenate the immune system, and restore immune homeostasis. We conclude that the future of immuno-oncology as a distinct discipline of cancer medicine has arrived.

INTRODUCTION

It is estimated that by 2035, one quarter of the global populations will be directly affected by cancers (https://cancerprogressreport.org/Pages/cpr18-cancer-in-2018.aspx). There are five main therapeutic modalities for cancer: surgery, radiation, chemotherapy, targeted therapy, and immunotherapy. With a few exceptions, the first four modalities are focused squarely on cancer itself. Immunotherapy represents conceptually a unique way of dealing with cancer which is to focus on eliminating cancer indirectly by harnessing the power of the host's immune system. The concept of cancer immunotherapy has been there for more than a century [1]. But it is only after the turn of this century that it has gained traction thanks to advancements in both basic immunology research [2] and the birth of immuno-oncology (IO) [3]. It is now established that as a genetically altered entity, cancer triggers both innate and adaptive immune response of the host during its evolution. Immune escape is recognized as one of the key hallmarks of cancer [4]. The implication of this fundamental and conceptual shift is significant because it inspires strategies to restore immunity to keep cancer permanently at bay, i.e., cure. Indeed, the discovery of both cellular and molecular mechanisms of cancer immune evasion fuels the development of IO agents, including immune checkpoint blockers against CTLA4, PD-1, and PD-L1 [5–7]. Importantly, the IO field is still at its early stage. There are more questions than answers. For example, less than one quarter of patients overall respond to PD-1/PD-L1 blockers. Frustratingly, there is a lack of biomarkers to predict who will respond and who will not to these agents. There has been no clear breakthrough to enhance efficacy of immune checkpoint inhibitors (ICIs). Furthermore, IO is shaking up the field of cancer medicine, but there is no clear and effective strategy to integrate immunotherapy into the conventional strategies for treating a majority of cancer types. Whereas ICIs have enjoyed unprecedented success, other immunotherapeutic strategies are not there yet in prime time. There are still no effective therapeutic vaccines. Approved cell therapy is also limited to B cell malignancies. The challenges IO field imposes to cancer medicine also include lack of adequate healthcare providers in this emerging field, and struggles of the regulatory agencies in crafting guidelines in steering and accelerating the clinical development of unconventional immune-regulatory agents. In light of these excitement and challenges, a much anticipated 2018 China Cancer Immunotherapy workshop was held in Beijing on June 30th and July 1st. This two-full-day meeting brought together IO experts from academia, industry, and government regulatory agencies around the world. This was the fourth time CAHON has partnered with the China FDA (joined also by Tsinghua University since 2017) to provide a high-level IO education conference annually to physicians, scientists, and drug developers in the industry to help advance IO in China and beyond.

Clinical Updates on Checkpoint Inhibitors

Two sessions of the conference were focused on clinical updates of ICIs. At the time of the conference (June 30–July 1, 2018), one CTLA4 blocker (Ipilumimab), two PD-1 inhibitors (Nivolumab and Pembrolizumab), and three PD-L1 antagonists (Durvalumab, Atezolizumab, and Avelumab) were approved by the US FDA for various indications (Table 1). Subsequently, the third PD-1 blocker Cemiplimab was approved for the treatment of patients with metastatic cutaneous squamous cell carcinoma (CSCC) or locally advanced CSCC who are not candidates for curative surgery or curative radiation. This is based on encouraging clinical study including the positive study by Migden et al. who performed an expansion phase I study as well as the pivotal phase 2 study for patients with metastatic disease CSCC [8]. Patients received cemiplimab i.v. at 3 mg/kg of body weight every 2 weeks and were assessed for clinical response every 8 weeks. Deep response in the phase 1 expansion cohort of patients was observed in 50% of patients ($n=26$), which was reproduced in the phase 2 study, with response rate in 28 of 59 patients (47%; 95% CI, 34 to 61). This response appeared to be durable, exceeding 6 months in most patients without observed new immune-related adverse events (irAEs).

Table 1: US FDA approved immune checkpoint blockers for cancer immunotherapy as of Oct 2018

Medicine (target, Ig isotype)	**Indication** ● Melanoma ● RCC ● MSI ● NSCLC ● HNSCC ● SCLC ● HD ● UC ● HCC ● PMBCL ● GC/GEJC ● MCC ● CSCC
Ipilimumab (CTLA-4, IgG1↗)	● ● ● *Full Approval:* • Unresectable or metastatic melanoma in adults and pediatric patients (12 years and older). • Adjuvant treatment of patients with cutaneous melanoma with pathologic involvement of regional lymph nodes of more than 1 mm who have undergone complete resection, including total lymphadenectomy. • Treatment of patients with intermediate or poor risk, previously untreated advanced renal cell carcinoma, in combination with nivolumab. *Accelerated Approval:* • Treatment of adult and pediatric patients 12 years of age and older with microsatellite instability-high (MSI-H) or mismatch repair deficient (dMMR) metastatic colorectal cancer that has progressed following treatment with a fluoropyrimidine, oxaliplatin, and irinotecan, in combination with nivolumab.
Nivolumab (PD-1, IgG4↗)	● ● ● ● ● ● ● ● ● *Full Approval:* • BRAF V600 wild-type unresectable or metastatic melanoma, as a single agent. • Melanoma with lymph node involvement or metastatic disease who have undergone complete resection, in the adjuvant setting. • Metastatic non-small cell lung cancer and progression on or after platinum-based chemotherapy. Patients with EGFR or ALK genomic tumor aberrations should have disease progression on FDA-approved therapy for these aberrations • Advanced renal cell carcinoma who have received prior antiangiogenic therapy. • Recurrent or metastatic squamous cell carcinoma of the head and neck with disease progression on or after a platinum-based therapy. *Accelerated Approval:* • BRAF V600 mutation-positive unresectable or metastatic melanoma, as a single agent. • Unresectable or metastatic melanoma, in combination with ipilimumab. • Metastatic small cell lung cancer with progression after platinum-based chemotherapy and at least one other line of therapy. • Intermediate or poor risk, previously untreated advanced renal cell carcinoma, in combination with ipilimumab. • Adult patients with classical Hodgkin lymphoma that has relapsed or progressed after autologous hematopoietic stem cell transplantation (HSCT) and brentuximab vedotin, or 3 or more lines of systemic therapy that includes autologous HSCT. • Locally advanced or metastatic urothelial carcinoma who have disease progression during or following platinum-containing chemotherapy, or have disease progression within 12 months of neoadjuvant or adjuvant treatment with platinum-containing chemotherapy. • Adult and pediatric (12 years and older) patients with microsatellite instability-high (MSI-H) or mismatch repair deficient (dMMR) metastatic colorectal cancer that has progressed following treatment with a fluoropyrimidine, oxaliplatin, and irinotecan, as a single agent or in combination with ipilimumab. • Hepatocellular carcinoma who have been previously treated with sorafenib.
Pembrolizumab (PD-1, IgG4↗)	● ● ● ● ● ● ● ● ● ● *Full Approval:* • Unresectable or metastatic melanoma. • First-line treatment of patients with metastatic non-squamous non–small cell lung cancer (NSCLC), with no EGFR or ALK genomic tumor aberrations, in combination with pemetrexed and platinum chemotherapy. • First-line treatment of patients with metastatic NSCLC whose tumors have PD-L1 expression [tumor proportion score (TPS) ≥50%], with no EGFR or ALK genomic tumor aberrations, as a single agent. • Single agent indication for patients with metastatic NSCLC whose tumors express PD-L1 (TPS ≥1%), with disease progression on or after platinum-containing chemotherapy. Patients with EGFR or ALK genomic tumor aberrations should have disease progression on FDA-approved therapy for these aberrations prior to receiving KEYTRUDA. • Locally advanced or metastatic urothelial carcinoma (mUC) who have disease progression during or following platinum-containing chemotherapy or within 12 months of neoadjuvant or adjuvant treatment with platinum-containing chemotherapy. *Accelerated Approval:* • Recurrent or metastatic head and neck squamous cell carcinoma (HNSCC) with disease progression on or after platinum-containing chemotherapy. • Adult and pediatric patients with refractory classical Hodgkin lymphoma (cHL), or who have relapsed after 3 or more prior lines of therapy. • Adult and pediatric patients with refractory primary mediastinal large B-cell lymphoma (PMBCL), or who have relapsed after 2 or more prior lines of therapy. • Locally advanced or metastatic urothelial carcinoma (mUC) who are not eligible for cisplatin-containing chemotherapy and whose tumors express PD-L1 [combined positive score (CPS) ≥10], or in patients who are not eligible for any platinum-containing chemotherapy regardless of PD-L1
	status. • Adult and pediatric patients with unresectable or metastatic microsatellite instability-high (MSI-H) or mismatch repair deficient (dMMR) solid tumors that have progressed following prior treatment and who have no satisfactory alternative treatment options, or colorectal cancer that has progressed following treatment with fluoropyrimidine, oxaliplatin, and irinotecan. • Recurrent locally advanced or metastatic gastric or gastroesophageal junction (GEJ) adenocarcinoma whose tumors express PD-L1 [combined positive score (CPS) ≥1], with disease progression on or after two or more prior lines of therapy including fluoropyrimidine- and platinum-containing chemotherapy and, if appropriate, HER2/neu-targeted therapy. • Recurrent or metastatic cervical cancer with disease progression on or after chemotherapy whose tumors express PD-L1 (CPS ≥1).
Cemiplimab (PD-1, IgG4)	● • Metastatic cutaneous squamous cell carcinoma (CSCC) or locally advanced CSCC who are not candidates for curative surgery or curative radiation.
Durvalumab (PD-L1, IgG1↗)	● ● *Full Approval:* • Unresectable Stage III non-small cell lung cancer (NSCLC) whose disease has not progressed following concurrent platinum-based chemotherapy and radiation therapy. *Accelerated Approval:* • Locally advanced or metastatic urothelial carcinoma who have disease progression during or following platinum-containing chemotherapy, or who have disease progression within 12 months of neoadjuvant or adjuvant treatment with platinum-containing chemotherapy
Atezolizumab (PD-L1, IgG1↗)	● ● *Full Approval:* • Metastatic non-small cell lung cancer who have disease progression during or following platinum-containing chemotherapy. Patients with EGFR or ALK genomic tumor aberrations should have disease progression on FDA approved therapy for these aberrations *Accelerated Approval:* • Locally advanced or metastatic urothelial carcinoma who are not eligible for cisplatin-containing chemotherapy and whose tumors express PD-L1 (PD-L1 stained tumor-infiltrating immune cells [IC] covering ≥ 5% of the tumor area), are not eligible for any platinum-containing chemotherapy regardless of PD-L1 status, or have disease progression during or following any platinum-containing chemotherapy, or within 12 months of neoadjuvant or adjuvant chemotherapy.
Avelumab (PD-L1, IgG1↙)	● ● *Accelerated Approval:* • Adults and pediatric patients 12 years and older with metastatic Merkel cell carcinoma (MCC). • Patients with locally advanced or metastatic urothelial carcinoma (UC) who: have disease progression during or following platinum-containing chemotherapy, or have disease progression within 12 months of neoadjuvant or adjuvant treatment with platinum-containing chemotherapy.

Altogether at the time of writing this report (Oct 27, 2018), seven ICIs have been approved collectively for the standard treatment of a total of 13 cancer types. Excitingly, the US FDA has also granted accelerated approval for Nivolumab (with and without Ipilumimab) and Pembrolizumab for the management of advanced cancers with DNA mismatch repair deficiency, regardless of the histology of the cancer types, marking for the first time the approval of cancer medicine in a tissue-agnostic fashion. The clinical experiences with these agents were highlighted in designated talks by Weijing Sun (gastrointestinal cancer, University of Kansas), Yiping Yang (hematological malignancy, Duke University), Jun Zhu (lymphoma, Beijing University Cancer Hospital), Mario Sznol (melanoma, Yale University), Jun Guo (melanoma, Beijing University Cancer Hospital), Yilong Wu (lung cancer, Guangdong General Hospital), Shukui Qin (hepatocellular carcinoma, Nanjing PLA Hospital), and Jingshong Zhang (genitourinary cancer, Lee Moffitt Cancer Center). In addition to the agents approved in the USA, researchers from China also presented exciting data regarding PD-1 inhibitors and other IO agents developed in China, by the following companies: Hengrui, Innovent, Beigene, Jun Shi, 3DMed, Zai Lab, and I-Mab.

Of note, clear differences do exist in both the distribution and biology of cancers between the West and the East, underscoring the importance of conducting IO trials in China rather than totally depending on clinical experience in other parts of the world, for guiding the IO approval process. In Asia, liver and upper gastrointestinal cancers are epidemics which may have different underlying biology. Whereas both acral and mucosal melanoma are exceedingly rare in the USA at 5% and 1–2% of all melanomas, Jun Guo pointed out that in China these two subsets could be 49.4% and 22.6% respectively [9]. Sznol highlighted the experience with stage IV melanoma with ICIs. Nivolumab plus Ipilimumab is an approved strategy in this setting. Among all the patients treated with this combination ($N=94$) in the initial phase I trial at a follow-up of 30.3 to 55.0 months, the 3-year overall survival rate was 63% and median overall survival had not been reached at the time of the publication of the analysis [10]. The investigators reported 42% objective response rate by modified WHO criteria, and median duration of response was 22.3 months. Unfortunately, the improved efficacy is also accompanied by the increased incidence of severe (grade 3 and 4) treatment-related adverse events at 59%. Nonetheless, the 3-year OS rate of 63% in advanced melanoma highlighted the significant clinical utility and efficacy of ICIs. Interestingly, the appearance of $CD21^{low}$ B cells in the peripheral blood in a study with a small cohort of patients appears to predict immune-related adverse events (irAEs) without affecting efficacy [11]. Sznol also outlined practical principle in the management of irAEs, by recommending the following: (a) ruling out the possibility of disease progression or infection, (b) following established guidelines [12–14], (c) having low threshold to start corticosteroids and admit the patients to the hospital for inpatient care, (d) maintaining high dose steroids for at least 1 week and tapering slowly over 30–40 days, and (e) discontinuing IO agents permanently for grade IV irAEs. These points were further underscored by Helen Chen (US National Cancer Institute), who cautioned of risks of enhanced toxicities with immunotherapy combinations with targeted agents. Several interesting combinations have since been discontinued due to increased toxicities, including durvalumab plus osimertinib (pneumonitis), tremelimumab plus suninitinib (renal failure), crizotinib plus nivolumab (hepatic toxicities) [15], and nivolumab plus pazopanib (hepatic toxicities).

Cell Therapy

2018 marked the year when IO enjoys unprecedented growth at many fronts. In a comprehensive analysis of the global IO landscape, Tang and colleagues found that in the span of just 1 year (September 2017 to September 2018), there was a 67% increase in the number of active agents in the global IO pipeline (2031 versus 3394) [16]. Impressively, the cell therapy class had the largest growth—a whopping 113% increase in the number of active agents. While it may be argued that bone marrow or hematopoietic stem cell transplantation represents the best-established cell therapy for human malignancy, CD19-targeted CAR-T cells for B cell neoplasms open up the imagination of scientists in the field in perhaps signaling what more could come in this extraordinary space. There are two approved CD19-CAR-T cell platforms: Tisagenlecleucel (Kymriah) and Axicabtagene ciloleucel (Yescarta), which have similarities and differences (Table 2). Both agents are autologous peripheral T cells engineered ex vivo to express a transmembrane chimeric antigen receptor composed of an extracellular antigen-specific single chain antibody and an intracellular T cell signaling domain. Both agents utilize single chain anti-CD19 antibody to target B cells, and CD3ζ intracellular signaling motif to deliver primary activating signals to T cells. However, tisagenlecleucel employs additional CD137 (4-1BB) signaling for co-stimulation as opposed to axicabtagene which does so with a CD28 signaling cassette. Both agents have been approved in the USA for the treatment of relapsed or refractory large B cell lymphoma after two or more lines of systemic therapy. Tisagenlecleucel is additionally approved for the treatment of patients up to 25 years of age with B cell precursor acute lymphoblastic leukemia (ALL) that is refractory or in second or later relapse.

Table 2: Comparison of two US FDA approved CAR-T products for B cell malignancies

Medicine	Signaling motifs	Dosage	Indication
YESCARTA Axicabtagene ciloleucel (Yescarta)	CD28 and CD3ζ	2×10^6 CAR-positive viable T cells per kg body weight, with a maximum of 2×10^8 CAR-positive viable T cells.	• Adult patients with relapsed or refractory large B cell lymphoma after two or more lines of systemic therapy, including diffuse large B cell lymphoma (DLBCL) not otherwise specified, primary mediastinal large B cell lymphoma, high grade B cell lymphoma, and DLBCL arising from follicular lymphoma.
KYMRIAH™ Tisagenlecleucel (Kymriah)	CD137 (4-1BB) and CD3ζ	Pediatric and young adult B cell ALL (up to 25 years of age): • For patients 50 kg or less, administer 0.2 to 5.0×10^6 CAR-positive viable T cells per kg body weight intravenously. • For patients above 50 kg, administer 0.1 to 2.5×10^8 total CAR-positive viable T cells (non-weight based) intravenously. Adult relapsed or refractory diffuse large B cell lymphoma: •Administer 0.6 to 6.0×10^8 CAR-positive viable T cells intravenously.	• Patients up to 25 years of age with B cell precursor acute lymphoblastic leukemia (ALL) that is refractory or in second or later relapse. • Adult patients with relapsed or refractory (r/r) large B cell lymphoma after two or more lines of systemic therapy including diffuse large B cell lymphoma (DLBCL) not otherwise specified, high grade B cell lymphoma and DLBCL arising from follicular lymphoma, excluding primary central nervous system lymphoma.

The presentation by Patrick Hwu (MD Anderson Cancer Center), Ke Liu (US FDA), Weidong Han (Army Hospital in Beijing), Sen Zhuang

(Johnson & Johnson), and Chunyan Gao (China National Medical Product Administration) discussed a number of important issues about cell therapy, as follows:

Flavors of Cell Therapy

Cellular products in the clinical application and testing including hematopoietic stem cells, CAR-T cells against CD19, and other targets, T cells engineered to express T cell receptor with known specificity (TCR-T), tumor-reactive or tumor-infiltrating T cells isolated and expanded from cancer patients (otherwise known as endogenous T cells, or ETC), polyclonal tumor-reactive T cells (tumor-infiltrating T cells, or TILs) isolated from the tumor, NK cells, NKT cells, dendritic cells, etc. Patrick Hwu summarized the MDACC experience in their TIL therapy program for 74 metastatic melanoma patients from 2007 to 2017 [17]. They found that the best overall response for the entire cohort was 42%: 47% in 43 ICIs-naïve patients, 38% when patients were exposed to anti-CTLA4 alone (21 patients) and 33% if also exposed to anti-PD1 (9 patients) prior to TIL therapy. Median overall survival was 17.3 months; 24.6 months in CTLA4-naïve patients and 8.6 months in patients with prior CTLA4 blockade. The latter patients were infused with fewer TILs and experienced a shorter duration of response. They found that infusion of higher numbers of TIL with CD8 predominance and expression of BTLA (B And T Lymphocyte Associated) by the tumor cells correlated with improved response in anti-CTLA4 naïve patients, but not in anti-CTLA4 refractory patients. Baseline serum levels of IL9 predicted response to TIL therapy, while curiously TIL persistence, tumor recognition, and mutation burden did not correlate with outcome. They concluded that there are deleterious effects of prior exposure to anti-CTLA4 on TIL therapy response. Hwu discussed a number of strategies to improve TIL cell therapy based on rational thinking and preclinical data including stably expressing dominant negative TGFβ receptor II in the TIL products to overcome immune suppression in the tumor microenvironment [18] and transduction of T cells with CXCR2 to allow them to better migrate to the tumor sites [19]. Importantly, recent breakthroughs in genomic medicine and informatics enable the detection of neoantigen epitopes and subsequent expansion of antigen-specific TILs using these antigens in the context of appropriate HLA. Adoptive transfer with neoantigen-specific T cells has been shown to mediate objective clinical responses in patients with metastatic bile duct, colon, and cervical cancers, as well as triple negative breast cancers [20–23]. The practical challenge of this approach is similar to what CD19-CAR-T technology faced almost 10 years ago [24], which is to determine how to move exciting proof-of-principle science from the academic settings to real world clinical practice.

Targeting Antigens of CAR-T cells

Without doubt, the bottle neck to prevent CAR-T technology to be widely used clinically is the lack of optimal target antigens for a majority of cancers like CD19 for B cell malignancies. Patients with B cell aplasia can live relatively healthy with maintenance therapy of intravenous immunoglobulins from normal donors. In comparison, life cannot be sustained with lack of myeloid cells which is why CAR-T based strategy has not found significant success for the treatment of myelodysplastic syndrome or acute myelogenous leukemia. To circumvent this problem, Kim et al. deleted CD33 from the normal human hematopoietic stem cells and transplanted into rhesus macaques with long-term multilineage engraftment with normal myeloid function [25]. These CD33-deficient cells then allow CD33-targeted CAR-T therapy for efficient elimination of $CD33^+$ leukemia without myelotoxicity. For plasma cell disorder, it is a different story. One can afford the ablation of normal plasma cells in order to eradicate malignant plasma clone with CAR-T based strategy. In this regard, BCMA (B cell maturation antigen)-CAR-T-based strategy, LCAR-B38M, was discussed by Sen Zhuang (Johnson and Johnson). A confirmation clinical trial has started in the USA, followed by the original encouraging data in China with 35 patients who participated in the study. In that study, all patients responded to the therapy, with 94% showing sustained complete or near-complete remission [26, 27]. As of July 2018, a total of 74 patients have been treated with LCAR-B38M, updated by Frank Fan (Legend Biotech). In various phases of clinical trials are also CAR-T cells targeting other cell surface antigens including GD2, HER2, CD20, EBV antigen, mesothelin, CD33, CD22, CD30, CD123, EGFR, PSMA, WT1, GPC3, CD38, EGFRvIII, MUC1, PDL1, and neoantigens [28].

Dual, Switchable, Off-the-shelf, SUPRA CAR-T, etc.

Weidong Han discussed multiple efforts in designing safer and more effective CAR-T strategies [29]. By gene editing methodology, genes encoding human leukocyte antigen (HLA) molecules and endogenous T cell receptors (TCRs) can be deleted and these T cells will then be transduced to express CAR-T construct, followed by expansion in vitro, cryopreservation, and aliquoting. These products can then be used for any patients whose cancer express the target of the CAR-T cells. This effort is ongoing for $CD19^+$ B cell malignancies. One switchable CAR-T cell strategy is to make these T cells to bind to a specific peptide that is genetically engrafted onto a tumor-binding Fab molecule. The "switch" acts as a bridge between target and effector cells, which can be titrated due to the relatively short half-life of the Fab fragment. It was found that such a strategy worked well in a preclinical model against human $Her2^+$ cancer in a mouse xenograft system [30]. Multiple other strategies have been developed to control CAR-T activity including using combinatorial antigen-sensing system [31], or engineering a built-in suicide system in the CAR to allow physicians to switch off CAR-T when unwanted toxicities emerge. Another exciting strategy was the so-called SUPRA CAR, which is a split, universal, and programmable system [32]. It has a two-component receptor system composed of a universal receptor (zipCAR) expressed on T cells and a tumor-targeting scFv adaptor (zipFv). Both the receptor and scFv adaptor contains leucine zipper, allowing targeting of multiple antigens without further genetic manipulations of a patient's T cells. This strategy had remarkable successes in preclinical models against several types of cancer by simultaneously targeting multiple antigens using one batch of engineered zipCAR-T cells.

Regulatory Challenges

Ke Liu (US FDA) and Chenyan Gao (CFDA) discussed the regulatory challenges imposed by the intense interests of the public in CAR-T technology. Like other products, the regulatory agencies uphold three basic principles when it comes to evaluate cell therapy products for approval: substantial evidence of efficacy, acceptable safety, and appropriate patient population. Ke Liu cautioned that both CD19-CAR-T products on the market carry black box warning for cytokine release syndrome and neurotoxicity. He emphasized that much work needs to be done in solid tumor space with focus on target identification, understanding and enhancing CAR-T cell tracking and homing to tumor site, to maximize the clinical benefit.

Emerging Immunotherapeutic Strategies

A number of exciting progresses have been made to usher the field of IO into the next phase, which is beyond ICIs against PD-1, PD-L1, and CTLA4. Space is limited to cover all of the new developments. What were highlighted in 2018 China Cancer Immunotherapy Workshop included the following:

Search for other surface-bound immune checkpoint molecules

Mounting evidence suggest there are additional immune checkpoint molecules to constrain tumor-reactive T cells. Through single-cell RNAseq and proteomics approach, a recent work from Anderson Regev, Kuchroo and colleagues discovered a module of co-inhibitory receptors in both $CD4^+$ and $CD8^+$ T cells that includes PD-1, TIM-3, LAG-3, TIGIT, activated protein C receptor (PROCR), and podoplanin (PDPN) [33]. The module of co-inhibitory receptors is shared by non-responsive T cells in several physiological contexts and is driven by the immunoregulatory cytokine IL-27. Importantly, they found that PRDM1 and c-MAF serve as cooperative transcription regulators of the co-inhibitory module. Chen Dong (Tsinghua University) updated his work on B7 superfamily member 1 (B7S1), also called B7-H4, B7x, or VTCN1. They found that the increased B7S1 expression on myeloid cells from patients with hepatocellular carcinoma correlated with $CD8^+$ T cell dysfunction [34]. The receptor of B7S1, yet to be defined, is co-expressed with PD-1 but not Tim-3 on T cells during activation, which promotes T cell exhaustion. Intriguingly, blocking of both B7S1 and PD-1 synergistically enhanced anti-tumor immune responses. Using a high throughput functional screening strategy, the team of Lieping Chen (Yale) discovered a cell surface molecule that is expressed by a subset of myeloid cells and tumor cells (ovarian, lung, bladder, pancreas, head, and neck cancer) called Siglec15 (unpublished). Although the receptor for Siglec15 on T cells has not been molecularly defined yet, Siglec clearly plays negative roles for T cell activation and function by inducing suppressive myeloid cells. In an unprecedented pace, NC318, a Siglec15 targeting antibody, has already entered a phase 1/2 clinical trial in patients with advanced or metastatic solid tumors.

Immunogenomics and precision immunotherapy

Precision immunotherapy requires understanding of both tumor microenvironment (the tumor) and macroenvironment (the host, i.e., the patient). A comprehensive presentation was delivered by Elizabeth Jaffee (Johns Hopkins), Tim Chan (Memorial Sloan-Kettering), Drew Pardoll (Johns Hopkins), and Siwen Hu-Lieskovan (UCLA). Immunogenomics is a rapid expanding area that allows researchers to interrogate and understand how changes of the cancer genome affect immunity or treatment responsiveness. For example, understanding tumor mutation burden (TMB), immunoediting score etc. will enable researchers and physicians to guide ICI therapy [35, 36]. Understanding TCR repertoire, neoantigen epitopes and HLA haplotypes will facilitate effort in neoantigen vaccine development and cell therapy. Jaffee discussed their meta-analysis results of patients on anti-PD-1/PD-L1 agents whose exome sequencing information were available [37]. They found a strong relationship between the tumor mutational burden and the activity of anti–PD-1 therapies across multiple cancer types. Their analysis allowed them to calculate objective response rate (ORR) with a linear correlation formula: $ORR = 10.8 \times \log_e(X) - 0.7$, where "*X*" is the number of coding somatic mutations per megabase of DNA. Validation of this finding with future prospective trials shall be helpful to guide the selection of patients for ICIs. Catherine Wu and her colleagues have identified a subcluster of MAGE-A cancer-germline antigens, located within a narrow 75 kb region of chromosome Xq28, that predicts resistance uniquely to blockade of CTLA4, but not PD-1 [38]. Tim Chan discussed the exciting study from his group that highlighted the importance of mutation of specific genes correlating to ICI responsiveness. They reported that somatic mutations in SERPINB3 and SERPINB4 are associated with survival after anti-CTLA4 immunotherapy in two independent cohorts of patients with melanoma ($n = 174$), although the underlying mechanism is unclear [39]. Furthermore, Tim Chan's group determined the HLA class I genotype of 1535 advanced cancer patients treated with ICIs. They found that maximal heterozygosity at HLA class loci correlated with improved overall survival compared with patients who were homozygous for at least one HLA locus. Curiously, in two independent melanoma cohorts, patients with the HLA-B44 had extended survival, whereas the HLA-B62 supertype (including HLA-B*15:01) or somatic loss of heterozygosity at HLA class I was associated with poor outcome [40]. Hu-Lieskovan discussed several lines of work in UCLA, including a remarkable 70% clinical response of patients with desmoplastic melanoma to PD-1 blockers, which correlated with high tumor mutation burden and frequent NF1 mutations in this unique subset of melanoma patients [41]. PD-1 blocker-based therapy ultimately depends on $CD8^+$ T cells and IFNγ for cancer eradication. Not surprisingly, loss of function mutations of MHC class I (e.g., loss of β2m) and key IFNγ signaling molecules JAK1/2 in the cancer are associated with intrinsic resistance to anti-PD-1 therapy [42, 43]. Perhaps, a more striking example of impact of cancer genomics on ICI treatment is the status of microsatellite instability-high (MSI-H) or DNA mismatch repair deficiency (dMMR) in the tumors [44–47]. About ~ 50% patients with advanced cancers and the defect in the mismatch repair pathway will derive clinical benefit in response to nivolumab or pembrolizumab. Genomics study of cancer can also shed light on the mechanism of immune evasion. For example, a multi-omic analysis of 1211 colorectal cancer primary tumors reveals that it should be possible to better monitor resistance in the 15% of cases that respond to ICI therapy and also to use WNT signaling inhibitors to reverse immune exclusion in the 85% of cases that currently do not [48]. Genomic and immunologic studies have also uncovered specific driver mutations correlated with lower (CTNNB1, NRAS, or IDH1) or higher (BRAF, TP53, or CASP8) leukocyte levels across all cancers [49]. The oncogenic pathways [50], such as PTEN loss [51, 52], and activation of the WNT/β-catenin signaling pathway [53] have been shown to lead to poor T cell infiltration and function in the tumor microenvironment.

In the field of personal neoantigen vaccines [54], there have been several high profile proof-of-principle studies. Ott et al. demonstrated the feasibility, safety, and immunogenicity of a neoantigen vaccine platform (up to 20 personized HLA-A/B-restricted peptides plus poly-ICLC as adjuvant) that targets advanced melanoma [55]. Evidence for T cells discriminating mutated from wild-type antigens was shown for some patients. Another group tested RNA-based poly-neo-epitope approach for patients with melanoma [56]. They found evidence suggesting that patients developed T cell responses against multiple vaccine neo-epitopes and increased T cell infiltration and neo-epitope-specific killing of autologous tumor cells in post-vaccination resected metastases. Although the sample size is too low to conclude the clinical utility for all of these studies, the neoantigen-based approach may prove to be useful in the adjuvant setting, particularly in combination with ICIs. Pardoll discussed their allele-integrated deep learning framework for improving class I and class II HLA-binding predictions, which may be useful for future neoantigen vaccine effort and also the expansion of tumor antigen-specific T cells [57]. Jaffee also discussed the Hopkins experience on the combination of neoantigen vaccine and ICIs and other IO agents such as CD40 agonist, CXCR4 inhibitor, and agents that target CD47, CSF1R, IDO, TGF-β, A2A, etc. But these studies are mostly at the preclinical stage. Undoubtedly, effective cancer immunotherapy depends on robust priming of tumor-specific T cells, enabling T cells to infiltrate the tumors and ensuring effective mechanism to prevent T cell dysfunction due to hostile tumor microenvironment.

Targeting soluble immune checkpoint

Besides cell surface immune checkpoint molecules, there are multiple soluble immune suppressive factors that play important roles in maintaining immune homeostasis. These factors include but, are not limited to, prostaglandins, nitric oxide, IL-10, TGF-β, IL-33, IL-35, IL-4, IL-13, IL-37, and VEGF. Thorsson et al. performed an extensive immunogenomic analysis of more than 10,000 tumors comprising 33 diverse cancer types by mining the TCGA data [49]. They identified six immune subtypes, including wound healing, IFNγ dominant, inflammatory, lymphocyte depleted, immunologically quiet, and TGF-β dominant. The importance of TGF-β in driving immune suppression and its place in targeted cancer immunotherapy was discussed by Zihai Li (Medical University of South Carolina). Accumulating evidence suggest that TGF-β is a key mechanism for resistance to blockade to PD-1/PD-L1 in multiple cancer types including bladder cancer [58], colorectal cancer [59], and others. However, TGF-β targeting alone, either with small molecule inhibitors of the signaling pathway or anti-TGF-β antibody, has met with limited clinical success due to narrow therapeutic window and heterogeneity of cancer biology in patient populations [60]. Recently, a bifunctional molecule targeting both PD-L1 and TGF-β, called M7824, has been developed [61]. M7824 is a chimeric molecule containing the N-terminal region of fully human IgG1 against human PD-L1 and the C-terminal TGF-β neutralizing trap component from the extracellular domain of the human TGF-β receptor 2. Preclinically, M7824 efficiently binds PD-L1 and TGF-β in vivo and suppressed tumor growth and metastasis more effectively than treatment with either an anti-PD-L1 antibody or TGF-β trap alone in syngeneic mouse models. Encouragingly, M7824 treatment resulted in activation of both the innate and adaptive immune systems, and synergize with radiotherapy or chemotherapy in mouse models. Gulley and his colleagues conducted a phase I open-label trial of M7824 in 19 heavily pretreated patients with advanced solid tumors [62]. They found that M7824 hit and saturated the targets at >1 mg/kg. Clinical efficacy was seen across all dose levels, including one ongoing confirmed complete response (cervical cancer), two durable confirmed partial responses (PR; pancreatic cancer, anal cancer), one near-PR (cervical cancer), and two cases of prolonged stable disease at study entry (pancreatic cancer, carcinoid). Ongoing clinical studies of M7824 include treatment of patients with colorectal cancer, HPV^+ malignancies, and a planned trial to compare M7824 with pembrolizumab as a first-line treatment in patients with PD-L1-expressing advanced non-small cell lung cancer (NSCLC).

Another development in the TGF-β field is the discovery of a cell surface dock receptor for activation of latent TGF-β, called Glycoprotein A Repetitions Predominant (GARP) [63]. Encoded by *LRRC32*, GARP has its restricted expression by regulatory T cells [64, 65] and platelets [66] in normal individuals. Whether GARP is expressed by cancer cells and how it impacts cancer have been investigated. It was found that GARP promotes oncogenesis and immune tolerance by enriching and activating latent TGF-β in the tumor microenvironment [67]. GARP expression and folding depends on a pro-oncogenic molecular chaperone gp96 in the endoplasmic reticulum [68]. Importantly, by both gain- and loss-of-function studies using normal mammary gland epithelial cells and carcinomas, GARP was found to increase the bioactivity of TGF-β and promote malignant transformation in immune-deficient mice [67]. In immune-intact mice, over-expression of GARP in mammary carcinomas drives expansion of regulatory T cells, which contributes to enhanced cancer progression and metastasis [67]. Intriguingly, Rachidi et al. discovered that constitutive expression of GARP on platelets is the most important mechanism of TGF-β activation in vivo, placing platelets squarely in the immune suppressive workforce [69]. Finally, several GARP-specific monoclonal antibodies have been reported. In one case, GARP-targeted antibody was shown to reduce regulatory T cell function in vivo [70]. In another case, a competitive anti-GARP antibody to block the binding between GARP and LTGF-β showed significant activity to perturb metastasis in an orthotopic breast cancer model [67]. Thus, a gp96-GARP-TGF-β switch is a novel oncogenic mechanism that can be exploited for both diagnostic and therapeutic purposes.

Rational combination therapy

The success of ICIs against the broad spectrum of cancers has now reset the baseline of IO. The focus of the IO field for the last 5 years has not been on replacing ICIs but on how to improve their efficacy for a greater proportion of patients. This topic became the central theme of the conference and was touched upon by almost all the speakers especially Lei Zheng (Johns Hopkins), Yang-Xin Fu (UT Southwestern), and Elizabeth Jaffee (Johns Hopkins). There are existing approved combination therapies with nivolumab and ipilimumab for treatment of advanced melanoma, renal cell carcinoma, MSI high tumors, etc. (Table 1). The first-line treatment of patients with metastatic NSCLC, without EGFR or ALK genomic tumor aberrations, is also in combination with pemetrexed and platinum chemotherapy. Not surprisingly, there has been an impressive increase in new combination studies in the past

5 years. Analyses of the Cancer Research Institute database by Tang and his colleagues show that in 2017 alone, 469 new studies were started, with a target enrollment of 52,539 patients, principally being combined with anti-PD-1/L1 agents [71]. For example, a phase 1b clinical trial was conducted to study the impact of oncolytic virotherapy with talimogene laherparepvec in combination with pembrolizumab for advanced melanoma [72]. Confirmed objective response rate was 62%, with a complete response rate of 33% per immune-related response criteria. Responders had increased $CD8^+$ T cells, elevated PD-L1 protein expression, as well as IFN-γ gene expression on several cell subsets in the tumors [72]. Excitingly, during the 2018 European Society for Medical Oncology (ESMO) annual meeting, a positive result of Phase 3 KEYNOTE-426 trial was announced by the study sponsors. This study tests pembrolizumab plus axitinib versus sunitinib alone in treatment-naive advanced/metastatic renal cell carcinoma (mRCC) (NCT02853331). A total of 861 patients with advanced or metastatic RCC were randomized to receive frontline treatment with pembrolizumab (200 mg IV every 3 weeks) plus axitinib (5 mg orally twice daily) for up to 24 months, or sunitinib (50 mg orally once daily for 4 weeks followed by no treatment for 2 weeks, continuously). No new safety concerns were raised. Although the final data is not available yet, the earlier study leading to the trial indeed offered encouraging results to potentially change the standard of practice for the treatment of advanced RCC [73]. There are also interests in combining cytokine-directed therapy with ICIs, as in the case of M7824 mentioned above to block TGFβ and PD-L1 simultaneously. The roles of common γ-chain cytokines including second generation IL-2 and IL-15 in boosting ICIs have also gained attention. For example, one encouraging phase Ib study has shown the utility of the combination of nivolumab and ALT-803 for patients with metastatic NSCLC [74]. ALT-803 is a homo-dimer of IL-15Rα-Fc (IgG1) bound with recombinant IL-15^{N72D} [75]. A pegylated IL-2, NKTR-214, which is a pro-drug and has the preferential release of the active IL-2 in the tumor microenvironment, has an excellent preclinical activity [76] and is now being tested in combination with ICIs for multiple malignancies in multiple settings. However, abundant evidence also sends a cautionary note to the field that the effective combination therapy is easy said than done. Indoleamine 2,3-dioxygenase 1 (IDO1) is a rate-limiting enzyme in the tryptophan catabolism and plays important roles in immune suppression [77]. It makes rational sense to combine inhibitors of PD-1 and IDO for cancer immunotherapy. However, despite the encouraging early phase data [78, 79], a recent phase III ECHO 301 trial testing the combination

of epacadostat (an orally bioavailable IDO inhibitor) with pembrolizumab in melanoma did not show superior outcome compared to pembrolizumab alone [80].

Lei Zheng (Johns Hopkins) discussed rational thought process in designing combination therapy. Ideally, the two combined agents or modalities shall have single agent efficacy (such as PD-1 and CTLA4 inhibitors), non-overlapping mechanism of actions and toxicities (e.g., ICIs and cytotoxic agents), and being used for the right populations of patients selected carefully based on precision biomarkers. The last point is important for IO agents because, for example, one would not want to treat T cell excluded tumors with agents that reverse T cell exhaustion only [81]. In patients when the frequency of tumor-reactive T cells is low, strategies need to be brought forward with vaccinations (proper antigens with new generation of adjuvants), adoptive transfer of tumor-reactive T cells, and mechanisms to amplify T cell responses with co-stimulatory agents (such as CD137 agonist), survival cytokines, and means to tame immune tolerance mechanisms such as turning off regulatory T cells.

Yang-Xin Fu (UT Southwestern) discussed several novel agents and their application preclinically by targeting both innate and adaptive immunity, which highlighted a number of important principles for developing future IP agents. LIGHT (TNFSF14) is immune stimulatory cytokine. A bifunctional molecule has been generated to link anti-EGFR antibody on the one arm with a three tandem LIGHT fused with Fc domain on the other arm. This α-EGFR-LIGHT fusion protein was shown to be able to overcome resistance to anti-PD-1 and convert non-T cell infiltrating ("cold") tumor to tumors with increased infiltrating T cells ("hot") tumor. Interestingly, a series of works from Fu and his colleagues showed that therapeutic roles of commonly used antibodies in oncology (against Her2, EGFR and CD20 for example) are dependent on T cells [82–84], providing a rationale for combining these antibodies with ICIs for cancer immunotherapy. Another intriguing strategy is targeting CD47, a "do not-eat-me" signal on macrophages and other antigen-presenting cells for cancer immunotherapy [85, 86]. A humanized anti-CD47 antibody, Hu5F9-G4, has demonstrated therapeutic efficacy in vitro and in vivo in patient-derived orthotopic xenograft models on five aggressive pediatric brain tumors [87]. The roles of CD47-targeting monotherapy might be problematic due to the significant side effect of causing red blood cell destruction and lack of preference of targeting tumor-infiltrating macrophages. However, by priming (1 mg/kg) and maintenance (10–30 mg/kg weekly starting week 2) dosing, the anemia induced by Hu5F9-G4 can be mitigated. When it was combined with rituximab, promising activity was seen in patients with refractory B cell lymphoma in a phase 1b study involving 22 patients [88]. Liu et al. recently revealed that CD47 and PD-L1 on tumor cells coordinately suppress innate and adaptive sensing to evade immune control. Targeted blockade of both CD47 and PD-L1 on tumor cells with a bispecific anti-PD-L1-SIRPα agent showed significantly enhanced tumor targeting and therapeutic efficacy comparing with monotherapy [89]. This finding makes sense because the cancer therapeutic effect of targeting CD47 also depends on $CD8^+$ T cells [90].

Expediting oncology drug development in an era of breakthrough therapies

Richard Pazdur (US FDA) provided unique perspectives on oncology drug development including in the area of IO. The FDA oversees medical and food industries that are a quarter of the America's expenditures. It is responsible for assurance of the safety, efficacy, and security of these products. The hematology and oncology division has established disease-specific structure that is akin to current academic models, including Division of Oncology Products 1 (dealing with genitourinary, breast, and gynecologic cancer), Division of Oncology Products 2 (thoracic, head and neck, gastrointestinal, melanoma-sarcoma, pediatric/neuroendocrine/rare tumors), Division of Hematology Products (benign hematology products, lymphoma, leukemias, and transplant), and Division of Hematology and Oncology Toxicology (toxicologists supporting each division). Oncology drug development and approval are unique comparing with other therapeutic areas in that cancer deals with severe and life-threatening diseases, it has a large public interest which needs to expedite drugs, the area has different risk tolerance for side effects, there are strong active advocacy groups, the area enjoys one of the most active biomedical research, 50% of breakthrough therapies are in oncology space, and the oncology drug approval often utilizes biomarkers for subgroup patient selection. Regarding efficacy endpoints, FDA has moved away from overall response rate and transitioned to more emphasis on overall survival which means putting more weight on how patients "feel, function, or survive." The explosion of IO field also coincides with the introduction of FDA expedited programs, leading from fast track to breakthrough therapy, to priority review, and eventually to accelerated approval. All ICIs now have indications based on the accelerated approval mechanism which requires post-marketing clinical trials to be underway at the time of approval. FDA also welcomes novel seamless trial design in drug development as opposed to the traditional rigid three discrete phases of clinical trials (I, II, and III). This is especially important for IO drug such as CD19-CAR-T which cannot be ethically tested in the phase III trial setting against relapsed and refractory B cell leukemia and lymphoma because the standard care offers negligible hope for controlling these diseases. Yangmin Ning (US FDA) discussed in details the nuts and bolts of the "Breakthrough Therapy Designation" program which started in 2012. It is designed to accelerate the approval of life-saving drugs with confirmed evidence that likely changes the standard of care for patients. There are two requirements for designating Breakthrough Therapy: (a) life-threatening diseases with unmet medical needs and (b) preliminary clinical evidence showing substantial improvement over available or existing therapies. It is important to keep in mind that such a designation does not mean an approval for marketing and implication of cure and is not restricted to oncology.

Zhimin Yang (China's NMPA) discussed the oncology drug approval approach in China which mirrors the practice in the USA. Leading up to June 25, 2018, there were 193 trials with PD-1 blockers in China that were listed in Clinicaltrials.gov. She discussed a number of issues that are not new but made more prominent in the approval consideration for IO medicine: patient selection (cancer types, histology, biomarkers, upfront vs salvage therapy, etc.), efficacy, monotherapy vs combination therapy, and manageable toxicity. Regarding clinical trial design, for cancer types or stages that do not have a standard care option, NMPA also allows single-arm trial to gain regulatory approval. Undoubtedly, the future of regulation of IO development will be more dependent on bio-marker selection of patients, rather than histological types of diseases. It will also be based on mechanistic insights of the medicine rather than empiric reasoning. All of these considerations will hopefully lead to the launch of much more effective and less toxic IO medicine into the clinics.

New Concept in Cancer Immunology and Immunotherapy

Cancer immunotherapy has come a long way. It has been fueled by the basic understanding of the immune system and the unveiling of the dynamic interaction between the host immunity and the transformed cells during oncogenesis. Experimental data coupled with human epidemiology studies have established that during the ontogeny of cancer, immune response against cancer undergoes three functional phases, namely elimination of the cancer cells, equilibrium between cancer and the host immunity, and escape of cancers from the immunological attack [2] (Fig. 1a). This three Es model is helpful for guiding the development of immunotherapeutic strategies which deal primarily with cancer immune escape. Accordingly, in principle, cancer immunotherapy can be summarized with a framework of three Rs, which are to reverse tolerance, rejuvenate the immune system, and restore the immune homeostasis (Fig. 1b). Each of the modalities has its unique characteristics with the first 2 Rs associated with significant toxicities and a limited scope of application at present. The Holy Grail of cancer immunotherapy is the third R, as argued and championed by Lieping Chen (Yale University) to be the process of normalizing the immune response (i.e., dial back the immune editing to the elimination phase). This idea, presented at the conference and further elaborated elegantly in a recent publication by Sanmamed and Chen [3], emphasizes the concept of normalization of anti-tumor immunity in the tumor microenvironment that has aberrant expression of tumor-associated immune regulatory molecules. We would like to coin the term TAICHI for tumor-associated immune checkpoint inhibitory molecules to describe these molecular entities. PD-L1 is a prime example of TAICHI. It is important to point out that the Three R strategies may need to be deployed at the same time, or given sequentially in order to maximize the chance of cancer cure.

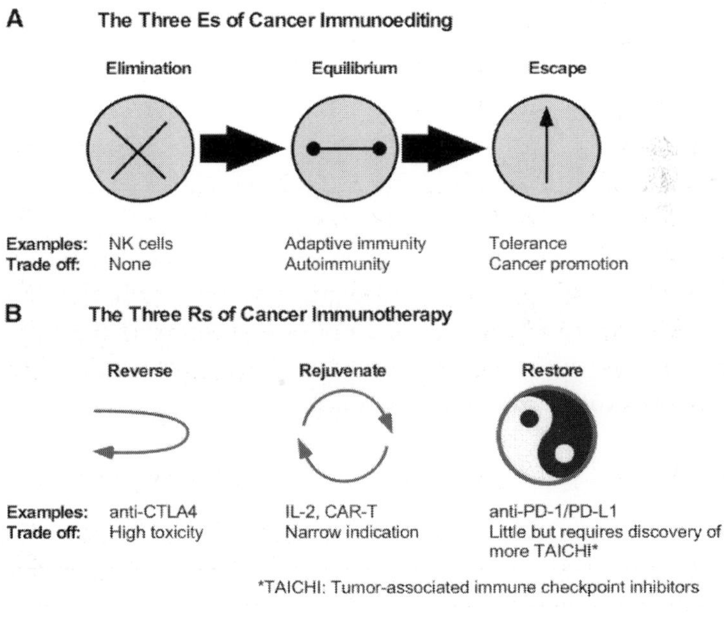

Figure 1: Principles of immunoediting and immunotherapy of cancer. a The 3Es model of cancer immunoediting is schematically shown, along with examples of the immune response and the trade-offs in each phase. b The 3Rs model

of cancer immunotherapy divides treatment modalities into three distinct mode of actions: to reverse, rejuvenate, and restore anti-cancer immunity.

In an attempt to discover more TAICHI for cancer immunotherapy, Lieping Chen and his colleagues performed functional screening for cell surface molecules that inhibit T cell activation. As discussed above, they found that Siglec-15 has previously unknown immunosuppressive roles through promoting the survival and differentiation of suppressive myeloid cells and negatively regulates T cell function. Anti-Siglec antibody has already entered the clinical trial for the dose-defining study in patients with advanced solid tumors.

ICIs, CAR-T cells, etc., are currently used primarily in patients with advanced cancers. Drew Pardoll (Johns Hopkins) argued that the maximal benefit of these agents has not been realized. Immunotherapy of early stage of cancers before intervention by conventional strategies might induce the best benefit and shall be considered. This concept is supported by encouraging results with ICIs used in the neoadjuvant settings, for the treatment of cancers such as melanoma [91], bladder cancer [92], and head and neck cancer [93].

To determine what anti-PD-1 agents do to the tumor microenvironment in early-stage diseases, Forde et al. tested the roles of nivolumab in the neoadjuvant setting for adults with untreated, surgically resectable early (stages I, II, or IIIA) NSCLC. Nivolumab was associated with few side effects, did not delay surgery, and induced a major pathological response in 45% of resected tumors. As predicted, the tumor mutational burden correlates with the pathological response to PD-1 blockade, and the treatment induced expansion of neoantigen-specific T-cell clones in peripheral blood [94]. Future studies will need to address if upfront immunotherapy can change the natural history of the diseases and if so what will be the roles (or lack of) of surgery if pathological complete responses can be accomplished. Conceptually, studies like this will push IO experts and the regulatory agencies to move IO medicine much earlier in the management of cancer rather than using it as the last reserve of treatment.

Weiping Zou (University of Michigan) discussed the holistic approach in cancer immunotherapy, by examining not only tumor microenvironment for genomic alterations and changes of immune infiltration pattern, but also looking at the macroenvironment of the patients, including metabolism, microbiome, and other co-morbidities. Regarding PD-1/L-1-based immunotherapy, work from Arlene Sharpe, Weiping Zou, Yang-Xin Fu, and others showed that PD-L1 expression on tumor-associated professional antigen-presenting cells, as opposed tumor cells, could be the major target for anti-PD-1/L1 responsiveness for some cancers [95–97]. Zou and his colleagues also asked a provocative question regarding the roles of the immune system in chemoresistance. They found that $CD8^+$ T cells, via the JAK/STAT1 pathway, can abrogate fibroblast-mediated chemoresistance in ovarian cancer model through upregulation of gamma-glutamyltransferases and repression of system xc(−) cystine and glutamate antiporter [98]. In colon cancer, it was found that *Fusobacterium (F.) nucleatum* in the gut was able to promote colorectal cancer resistance to chemotherapy by activating innate immunity and the autophagy pathway and thereby altering colorectal cancer chemotherapeutic response. Thus, how chemotherapy and microbiome contribute to reducing cancer burden and death shall also be re-examined in the era of IO. Finally, a case was made to effectively target regulatory T cells (Tregs) as a major path forward for immunotherapy [99]. Multiple agents have been tested for depleting Tregs or inactivating Treg function, including antibodies that block CTLA-4, GITR, GARP, CCR4, CD25, PD-1, OX-40, and LAG3, and small molecule inhibitors against PI3Kδ, PTEN, IDO, EZH2, and ZAP70 [99, 100]. However, a cautionary note was provided that even apoptotic Tregs can release high levels of ATP, which are then converted to adenosine via CD39 and CD73 to suppress T cell immunity [101].

Finally, the unique and distinct role of non-profit organizations in promoting and supporting IO development was shared by representatives from American Associaton of Cancer Research (AACR) (Elizabeth Jaffee), Society of Immunotherapy of Cancer (SITC) (Mario Sznol), Cancer Drug Development Forum (CDDF) (Heinz Zwierzina), Parker Institute for Cancer Immunotherapy (PICI) (Ramy Ibrahim), Chinese Society of Clinical Oncology (CSCO) (Jin Li), and National Foundation for Cancer Research (NFCR) (Sajuan Ba).

Conclusive Remarks

Without a doubt, the era of immuno-oncology is upon us. The true significance of IO medicine in the battle of mankind against cancer may still not be fully appreciated until a decade or so later. The broad activity of PD-1/ PD-L1 agents against cancer has cemented the notion that immune escape is indeed a fundamental hallmark of cancer. Such a revelation raises hope and lifts the cloud of years' frustration and failure over the field of cancer immunology. Thus, the Nobel Committee is right to acknowledge that the work of blocking inhibitory signals for treatment of cancer is Nobel-worthy. It is the long-term and painstaking basic research in immunology that has made this feat possible. James Allison has relentlessly pursued anti-CTLA4 antibody for cancer immunotherapy [102, 103] and has been the champion leading the current IO revolution. Studying basic mechanism of activation-induced cell death of lymphocytes, Tasuko Honjo cloned PD-1 [104] and showed later the importance of PD-1 pathway in negatively regulating T cell function [105–107]. However, the list of mavericks and pioneers of IO who have contributed to the Nobel-worthy work is long, including Lieping Chen who first cloned PD-L1 (also known as B7-H1) [108] and showed its inhibitory function [109, 110], IFNγ-inducibility [109], and its roles in constraining T cell immunity against cancer [109]; Gordon Freeman who collaborated with Honjo to establish the receptor-ligand interaction between PD-1 and PD-L1 [106]; Pierre Goldstein who first cloned CTLA4 [111]; and Jeffrey Bluestone [112, 113], Tak Mak [114], and Arlene Sharpe [113] who demonstrated the inhibitory function of CTLA4. Thanks to the work of those and others, the stage of IO has been set and the script has been written. We are here for a remarkable thriller which we hope will put smiles on face of all of our patients.

ACKNOWLEDGEMENTS

The authors are indebted to other 2018 China Cancer Immunotherapy Workshop organizing committee members from CAHON (Ke Liu, Lei Zheng), CFDA (Jin Cui, Chenyan Gao, Zhimin Yang), and Tsinghua University (Chen Dong, Xin Lin), as well as all the invited speakers for their contributions to the success of the meeting. The authors apologize to colleagues whose work could not be cited due to space limitations and restrictions on the number of references. The authors also wish to acknowledge excellent editorial assistance from Mr. James Fant and Ms. Julia Singleton.

AUTHORS' CONTRIBUTIONS

ZL drafted the manuscript and finalized it with input from WS, MR, and DL. All authors approved the final version.

REFERENCES

1. Old LJ. Immunotherapy for cancer. Sci Am. 1996;275(3):136–143. doi: 10.1038/scientificamerican0996-136.
2. Schreiber RD, Old LJ, Smyth MJ. Cancer immunoediting: integrating immunity's roles in cancer suppression and promotion. Science. 2011;331(6024):1565–1570. doi: 10.1126/science.1203486.
3. Sanmamed MF, Chen L. A paradigm shift in cancer immunotherapy: from enhancement to normalization. Cell. 2018;175(2):313–326. doi: 10.1016/j.cell.2018.09.035.
4. Hanahan D, Weinberg RA. Hallmarks of cancer: the next generation.Cell.2011;144(5):646–674.doi:10.1016/j.cell.2011.02.013.
5. Ribas A, Wolchok JD. Cancer immunotherapy using checkpoint blockade. Science. 2018;359(6382):1350–1355. doi: 10.1126/science.aar4060.
6. Wang J, Yuan R, Song W, Sun J, Liu D, Li Z. PD-1, PD-L1 (B7-H1) and tumor-site immune modulation therapy: the historical perspective. J Hematol Oncol. 2017;10(1):34. doi: 10.1186/s13045-017-0403-5.
7. Sharma P, Allison JP. Immune checkpoint targeting in cancer therapy: toward combination strategies with curative potential. Cell. 2015;161(2):205–214. doi: 10.1016/j.cell.2015.03.030.
8. Migden MR, Rischin D, Schmults CD, Guminski A, Hauschild A, Lewis KD, Chung CH, Hernandez-Aya L, Lim AM, Chang ALS, et al. PD-1 blockade with cemiplimab in advanced cutaneous squamous-cell carcinoma. N Engl J Med. 2018;379(4):341–351. doi: 10.1056/NEJMoa1805131.
9. Chi Z, Li S, Sheng X, Si L, Cui C, Han M, Guo J. Clinical presentation, histology, and prognoses of malignant melanoma in ethnic Chinese: a study of 522 consecutive cases. BMC Cancer. 2011;11:85. doi: 10.1186/1471-2407-11-85.
10. Callahan MK, Kluger H, Postow MA, Segal NH, Lesokhin A, Atkins MB, Kirkwood JM, Krishnan S, Bhore R, Horak C, et al. Nivolumab plus ipilimumab in patients with advanced melanoma: updated survival, response, and safety data in a phase I dose-escalation study. J Clin Oncol. 2018;36(4):391–398. doi: 10.1200/JCO.2017.72.2850.
11. Das R, Bar N, Ferreira M, Newman AM, Zhang L, Bailur JK, Bacchiocchi A, Kluger H, Wei W, Halaban R, et al. Early B cell changes

predict autoimmunity following combination immune checkpoint blockade. J Clin Invest. 2018;128(2):715–720. doi: 10.1172/JCI96798.

12. Puzanov I, Diab A, Abdallah K, Bingham CO, 3rd, Brogdon C, Dadu R, Hamad L, Kim S, Lacouture ME, LeBoeuf NR, et al. Managing toxicities associated with immune checkpoint inhibitors: consensus recommendations from the Society for Immunotherapy of Cancer (SITC) Toxicity Management Working Group. J Immunother Cancer. 2017;5(1):95. doi: 10.1186/s40425-017-0300-z.
13. Brahmer JR, Lacchetti C, Schneider BJ, Atkins MB, Brassil KJ, Caterino JM, Chau I, Ernstoff MS, Gardner JM, Ginex P, et al. Management of immune-related adverse events in patients treated with immune checkpoint inhibitor therapy: American Society of Clinical Oncology Clinical Practice Guideline. J Clin Oncol. 2018;36(17):1714–1768. doi: 10.1200/JCO.2017.77.6385.
14. Thompson JA. New NCCN guidelines: recognition and management of immunotherapy-related toxicity. J Natl Compr Canc Netw. 2018;16(5s):594–596. doi: 10.6004/jnccn.2018.0047.
15. Spigel DR, Reynolds C, Waterhouse D, Garon EB, Chandler J, Babu S, Thurmes P, Spira A, Jotte R, Zhu J, et al. Phase 1/2 study of the safety and tolerability of nivolumab plus crizotinib for the first-line treatment of anaplastic lymphoma kinase translocation - positive advanced non-small cell lung cancer (CheckMate 370) J Thorac Oncol. 2018;13(5):682–688. doi: 10.1016/j.jtho.2018.02.022.
16. Tang J, Pearce L, O'Donnell-Tormey J, Hubbard-Lucey VM. Trends in the global immuno-oncology landscape. Nat Rev Drug Discov. 2018. 10.1038/nrd.2018.202.
17. Forget MA, Haymaker C, Hess KR, Meng YJ, Creasy C, Karpinets T, Fulbright OJ, Roszik J, Woodman SE, Kim YU, et al. Prospective analysis of adoptive TIL therapy in patients with metastatic melanoma: response, impact of anti-CTLA4, and biomarkers to predict clinical outcome. Clin Cancer Res. 2018;24(18):4416–4428. doi: 10.1158/1078-0432.CCR-17-3649.
18. Zhang L, Yu Z, Muranski P, Palmer DC, Restifo NP, Rosenberg SA, Morgan RA. Inhibition of TGF-beta signaling in genetically engineered tumor antigen-reactive T cells significantly enhances tumor treatment efficacy. Gene Ther. 2013;20(5):575–580. doi: 10.1038/gt.2012.75.
19. Peng W, Ye Y, Rabinovich BA, Liu C, Lou Y, Zhang M, Whittington M, Yang Y, Overwijk WW, Lizee G, et al. Transduction of tumor-specific

T cells with CXCR2 chemokine receptor improves migration to tumor and antitumor immune responses. Clin Cancer Res. 2010;16(22):5458–5468. doi: 10.1158/1078-0432.CCR-10-0712.

20. Tran E, Ahmadzadeh M, Lu YC, Gros A, Turcotte S, Robbins PF, Gartner JJ, Zheng Z, Li YF, Ray S, et al. Immunogenicity of somatic mutations in human gastrointestinal cancers. Science. 2015;350(6266):1387–1390. doi: 10.1126/science.aad1253.
21. Tran E, Robbins PF, Lu YC, Prickett TD, Gartner JJ, Jia L, Pasetto A, Zheng Z, Ray S, Groh EM, et al. T-cell transfer therapy targeting mutant KRAS in Cancer. N Engl J Med. 2016;375(23):2255–2262. doi: 10.1056/NEJMoa1609279.
22. Tran E, Turcotte S, Gros A, Robbins PF, Lu YC, Dudley ME, Wunderlich JR, Somerville RP, Hogan K, Hinrichs CS, et al. Cancer immunotherapy based on mutation-specific CD4+ T cells in a patient with epithelial cancer. Science. 2014;344(6184):641–645. doi: 10.1126/science.1251102.
23. Zacharakis N, Chinnasamy H, Black M, Xu H, Lu YC, Zheng Z, Pasetto A, Langhan M, Shelton T, Prickett T, et al. Immune recognition of somatic mutations leading to complete durable regression in metastatic breast cancer. Nat Med. 2018;24(6):724–730. doi: 10.1038/s41591-018-0040-8.
24. Porter DL, Levine BL, Kalos M, Bagg A, June CH. Chimeric antigen receptor-modified T cells in chronic lymphoid leukemia. N Engl J Med. 2011;365(8):725–733. doi: 10.1056/NEJMoa1103849.
25. Kim MY, Yu KR, Kenderian SS, Ruella M, Chen S, Shin TH, Aljanahi AA, Schreeder D, Klichinsky M, Shestova O, et al. Genetic inactivation of CD33 in hematopoietic stem cells to enable CAR T cell immunotherapy for acute myeloid leukemia. Cell. 2018;173(6):1439–1453. doi: 10.1016/j.cell.2018.05.013.
26. Dolgin E. CAR-Ts move beyond B-cell cancers to myeloma. Nat Biotechnol. 2017;35(7):599–601. doi: 10.1038/nbt0717-599.
27. Cho SF, Anderson KC, Tai YT. Targeting B cell maturation antigen (BCMA) in multiple myeloma: potential uses of BCMA-based immunotherapy. Front Immunol. 2018;9:1821. doi: 10.3389/fimmu.2018.01821.
28. Tang J, Hubbard-Lucey VM, Pearce L, O'Donnell-Tormey J, Shalabi A. The global landscape of cancer cell therapy. Nat Rev Drug

Discov. 2018;17(7):465–466. doi: 10.1038/nrd.2018.74.

29. Wang Z, Wu Z, Liu Y, Han W. New development in CAR-T cell therapy. J Hematol Oncol. 2017;10(1):53. doi: 10.1186/s13045-017-0423-1.
30. Raj D, Yang MH, Rodgers D, Hampton EN, Begum J, Mustafa A, Lorizio D, Garces I, Propper D, Kench JG, et al. Switchable CAR-T cells mediate remission in metastatic pancreatic ductal adenocarcinoma. Gut. 2018. 10.1136/gutjnl-2018-316595.
31. Roybal KT, Rupp LJ, Morsut L, Walker WJ, McNally KA, Park JS, Lim WA. Precision tumor recognition by T cells with combinatorial antigen-sensing circuits. Cell. 2016;164(4):770–779. doi: 10.1016/j.cell.2016.01.011.
32. Cho JH, Collins JJ, Wong WW. Universal chimeric antigen receptors for multiplexed and logical control of T cell responses. Cell. 2018;173(6):1426–1438. doi: 10.1016/j.cell.2018.03.038.
33. Chihara N, Madi A, Kondo T, Zhang H, Acharya N, Singer M, Nyman J, Marjanovic ND, Kowalczyk MS, Wang C, et al. Induction and transcriptional regulation of the co-inhibitory gene module in T cells. Nature. 2018;558(7710):454–459. doi: 10.1038/s41586-018-0206-z.
34. Li J, Lee Y, Li Y, Jiang Y, Lu H, Zang W, Zhao X, Liu L, Chen Y, Tan H, et al. Co-inhibitory molecule B7 superfamily member 1 expressed by tumor-infiltrating myeloid cells induces dysfunction of anti-tumor CD8(+) T cells. Immunity. 2018;48(4):773–786. doi: 10.1016/j.immuni.2018.03.018.
35. Liu XS, Mardis ER. Applications of immunogenomics to cancer. Cell. 2017;168(4):600–612. doi: 10.1016/j.cell.2017.01.014.
36. Havel JJ, Chan TA. High-resolution genomic analysis: the tumor-immune interface comes into focus. Genome Biol. 2015;16:65. doi: 10.1186/s13059-015-0631-3.
37. Yarchoan M, Hopkins A, Jaffee EM. Tumor mutational burden and response rate to PD-1 inhibition. N Engl J Med. 2017;377(25):2500–2501. doi: 10.1056/NEJMc1713444.
38. Shukla SA, Bachireddy P, Schilling B, Galonska C, Zhan Q, Bango C, Langer R, Lee PC, Gusenleitner D, Keskin DB, et al. Cancer-germline antigen expression discriminates clinical outcome to CTLA-4

blockade. Cell. 2018;173(3):624–633. doi: 10.1016/j.cell.2018.03.026.

39. Riaz N, Havel JJ, Kendall SM, Makarov V, Walsh LA, Desrichard A, Weinhold N, Chan TA. Recurrent SERPINB3 and SERPINB4 mutations in patients who respond to anti-CTLA4 immunotherapy. Nat Genet. 2016;48(11):1327–1329. doi: 10.1038/ng.3677.

40. Chowell D, Morris LGT, Grigg CM, Weber JK, Samstein RM, Makarov V, Kuo F, Kendall SM, Requena D, Riaz N, et al. Patient HLA class I genotype influences cancer response to checkpoint blockade immunotherapy. Science. 2018;359(6375):582–587. doi: 10.1126/science.aao4572.

41. Eroglu Z, Zaretsky JM, Hu-Lieskovan S, Kim DW, Algazi A, Johnson DB, Liniker E, Ben K, Munhoz R, Rapisuwon S, et al. High response rate to PD-1 blockade in desmoplastic melanomas. Nature. 2018;553(7688):347–350. doi: 10.1038/nature25187.

42. Zaretsky JM, Garcia-Diaz A, Shin DS, Escuin-Ordinas H, Hugo W, Hu-Lieskovan S, Torrejon DY, Abril-Rodriguez G, Sandoval S, Barthly L, et al. Mutations associated with acquired resistance to PD-1 blockade in melanoma. N Engl J Med. 2016;375(9):819–829. doi: 10.1056/NEJMoa1604958.

43. Shin DS, Zaretsky JM, Escuin-Ordinas H, Garcia-Diaz A, Hu-Lieskovan S, Kalbasi A, Grasso CS, Hugo W, Sandoval S, Torrejon DY, et al. Primary resistance to PD-1 blockade mediated by JAK1/2 mutations. Cancer Discov. 2017;7(2):188–201. doi: 10.1158/2159-8290.CD-16-1223.

44. Bever KM, Le DT. DNA repair defects and implications for immunotherapy. J Clin Invest. 2018;128(10):4236–4242. doi: 10.1172/JCI122010.

45. Le DT, Durham JN, Smith KN, Wang H, Bartlett BR, Aulakh LK, Lu S, Kemberling H, Wilt C, Luber BS, et al. Mismatch repair deficiency predicts response of solid tumors to PD-1 blockade. Science. 2017;357(6349):409–413. doi: 10.1126/science.aan6733.

46. Diaz LA, Jr, Le DT. PD-1 blockade in tumors with mismatch-repair deficiency. N Engl J Med. 2015;373(20):1979. doi: 10.1056/NEJMc1510353.

47. Le DT, Uram JN, Wang H, Bartlett BR, Kemberling H, Eyring AD,

Skora AD, Luber BS, Azad NS, Laheru D, et al. PD-1 blockade in tumors with mismatch-repair deficiency. N Engl J Med. 2015;372(26):2509–2520. doi: 10.1056/NEJMoa1500596.

48. Grasso CS, Giannakis M, Wells DK, Hamada T, Mu XJ, Quist M, Nowak JA, Nishihara R, Qian ZR, Inamura K, et al. Genetic mechanisms of immune evasion in colorectal cancer. Cancer Discov. 2018;8(6):730–749. doi: 10.1158/2159-8290.CD-17-1327.
49. Thorsson V, Gibbs DL, Brown SD, Wolf D, Bortone DS, Ou Yang TH, Porta-Pardo E, Gao GF, Plaisier CL, Eddy JA, et al. The immune landscape of cancer. Immunity. 2018;48(4):812–830. doi: 10.1016/j.immuni.2018.03.023.
50. Spranger S, Gajewski TF. Impact of oncogenic pathways on evasion of antitumour immune responses. Nat Rev Cancer. 2018;18(3):139–147. doi: 10.1038/nrc.2017.117.
51. Peng W, Chen JQ, Liu C, Malu S, Creasy C, Tetzlaff MT, Xu C, McKenzie JA, Zhang C, Liang X, et al. Loss of PTEN promotes resistance to T cell-mediated immunotherapy. Cancer Discov. 2016;6(2):202–216. doi: 10.1158/2159-8290.CD-15-0283.
52. George S, Miao D, Demetri GD, Adeegbe D, Rodig SJ, Shukla S, Lipschitz M, Amin-Mansour A, Raut CP, Carter SL, et al. Loss of PTEN is associated with resistance to anti-PD-1 checkpoint blockade therapy in metastatic uterine leiomyosarcoma. Immunity. 2017;46(2):197–204. doi: 10.1016/j.immuni.2017.02.001.
53. Spranger S, Bao R, Gajewski TF. Melanoma-intrinsic beta-catenin signalling prevents anti-tumour immunity. Nature. 2015;523(7559):231–235. doi: 10.1038/nature14404.
54. Schumacher TN, Schreiber RD. Neoantigens in cancer immunotherapy. Science. 2015;348(6230):69–74. doi: 10.1126/science.aaa4971.
55. Ott PA, Hu Z, Keskin DB, Shukla SA, Sun J, Bozym DJ, Zhang W, Luoma A, Giobbie-Hurder A, Peter L, et al. An immunogenic personal neoantigen vaccine for patients with melanoma. Nature. 2017;547(7662):217–221. doi: 10.1038/nature22991.
56. Sahin U, Derhovanessian E, Miller M, Kloke BP, Simon P, Lower M, Bukur V, Tadmor AD, Luxemburger U, Schrors B, et al. Personalized RNA mutanome vaccines mobilize poly-specific therapeutic immunity

against cancer. Nature. 2017;547(7662):222–226. doi: 10.1038/nature23003.

57. Danilova L, Anagnostou V, Caushi JX, Sidhom JW, Guo H, Chan HY, Suri P, Tam A, Zhang J, Asmar ME, et al. The mutation-associated neoantigen functional expansion of specific T cells (MANAFEST) assay: a sensitive platform for monitoring antitumor immunity. Cancer Immunol Res. 2018;6(8):888–899. doi: 10.1158/2326-6066.CIR-18-0129.

58. Mariathasan S, Turley SJ, Nickles D, Castiglioni A, Yuen K, Wang Y, Kadel EE, III, Koeppen H, Astarita JL, Cubas R, et al. TGFbeta attenuates tumour response to PD-L1 blockade by contributing to exclusion of T cells. Nature. 2018;554(7693):544–548. doi: 10.1038/nature25501.

59. Tauriello DVF, Palomo-Ponce S, Stork D, Berenguer-Llergo A, Badia-Ramentol J, Iglesias M, Sevillano M, Ibiza S, Canellas A, Hernando-Momblona X, et al. TGFbeta drives immune evasion in genetically reconstituted colon cancer metastasis. Nature. 2018;554(7693):538–543. doi: 10.1038/nature25492.

60. Herbertz S, Sawyer JS, Stauber AJ, Gueorguieva I, Driscoll KE, Estrem ST, Cleverly AL, Desaiah D, Guba SC, Benhadji KA, et al. Clinical development of galunisertib (LY2157299 monohydrate), a small molecule inhibitor of transforming growth factor-beta signaling pathway. Drug Des Devel Ther. 2015;9:4479–4499.

61. Lan Y, Zhang D, Xu C, Hance KW, Marelli B, Qi J, Yu H, Qin G, Sircar A, Hernandez VM, et al. Enhanced preclinical antitumor activity of M7824, a bifunctional fusion protein simultaneously targeting PD-L1 and TGF-beta. Sci Transl Med. 2018;10(424).

62. Strauss J, Heery CR, Schlom J, Madan RA, Cao L, Kang Z, Lamping E, Marte JL, Donahue RN, Grenga I, et al. Phase I trial of M7824 (MSB0011359C), a bifunctional fusion protein targeting PD-L1 and TGFbeta, in advanced solid tumors. Clin Cancer Res. 2018;24(6):1287–1295. doi: 10.1158/1078-0432.CCR-17-2653.

63. Metelli A, Salem M, Wallace CH, Wu BX, Li A, Li X, Li Z. Immunoregulatory functions and the therapeutic implications of GARP-TGF-beta in inflammation and cancer. J Hematol Oncol. 2018;11(1):24. doi: 10.1186/s13045-018-0570-z.

64. Wang R, Kozhaya L, Mercer F, Khaitan A, Fujii H, Unutmaz D. Expression of GARP selectively identifies activated human FOXP3+

regulatory T cells. Proc Natl Acad Sci U S A. 2009;106(32):13439–13444. doi: 10.1073/pnas.0901965106.

65. Wang R, Wan Q, Kozhaya L, Fujii H, Unutmaz D. Identification of a regulatory T cell specific cell surface molecule that mediates suppressive signals and induces Foxp3 expression. PLoS One. 2008;3(7):e2705. doi: 10.1371/journal.pone.0002705.
66. Tran DQ, Andersson J, Wang R, Ramsey H, Unutmaz D, Shevach EM. GARP (LRRC32) is essential for the surface expression of latent TGF-beta on platelets and activated FOXP3+ regulatory T cells. Proc Natl Acad Sci U S A. 2009;106(32):13445–13450. doi: 10.1073/pnas.0901944106.
67. Metelli A, Wu BX, Fugle CW, Rachidi S, Sun S, Zhang Y, Wu J, Tomlinson S, Howe PH, Yang Y, et al. Surface expression of TGFbeta docking receptor GARP promotes oncogenesis and immune tolerance in breast cancer. Cancer Res. 2016;76(24):7106–7117. doi: 10.1158/0008-5472.CAN-16-1456.
68. Zhang Y, Wu BX, Metelli A, Thaxton JE, Hong F, Rachidi S, Ansa-Addo E, Sun S, Vasu C, Yang Y, et al. GP96 is a GARP chaperone and controls regulatory T cell functions. J Clin Invest. 2015;125(2):859–869. doi: 10.1172/JCI79014.
69. Rachidi S, Metelli A, Riesenberg B, Wu BX, Nelson MH, Wallace C, Paulos CM, Rubinstein MP, Garrett-Mayer E, Hennig M, et al. Platelets subvert T cell immunity against cancer via GARP-TGFbeta axis. Sci Immunol. 2017;2(11).
70. Cuende J, Lienart S, Dedobbeleer O, van der Woning B, De Boeck G, Stockis J, Huygens C, Colau D, Somja J, Delvenne P, et al. Monoclonal antibodies against GARP/TGF-beta1 complexes inhibit the immunosuppressive activity of human regulatory T cells in vivo. Sci Transl Med. 2015;7(284):284ra256. doi: 10.1126/scitranslmed.aaa1983.
71. Tang J, Shalabi A, Hubbard-Lucey VM. Comprehensive analysis of the clinical immuno-oncology landscape. Ann Oncol. 2018;29(1):84–91. doi: 10.1093/annonc/mdx755.
72. Ribas A, Dummer R, Puzanov I, VanderWalde A, Andtbacka RHI, Michielin O, Olszanski AJ, Malvehy J, Cebon J, Fernandez E, et al. Oncolytic virotherapy promotes intratumoral T cell infiltration and improves anti-PD-1 immunotherapy. Cell. 2017;170(6):1109–1119. doi: 10.1016/j.cell.2017.08.027.

73. Atkins MB, Plimack ER, Puzanov I, Fishman MN, McDermott DF, Cho DC, Vaishampayan U, George S, Olencki TE, Tarazi JC, et al. Axitinib in combination with pembrolizumab in patients with advanced renal cell cancer: a non-randomised, open-label, dose-finding, and dose-expansion phase 1b trial. Lancet Oncol. 2018;19(3):405–415. doi: 10.1016/S1470-2045(18)30081-0.
74. Wrangle JM, Velcheti V, Patel MR, Garrett-Mayer E, Hill EG, Ravenel JG, Miller JS, Farhad M, Anderton K, Lindsey K, et al. ALT-803, an IL-15 superagonist, in combination with nivolumab in patients with metastatic non-small cell lung cancer: a non-randomised, open-label, phase 1b trial. Lancet Oncol. 2018;19(5):694–704. doi: 10.1016/S1470-2045(18)30148-7.
75. Robinson TO, Schluns KS. The potential and promise of IL-15 in immuno-oncogenic therapies. Immunol Lett. 2017;190:159–168. doi: 10.1016/j.imlet.2017.08.010.
76. Charych DH, Hoch U, Langowski JL, Lee SR, Addepalli MK, Kirk PB, Sheng D, Liu X, Sims PW, VanderVeen LA, et al. NKTR-214, an engineered cytokine with biased IL2 receptor binding, increased tumor exposure, and marked efficacy in mouse tumor models. Clin Cancer Res. 2016;22(3):680–690. doi: 10.1158/1078-0432.CCR-15-1631.
77. Munn DH, Mellor AL. IDO in the tumor microenvironment: inflammation, counter-regulation, and tolerance. Trends Immunol. 2016;37(3):193–207. doi: 10.1016/j.it.2016.01.002.
78. Beatty GL, O'Dwyer PJ, Clark J, Shi JG, Bowman KJ, Scherle PA, Newton RC, Schaub R, Maleski J, Leopold L, et al. First-in-human phase I study of the oral inhibitor of Indoleamine 2,3-dioxygenase-1 epacadostat (INCB024360) in patients with advanced solid malignancies. Clin Cancer Res. 2017;23(13):3269–3276. doi: 10.1158/1078-0432.CCR-16-2272.
79. Mitchell TC, Hamid O, Smith DC, Bauer TM, Wasser JS, Olszanski AJ, Luke JJ, Balmanoukian AS, Schmidt EV, Zhao Y, et al. Epacadostat plus pembrolizumab in patients with advanced solid tumors: phase I results from a multicenter, open-label phase I/II trial (ECHO-202/KEYNOTE-037). J Clin Oncol. 2018;36(32):3223–30.
80. Muller AJ, Manfredi MG, Zakharia Y, Prendergast GC. Inhibiting IDO pathways to treat cancer: lessons from the ECHO-301 trial and beyond. Semin Immunopathol. 2018. 10.1007/s00281-018-0702-0. .
81. Foley K, Kim V, Jaffee E, Zheng L. Current progress in immunotherapy

for pancreatic cancer. Cancer Lett. 2016;381(1):244–251. doi: 10.1016/j.canlet.2015.12.020.

82. Yang X, Zhang X, Mortenson ED, Radkevich-Brown O, Wang Y, Fu YX. Cetuximab-mediated tumor regression depends on innate and adaptive immune responses. Mol Ther. 2013;21(1):91–100. doi: 10.1038/mt.2012.184.

83. Ren Z, Guo J, Liao J, Luan Y, Liu Z, Sun Z, Liu X, Liang Y, Peng H, Fu YX. CTLA-4 limits anti-CD20-mediated tumor regression. Clin Cancer Res. 2017;23(1):193–203. doi: 10.1158/1078-0432.CCR-16-0040.

84. Park S, Jiang Z, Mortenson ED, Deng L, Radkevich-Brown O, Yang X, Sattar H, Wang Y, Brown NK, Greene M, et al. The therapeutic effect of anti-HER2/neu antibody depends on both innate and adaptive immunity. Cancer Cell. 2010;18(2):160–170. doi: 10.1016/j.ccr.2010.06.014.

85. Weissman I. How one thing led to another. Annu Rev Immunol. 2016;34:1–30. doi: 10.1146/annurev-immunol-032414-112003.

86. Liu X, Kwon H, Li Z, Fu YX. Is CD47 an innate immune checkpoint for tumor evasion? J Hematol Oncol. 2017;10(1):12. doi: 10.1186/s13045-016-0381-z.

87. Gholamin S, Mitra SS, Feroze AH, Liu J, Kahn SA, Zhang M, Esparza R, Richard C, Ramaswamy V, Remke M, et al. Disrupting the CD47-SIRPalpha anti-phagocytic axis by a humanized anti-CD47 antibody is an efficacious treatment for malignant pediatric brain tumors. Sci Transl Med. 2017;9(381).

88. Advani R, Flinn I, Popplewell L, Forero A, Bartlett NL, Ghosh N, Kline J, Roschewski M, LaCasce A, Collins GP, et al. CD47 blockade by Hu5F9-G4 and rituximab in non-Hodgkin's lymphoma. N Engl J Med. 2018;379(18):1711–1721. doi: 10.1056/NEJMoa1807315.

89. Liu X, Liu L, Ren Z, Yang K, Xu H, Luan Y, Fu K, Guo J, Peng H, Zhu M, et al. Dual targeting of innate and adaptive checkpoints on tumor cells limits immune evasion. Cell Rep. 2018;24(8):2101–2111. doi: 10.1016/j.celrep.2018.07.062.

90. Liu X, Pu Y, Cron K, Deng L, Kline J, Frazier WA, Xu H, Peng H, Fu YX, Xu MM. CD47 blockade triggers T cell-mediated destruction of immunogenic tumors. Nat Med. 2015;21(10):1209–1215. doi: 10.1038/nm.3931.

91. Blank CU, Rozeman EA, Fanchi LF, Sikorska K, van de Wiel B, Kvistborg P, Krijgsman O, van den Braber M, Philips D, Broeks A, et al. Neoadjuvant versus adjuvant ipilimumab plus nivolumab in macroscopic stage III melanoma. Nat Med. 2018;24(11):1655–61.

92. Kim HS, Seo HK. Immune checkpoint inhibitors for urothelial carcinoma. Investig Clin Urol. 2018;59(5):285–296. doi: 10.4111/icu.2018.59.5.285.

93. Saleh K, Eid R, Haddad FG, Khalife-Saleh N, Kourie HR. New developments in the management of head and neck cancer - impact of pembrolizumab. Ther Clin Risk Manag. 2018;14:295–303. doi: 10.2147/TCRM.S125059.

94. Forde PM, Chaft JE, Smith KN, Anagnostou V, Cottrell TR, Hellmann MD, Zahurak M, Yang SC, Jones DR, Broderick S, et al. Neoadjuvant PD-1 blockade in resectable lung cancer. N Engl J Med. 2018;378(21):1976–1986. doi: 10.1056/NEJMoa1716078.

95. Lin H, Wei S, Hurt EM, Green MD, Zhao L, Vatan L, Szeliga W, Herbst R, Harms PW, Fecher LA, et al. Host expression of PD-L1 determines efficacy of PD-L1 pathway blockade-mediated tumor regression. J Clin Invest. 2018;128(4):1708. doi: 10.1172/JCI120803.

96. Tang H, Liang Y, Anders RA, Taube JM, Qiu X, Mulgaonkar A, Liu X, Harrington SM, Guo J, Xin Y, et al. PD-L1 on host cells is essential for PD-L1 blockade-mediated tumor regression. J Clin Invest. 2018;128(2):580–588. doi: 10.1172/JCI96061.

97. Juneja VR, McGuire KA, Manguso RT, LaFleur MW, Collins N, Haining WN, Freeman GJ, Sharpe AH. PD-L1 on tumor cells is sufficient for immune evasion in immunogenic tumors and inhibits CD8 T cell cytotoxicity. J Exp Med. 2017;214(4):895–904. doi: 10.1084/jem.20160801.

98. Wang W, Kryczek I, Dostal L, Lin H, Tan L, Zhao L, Lu F, Wei S, Maj T, Peng D, et al. Effector T cells abrogate stroma-mediated chemoresistance in ovarian cancer. Cell. 2016;165(5):1092–1105. doi: 10.1016/j.cell.2016.04.009.

99. Tanaka A, Sakaguchi S. Regulatory T cells in cancer immunotherapy. Cell Res. 2017;27(1):109–118. doi: 10.1038/cr.2016.151.

100. Munn DH, Sharma MD, Johnson TS. Treg destabilization and reprogramming: implications for cancer immunotherapy. Cancer Res. 2018;78(18):5191–5199. doi: 10.1158/0008-5472.CAN-18-1351.

101. Maj T, Wang W, Crespo J, Zhang H, Wang W, Wei S, Zhao L, Vatan L, Shao I, Szeliga W, et al. Oxidative stress controls regulatory T cell apoptosis and suppressor activity and PD-L1-blockade resistance in tumor. Nat Immunol. 2017;18(12):1332–1341. doi: 10.1038/ni.3868.

102. Leach DR, Krummel MF, Allison JP. Enhancement of antitumor immunity by CTLA-4 blockade. Science. 1996;271(5256):1734–1736. doi: 10.1126/science.271.5256.1734.

103. Kwon ED, Hurwitz AA, Foster BA, Madias C, Feldhaus AL, Greenberg NM, Burg MB, Allison JP. Manipulation of T cell costimulatory and inhibitory signals for immunotherapy of prostate cancer. Proc Natl Acad Sci U S A. 1997;94(15):8099–8103. doi: 10.1073/pnas.94.15.8099.

104. Ishida Y, Agata Y, Shibahara K, Honjo T. Induced expression of PD-1, a novel member of the immunoglobulin gene superfamily, upon programmed cell death. EMBO J. 1992;11(11):3887–3895. doi: 10.1002/j.1460-2075.1992.tb05481.x.

105. Nishimura H, Nose M, Hiai H, Minato N, Honjo T. Development of lupus-like autoimmune diseases by disruption of the PD-1 gene encoding an ITIM motif-carrying immunoreceptor. Immunity. 1999;11(2):141–151. doi: 10.1016/S1074-7613(00)80089-8.

106. Freeman GJ, Long AJ, Iwai Y, Bourque K, Chernova T, Nishimura H, Fitz LJ, Malenkovich N, Okazaki T, Byrne MC, et al. Engagement of the PD-1 immunoinhibitory receptor by a novel B7 family member leads to negative regulation of lymphocyte activation. J Exp Med. 2000;192(7):1027–1034. doi: 10.1084/jem.192.7.1027.

107. Iwai Y, Terawaki S, Honjo T. PD-1 blockade inhibits hematogenous spread of poorly immunogenic tumor cells by enhanced recruitment of effector T cells. Int Immunol. 2005;17(2):133–144. doi: 10.1093/intimm/dxh194.

108. Dong H, Zhu G, Tamada K, Chen L. B7-H1, a third member of the B7 family, co-stimulates T-cell proliferation and interleukin-10 secretion. Nat Med. 1999;5(12):1365–1369. doi: 10.1038/70932.

109. Dong H, Strome SE, Salomao DR, Tamura H, Hirano F, Flies DB, Roche PC, Lu J, Zhu G, Tamada K, et al. Tumor-associated B7-H1 promotes T-cell apoptosis: a potential mechanism of immune evasion. Nat Med. 2002;8(8):793–800. doi: 10.1038/nm730.

110. Dong H, Zhu G, Tamada K, Flies DB, van Deursen JM, Chen L. B7-H1 determines accumulation and deletion of intrahepatic CD8(+) T

lymphocytes. Immunity. 2004;20(3):327–336. doi: 10.1016/S1074-7613(04)00050-0.

111. Brunet JF, Denizot F, Luciani MF, Roux-Dosseto M, Suzan M, Mattei MG, Golstein P. A new member of the immunoglobulin superfamily--CTLA-4. Nature. 1987;328(6127):267–270. doi: 10.1038/328267a0.

112. Walunas TL, Lenschow DJ, Bakker CY, Linsley PS, Freeman GJ, Green JM, Thompson CB, Bluestone JA. CTLA-4 can function as a negative regulator of T cell activation. Immunity. 1994;1(5):405–413. doi: 10.1016/1074-7613(94)90071-X.

113. Tivol EA, Borriello F, Schweitzer AN, Lynch WP, Bluestone JA, Sharpe AH. Loss of CTLA-4 leads to massive lymphoproliferation and fatal multiorgan tissue destruction, revealing a critical negative regulatory role of CTLA-4. Immunity. 1995;3(5):541–547. doi: 10.1016/1074-7613(95)90125-6.

114. Waterhouse P, Penninger JM, Timms E, Wakeham A, Shahinian A, Lee KP, Thompson CB, Griesser H, Mak TW. Lymphoproliferative disorders with early lethality in mice deficient in Ctla-4. Science. 1995;270(5238):985–988. doi: 10.1126/science.270.5238.985.

3

Next-generation immuno-oncology Agents: Current Momentum Shifts in Cancer Immunotherapy

Chongxian Pan[1,2], Hongtao Liu[1,3], Elizabeth Robins[4] , Wenru Song[1,5], Delong Liu[1,6], Zihai Li[1,4] and Lei Zheng[1,7]

[1] Chinese American Hematologist and Oncologist Network, New York, NY, USA
[2] University of California, Davis, CA, USA
[3] University of Chicago, Chicago, IL, USA
[4] Pelotonia Institute for Immuno-Oncology, The Ohio State University, Columbus, OH, USA
[5] Kira Pharmaceuticals, Cambridge, MA, USA
[6] New York Medical College, Valhalla, NY, USA
[7] Johns Hopkins University, Baltimore, MD, USA

Citation: Pan C, Liu H, Robins E, et al. Next-generation immuno-oncology agents: current momentum shifts in cancer immunotherapy. J Hematol Oncol. 2020;13(1):29. Published 2020 Apr 3. doi:10.1186/s13045-020-00862-w

ABSTRACT

Cancer immunotherapy has reached a critical point, now that immune checkpoint inhibitors and two CAR-T products have received market approval in treating 16 types of cancers and 1 tissue-agnostic cancer indication. Accompanying these advances, the 2018 Nobel Prize was awarded for the discovery of immune checkpoint pathways, which has led to the revolution of anti-cancer treatments. However, expanding the indications of immuno-oncology agents and overcoming treatment resistance face mounting challenges. Although combination immunotherapy is an obvious strategy to pursue, the fact that there have been more failures than successes in this effort has served as a wake-up call, placing emphasis on the importance of building a solid scientific foundation for the development of next-generation immuno-oncology (IO) agents. The 2019 China Cancer Immunotherapy Workshop was held to discuss the current challenges and opportunities in IO. At this conference, emerging concepts and strategies for IO development were proposed, focusing squarely on correcting the immunological defects in the tumor microenvironment. New targets such as Siglec-15 and new directions including neoantigens, cancer vaccines, oncolytic viruses, and cytokines were reviewed. Emerging immunotherapies were discussed in the areas of overcoming primary and secondary resistance to existing immune checkpoint inhibitors, activating effector cells, and targeting immunosuppressive mechanisms in the tumor microenvironment. In this article, we highlight old and new waves of IO therapy development, and provide perspectives on the latest momentum shifts in cancer immunotherapy.

Keywords: Neoantigen, Immune checkpoint inhibitor, Tumor microenvironment, PD-1, PD-L1, CTLA-4, CAR-T

INTRODUCTION

Cancer immunotherapy has been a game changer in cancer treatment since the approval of the immune checkpoint inhibitor (ICI) ipilimumab in 2011. Currently, 11 immune checkpoint inhibitors (Table 1) and 2 chimeric antigen receptor T cell (CAR-T) products have been approved in treating 16 types of malignant diseases and 1 tissue-agnostic indication. In 2018, one half of the Nobel Prize in Physiology or Medicine was awarded to James Allison, who conceptualized cancer immunotherapy by targeting the immunosuppressive signal mediated by Cytotoxic T Lymphocyte-Associated Protein 4 (CTLA-

4) [1, 2]. This conceptual breakthrough led to the subsequent revolutionary development of immune checkpoint inhibitors (ICIs). In addition, co-Nobel Prize awardee Tasuko Honjo showed that a basic mechanism of activation-induced cell death in lymphocytes is mediated by Programmed Cell Death 1 (PD-1) [3]. Honjo subsequently demonstrated that the PD-1 pathway is an important negative regulator of T cell function [4–6].

Table 1: Immune checkpoint inhibitors and their US FDA/EMA/China NMPA approved indications

Immune checkpoint inhibitor	Targets	US FDA/EMA approved indications	China NMPA approved indications
Pembrolizumab	PD-1	Melanoma, non-small cell lung cancer, head and neck cancer, Hodgkin's lymphoma, urothelial carcinoma, MSI-H/dMMR* colorectal cancer, MSI-H/dMMR cancers, gastric cancer, cervical cancer, hepatocellular carcinoma, Merkel cell carcinoma, renal cell carcinoma, small cell lung cancer, esophageal carcinoma, endometrial cancer	Melanoma, non-small cell lung cancer
Nivolumab	PD-1	Melanoma, non-small cell lung cancer, renal cell carcinoma, Hodgkin's lymphoma, head and neck cancer, urothelial carcinoma, MSI-H/dMMR colorectal cancer, hepatocellular carcinoma, small cell lung cancer	Non-small cell lung cancer
Atezolizumab	PD-L1	Urothelial cancer, non-small cell lung cancer, breast cancer, small cell lung cancer	Non-small cell lung cancer
Durvalumab	PD-L1	Urothelial carcinoma, non-small cell lung cancer	
Avelumab	PD-L1	Merkel cell carcinoma, urothelial carcinoma, renal cell carcinoma	
Cemiplimab	PD-1	Cutaneous squamous cell carcinoma	
Ipilimumab	CTLA4	Melanoma, metastatic, renal cell carcinoma, MSI-H/dMMR colorectal cancer	
Toripalimab	PD-1		Melanoma
Sintilimab	PD-1		Hodgkin's lymphoma

Camrelizumab	PD-1		Hodgkin's lymphoma
Tislelizumab	PD-1		Hodgkin's lymphoma

*Microsatellite instability high (MSI-H) or mismatch repair deficient (dMMR)

While the discoveries made by Allison and Honjo are truly seminal, the IO revolution, like any other scientific breakthroughs, has taken a "village". Lieping Chen, for example, first cloned Programmed Cell Death 1 Ligand 1 (PD-L1, also known as B7-H1) [7], showed its inhibitory function [8, 9], and indicated that blocking this pathway may have therapeutic potential [9]. Other notable scientific contributors include Gordon Freeman, who collaborated with Honjo to establish the receptor-ligand relationship between PD-1 and PD-L1 [5]; Pierre Goldstein, who first cloned CTLA-4 [10]; and Jeffrey Bluestone [11, 12], Tak Mak [13], and Arlene Sharpe [12], who demonstrated the inhibitory function of CTLA-4. The IO village has also included many clinical investigators, who masterfully designed and completed ICI clinical trials and taken ICIs into today's standard clinical practice [14]. Having witnessed the flourishing of cancer immunotherapy, the Chinese American Hematologist and Oncologist Network (CAHON), in partnership with the China National Medical Product Administration (NMPA), and later joined by Tsinghua University, have organized the annual China Cancer Immunotherapy Workshop since 2017 to provide update and education to physicians, scientists, and drug developers [15]. The fifth China Cancer Immunotherapy Workshop was held in Tianjin on June 2930, 2019, and it proved again to be an international forum on the discussion of the cutting edge of cancer immunotherapy. There were 4 major themes in the 2019 conference. The first theme was centered on current challenges in ICI development, and new visions for the future of this field. The second one focused on the development and application of cell therapy, where new IO agents continue to rapidly emerge. The third theme featured new immunotherapy strategies that are driven by advancements in basic immunology research. The last theme highlighted regulatory challenges and solutions in clinical research and development of cancer immunotherapeutics by experts from the China NMPA, the United States Food and Drug Administration (US FDA), and the European Medicines Agency (EMA).

Herein, we use the program of the 2019 China Cancer Immunotherapy Workshop as the general framework to critically review the most recent conceptual shifts and therapeutic advancements in the increasingly exciting and complex IO field.

A CURRENT PERSPECTIVE ON THE ANTI-CANCER IMMUNE RESPONSE

Challenges in improving the efficacy of existing immunotherapies, and the development of new ones, have led to a deeper appreciation of understanding the mechanisms underlying an effective anti-cancer immune response, as well as the "defects" that are responsible for the lack of an effective anti-cancer immune response in cancer patients.

The Cancer-immunity Cycle

We present a model of the anti-cancer immunity "cycle" (Fig. 1, innermost circle) [16], which provides a summary of our scientific knowledge on each step of an effective anti-cancer immune response. The cycle starts when tumor antigens are recognized by the immune system. Genomic instability/mutation is 1 of the 2 enabling characteristics of cancer [17]. All cancers, regardless of their tissue origin(s), harbor genetic alterations that range from a few mutations in pediatric malignancies to dozens or hundreds in adult cancers [18]. These non-synonymous DNA alterations can give rise to proteins that differ from the proteins expressed in normal cells, i.e., tumor antigens. As a second enabling characteristic, some cancers express non-mutation-associated tumor antigens, such as proteins normally expressed in immune-privileged sites, viral proteins, or proteins encoded by endogenous retroviral genes. When these antigens are taken up and processed by professional antigen-presenting cells (APCs), the APCs migrate to secondary lymphoid organs and activate naïve T cells in concert with a highly-coordinated hierarchy of co-stimulatory signals, such as the CD28/B7-1/2-mediated signal. To achieve homeostasis and prevent over-reaction to non-self antigens, the immune system has also developed highly coordinated negative feedback circuits. CTLA-4 is one of the major negative regulators of the T cell-mediated immune response. CTLA-4 expression is rapidly upregulated upon T cell receptor (TCR) engagement [19], allowing it to outcompete CD28 for ligation by B7-1/2, and thereby negatively regulate T cell activation and effector function.

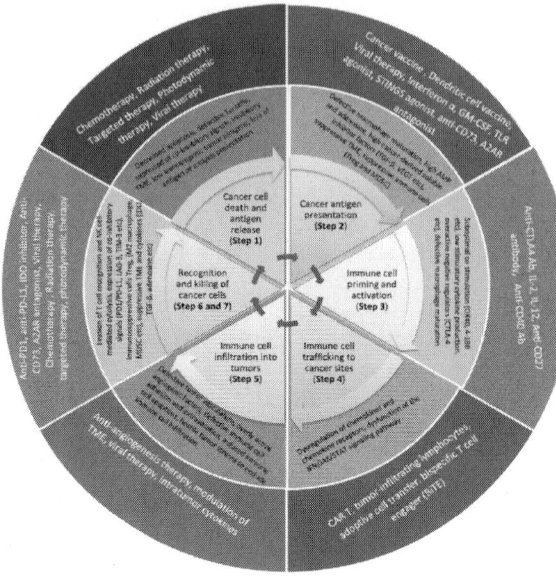

Figure 1: The cancer-immunity cycle, immune-resistant mechanisms and strategies for anti-cancer immunotherapy. The anti-cancer immunity cycle (innermost circle) starts when cancer cells release tumor antigens. Antigen-presenting cells take up tumor antigens and present antigen-derived peptides to immune cells, which in turn activate the immune cells to migrate through the circulation, enter tumor sites, and kill cancer cells. The death of cancer cells induces the release of additional tumor antigens, which initiates another cancer-immunity cycle. The immune system has developed complex negative feedback loops to rein in the anti-pathogen response. These negative feedback loops have been exploited by cancer cells to evade anti-cancer immunity (middle circle). Current anti-cancer immunotherapy approaches (outermost circle) have been targeting and harnessing various mechanisms along this cancer-immunity circle. There are two major approaches for cancer immunotherapy: (1) the enhancement approach, which aims to augment "normal" anti-cancer immune mechanisms. Strategies in this category range from the traditional non-specific enhancement of IL-2 signaling to the more recent cancer-specific CAR-T cell therapy; and (2) the normalization approach, which aims to restore defective anti-cancer immunity in the tumor microenvironment. Strategies include FDA-approved immune checkpoint inhibitors and other drugs in development (e.g., inhibitors of the adenosine pathway).

Once activated, effector T cells traffic into the body systemically, infiltrate the cancer site(s), recognize cancer cells expressing tumor antigen-derived peptides presented by Major Histocompatibility Complex (MHC), and kill target cancer cells. In turn, cancer cells release neoantigen(s) which are cross-presented by APCs, leading to further amplification of the anti-cancer immune response by allowing priming and activation of more T cells to recognize and attack the tumor [20]. Just as in any immune response, the final stage of the anti-cancer response is regulated by a complex network of stimulatory and inhibitory accessory pathways. The PD-1/PD-L1 pathway is one of the major inhibitory pathways. Engagement of the TCR with its cognate antigen-MHC complex, together with cytokine stimulation (e.g., IL-2 stimulation), induces the expression of PD-1. Engagement of PD-1 with PD-L1 on target cells inhibits T cell proliferation and IL-2 production, dampening the immune response. Thus, a rational combination immunotherapy must be aimed at coordinated facilitation of T cell activation and effector function, along with coordinated suppression of inhibitory T cell mechanisms.

The Immune Microenvironment of the Tumor

Study of a Tumor Immunity in the MicroEnvironment (TIME) classification system can be used as the first step in assessing anti-cancer immunity and determining underlying tumor resistance mechanisms. TIME classification is based on two major factors: (1) tumor expression of PD-L1, and (2) the presence of immune cell infiltration, mainly tumor-infiltrating lymphocytes (TIL) (Fig. 2) [21–23]. Correspondingly, 4 distinct TIME subtypes can be described [21]: T1 (PD-L1$^-$, TIL$^-$), T2 (PD-L1$^+$, TIL$^+$), T3 (PD-L1$^-$, TIL$^+$), and T4 (PD-L1$^+$, TIL$^-$).

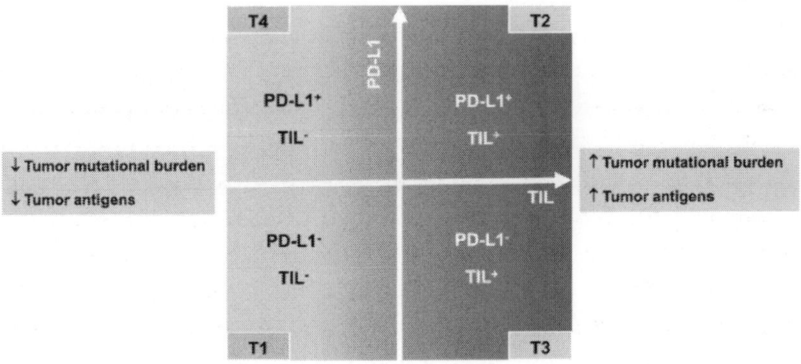

Figure 2: TIME classification based on PD-L1 expression and infiltration of immune cells, mainly tumor-infiltrating lymphocytes (TIL), in the tumor microenvironment. TIME, tumor immunity in the microenvironment.

In cancers with no immune cell infiltration (T1 or T4 TIME), no anti-cancer immunity exists at the cancer site(s), suggesting defects in cancer antigen release (Cancer-Immunity Cycle Step 1), presentation (Cancer-Immunity Cycle Step 2), immune cell priming and activation (Cancer-Immunity Cycle Step 3), or trafficking of immune cells into cancer sites (Cancer-Immunity Cycle Step 4). In these cases, normalization of cancer immunity using anti-PD1/PD-L1 therapy may not work, since no cancer immunity exists to be de-repressed. On the other hand, the majority of solid tumors (approximately 70%) have a T4 TIME, which underscores the importance of developing rational IO combinations to address both a lack of effector cell infiltration and the presence of non-PD-L1/PD-1 immunosuppressive components. Furthermore, T1 or T4 TIME tumors often exhibit low levels of tumor mutation burden and tumor antigens. For example, androgen-dependent prostate cancer usually presents with a T1 or T4 TIME, with little lymphocyte infiltration. In other cases, physical barriers can inhibit TIL infiltration such as in pancreatic cancer, even though an anti-cancer immune response emerges in some tumors, an immune-excluded phenotype is commonly observed because the desmoplastic stroma precludes the immune cells from penetrating into the tumor.

In cancers with immune cell infiltration into the TIME (T2 and T3), an anti-cancer immune response exists. However, the immunosuppressive TIME inhibits the activity of effector immune cells from killing cancer cells. Absence of PD-L1 in T3 ($PD\text{-}L1^{-}$, TIL^{+}) suggests that the suppression of anti-cancer immunity is largely mediated by mechanisms other than the PD1/PD-L1 pathway. Interestingly, the T3 TME only exists in approximately 10% of solid tumors. Thus, agents that target alternative co-inhibitory pathways, such as anti-CTLA-4 antibody and anti-LAG3 antibody, may be effective in cancers with a T3 TIME. On the other hand, while TIL are present in T2 and T3 TIMEs, their location and functional capacity may be crucial [24–27]. The immune-inflamed phenotype of T lymphocytes is often accompanied by myeloid and monocytic cells and TIL infiltration into the tumor sites, while the immune-excluded phenotype is characterized by immune cell retention in the stroma that surrounds nests of tumor cells, but does not penetrate into the tumor parenchyma. Cancers with high mutation burdens, such as melanoma, lung, and bladder cancer, can induce a strong anti-cancer immune response, and therefore present with T2 or T3 TIME phenotypes with abundant lymphocyte infiltration. In this manner, TIME classification complements the cancer immunity cycle, stratifying the complex milieu of cancer immunophenotypes into therapeutically-meaningful compartments that can serve as a guide for the study of cancer immuno-resistant mechanisms.

From Enhancing Immunity to Normalizing TIME

Historically, cancer immunotherapy has focused on amplifying tumor immunity above physiological levels, which is associated with clinical response in a minority of patients, in highly selected cancers (e.g., kidney and melanoma), and with off-tumor toxicities. It is becoming increasingly appreciated that many cancer patients have anti-cancer T cells, but the TIME can effectively suppress their immune response by harnessing immune homeostasis mechanisms to negatively regulate anti-cancer immunity or cell survival. As a result, cancer cells that can evade immune attack are naturally selected for survival. Hence, Lieping Chen and his colleagues have emphasized that, instead of enhancing the immune system, it is important to restore the function of the TIME [28]. The lessons we have learned from the failure of boosting immunity and the success of ICI development substantiate this notion of TIME normalization [29–31]. It is now crucial that we determine how to normalize the defects in TIME. In particular, targets for normalizing T1 (PD-L1$^-$, TIL$^-$) TIME remain to be discovered and validated. Searching for and defining such targets from T1 tumors are anticipated by Chen to be the next game changer in cancer immunotherapy. Chen reported one result that has already been gained from such efforts. He indicated that Siglec-15, mainly expressed by myeloid cells, may be a newly-defined T cell immune checkpoint target [32]. A first-in-class, first-in-human clinical trial of anti-Siglec-15 monoclonal antibody NC318 for the treatment of advanced cancers resistant to the current ICS has been ongoing since October 2018. Additional strategies aimed at normalizing the TIME by targeting myeloid cells and regulatory T cells are anticipated.

CHALLENGES AND NEW PERSPECTIVES IN ICI DEVELOPMENT

Simple Addition: the Traditional Method of Combination ICI Development

To date, the standard approach taken by pharmaceutical companies to develop new ICI therapeutic regimens has been to combine two agents that each has shown single-agent activity. This approach has, in fact, yielded progress in some studies that have combined a chemotherapy agent with an ICI agent. As discussed by Roy Herbst from Yale University, the combination of chemotherapy with PD-1 inhibitor pembrolizumab as a first-line therapy for metastatic non-small cell lung cancer (NSCLC) was effective. Promising results were also yielded by the IMpassion130 study, which tested the combination of PD-L1 inhibitor atezolizumab and nano-particle albumin-bound chemotherapy agent nab-paclitaxel in metastatic triple negative breast cancer (TNBC) [33]. A similar conclusion can be drawn from the outcome of the Keynote-552 study, which tested the combination of pembrolizumab and chemotherapy in the neoadjuvant and adjuvant setting for resectable advanced triple negative breast cancer (TNBC) (ESMO 2019. Abstract LBA8_PR).

However, time will tell whether other chemotherapy/ICI pairing strategies will succeed or fail. Four anti-PD-1 antibodies that were developed by China-based biopharmaceuticals have been approved in China for two disease indications (Table 1). They have demonstrated similar anti-cancer activities as those developed in the USA although some of them demonstrated higher in vitro binding affinity to the PD-1. Three of them have been approved for treating classical Hodgkin's lymphoma whereas one has been approved for treating melanoma [34–40]. None of them have been approved for more common types of malignancies such as NSCLC. By contrast, three anti-PD-1/PD-L1 antibodies that were developed by global pharmaceuticals have been approved to treat NSCLC. Although China's anti-PD-1 antibodies lag behind in clinical developments as single agents, a strategic emphasis on developing anti-PD-1 antibody based combination immunotherapy has been made for these anti-PD-1 antibodies. In combination with chemotherapy agents, these IO agents are being tested as candidate first-line therapies for nasopharyngeal carcinoma, which is endemic in southern China. An investigator-initiated clinical trial testing the combination of chemotherapy and anti-PD-1 antibody camrelizumab in nasopharyngeal carcinoma showed an impressive response rate in 22 evaluable patients [41]. In this study, the combination of chemotherapy with camrelizumab yielded an overall response rate (ORR) of 91%, compared to the phase 1 ORR of 34% with camrelizumab monotherapy in chemotherapy-refractory nasopharyngeal carcinomas. These results provide a contrast to the findings of the Keynote-062 study, as discussed by Andrew Zhu from Massachusetts General Hospital and Lei Zheng from Johns Hopkins University. In Keynote-062, the combination of chemotherapy and pembrolizumab has failed to demonstrate clinical benefit in gastric and gastroesophageal carcinoma patients.

Another strategy for combination immunotherapy development is to combine an ICI and a tyrosine kinase inhibitor (TKI). Lawrence Fong from University of California at San Francisco and Lei Zheng from Johns Hopkins University discussed the results of the KEYNOTE-426 trial, which combined pembrolizumab and vascular endothelial growth factor receptor (VEGFR) inhibitor axitinib. This dual regimen enhanced progression-free survival (PFS), overall survival (OS), and overall response rate (ORR) over single-agent therapy with platelet-derived growth factor receptor (PDGFR)/ VEGFR inhibitor sunitinib in treatment-naive advanced/metastatic renal cell carcinoma (mRCC) [42]. In another phase 3 study, PD-L1 inhibitor avelumab plus axitinib produced significantly prolonged PFS, OS, and ORR over sunitinib alone in treatment-naïve mRCC, irrespective of PD-L1 status [43]. Thus, the pembrolizumab/axitinib and avelumab/axitinib combinations have become the standard-of-care, first-line treatment option for mRCC. Another combination ICI/TKI therapy that has made it to first-line treatment status is the combination of pembrolizumab and multi-kinase inhibitor lenvatinib for advanced endometrial cancer [44].

Other ICI/TKI therapies have also shown potential. As reported at the 2019 Annual Meeting of the American Society of Clinical Oncology, the combination of anti-PD-1 monoclonal antibody nivolumab and angiopoietin receptor/VEGFR inhibitor regorafenib was found to yield promising objective responses in chemotherapy-refractory gastric adenocarcinoma and colorectal adenocarcinoma. Similarly, camrelizumab combined with VEGFR2 inhibitor apatinib in advanced HCC, gastric junction cancer, or esophagogastric junction cancer produced a 50% ORR, a 93.8% disease control rate (DCR), and a PFS period of 7.2 months in HCC [45]. All of these metrics were better than those produced by camrelizumab alone. Furthermore, these results are comparable to, if not better than, those of a clinical trial testing the combination of pembrolizumab and VEGFR inhibitor lenvatinib in HCC, as well as those of a clinical trial testing the combination of nivolumab and regorafenib in gastric and gastroesophageal carcinoma. The combination of anti-PD-1 antibody toripalimab and axitinib in treating mucosal melanoma has also shown encouraging results [46]. Taken together, these studies indicate that combination ICI/TKI therapy may be an effective strategy, even though its mechanism of action remains unknown. It is possible that one of the two treatments stablizes disease, allowing more patients to respond to the other treatment. Nevertheless, combination ICI/TKI therapy has indeed been built on strong evidence of the clinical benefit of ICI and TKI, either as single-agents or in combination. Whether this strategy is able to overcome the challenge of ICI-resistant tumors remains to be investigated.

A third combination therapy strategy is to combine agents targeting different immune checkpoint pathways. The combination of anti-CTLA-4 and anti-PD-1/PD-L1 antibodies has shown augmented efficacy versus monotherapy in several cancers, but mainly in those that are known to have responded to anti-PD-1/PD-L1 antibodies as a single agent. As discussed by Andrew Zhu, this approach is standard-of-care for melanoma [47] and renal cell carcinoma [48], and it is considered to be a promising treatment option for hepatocellular carcinoma. More recently, the combination of nivolumab and CTLA-4 inhibitor ipilimumab was shown to significantly enhance OS when compared to first-line chemotherapy in metastatic NSCLC, regardless of PD-L1 status [49]. Whether combination nivolumab/ipilimumab is effective in cancer types that are primarily resistant to anti-PD-1/PD-L1 ICIs remain questionable.

Lessons Learned from the Development and Application of Immune Checkpoint Inhibitors

ICIs continue to receive approval for new indications; however, these indications are underpinned by clinical proof-of-concept data that were produced before 2016. Besides CAR-T therapies, no new immunotherapy strategies have been brought into standard clinical practice after the initial emergence of ICIs. While numerous clinical trials on new immunotherapeutic agents have been conducted, including many that combine new agents with existing ICIs, the majority of these clinical trials have not shown promising results, and some have shown disappointing results. For example, the field learned a lesson from the results of the Phase-III ECHO-301 trial, which tested the combination of epacadostat with pembrolizumab in metastatic melanoma [50]. Epacadostat is an orally-bioavailable inhibitor of indoleamine 2,3-dioxygenase 1 (IDO1), a rate-limiting enzyme in tryptophan catabolism that plays important roles in immune suppression [51]. Despite encouraging early-phase data [52, 53], the ECHO-301 trial did not demonstrate that the combination of epacadostat and pembrolizumab yielded a superior outcome to pembrolizumab alone [50]. The failure of the ECHO-301 trial calls for a better rationale when designing trials for combination immunotherapy. Several strategies for improvement were discussed at the 2019 China Cancer Immunotherapy Workshop.

Roy Herbst presented the long-term follow-up results from both the CA209-003 and Keynote-001 studies, which have clearly documented the long-term survival benefit of nivolumab and pembrolizumab, respectively, in NSCLC [54, 55]. Herbst reviewed the studies that have demonstrated the correlation between higher frequencies of tumor-infiltrating lymphocytes (TIL) and better survival following ICI treatment. However, more specific biomarkers to define TIL subtypes are needed to better predict tumor sensitivity to ICIs. Concerning combination immunotherapy for NSCLC, Herbst highlighted the success of combining chemotherapy and pembrolizumab as a first-line therapy for NSCLC that is not a candidate for targeted therapies [56]. He suggested that future strategies to enhance the efficacy of combination therapy should include the following: (1) reducing tumor bulk to improve the T cell:tumor target ratio, (2) reducing T cell inhibitory substances produced by the tumor, (3) altering tumor barriers (i.e., vasculature/pressure) to T cell penetration, (4) using cancer vaccines to sensitize T cells and antigen-presenting cells (APCs), and (5) altering T cell signaling/gene expression to enhance the production of T cell attractants. In addition, Herbst noted that biomarker-driven studies such as the database-rich Lung-MAP study, the Keynote-495/KeyImPaCT NSCLC "umbrella" study of multiple pembrolizumab-based combinations, and a durvalumab (PD-L1 inhibitor)-based study of locally advanced NSCLC with biomarker correlation, could open the door for individualized combination immunotherapy for NSCLC patients.

In contrast, Andrew Zhu pointed out that ICIs are only indicated for a small percentage of gastrointestinal cancer patients. The Checkmate-040 study led to the approval of nivolumab [57], and the Keynote-224 study led to the approval of pembrolizumab [58], as second-line therapies for HCC. Both antibodies produced a durable response in the majority of treatment-responsive patients. However, the Phase-III CheckMate-459 study comparing nivolumab to multi-kinase inhibitor sorafenib as a first-line therapy for HCC failed to meet the primary endpoint. Nevertheless, the combinations of durvalumab with CTLA-4 inhibitor tremelimumab, ipilimumab with nivolumab, and atezolizumab with VEGF-A inhibitor bevacizumab have all demonstrated promising ORRs. The Phase-Ib study of pembrolizumab and lenvatinib as a first-line therapy for HCC showed an impressive 42.2% ORR. All considered, these studies suggest that combination immunotherapy may become a major platform for treating gastrointestinal cancer patients. After the initial approval of nivolumab and pembrolizumab as third-line therapies for gastric or gastroesophageal cancers, however, single-agent ICI therapies have encountered multiple failures in attempting to expand their indications to first-line and second-line settings. Nevertheless, both pembrolizumab and nivolumab can prolong OS when compared to second-line chemotherapy in patients with squamous cell carcinoma of the esophagus [59], which would be anticipated in this relatively high mutational-burden tumor. Earlier, ICIs had been approved for microsatellite instability (MSI)-high colorectal cancers, which constitute approximately 5% of metastatic colorectal cancers, as well as other MSI-high gastrointestinal malignancies which constitute less than 12% of their respective types of tumors [60, 61]. Therefore, the majority of gastrointestinal cancers still do not respond to ICIs as a single agent, making combination immunotherapy strategies highly desirable.

From an immunology perspective, Lei Zheng shared his viewpoints on the lessons learned from failed IO clinical trials. He summarized five major reasons for the failures, based on the tendency of the field to make unsubstantiated assumptions. The first reason is the assumption that two ICIs are better than one. A lesson in this reason for failure came from clinical trials that combined an anti-PD-1 antibody with an IDO inhibitor or other ICIs (such as anti-LAG3 antibody) as discussed above. The second reason is the assumption that there is synergy between ICIs and chemotherapy. This was exemplified in the outcome of the Keynote-062 study, where pembrolizumab in combination with chemotherapy failed to demonstrate a clinical advantage in gastric and gastroesophageal cancers. The third reason is the assumption that if a combination works for one cancer type, it would also work for other cancer types. The lack of treatment efficacy in Keynote-062 is disparate from the success of using combination chemotherapy and pembrolizumab in NSCLC. The fourth reason is premature conclusions based on results from studies with small sample sizes. This is best exemplified by a Phase-Ib study of nivolumab in combination with NKTR-214, an IL-2 receptor (IL-2R) agonist with biased activation through the IL-2Rβ/γ subunits [62]. This early study had shown a high ORR, which decreased to a lower ORR with more patients enrolled. A study with a small sample size must be interpreted with caution. The only way to prove the effectiveness of a combination therapy is to conduct a well-controlled, randomized study with a sufficiently large sample size to draw the definitive statistical conclusion. The fifth reason is underestimation of the number of immune "defects" that would need to be normalized. Although inflamed tumors may only have a single "defect", such as T cell exhaustion due to activation of immune checkpoints, non-inflamed tumors often have multiple "defects" in the cancer-immunity cycle [63].

FAST AND FURIOUS DEVELOPMENT OF CELL THERAPY

Although a paradigm shift from immunity enhancement to normalization of TIME may be advisable in IO development for solid tumors, immunity enhancement remains a mainstay therapeutic strategy for hematologic malignancies. CD19-targeted CAR-T cells for B cell neoplasms have opened up a new era in synthetic cancer immunotherapy [64]. There are two approved CD19-CAR-T cell platforms: tisagenlecleucel (Kymriah) and axicabtagene ciloleucel (Yescarta). Besides CD3ζ chain, Tisagenlecleucel uses CD137 (4-1BB) as additional co-stimulating signal (COS), while axicabtagene ciloleucel uses CD28 a COS. Both agents utilize a single chain anti-CD19 fragment to target malignant B cells. Tisagenlecleucel is approved for the treatment of patients up to 25 years of age with B cell precursor acute lymphoblastic leukemia (ALL) that is refractory to standard therapy or in at least second relapse. Both agents are indicated for the treatment of relapsed or refractory large B cell lymphoma.

Cutting-edge Developments in Adoptive Cell Therapy in the USA

Cassian Yee from the University of Texas MD Anderson Cancer Center reviewed more than 20 types of antigen-specific T cells from the peripheral blood of cancer patients for adoptive transfer therapy [65]. Yee presented the experience of his research group to use endogenous T cell (ETC) therapy alone to put advanced melanoma into long term remission. Yee's group was also the first to combine ICI therapy with human central memory T cell ETC therapy to treat advanced solid tumors, which yielded astonishingly good responses in some patients [66]. In addition, Yee reported on several investigator-initiated clinical trials in multiple solid tumor types.

Elizabeth Budde from the City of Hope National Medical Center presented long-term follow-up data from several major CAR-T trials in lymphoma, showing approximately 40% PFS at 2 years after CAR-T treatment [67]. Budde also provided an update on the real-world data concerning the two FDA-approved CD19 CAR-T products, which reflects a large fraction of patients who would have been ineligible for the original CAR-T clinical trials. It is reassuring that the real-world response rate, the severity of cytokine release syndrome (CRS), and neurotoxicity were comparable with the results from the clinical trials, indicating that CAR-T can be successfully used to provide clinical benefit in more patients at CAR-T certified medical centers [68]. Multiple myeloma is no doubt one of the most active areas of CAR-T development. CAR-T therapy targets for multiple myeloma include BCMA, CD19, SLAMF7 (CS1), NKG2D, CD56, CD70, CD38, CD138, CD44v6, and IgκλC. Among them, BCMA is the most studied [69, 70]. Two BCMA-targeted CAR-T clinical trials (bb2121 and LCAR-B38M) evidenced a 5074% CR rate, a 10.816-month duration of response, and an 11.815-month PFS [71, 72]. On the other hand, targets for CAR-T therapy in AML include CD123, CD33, NKG2D, Lewis Y, FLT-3, CLL-1, CD44v6, IL1-RAP, and TIM-3 [73, 74]. CD123 is the most-studied target. Budde is conducting CAR-T cell therapy targeting CD123 in AML patients at the City of Hope, and promising outcomes have already been observed in some patients [75, 76].

Yangbin Zhao from the University of Pennsylvania discussed his group's efforts with genetic engineering to improve T cell therapy for solid tumors. Zhao pointed out that the ideal T cells for cellular therapy might be effector memory T (T_{EM})-like cells with excellent proliferation and long-term in vivo persistance [77, 78]. Zhao's update focused on the ongoing first US trial with universal CAR-T cells that were genetically edited using CRISPR/Cas9 technology to eliminate their expression of endogenous TCRs and PD-1. This universal CAR-T cell product targets ESO-1 in solid tumors. Innovative designs of CAR-T cells to enhance their efficacy and function were described, including CAR-T cells with co-expression of a fusion protein between extracellular domain of PD-1 and transmembrane/signaling domain of CD28 to rescue them from hypofunction and enhance their tumor-killing effects [79]. Zhao also provided a briefing on a new strategy to ectopically express a dominant-negative transforming growth factor-beta receptor (TGFβR), thereby removing a negative control mechanism of CAR-T function [80].

Finally, Hongtao Liu from the University of Chicago discussed the management of CAR-T toxicities. Liu discussed the newly-published American Society for Transplantation and Cellular Therapy (ASTCT) consensus grading system for CRS and neurologic toxicity associated with immune effector cells. The new grading relies heavily on clinical parameters and presentation, which could be easily used at the bedside [81]. In addition, Liu discussed the risk factors for CRS and neurologic toxicities such as immune effector cell-associated neurotoxicity syndrome (ICANS), as well as recent studies that might change the management of these toxicities. Liu presented new findings from two separate groups on the critical roles of IL-1 receptor (IL-1R) and granulocyte-macrophage colony-stimulating factor

(GM-CSF) in CRS and ICANS, providing a rationale to block the IL-1R pathway using the IL-1R antagonist, Anakinra [82, 83], and to neutralize GM-CSF, in order to control and prevent CRS and ICANS during CAR-T treatment [84].

China Update: Rapid Catch-up in the Space of Cell Therapy

Investigators in China have also made marked progress in immunotherapy, particularly in the realm of adoptive cell transfer therapy. Jun Ren from Beijing Shijitan Hospital of Capital Medical University discussed autologous dendritic cell and cytokine-induced killer cell (DC-CIK) cellular immunotherapy, as well as combination DC-CIK and chemotherapy [85, 86]. Correlation of immunological biomarkers with clinical efficacy and clonal diversity of the TCRβ repertoires was reported [87]. The combination of tumor vaccines and hyperthermia therapy with cellular therapy also has generated promising clinical responses [88]. Several clinical trials that combine anti-PD-1 antibody with cellular immunotherapy have been initiated for several tumor types, including metastatic NSCLC, SCLC, HCC, and other advanced solid tumors.

Weidong Han from Chinese PLA General Hospital presented his group's clinical results from therapy using $CD19^+CD22^+$ dual-targeted CAR-T cells ($CD19^+CD22^+$ CAR-T), which suggests a general rationale for building dual-targeted CAR-T cells. One ALL patient who failed $CD19^+$CAR-T therapy was able to enter long term CR with $CD19^+CD22^+$CAR-T therapy [89]. $CD19^+CD22^+$CAR-T therapy has strong potency with limited CRS and neurotoxicity. In 46 evaluable refractory/resistant NHL patients, the ORR (CR+PR) reached 97.8% and the CR reached 82.3% after 1 month post-CAR-T infusion; at 6 months post-CAR-T infusion, the OOR was 69.2%, which was favorable in comparison with the results from large historical studies using single-target $CD19^+$ CAR-T. The main reasons for disease relapse/progression included the following: (1) loss of the target antigen on the leukemia cells, (2) low level or loss of CAR-T cells after infusion, (3) CAR-T suppression by the negative regulators in the TIME, and (4) acquired resistance to weak CAR-T cell efficacy. Han also noted that some initial PR patients could exhibit delayed CR with longer follow-up.

Jun Zhu from Peking University Cancer Hospital provided an update on his group's CAR-T product (MC-19PD1 CAR-T), which contains a PD-1/CD28 common gene chimerism for switching the suppressive signal of PD-1 to an activating signal that enhances CAR-T efficacy. Their pilot trial enrolled

17 relapsed/refractory lymphoma patients, including 15 patients with PD-L1⁺ lymphoma, which typically had poor or no response to the CAR-T products from Kite and Novartis. The ORR was 58.8% (CR rate of 41.2%), which were associated only with grade 2 CRS and grade 1 neurotoxicity. Zhu also discussed approaches to modify the hinge and transmembrane regions of CD8α in a CAR-T construct, in order to decrease the risk of CRS without compromising the anti-tumor efficacy of the modified CAR-T cell [90].

NEOANTIGEN-BASED VACCINATION: A PROMISING STRATEGY TO BOOST IMMUNOTHERAPY

An effective cancer-immunity cycle needs to be initiated by an innate immune response to drive the antigen presentation and priming process. Thus, an IO strategy that has emerged is to stimulate a strong innate immune response that will support robust antigen presentation and innate immunity. Neoantigen-based vaccination is one approach to this strategy, and it was discussed by several investigators at the Workshop.

In Situ Vaccination can Potentiate the Clinical Response to ICI Therapy

Tony Ribas from University of California at Los Angeles elaborated on the mechanisms of acquired resistance to ICIs: (1) defects in antigen presentation, and (2) defects in the IFN-γ signaling pathway [91, 92]. In situ vaccination has emerged as a candidate method for overcoming these defects, by facilitating TIME priming. Specifically, toll-like receptor 9 (TLR9) agonists [93], oncolytic virus [94], and IL-2 receptor agonists [95], represent three major targets for in situ vaccination. Clinical trials using these approaches to prime the melanoma TIME and enhance the sensitivity of anti-PD-1 antibodies have been conducted. Preliminary results from these studies support using this novel strategy in overcoming the resistance to ICIs through activating T cells directly by activating type-I cytokine receptors or indirectly through activating innate immune responses.

Ronald Levy from Stanford University shared with the audience his recent research on the in situ vaccination strategy. Levy discussed intratumoral injection of the CpG oligodeoxynucleotide (CpG) as an in situ therapeutic vaccination to boost anti-cancer immunity. Unmethylated CpG commonly exists in microbial genomes, but rarely in vertebrates [96]. Therefore, CpG is recognized via TLR-9 expressed by APCs, including dendritic cells and B cells [97], which activates both innate and adaptive immune responses. In tumor-bearing mice, CpG induced an anti-tumor response only after direct intratumoral injection. When intratumoral vaccination with CpG was combined with low dose radiation during a multi-center Phase-I/II clinical trial in patients with low-grade B-cell lymphoma, durable responses were observed at distant tumor sites [98]. With this success, several other combinations were tested both in the preclinical murine setting as well as in clinical trials. For example, Ibrutinib, a BTK and ITK inhibitor that can suppress myeloid-derived suppressor cells and regulatory T (T_{reg}) cells, produced synergistic anti-tumor activity when combined with CpG and low-dose radiation. In some cases, complete remission was observed in the treated lesions as well as distant lesions that were not treated with CpG. Flow cytometry and single-cell sequencing with paired biopsy specimens obtained pre- and post-treatment showed a decrease in tumor B cells post-treatment, while normal NK, B, and T cells increased. In addition, CpG vaccination not only stimulated an immune response but also induced the expression of OX40, also known as tumor necrosis factor receptor superfamily member 4, and a secondary co-stimulatory molecule expressed on T_{regs} and activated T cells. The combination of CpG vaccination and anti-OX40 therapy enhanced the anti-tumor immune response and eliminated established lymphoma as well as solid tumors in mice. In fact, this combination was more effective than the combination of CpG and anti-PD-L1 antibody [99]. With these promising results in hand, two clinical trials are currently ongoing: (1) combination therapy with CpG, OX40 agonist, and low-dose radiation for non-Hodgkin's lymphoma and (2) combination therapy with CpG and OX40 agonist in all cancer types.

Liang Deng from Memorial Sloan Kettering Cancer Center discussed a novel virotherapy based on vaccinia virus, another approach to the neoantigen-based in situ vaccination strategy. Similar to CpG, oncolytic virus is another kind of in situ therapy that can stimulate cancer immunotherapy as outlined in Fig. 1 (outermost circle). Oncolytic virus triggers an antitumor immune response through induction of immunogenic cell death, release of tumor-associated antigens (including damage-associated molecular patterns (DAMPs)), alteration of an immunosuppressive TIME, and promotion of dendritic cell maturation and antigen presentation. Hence, localized oncolytic virus can convert non-immunogenic "cold" tumors into immunogenic "hot" ones, inducing tumor infiltration by immune cells and overcoming resistance to ICIs [100]. T-VEC is a replication-competent Herpes Simplex 1 (HSV-

1) virus that expresses human GM-CSF (hGM-CSF). It was approved in the USA for treatment of advanced melanoma in 2015, making it the first oncolytic virus approved for this indication. However, compared to hGM-CSF control therapy, intratumoral injection of T-VEC only improved OS by 4.4 months [101]. To further improve clinical efficacy, T-VEC is being tested in combination with immune checkpoint inhibitors, such as anti-CTLA-4 antibody [102]. On the other hand, modified, attenuated vaccinia virus Ankara (MVA) is a new generation of smallpox vaccine that serves as a promising vaccine vector for infectious diseases and cancer. It has a deletion of 30 Kb from the parental vaccinia genome, which inhibits replication of the virus in mammals [103]. Intratumoral injection of heat-inactivated MVA induces innate immunity via the STING pathway, which enhances tumor antigen presentation, promotes dendritic cell maturation, stimulates naïve T cell priming, increases tumor-specific T cell expansion and migration, and boosts cytotoxic T lymphocyte (CTL)-mediated killing of cancer cells [104–106]. The anti-tumor effect of heat-inactivated MVA requires $CD8^+$ T cells, and the long-term anti-tumor memory response requires $CD4^+$ T cells. To generate the next generation of MVA with still greater efficacy, MVA with deletion of C7L (MVAΔC7L) was generated. MVAΔC7L can induce much higher levels of type I interferon, proinflammatory cytokines, and chemokines [105, 107]. Additionally, MVA can be further engineered to express Flt3L, which is a growth factor for $CD103^+$ and plasmacytoid dendritic cells, and OX40L, which serves as co-stimulatory ligand for OX40 expressed by T cells. After intratumoral injection, Flt3L- and OX40L-expressing MVAΔC7L induced more $CD8^+$ and $CD4^+$ T cells responding in distant, non-injected tumors, and synergized with anti-PD-L1 antibody in multiple mouse tumor models as compared to heat-inactivated parental MVA. Currently, the Memorial Sloan Kettering Cancer Center has two vaccinia-based vectors: (1) recombinant MVA that expresses a non-replicative, safe immune activator, activates multiple innate immune pathways (including cGAS/STING), and can be used for intratumoral injection as a monotherapy or in combination with ICIs; (2) an oncolytic vaccinia platform that is replication-competent, has the capability to express large proteins (e.g., antibodies against immune checkpoint molecules), enhances anti-tumor activity and reduces immune-related toxicities.

Personalized Neoantigen-based Vaccination as a Novel Combination Immunotherapy in Cancer

David Reardon from Dana-Farber Cancer Institute and Harvard Medical School discussed peptide-based neoantigen vaccination as a personalized cancer immunotherapy. All cancer cells harbor genetic alterations, including missense, deletion, frame shift, and gene fusion mutations, which can generate tumor neoepitopes. If these neoepitopes can be presented to MHC molecules for T cell recognition, in theory, they can be formulated to make neoantigen-specific cancer vaccines [18]. Previous studies have shown that high neoantigen load is associated with better immunotherapy outcomes. Most investigators use three parallel steps to generate a personalized neoantigen vaccine: (1) DNA from normal cells is sequenced to determine the human leukocyte antigen (HLA) type; (2) paired normal and tumor DNAs are sequenced to detect somatic mutations; and (3) tumor mRNA is sequenced to determine gene expression. Information from these three approaches is combined to predict personal HLA-binding peptides which will guide the manufacturing of candidate neoantigens for personalized vaccine. In the first study with six patients with melanoma as reported by Reardon, four had no cancer recurrence at 25 months after vaccination while the remaining two patients had recurrence, but achieved complete tumor regression after subsequent anti-PD-1 therapy [108]. Results using the enzyme-linked immune absorbent spot (ELISPOT) assay showed significant ex vivo IFN-γ responses against a neoepitope pool. Neoantigen vaccination elicited polyfunctional de novo $CD4^+$ and $CD8^+$ T cell responses against neoepitope pools and induced an anti-tumor T cell response that discriminated between mutated antigens and the corresponding wild type epitopes. Having obtained these exciting results with neoantigen vaccination in melanoma, a glioblastoma NeoVax trial was designed that included approximately 20 synthetic long (i.e., 20-30-mer) peptides per patient [109]. This was an open-label, Phase-I study that included newly-diagnosed glioblastoma patients with MGMT-unmethylated tumors. The vaccination schedule included five injections of high antigen exposure during the first 4 weeks, followed by two boost vaccinations at weeks 12 and 20. The therapy was very well tolerated with no dose-limiting toxicity, and there were no dose delays. Multiple de novo, polyfunctional T cell responses were noted primarily against mutant and not wild-type peptides among patients not treated with dexamethasone. When relapsed tumor tissue was analyzed to assess post-vaccination changes, increased $CD4^+$ and $CD8^+$ T cell infiltration was observed in dexamethasone-untreated patients, but not in dexamethasone-treated patients. Infiltrating T cells detected at the intracranial tumor site were neoantigen-specific. With a median follow-up of 31.5 months, the median PFS was 14.2 months and the median OS was 29.0 months. It can therefore be concluded that the generation of personalized vaccines based on individual patient mutation profiles and HLA types is feasible, and that these vaccines are well-tolerated and can induce neoepitope-specific immune responses. A new clinical trial has been planned, which will combine NeoVax with pembrolizumab for patients with newly-diagnosed GBM.

Finally, Robert Schreiber from Washington University discussed approaches in characterizing $CD8^+$ and $CD4^+$ T cell epitopes and TCR repertoire in murine tumor models [110]. Identification and characterization of neoantigen-specific T cells within tumors is anticipated to lead to a new wave of adoptive T cell therapies, which can be actively developed as a personalized immunotherapy strategy in tandem with neoantigen-based vaccine therapy.

METABOLIC REPROGRAMMING: THE KEY FOR SUSTAINED T CELL EFFECTOR FUNCTION IN CANCER IMMUNOTHERAPY

Although the innate immune response and antigen presentation are crucial for the initiation of an anti-tumor immune response, effective antitumor immunity is often not sustained due to the dysfunctional status of effector T cells due to both T cell intrinsic and extrinsic mechanisms. In the 2019 China Cancer Immunotherapy Workshop, several cutting-edge unpublished preclinical and clinical studies were reported, which demonstrated the role of metabolism in anti-tumor immune modulation.

Reprogramming of Amino Acid Metabolism to Sensitize Tumors for Immunotherapy

Weiping Zou from University of Michigan elucidated how ferroptosis, an iron-dependent, peroxidation-induced type of cell death, regulates the immune system [111, 112] and functions as an important effector T cell-mediated pathway of cancer cell death. Zou's research has found that IFN-γ sensitizes and promotes tumor cell ferroptosis, whereas ferroptosis inhibition attenuates anti-tumor immunity. Zou's group further dissected the mechanisms of IFN-γ-regulated tumor cell ferroptosis and found that IFN-γ regulates tumor ferroptosis by targeting SLC7A11 and SLC3A2. Both

SLC7A11 and SLC3A2 are cystine transporters that are responsible for cystine uptake, glutamate release, and glutathione maintenance. Thus, the reprogramming of amino acid metabolism is crucial for immune-mediated tumor cell death, and targeting ferroptosis may sensitize tumors to immune attack.

Targeting the Adenosine Pathway to Potentiate Cancer Immunotherapy

Extracellular adenosine has significant immunosuppressive effects on both effector immune cells and immunosuppressive regulatory cells [113, 114]. Adenosine is metabolized through dephosphorylation of ATP by CD39 and CD73, which are highly expressed on stromal and immune cells in the TIME [115]. Interaction of adenosine with its receptor blocks T cell activation and promotes myeloid suppression. Hence, targeting adenosine and other molecules upstream and downstream of its pathway may restore anti-cancer immunity (Fig. 1, outermost circle). Lawrence Fong from University of California San Francisco discussed strategies to potentiate cancer immunotherapy through targeting the adenosine pathway. CPI-444 is an oral small molecule antagonist of the adenosine 2A receptor (A2AR) [116]. A Phase-I/Ib clinical trial with oral CPI-444 alone or CPI-444 in combination with atezolizumab is currently ongoing in patients with renal cell carcinoma, non-small cell lung cancer, melanoma, triple-negative breast cancer, or other cancer types (unpublished data). These patients have exhibited progressive disease on prior therapy, including immunotherapy, and they were not selected based on PD-L1 expression. So far, this regimen has been tolerated very well. A2A inhibition alone or in combination with atezolizumab has led to clinical responses in both atezolizumab-naïve as well as atezolizumab-resistant/refractory patients. CPI-444 treatment induces $CD8^+$ T cell infiltration into tumor sites, and the adenosine gene signature is associated with tumor response to therapy. Since adenosine is converted from AMP by CD73, another clinical trial is currently ongoing with the humanized anti-CD73 antibody CPI-006, which blocks CD73 catalytic activity and has agonistic immunomodulatory activity on $CD73^+$ cells. CPI-006 has been well-tolerated at the dose of 12 mg/kg, can completely inhibit CD73 enzymatic activity in tumor biopsies, and can induce serum pro-inflammatory cytokines. In conclusion, targeting the adenosine pathway can potentially reverse an immunosuppressive TIME and potentiate the anti-tumor response to ICIs.

Discovering and Targeting the RNA Metabolic Switch between Regulatory and Effector T Cells

RNA-binding proteins are important regulators of RNA biology (splicing, stability, etc.) and protein translation. Recently, it was found that these proteins play critical roles in the regulation of gene expression upon T cell activation. One member of the RNA-binding proteins, the poly-C-binding protein 1 (PCBP1), binds to DNA and RNA and controls protein expression. In addition, its phosphorylation status can determine its RNA binding function [117]. In a chronic immunosuppressive setting such as cancer, the phosphorylation of PCBP1 induced by TGF-β can affect expression of proteins such as moesin [118], and it can regulate T cell differentiation and the immune response. Zihai Li from The Ohio State University reported his group's unpublished data on the roles of PCBP1 in stabilizing effector T cell (T_{eff}) function and suppressing the T_{eff}-T_{reg} commitment program (Ephraim Ansa-Addo and Zihai Li, unpublished). However, once T_{regs} have differentiated, PCBP1 knockdown can no longer reverse this program back to the T_{eff} program. Consequently, genetic disruption of PCBP1 in T cells exacerbates tumor growth. However, melanoma with low PCBP1 expression responds better to anti-PD-1 therapy, suggesting that, in this ICI-sensitive tumor, the T_{eff}-T_{reg} commitment program sensitizes tumors for ICI treatment. Li also showed that high PCBP1 expression is associated with low expression of immune checkpoint signals in melanoma. Therefore, it will be interesting to see the expression levels of PCBP1 in ICI-resistant, cold tumors, and examine how this plays a role in the T_{eff}-T_{reg} commitment program and ICI resistance.

CONCLUSIONS AND FUTURE PERSPECTIVES

The development of cancer immunotherapy for ICI-resistant cancers has been a challenge. Current ICI-based combination therapy strategies have achieved some, albeit limited, success. A deep understanding of TIME biology in the IO field is necessary for generating next generation immuno-oncology therapeutic strategies. ICIs, CAR-T therapy, adoptive cell therapy, and other anti-tumor immunity enhancement approaches, will continue to lead the way in the clinical IO space. However, new classes of immunotherapy are emerging. These new classes aim at normalizing the "defective" TIME by targeting immunosuppressive components unique to the tumors, priming effector T cells by in situ oncolytic therapy, broadening effective T cell repertoire with multi-valent neoantigen-based vaccines, modulating metabolic programming for sustained T cell function and promoting effective immune-mediated cell death. All together, these emerging strategies point towards a promising new wave of cancer immunotherapies that may allow us to surmount the limitations of previous ones.

ACKNOWLEDGEMENTS

The authors are indebted to all organizing committee members of 2019 China Cancer Immunotherapy Workshop, including those from CAHON (Ke Liu), China NMPA (Jin Cui, Chenyan Gao, Zhimin Yang), and Tsinghua University (Chen Dong, Xin Lin), as well as all the invited speakers for their contributions to the success of the meeting. The authors apologize to colleagues whose work could not be cited due to space limitations.

AUTHORS' CONTRIBUTIONS

LZ, CP, HL drafted the manuscript and finalized it with input from ER, WS, MR, DL and ZL. All authors approved the final manuscript.

REFERENCES

1. Leach DR, Krummel MF, Allison JP. Enhancement of antitumor immunity by CTLA-4 blockade. Science. 1996;271:1734–1736. doi: 10.1126/science.271.5256.1734.
2. Kwon ED, Hurwitz AA, Foster BA, et al. Manipulation of T cell costimulatory and inhibitory signals for immunotherapy of prostate cancer. Proc Natl Acad Sci U S A. 1997;94:8099–8103. doi: 10.1073/pnas.94.15.8099.
3. Ishida Y, Agata Y, Shibahara K, Honjo T. Induced expression of PD-1, a novel member of the immunoglobulin gene superfamily, upon programmed cell death. EMBO J. 1992;11:3887–3895. doi: 10.1002/j.1460-2075.1992.tb05481.x.
4. Nishimura H, Nose M, Hiai H, et al. Development of lupus-like autoimmune diseases by disruption of the PD-1 gene encoding an ITIM motif-carrying immunoreceptor. Immunity. 1999;11:141–151. doi: 10.1016/S1074-7613(00)80089-8.
5. Freeman GJ, Long AJ, Iwai Y, et al. Engagement of the PD-1 immunoinhibitory receptor by a novel B7 family member leads to negative regulation of lymphocyte activation. J Exp Med. 2000;192:1027–1034. doi: 10.1084/jem.192.7.1027.
6. Iwai Y, Terawaki S, Honjo T. PD-1 blockade inhibits hematogenous spread of poorly immunogenic tumor cells by enhanced recruitment of effector T cells. Int Immunol. 2005;17:133–144. doi: 10.1093/intimm/dxh194.
7. Dong H, Zhu G, Tamada K, Chen L. B7-H1, a third member of the B7 family, co-stimulates T-cell proliferation and interleukin-10 secretion. Nat Med. 1999;5:1365–1369. doi: 10.1038/70932.
8. Dong H, Strome SE, Salomao DR, et al. Tumor-associated B7-H1 promotes T-cell apoptosis: a potential mechanism of immune evasion. Nat Med. 2002;8:793–800. doi: 10.1038/nm730.
9. Dong H, Zhu G, Tamada K, et al. B7-H1 determines accumulation and deletion of intrahepatic CD8(+) T lymphocytes. Immunity. 2004;20:327–336. doi: 10.1016/S1074-7613(04)00050-0.
10. Brunet JF, Denizot F, Luciani MF, et al. A new member of the immunoglobulin superfamily--CTLA-4. Nature. 1987;328:267–270. doi: 10.1038/328267a0.

11. Walunas TL, Lenschow DJ, Bakker CY, et al. CTLA-4 can function as a negative regulator of T cell activation. Immunity. 1994;1:405–413. doi: 10.1016/1074-7613(94)90071-X.
12. Tivol EA, Borriello F, Schweitzer AN, et al. Loss of CTLA-4 leads to massive lymphoproliferation and fatal multiorgan tissue destruction, revealing a critical negative regulatory role of CTLA-4. Immunity. 1995;3:541–547. doi: 10.1016/1074-7613(95)90125-6.
13. Waterhouse P, Penninger JM, Timms E, et al. Lymphoproliferative disorders with early lethality in mice deficient in Ctla-4. Science. 1995;270:985–988. doi: 10.1126/science.270.5238.985.
14. Pardoll D. Immunotherapy: it takes a village. Science. 2014;344:149. doi: 10.1126/science.344.6180.149-a.
15. Li Z, Song W, Rubinstein M, Liu D. Recent updates in cancer immunotherapy: a comprehensive review and perspective of the 2018 China Cancer Immunotherapy Workshop in Beijing. J Hematol Oncol. 2018;11:142. doi: 10.1186/s13045-018-0684-3.
16. Chen DS, Mellman I. Oncology meets immunology: the cancer-immunity cycle. Immunity. 2013;39:1–10. doi: 10.1016/j.immuni.2013.07.012.
17. Hanahan D, Weinberg RA. Hallmarks of cancer: the next generation. Cell. 2011;144:646–674. doi: 10.1016/j.cell.2011.02.013.
18. Vogelstein B, Papadopoulos N, Velculescu VE, et al. Cancer genome landscapes. Science. 2013;339:1546–1558. doi: 10.1126/science.1235122.
19. Perkins D, Wang Z, Donovan C, et al. Regulation of CTLA-4 expression during T cell activation. J Immunol. 1996;156:4154–4159.
20. Yi M, Dong B, Chu Q, Wu K. Immune pressures drive the promoter hypermethylation of neoantigen genes. Exp Hematol Oncol. 2019;8:32. doi: 10.1186/s40164-019-0156-7.
21. Taube JM, Anders RA, Young GD, et al. Colocalization of inflammatory response with B7-h1 expression in human melanocytic lesions supports an adaptive resistance mechanism of immune escape. Sci Transl Med. 2012;4:127ra137. doi: 10.1126/scitranslmed.3003689.
22. Chen L, Han X. Anti-PD-1/PD-L1 therapy of human cancer: past, present, and future. J Clin Invest. 2015;125:3384–3391. doi: 10.1172/JCI80011.

23. Zhang Y, Chen L. Classification of advanced human cancers based on tumor immunity in the microEnvironment (TIME) for cancer immunotherapy. JAMA Oncol. 2016;2:1403–1404. doi: 10.1001/jamaoncol.2016.2450.
24. Herbst RS, Soria JC, Kowanetz M, et al. Predictive correlates of response to the anti-PD-L1 antibody MPDL3280A in cancer patients. Nature. 2014;515:563–567. doi: 10.1038/nature14011.
25. Salmon H, Franciszkiewicz K, Damotte D, et al. Matrix architecture defines the preferential localization and migration of T cells into the stroma of human lung tumors. J Clin Invest. 2012;122:899–910. doi: 10.1172/JCI45817.
26. Spranger S, Spaapen RM, Zha Y, et al. Up-regulation of PD-L1, IDO, and T (regs) in the melanoma tumor microenvironment is driven by CD8(+) T cells. Sci Transl Med. 2013;5:200ra116. doi: 10.1126/scitranslmed.3006504.
27. Gajewski TF, Woo SR, Zha Y, et al. Cancer immunotherapy strategies based on overcoming barriers within the tumor microenvironment. Curr Opin Immunol. 2013;25:268–276. doi: 10.1016/j.coi.2013.02.009.
28. Sanmamed MF, Chen L. A paradigm shift in cancer immunotherapy: from enhancement to normalization. Cell. 2018;175:313–326. doi: 10.1016/j.cell.2018.09.035.
29. Lutz ER, Wu AA, Bigelow E, et al. Immunotherapy converts nonimmunogenic pancreatic tumors into immunogenic foci of immune regulation. Cancer Immunol Res. 2014;2:616–631. doi: 10.1158/2326-6066.CIR-14-0027.
30. Lutz ER, Kinkead H, Jaffee EM, Zheng L. Priming the pancreatic cancer tumor microenvironment for checkpoint-inhibitor immunotherapy. Oncoimmunology. 2014;3:e962401. doi: 10.4161/21624011.2014.962401.
31. Kleponis J, Skelton R, Zheng L. Fueling the engine and releasing the break: combinational therapy of cancer vaccines and immune checkpoint inhibitors. Cancer Biol Med. 2015;12:201–208.
32. Wang J, Sun J, Liu LN, et al. Siglec-15 as an immune suppressor and potential target for normalization cancer immunotherapy. Nat Med. 2019;25:656–666. doi: 10.1038/s41591-019-0374-x.
33. Schmid P, Adams S, Rugo HS, et al. Atezolizumab and nab-paclitaxel in advanced triple-negative breast cancer. N Engl J Med. 2018;379:2108–

2121. doi: 10.1056/NEJMoa1809615.

34. Keam SJ. Toripalimab: first global approval. Drugs. 2019;79:573–578. doi: 10.1007/s40265-019-01076-2.
35. Hoy SM. Sintilimab: first global approval. Drugs. 2019;79:341–346. doi: 10.1007/s40265-019-1066-z.
36. Shi Y, Su H, Song Y, et al. Safety and activity of sintilimab in patients with relapsed or refractory classical Hodgkin lymphoma (ORIENT-1): a multicentre, single-arm, phase 2 trial. Lancet Haematol. 2019;6:e12–e19. doi: 10.1016/S2352-3026(18)30192-3.
37. Tang B, Yan X, Sheng X, et al. Safety and clinical activity with an anti-PD-1 antibody JS001 in advanced melanoma or urologic cancer patients. J Hematol Oncol. 2019;12:7. doi: 10.1186/s13045-018-0693-2.
38. Markham A, Keam SJ. Camrelizumab: first global approval. Drugs. 2019;79:1355–1361. doi: 10.1007/s40265-019-01167-0.
39. Song Y, Wu J, Chen X, et al. A single-arm, multicenter, phase II study of camrelizumab in relapsed or refractory classical Hodgkin lymphoma. Clin Cancer Res. 2019;25:7363–7369. doi: 10.1158/1078-0432.CCR-19-1680.
40. Song Y, Gao Q, Zhang H, et al. Treatment of relapsed or refractory classical Hodgkin lymphoma with the anti-PD-1, tislelizumab: results of a phase 2, single-arm, multicenter study. Leukemia. 2020;34:533–542. doi: 10.1038/s41375-019-0545-2.
41. Fang W, Yang Y, Ma Y, et al. Camrelizumab (SHR-1210) alone or in combination with gemcitabine plus cisplatin for nasopharyngeal carcinoma: results from two single-arm, phase 1 trials. Lancet Oncol. 2018;19:1338–1350. doi: 10.1016/S1470-2045(18)30495-9.
42. Rini BI, Plimack ER, Stus V, et al. Pembrolizumab plus axitinib versus sunitinib for advanced renal-cell carcinoma. N Engl J Med. 2019;380:1116–1127. doi: 10.1056/NEJMoa1816714.
43. Motzer RJ, Penkov K, Haanen J, et al. Avelumab plus axitinib versus sunitinib for advanced renal-cell carcinoma. N Engl J Med. 2019;380:1103–1115. doi: 10.1056/NEJMoa1816047.
44. Makker V, Rasco D, Vogelzang NJ, et al. Lenvatinib plus pembrolizumab in patients with advanced endometrial cancer: an interim analysis of a multicentre, open-label, single-arm, phase 2 trial. Lancet

Oncol. 2019;20:711–718. doi: 10.1016/S1470-2045(19)30020-8.

45. Xu J, Zhang Y, Jia R, et al. Anti-PD-1 Antibody SHR-1210 Combined with apatinib for advanced hepatocellular carcinoma, gastric, or esophagogastric junction cancer: an open-label, dose escalation and expansion study. Clin Cancer Res. 2019;25:515–523. doi: 10.1158/1078-0432.CCR-18-2484.

46. Sheng X, Yan X, Chi Z et al. Axitinib in combination with toripalimab, a humanized immunoglobulin G4 monoclonal antibody against programmed cell death-1, in patients with metastatic mucosal melanoma: an open-label phase IB trial. J Clin Oncol 2019; JCO1900210.

47. Wolchok JD, Kluger H, Callahan MK, et al. Nivolumab plus ipilimumab in advanced melanoma. N Engl J Med. 2013;369:122–133. doi: 10.1056/NEJMoa1302369.

48. Motzer RJ, Tannir NM, McDermott DF, et al. Nivolumab plus ipilimumab versus sunitinib in advanced renal-cell carcinoma. N Engl J Med. 2018;378:1277–1290. doi: 10.1056/NEJMoa1712126.

49. Hellmann MD, Paz-Ares L, Bernabe Caro R, et al. Nivolumab plus ipilimumab in advanced non-small-cell lung cancer. N Engl J Med. 2019.

50. Muller AJ, Manfredi MG, Zakharia Y, Prendergast GC. Inhibiting IDO pathways to treat cancer: lessons from the ECHO-301 trial and beyond. Semin Immunopathol. 2019;41:41–48. doi: 10.1007/s00281-018-0702-0.

51. Munn DH, Mellor AL. IDO in the tumor microenvironment: inflammation, counter-regulation, and tolerance. Trends Immunol. 2016;37:193–207. doi: 10.1016/j.it.2016.01.002.

52. Beatty GL, O'Dwyer PJ, Clark J, et al. First-in-human phase I study of the oral inhibitor of indoleamine 2,3-dioxygenase-1 epacadostat (INCB024360) in patients with advanced solid malignancies. Clin Cancer Res. 2017;23:3269–3276. doi: 10.1158/1078-0432.CCR-16-2272.

53. Mitchell TC, Hamid O, Smith DC et al. Epacadostat plus pembrolizumab in patients with advanced solid tumors: phase I results from a multicenter, open-label phase I/II trial (ECHO-202/KEYNOTE-037). J Clin Oncol 2018; JCO2018789602.

54. Gettinger S, Horn L, Jackman D, et al. Five-year follow-up of nivolumab in previously treated advanced non-small-cell lung cancer:

results from the CA209-003 study. J Clin Oncol. 2018;36:1675–1684. doi: 10.1200/JCO.2017.77.0412.

55. Hamid O, Robert C, Daud A, et al. Five-year survival outcomes for patients with advanced melanoma treated with pembrolizumab in KEYNOTE-001. Ann Oncol. 2019;30:582–588. doi: 10.1093/annonc/mdz011.

56. Gandhi L, Rodriguez-Abreu D, Gadgeel S, et al. Pembrolizumab plus chemotherapy in metastatic non-small-cell lung cancer. N Engl J Med. 2018;378:2078–2092. doi: 10.1056/NEJMoa1801005.

57. El-Khoueiry AB, Sangro B, Yau T, et al. Nivolumab in patients with advanced hepatocellular carcinoma (CheckMate 040): an open-label, non-comparative, phase 1/2 dose escalation and expansion trial. Lancet. 2017;389:2492–2502. doi: 10.1016/S0140-6736(17)31046-2.

58. Zhu AX, Finn RS, Edeline J, et al. Pembrolizumab in patients with advanced hepatocellular carcinoma previously treated with sorafenib (KEYNOTE-224): a non-randomised, open-label phase 2 trial. Lancet Oncol. 2018;19:940–952. doi: 10.1016/S1470-2045(18)30351-6.

59. Shah MA, Kojima T, Hochhauser D, et al. Efficacy and safety of pembrolizumab for heavily pretreated patients with advanced, metastatic adenocarcinoma or squamous cell carcinoma of the esophagus: the phase 2 KEYNOTE-180 study. JAMA Oncol. 2019;5:546–550. doi: 10.1001/jamaoncol.2018.5441.

60. Le DT, Uram JN, Wang H, et al. PD-1 Blockade in tumors with mismatch-repair deficiency. N Engl J Med. 2015;372:2509–2520. doi: 10.1056/NEJMoa1500596.

61. Le DT, Durham JN, Smith KN, et al. Mismatch repair deficiency predicts response of solid tumors to PD-1 blockade. Science. 2017;357:409–413. doi: 10.1126/science.aan6733.

62. Bentebibel SE, Hurwitz ME, Bernatchez C, et al. A first-in-human study and biomarker analysis of NKTR-214, a novel IL2Rbetagamma-biased cytokine, in patients with advanced or metastatic solid tumors. Cancer Discov. 2019;9:711–721. doi: 10.1158/2159-8290.CD-18-1495.

63. Upadhrasta S, Zheng L. Strategies in developing immunotherapy for pancreatic cancer: recognizing and correcting multiple immune "defects" in the tumor microenvironment. J Clin Med. 2019;8.

64. June CH, O'Connor RS, Kawalekar OU, et al. CAR T cell immunotherapy for human cancer. Science. 2018;359:1361–1365. doi: 10.1126/science.aar6711.
65. Yee C, Lizee G, Schueneman AJ. Endogenous T-cell therapy: clinical experience. Cancer J. 2015;21:492–500. doi: 10.1097/PPO.0000000000000158.
66. Yee C. Adoptive T cell therapy: points to consider. Curr Opin Immunol. 2018;51:197–203. doi: 10.1016/j.coi.2018.04.007.
67. Schuster SJ, Bishop MR, Tam CS, et al. Tisagenlecleucel in adult relapsed or refractory diffuse large B-cell lymphoma. N Engl J Med. 2019;380:45–56. doi: 10.1056/NEJMoa1804980.
68. Nastoupil LJ, Jain MD, Spiegel JY, et al. Axicabtagene ciloleucel (Axi-cel) CD19 chimeric antigen receptor (CAR) T-cell therapy for relapsed/refractory large B-cell lymphoma: real world experience. Blood. 2018;132:91. doi: 10.1182/blood-2018-99-114152.
69. Lin Q, Zhao J, Song Y, Liu D. Recent updates on CAR T clinical trials for multiple myeloma. Mol Cancer. 2019;18:154. doi: 10.1186/s12943-019-1092-1.
70. Wu C, Zhang L, Brockman QR, et al. Chimeric antigen receptor T cell therapies for multiple myeloma. J Hematol Oncol. 2019;12:120. doi: 10.1186/s13045-019-0823-5.
71. Zhao WH, Liu J, Wang BY, et al. A phase 1, open-label study of LCAR-B38M, a chimeric antigen receptor T cell therapy directed against B cell maturation antigen, in patients with relapsed or refractory multiple myeloma. J Hematol Oncol. 2018;11:141. doi: 10.1186/s13045-018-0681-6.
72. Raje N, Berdeja J, Lin Y, et al. Anti-BCMA CAR T-cell therapy bb2121 in relapsed or refractory multiple myeloma. N Engl J Med. 2019;380:1726–1737. doi: 10.1056/NEJMoa1817226.
73. Liu D. Cancer biomarkers for targeted therapy. Biomark Res. 2019;7:25. doi: 10.1186/s40364-019-0178-7.
74. Liu D. CAR-T "the living drugs", immune checkpoint inhibitors, and precision medicine: a new era of cancer therapy. J Hematol Oncol. 2019;12:113. doi: 10.1186/s13045-019-0819-1.
75. Budde L, Song JY, Kim Y, et al. Remissions of acute myeloid leukemia and blastic plasmacytoid dendritic cell neoplasm following treatment with CD123-specific CAR T cells: a first-in-human clinical

trial. Blood. 2017;130:811.

76. Zhang W, Stevens BM, Budde EE, et al. Anti-CD123 CAR T-cell therapy for the treatment of myelodysplastic syndrome. Blood. 2017;130:1917.

77. Liu X, Jiang S, Fang C, et al. Novel T cells with improved in vivo anti-tumor activity generated by RNA electroporation. Protein Cell. 2017;8:514–526. doi: 10.1007/s13238-017-0422-6.

78. Liu X, Barrett DM, Jiang S et al. Improved anti-leukemia activities of adoptively transferred T cells expressing bispecific T-cell engager in mice. Blood Cancer J. 2016;6:e430.

79. Liu X, Ranganathan R, Jiang S, et al. A chimeric switch-receptor targeting PD1 augments the efficacy of second-generation CAR T cells in advanced solid tumors. Cancer Res. 2016;76:1578–1590. doi: 10.1158/0008-5472.CAN-15-2524.

80. Kloss CC, Lee J, Zhang A, et al. Dominant-negative TGF-beta receptor enhances PSMA-targeted human CAR T cell proliferation and augments prostate cancer eradication. Mol Ther. 2018;26:1855–1866. doi: 10.1016/j.ymthe.2018.05.003.

81. Lee DW, Santomasso BD, Locke FL, et al. ASTCT consensus grading for cytokine release syndrome and neurologic toxicity associated with immune effector cells. Biol Blood Marrow Transplant. 2019;25:625–638. doi: 10.1016/j.bbmt.2018.12.758.

82. Giavridis T, van der Stegen SJC, Eyquem J, et al. CAR T cell-induced cytokine release syndrome is mediated by macrophages and abated by IL-1 blockade. Nat Med. 2018;24:731–738. doi: 10.1038/s41591-018-0041-7.

83. Norelli M, Camisa B, Barbiera G, et al. Monocyte-derived IL-1 and IL-6 are differentially required for cytokine-release syndrome and neurotoxicity due to CAR T cells. Nat Med. 2018;24:739–748. doi: 10.1038/s41591-018-0036-4.

84. Sterner RM, Sakemura R, Cox MJ, et al. GM-CSF inhibition reduces cytokine release syndrome and neuroinflammation but enhances CAR-T cell function in xenografts. Blood. 2019;133:697–709. doi: 10.1182/blood-2018-10-881722.

85. Qiao G, Wang X, Zhou L, et al. Autologous dendritic cell-cytokine induced killer cell immunotherapy combined with S-1 plus cisplatin in patients with advanced gastric cancer: a prospective study. Clin Cancer Res. 2019;25:1494–1504. doi: 10.1158/1078-0432.CCR-18-2360.

86. Zhao Y, Qiao G, Wang X, et al. Combination of DC/CIK adoptive T cell immunotherapy with chemotherapy in advanced non-small-cell lung cancer (NSCLC) patients: a prospective patients' preference-based study (PPPS) Clin Transl Oncol. 2019;21:721–728. doi: 10.1007/s12094-018-1968-3.
87. Ren J, Gwin WR, Zhou X et al. Adaptive T cell responses induced by oncolytic herpes simplex virus-granulocyte macrophage-colony-stimulating factor therapy expanded by dendritic cell and cytokine-induced killer cell adoptive therapy. Oncoimmunology 6: e1264563.
88. Sun Y, Wang S, Yang H, et al. Impact of synchronized anti-PD-1 with Ad-CEA vaccination on inhibition of colon cancer growth. Immunotherapy. 2019;11:953–966. doi: 10.2217/imt-2019-0055.
89. Jia H, Wang Z, Wang Y, et al. Haploidentical CD19/CD22 bispecific CAR-T cells induced MRD-negative remission in a patient with relapsed and refractory adult B-ALL after haploidentical hematopoietic stem cell transplantation. J Hematol Oncol. 2019;12:57. doi: 10.1186/s13045-019-0741-6.
90. Ying Z, Huang XF, Xiang X, et al. A safe and potent anti-CD19 CAR T cell therapy. Nat Med. 2019;25:947–953. doi: 10.1038/s41591-019-0421-7.
91. Zaretsky JM, Garcia-Diaz A, Shin DS, et al. Mutations associated with acquired resistance to PD-1 blockade in melanoma. N Engl J Med. 2016;375:819–829. doi: 10.1056/NEJMoa1604958.
92. Shin DS, Zaretsky JM, Escuin-Ordinas H, et al. Primary resistance to PD-1 blockade mediated by JAK1/2 mutations. Cancer Discov. 2017;7:188–201. doi: 10.1158/2159-8290.CD-16-1223.
93. Ribas A, Medina T, Kummar S, et al. SD-101 in combination with pembrolizumab in advanced melanoma: results of a phase Ib, multicenter study. Cancer Discov. 2018;8:1250–1257. doi: 10.1158/2159-8290.CD-18-0280.
94. Ribas A, Dummer R, Puzanov I, et al. Oncolytic virotherapy promotes intratumoral T cell infiltration and improves anti-PD-1 immunotherapy. Cell. 2017;170:1109–1119. doi: 10.1016/j.cell.2017.08.027.
95. Sockolosky JT, Trotta E, Parisi G, et al. Selective targeting of engineered T cells using orthogonal IL-2 cytokine-receptor

complexes. Science. 2018;359:1037–1042. doi: 10.1126/science.aar3246.

96. Bauer S, Wagner H. Bacterial CpG-DNA licenses TLR9. Curr Top Microbiol Immunol. 2002;270:145–154.
97. Rothenfusser S, Tuma E, Endres S, Hartmann G. Plasmacytoid dendritic cells: the key to CpG. Hum Immunol. 2002;63:1111–1119. doi: 10.1016/S0198-8859(02)00749-8.
98. Brody JD, Ai WZ, Czerwinski DK, et al. In situ vaccination with a TLR9 agonist induces systemic lymphoma regression: a phase I/II study. J Clin Oncol. 2010;28:4324–4332. doi: 10.1200/JCO.2010.28.9793.
99. Sagiv-Barfi I, Czerwinski DK, Levy S, et al. Eradication of spontaneous malignancy by local immunotherapy. Sci Transl Med. 2018;10.
100. Zamarin D, Holmgaard RB, Subudhi SK, et al. Localized oncolytic virotherapy overcomes systemic tumor resistance to immune checkpoint blockade immunotherapy. Sci Transl Med. 2014;6:226ra232. doi: 10.1126/scitranslmed.3008095.
101. Andtbacka RH, Kaufman HL, Collichio F, et al. Talimogene laherparepvec improves durable response rate in patients with advanced melanoma. J Clin Oncol. 2015;33:2780–2788. doi: 10.1200/JCO.2014.58.3377.
102. Puzanov I, Milhem MM, Minor D, et al. Talimogene laherparepvec in combination with ipilimumab in previously untreated, unresectable stage IIIB-IV melanoma. J Clin Oncol. 2016;34:2619–2626. doi: 10.1200/JCO.2016.67.1529.
103. Pavot V, Sebastian S, Turner AV, et al. Generation and production of modified vaccinia virus Ankara (MVA) as a vaccine vector. Methods Mol Biol. 2017;1581:97–119. doi: 10.1007/978-1-4939-6869-5_6.
104. Dai P, Cao H, Merghoub T, et al. Myxoma virus induces type I interferon production in murine plasmacytoid dendritic cells via a TLR9/MyD88-, IRF5/IRF7-, and IFNAR-dependent pathway. J Virol. 2011;85:10814–10825. doi: 10.1128/JVI.00104-11.
105. Dai P, Wang W, Cao H, et al. Modified vaccinia virus Ankara triggers type I IFN production in murine conventional dendritic cells via a cGAS/STING-mediated cytosolic DNA-sensing pathway. PLoS Pathog. 2014;10:e1003989. doi: 10.1371/journal.ppat.1003989.
106. Dai P, Wang W, Yang N, et al. Intratumoral delivery of inactivated modified vaccinia virus Ankara (iMVA) induces systemic antitumor

immunity via STING and Batf3-dependent dendritic cells. Sci Immunol. 2017;2.
107. Deng L, Dai P, Ding W, et al. Vaccinia virus infection attenuates innate immune responses and antigen presentation by epidermal dendritic cells. J Virol. 2006;80:9977–9987. doi: 10.1128/JVI.00354-06.
108. Ott PA, Hu Z, Keskin DB, et al. An immunogenic personal neoantigen vaccine for patients with melanoma. Nature. 2017;547:217–221. doi: 10.1038/nature22991.
109. Keskin DB, Anandappa AJ, Sun J, et al. Neoantigen vaccine generates intratumoral T cell responses in phase Ib glioblastoma trial. Nature. 2019;565:234–239. doi: 10.1038/s41586-018-0792-9.
110. Alspach E, Lussier DM, Miceli AP, et al. MHC-II neoantigens shape tumour immunity and response to immunotherapy. Nature. 2019;574:696–701. doi: 10.1038/s41586-019-1671-8.
111. Wang W, Green M, Choi JE, et al. CD8(+) T cells regulate tumour ferroptosis during cancer immunotherapy. Nature. 2019;569:270–274. doi: 10.1038/s41586-019-1170-y.
112. Lang X, Green MD, Wang W, et al. Radiotherapy and immunotherapy promote tumoral lipid oxidation and ferroptosis via synergistic repression of SLC7A11. Cancer Discov. 2019.
113. Cekic C, Linden J. Purinergic regulation of the immune system. Nat Rev Immunol. 2016;16:177–192. doi: 10.1038/nri.2016.4.
114. Allard B, Longhi MS, Robson SC, Stagg J. The ectonucleotidases CD39 and CD73: Novel checkpoint inhibitor targets. Immunol Rev. 2017;276:121–144. doi: 10.1111/imr.12528.
115. Leone RD, Emens LA. Targeting adenosine for cancer immunotherapy. J Immunother Cancer. 2018;6:57. doi: 10.1186/s40425-018-0360-8.
116. Leone RD, Sun IM, Oh MH, et al. Inhibition of the adenosine A2a receptor modulates expression of T cell coinhibitory receptors and improves effector function for enhanced checkpoint blockade and ACT in murine cancer models. Cancer Immunol Immunother. 2018;67:1271–1284. doi: 10.1007/s00262-018-2186-0.
117. Guo J, Jia R. Splicing factor poly (rC)-binding protein 1 is a novel and distinctive tumor suppressor. J Cell Physiol. 2018;234:33–41. doi: 10.1002/jcp.26873.

118. Ansa-Addo EA, Zhang Y, Yang Y, et al. Membrane-organizing protein moesin controls Treg differentiation and antitumor immunity via TGF-beta signaling. J Clin Invest. 2017;127:1321–1337. doi: 10.1172/JCI89281.

4

Cancer Immunotherapy beyond Immune Checkpoint Inhibitors

Julian A. Marin-Acevedo[1], Aixa E. Soyano[2], Bhagirathbhai Dholaria[2,3], Keith L. Knutson[4] and Yanyan Lou[2]

[1] Department of Internal Medicine, Mayo Clinic, Jacksonville, FL, USA
[2] Department of Hematology and Oncology, Mayo Clinic, 4500 San Pablo Road, Jacksonville, FL 32224, USA
[3] Current address: Department of Blood and Marrow Transplant and Cellular Immunotherapy, Moffitt Cancer Center, Tampa, FL, USA
[4] Department of Immunology, Mayo Clinic, Jacksonville, FL, USA

ABSTRACT

Malignant cells have the capacity to rapidly grow exponentially and spread in part by suppressing, evading, and exploiting the host immune system. Immunotherapy is a form of oncologic treatment directed towards enhancing

Citation: Marin-Acevedo JA, Soyano AE, Dholaria B, Knutson KL, Lou Y. Cancer immunotherapy beyond immune checkpoint inhibitors. J Hematol Oncol. 2018;11(1):8. Published 2018 Jan 12. doi:10.1186/s13045-017-0552-6.

the host immune system against cancer. In recent years, manipulation of immune checkpoints or pathways has emerged as an important and effective form of immunotherapy. Agents that target cytotoxic T lymphocyte-associated molecule-4 (CTLA-4), programmed cell death receptor-1 (PD-1), and programmed cell death ligand-1 (PD-L1) are the most widely studied and recognized. Immunotherapy, however, extends beyond immune checkpoint therapy by using new molecules such as chimeric monoclonal antibodies and antibody drug conjugates that target malignant cells and promote their destruction. Genetically modified T cells expressing chimeric antigen receptors are able to recognize specific antigens on cancer cells and subsequently activate the immune system. Native or genetically modified viruses with oncolytic activity are of great interest as, besides destroying malignant cells, they can increase anti-tumor activity in response to the release of new antigens and danger signals as a result of infection and tumor cell lysis. Vaccines are also being explored, either in the form of autologous or allogenic tumor peptide antigens, genetically modified dendritic cells that express tumor peptides, or even in the use of RNA, DNA, bacteria, or virus as vectors of specific tumor markers. Most of these agents are yet under development, but they promise to be important options to boost the host immune system to control and eliminate malignancy. In this review, we have provided detailed discussion of different forms of immunotherapy agents other than checkpoint-modifying drugs. The specific focus of this manuscript is to include first-in-human phase I and phase I/II clinical trials intended to allow the identification of those drugs that most likely will continue to develop and possibly join the immunotherapeutic arsenal in a near future.

Keywords: Immunotherapy, Tumor-directed monoclonal antibodies, Antibody drug conjugates, Chimeric antigen receptor therapy, Oncolytic viruses, Tumor vaccines, Viral gene therapy

BACKGROUND

Immunotherapy consists in harnessing the body's own immune system to generate an anti-tumor response, which is often sustained after treatment has finished, suggesting a role of modulation and modification of the immune system [1, 2]. The most commonly used strategy is the modulation of immune checkpoints, particularly the cytotoxic T lymphocyte-associated molecule-4 (CTLA-4), programmed cell death receptor-1 (PD-1), or programmed cell death ligand-1 (PD-L1), although new inhibitory (e.g., lymphocyte activation gene/LAG-3, T cell immunoglobulin/TIM-3, V-domain Ig suppressor of T cell activation/VISTA) and stimulatory (e.g., inducible co-stimulator/ICOS, OX40, 4-1BB) pathways have also emerged as targets [3]. Unfortunately, some limitations with immune checkpoint therapy are of concern including a response heterogeneity where some patients achieve a complete response (CR) but others never do. Furthermore, there can be tumor relapse due to alternative immune escape mechanisms, and there is lack of optimal biomarkers to predict response and toxicity. Other major issues have been the emergence of new adverse events and autoimmune-like reactions, and the cost associated with this therapy. Thus, approaches that differ from manipulating immune checkpoints, as alternatives, are under investigation and are the focus of this review. Specifically, we discuss current molecules under phase I and I/II clinical investigation in the area of conjugated monoclonal antibodies (mAbs), chimeric antigen receptor (CAR) T cells, oncolytic viruses, vaccines, and other immune-based approaches that are under investigation. Agents in more advanced investigational stages (e.g., phase III) have not been included. Figure 1 summarizes the different strategies that will be discussed, and a summary of these agents is found in Table 1.

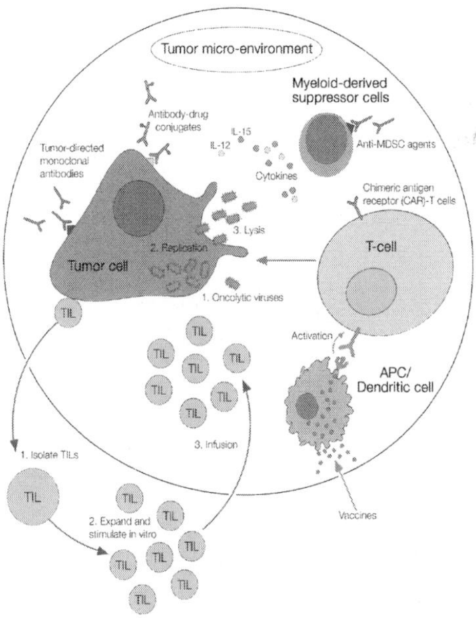

Figure 1: Multi-modality cancer immunotherapy approaches.

Table 1: Summary of non-immune checkpoint blockade agents

Category	Drug	Target	Trial	Phase	Type of tumor	Clinical efficacy	Safety	Comments
Tumor-directed monoclonal antibodies								
	Ensituximab	MUC5A	NCT01040000	I/II	CRC and pancreatic	OS improved from 5.0 to 6.8 months, 21/56 pts survived > 12 months	< 2% of patients with grade 3 toxicities and no grade 4 adverse events	Trial completed
	CEA CD3 TCB (RG7802, RO6958688)	CEA and CD3	NCT02324257	I	CEA(+) solid tumors	5% PR	16% patients developed grade 3 or more adverse events	In conjunction with obinutuzumab
			NCT02650713	I	CEA(+) solid tumors	20% PR		In conjunction with atezolizumab
	Blinatumomab	CD19 and CD3	NCT01741792	II	DLBCL	19% CR, PFS up to 20 months	Grade 3 neurologic events (9% encephalopathy, 9% aphasia)	Trial completed
	BAY2010112 (AMG212, MT112)	PSMA and CD3	NCT01723475	I	Castration-resistant prostate cancer	Not reported	–	Study is ongoing but not recruiting
	MOR209/ ES414	PSMA and CD3	NCT02262910	I				–
	AFM13	CD30 and CD16A	NCT01221571	I	CD30+ HL	3/26 PR, 13/26 SD, overall DCR 61.5%	Mild to moderate adverse events ranging from fever to infusion reactions	Trial completed
			NCT02321592	II	CD30+ HL	–	–	–
Antibody drug conjugates								
	ABBV-339	c-Met	NCT02099058	I	NSCLC	19% PR	All-grade adverse events in > 10% of patients	Used in conjunction with erlotinib

	Glembatu-mumab vedotin (GV, CDX-011)	gpNMB	NCT02302339	II	Melanoma	1/62 CR, 6/62 PR, and 33/62 SD	Alopecia, neuropathy, rash, fatigue and neutro-penia	–
	Losatuxi-zumab vedotin (ABBV-221)	EGFR	NCT02365662	I	EGFR-dependent tumors	38% SD and 1 patient had unconfirmed PR	Infusion reactions and fatigue	–
	Mirvetuximab soravtansine (IMGN853)	Folate receptor alpha (FRα)	NCT01609556 (FORWARD I)	I	Ovarian cancer	ORR 46%	1/37 CR and 16/37 PR	Monotherapy
			NCT02606305 (FORWARD II)	I	Ovarian cancer	–	Most adverse events were grade 2 or less	Used in com-bination with bevacizumab, carboplatin, liposomal doxorubicin, or pembroli-zumab
	Enfortumab vedotin (ASG-22CE; ASG-22ME)	Nectin-4	NCT02091999	I	Urothelial tumors	ORR 40%, CR 3/68, median duration of response was 18 weeks, and median PFS 17 weeks	85% devel-oped adverse events, but most were grade 2 or less	–
	Sacituzumab govitecan (IMMU-132)	Trop-2	NCT01631552	I/II	Epithelial cell tumors	30% ORR, 2/69 CR, 19/69 PR, median OS 16.6 months for triple negative breast cancer	Neutropenia, diarrhea, febrile neu-tropenia	–
						ORR 19%, median		
						PFS 5.2 months, and a median OS of 9.5 months in NSCLC		
	Inotuzumab ozogamicin (InO/CMC-544)	CD22	NCT01055496	I	CD22+ NHL	ORR 53%	85% throm-bocytopenia, 69% of neutropenia	Used in con-junction with rituximab, gemcitabine, dexametha-sone, and cisplatin Trial com-pleted
	Labetuzumab govitecan (IMMU-130)	CEACAM	NCT01605318	II	CRC	1/86 PR, 42/86 had SD, OS was 6.9 months, and PFS was 3.6 months	16% Neutro-penia), 9% anemia, and 7% diarrhea	–

	Lorvotuzumab mertansine (IMGN901)	CD56	NCT01237678	I/II	SCLC	94 patients (combination ADC with chemo-therapy) and 47 patients (no ADC) achieved a PFS of 6.2 and 6.7 months, respectively Median OS was 10 months in both cohorts	29% Periph-eral neuropa-thy , 21/94 patients had a treatment-emergent adverse event leading to death	Used in com-bination with carboplatin and etoposide
	Rovalpitu-zumab tesirine (Rova-T)	DLL3	NCT02674568	II	SCLC	18% ORR, and 54% SD	38% devel-oped serious adverse events (pleural and pericardial effusions)	Ongoing
			NCT03026166	I	SCLC	–	–	Used in com-bination with nivolumab or nivolumab and ipilim-umab
	ADCT-301	CD25	NCT02432235	I	HL and NHL	1/18 CR, 1/18	Rash, mucositis, enteritis, and elevated CPK	-
						PR, 6/18 SD		
			NCT02588092	I	AML and ALL	–	–	–
	TAK-264 (MLN0264)	GCC	NCT02202785	II	Pancreatic adenocarci-noma	3% ORR	All patients developed at least 1 ad-verse event	Trial was terminated
	MEDI-4276	HER2	NCT02576548	I/II	HER2+ solid tumors	–	–	–
	XMT-1522	HER2	NCT02952729	I	HER2+ breast cancer	–	–	–
	ARX-788	HER2	NCT02512237	I	HER2+ clancers	–	–	–
	DS-8201a	HER2	NCT02564900	I	Solid tumors	–	–	–
	SDY985	HER2	NCT02277717	I	Solid tumors	–	–	–
	ADCT-502	HER2	NCT03125200	I	HER2+ solid tumors	–	–	–

	Glembatumumab vedotin (CDX-011, CR-11-vc-MMAE)	GPNMB	NCT02302339	II	Melanoma	1/62 CR, 6 PR, 33 SD, and median OS was 9.8 months	Alopecia, neutropenia, and rash	–
			NCT01156753	II	Advanced GPNMB-expressing breast cancer	ORR 6%	Rash and pruritus	Trial completed
	SAR566658	CA-6	NCT01156870	II	Solid tumors expressing CA6	1/114 CR, 8 PR, and 39% SD	Mild toxicities: fatigue, neuropathy, neutropenia	Trial completed
	SGN-LIV1A	LIV-1	NCT01969643	I	Breast cancer	ORR 11% and SD or better achieved in 63% of patients	No dose-limiting toxicities	Used in combination with trastuzumab
	PF-06647020	Tyrosine kinase 7	NCT02222922	I	Advanced solid tumors	1/76 CR, 5 PR, 12 SD, and 4 PD	Most toxicities were mild: nausea, alopecia, neutropenia	–
	PF-6647263	Ephrin-A4	NCT02078752	I	Advanced solid tumors	5/48 PR	Dose-limiting toxicities were observed in 6 patients	Trial completed
	SAR428926	LAMP-1	NCT02575781	I	Solid tumors	–	–	–
	PCA062	P-cadherin 3	NCT02375958	I	Triple negative breast cancer, head and neck cancer, esophageal cancer	–	–	–
	U3-1402	HER3	NCT02980341	I/II	HER3+ metastatic breast Cancer	–	–	–
	HuMax-Axl	Axl	NCT02988817		Ovarian, cervical, endometrial, NSCLC, thyroid cancer, and melanoma	–	–	–
	MEDI3726	PMSA	NCT02991911	I	Metastatic castration-resistant prostate cancer	–	–	Used in combination with enzalutamide
CAR T cells								

	T4 immunotherapy	ErbB dimers, IL4	NCT01818323	I	HNSCC	DCR 44%	All grade 2 (or less) adverse events	Intratumoral T-4 therapy
	CART-19	CD19	NCT01044069	I	B-ALL	CR rates were 95 and 77% on patients with < 5% of blasts in the bone marrow and those with > 5%, respectively	CRS and neurotoxicity	–
			NCT02348216	I/II	NHL	ORR 82% and CR 39% after 8 months	31% Febrile neutropenia, 24% thrombocytopenia, 21% encephalopathy, and 13% CRS	Some patients received steroids and others tocilizumab
			NCT01865617	I/II	ALL, NHL, and CLL	31/33 ALL patients achieved CR, 6/12 CLL with CR, 84% ORR in NHL	16% CRS and 31% neurotoxicity	All CLL pts had received prior ibrutinib
	Anti-GPC3	GPC3	NCT02395250	I	GPC3+ HCC	1/13 with PR	No dose-limiting toxicities	–
	CART-133	CD133	NCT02541370	I	HCC, pancreas, CRC, cholangiocarcinoma	21/23 PFS ranging from 8 to 22 weeks	Hyperbilirubinemia and CRS	–
	bb2121	BCMA	NCT02658929	I	MM	6/11 ORR	Only grade 1–2 CRS	ORR seen in patients who received higher doses of T cells
	Anti-kappa light chain	Kappa light chains	NCT00881920	I	Kappa (+) CLL, NHL, and MM	2/9 CR, 1/9 PR. 4/7 patients with MM had SD	None	–
	Anti-CD30	CD30	NCT01316146	I	HL and NHL	2 patients with HL and 1 w/ ALCL achieved CR, 3 patients with HL achieved SD	None	–
			NCT02690545	Ib/II	CD30+ HL and NHL	–	–	–
	Anti-IL13	IL13Rα2	NCT02208362	I	Glioblastoma	CR in 1 patient	None	–

TCR gene-modified T cell therapy								
	NY-ESO-1c259t	NY-ESO-1 and HLA-A2	NCT01343043	I/II	Sarcoma	ORR 50%, 1 case of CR	96% Leukopenia, 79% anemia, 79% thrombocytopenia, 1 fatal bone marrow failure, and 11/34 cases with CRS	–
	Anti-E6	E6	NCT02280811	I/II	HPV 16+ carcinomas (e.g., cervical, anal, pharyngeal)	2/12 PR	No dose-limiting toxicity	Study completed
	Anti-MAGE A10	MAGE-A10	NCT02989064	I	Urothelial cancer, HNSCC, or melanoma	–	–	–
Tumor-infiltrating cell therapy								
	TIL	Varies depending on tumor	NCT01319565	II	Melanoma	CR24%	13/48 patients who received TBI developed thrombotic microangiopathy not seen in patients with no TBI	In combination with TBI
	MIL	Varies depending on tumor	NCT00566098	I/II	MM	27% CR, 27% PR, 23% SD, and 14% PD	Not mentioned	Study completed
Oncolytic viruses								
	Coxsackievirus A21 (CVA21–CAVATAK)	ICAM-1-expressing tumors	NCT02565992	Ib	Melanoma, NSCLC, bladder, and prostate cancer	ORR 73% and DCR 91% in melanoma	No dose-limiting toxicities	In combination with pembrolizumab
			NCT02043665					
			NCT02307149	I	Melanoma	ORR 38%	Minimal toxicity	In conjunction with ipilimumab
						DCR 88%		

	Pelareorep (Reolysin)	Different tumors	NCT00984464	II	Melanoma	ORR 21%, 1-year survival 43%, DCR85%	Fever was the most common toxicity	In combination with carboplatin and paclitaxel. Study was completed
			NCT01656538	II	Breast cancer	Median OS was 17.4 months for patients with both agents and 10.4 months for patients with paclitaxel alone	Fatigue, nausea, vomiting, diarrhea	In combination with paclitaxel, or paclitaxel alone
	DNX-2401	Glioblastoma	NCT01956734	I	Glioblastoma	1 patient was alive 30 months into treatment and 2 after 23 months	Related to temozolamide or underlying disease	In combination with temozolamide
	Enadenotucirev (EnAd)	Tumors of epithelial origin	NCT02636036	I	Epithelial tumors	–	–	In combination with nivolumab
Vaccines								
	HS-110 (Viagenpumatucel-L)	Lung adenocarcinoma cells	NCT02439450	I/IIb	Lung adenocarcinoma	ORR 50%	Injection site reactions, maculopapular rash	In combination with nivolumab
	gp96	Gastric cancer cells	NCT02317471	II	Gastric cancer	2-year OS was 81.9% in the vaccination arm vs. 67.9% with chemotherapy-alone arm	No clinically significant adverse events	In combination with oxaliplatin
	GM.CD40L	Lung adenocarcinoma cells	NCT02466568	I/II	Lung adenocarcinoma	Median OS was 9.4 months	No dose-limiting toxicities	Some patients had added CCL21 to GM.CD40L
	RNA-lipoplex (RNA(LIP))	Melanoma antigens	NCT02410733	I/II	Melanoma	–	No dose-limiting toxicities	–
	VXM01	VEGFR-2	EudraCT 2011-000222-29	I	Pancreatic cancer	OS was 9.3 months vs 8.4 months (placebo)	Lymphopenia and increased diarrhea	Oral vaccine
	INO-5150	Prostate cancer antigens	NCT02514213	I	Prostate cancer	10% PD	No dose-limiting toxicities	Used with or without IL-12. Study is not recruiting patients
	INVAC-1	Human telomerase	NCT02301754	I	Solid tumors	12/20 SD	Asthenia and local reaction at injection site	–

	pTVG-HP	PAP	NCT01341652	II	Prostate cancer	–	–	Ongoing
	ADXS11-001	E7 antigen	NCT02164461	I	Cervical cancer	–	Only 1 > grade 2 adverse event (hypotension)	Ongoing
	AdMA3	MAGE-A3	NCT02285816	I	Solid tumors	Evidence of induction of pro-inflammatory genes and subsequent anti-tumor activity	4/41 patient developed dose-limiting toxicities (hypoxia, dyspnea, vomiting, headache)	Used in conjunction with an oncolytic virus (MG1MA3)
	AdHER-2ECTM	Her2	NCT01730118	I	Her2(+) tumors	1/27 CR, 1 PR, and 5 SD	Local injection site reactions	–
	CMB305	NY-ESO-1	NCT02387125	I	NY-ESO-1-expressing solid tumors	Increase of anti-NY-ESO-1 T cells in 65% of patients and anti-NY-ESO-1 antibodies in 68% of patients	–	–
	MVA	Brachyury	NCT02179515	I	Advanced solid malignancies	82% of patients developed brachyury-specific immune responses	No dose-limiting toxicities were observed	Trial completed
	BPX101	Human prostate-specific membrane antigen	NCT00868595	I	Prostate cancer	1/18 with CR and 2/18 PR	No dose-limiting toxicities	Trial completed
	WT-1	WT-1-expressing tumors	UMIN000005248	II	Pancreatic adenocarcinoma	Increased OS from 21.5 (gemcitabine alone) to 34.2 (gemcitabine with WT-1 vaccine)	–	Used in conjunction with gemcitabine
	WT4869	WT-1-expressing tumors	JapicCTI-101374	I/II	MDS	ORR 18.2%, median OS 64.71 weeks	30.8% Neutropenia, 7.7% febrile neutropenia, and 7.7% elevated CPK	Trial completed
	Galinpepimut-S	WT-1-expressing tumors	NCT01265433	II	Pleural mesothelioma	PFS 45%, median OS 22.8 months	Mild and not clinically significant	Trial completed

	DPX-Survivac	Survivin-expressing tumors	NCT01416038	Ib	Ovarian, fallopian, and peritoneal cancer	Sustained immune responses of varying magnitude and duration	Skin ulceration	Trial completed
	AE37	Her2	NCT00524277	II	Her2(+) breast cancer	Disease-free survival improved from 51% (GM-CSF alone) to 89% (AE37+ GM-CSF)	Vaccines is safe and well tolerated	Used with GM-CSF
	Multi-HLA binding peptides	HSP70 and GPC3	UMIN000020440	I	Solid tumors	Decreased tumor-marker expression in 6/12 patients and disease control in 5/12 patients	No severe toxicities	–
	URLC10-CDCA1-KOC1	URLC10, CDCA1, KOC1	UMIN000003557	II	Esophageal squamous cell carcinoma	No significant difference of relapse-free survival compared to control group, but there was good immunological response	–	–
	Poly-ICLC	TLR-3	NCT01984892	I/II	Solid tumors including melanoma, breast, and HNSCC	1/8 SD for 41 weeks; the remainder of patients showed PD	Mild and limited to the site of application	Study was terminated
	BO-112	MDA-5 and NOXA	NCT02828098	I	Melanoma and breast cancer	–	1 case of reversible thrombocytopenia	–
	IVAC MUTANOME	Personal tumor neoantigens	NCT02035956	I	Melanoma	8/13 patients remained recurrence-free for the entire follow-up period (12–23 months)	No major adverse events	Ongoing
	Immunogenic personal neoantigen vaccine	Personal tumor neoantigens	NCT01970358	I	Melanoma	4/6 patients that had no recurrence of the disease at 25 months after vaccination	Mild flu-like symptoms, injection site reactions, rash, and fatigue	–
Targeting MDSCs								
	DS-8273a	TRAIL-R2 (DR-5)	NCT02076451	I	Solid tumors	–	No dose-limiting toxicities	–

Cytokine gene therapy								
	Ad-RTS-hIL-12	IL-12	NCT02026271	I	Glioblastoma	Median OS 12.5 months	Flu-like illness, grade 3 CRS, transaminitis	Used in with Veledimex
	NKTR-214	IL-2	NCT02983045	I/II	Solid malignancies	1 patient had unconfirmed CR	No dose-limiting toxicities	Used in conjunction with nivolumab
			NCT02869295	I/II	Solid malignancies	23% achieved tumor size reduction ranging from 10–30%	No dose-limiting toxicities	–
Agents targeting tumor micro-environment								
	BMS-986205	IDO	NCT02658890	I	Solid tumors	–	Hepatitis, rash	Used in conjunction with nivolumab
	Indoximod	IDO	NCT02073123	II	Melanoma	ORR 52%	No significant toxicities	Used in conjunction with ipilimumab, nivolumab, or pembrolizumab
			NCT02077881	II	Pancreatic	ORR 37%	One case of colitis	Used with both gemcitabine and nab-paclitaxel
			NCT01560923	II	Prostate	Median PFS increased from 4.1 to 10.3 months	No significant adverse events	–
	Epacadostat	IDO	NCT02327078 NCT02178722	I/II	Solid and hematologic malignancies	ORR of 75% (melanoma) and 4% (CRC)	No dose-limiting toxicities	–
	MEDI9197	TLR7/8	NCT02556463	I	Solid malignancies	–	Mild adverse events only	In combination with durvalumab and radiation therapy
	PG545 (pixatimod, pINN)	TLR9/ IL-12	NCT02042781	I	Solid malignancies	SD for 24 weeks, DCR of 38%	Dose-limiting toxicities in 3/23 patients	–
	Poly-ICLC	TLR3	NCT00553683	I	HCC	PFS 66% at 6 months, 28% at 24 months	Most grade I–II adverse events	In combination with radiation therapy

						OS 69% at 1 year, 38% at 2 years		
	CB-1158	Arginase	NCT02903914	I	Solid malignancies	–	No dose-limiting toxicities	In conjunction with nivolumab
Oncolytic peptides								
	LTX-315	Tumor mitochondrial membranes	NCT01986426	I	Melanoma and breast cancer	2/28 CR, 5 patients had a decreased of > 50% of the tumor size, and 8 patients achieved SD	Most common adverse events were mild local erythema, flushing, pruritus, and transient hypotension	In combination with ipilimumab or pembrolizumab

Abbreviations: *ALL* acute lymphocytic leukemia, *ALCL* anaplastic large cell lymphoma, *AML* acute myeloid leukemia, *B-ALL* B cell acute lymphocytic leukemia, *CAR* chimeric antigen receptor, *BCMA* B cell maturation antigen, *CEACAM* CEA cell adhesion molecule, *CLL* chronic lymphocytic leukemia, CPK creatine phosphokinase, *CR* complete response; *CRC* colorectal cancer, *DLBCL* diffuse large B cell lymphoma, *CRS* cytokine release syndrome, *DCR* disease control rate, *DLL3* delta-like protein 3, *GCC* guanylyl cyclase C, *GPC3* glypican-3, *gPNMB* glycoprotein non-metastatic B, *HCC* hepatocellular carcinoma, *HD* Hodgkin's disease, *HNSCC* head and neck squamous cell carcinoma, *IDO* indoleamine 2,3-dioxygenase, *MDS* myelodysplastic syndrome, *MDSCs* myeloid-derived suppressor cells, *MIL* marrow-infiltrating lymphocyte, *MM* multiple myeloma, *NHL* non-Hodgkin's lymphoma, *NSCLC* non-small cell lung carcinoma, *MVA* Modified Vaccinia Ankara, *OS* overall survival, *ORR* objective response rate, *PAP* prostatic acid phosphatase, *PD* progressive disease, PFS progression-free survival, PSMA prostate-specific membrane antigen, *Poly-ICLC* polyinosinic-polycytidylic acid polylysine carboxymethylcellulose, *PR* partial response, *SCLC* small cell lung cancer, *SD* stable disease, *TBI* total body irradiation, *TIL* tumor infiltrating lymphocyte, *TLPLDC* tumor lysate, particle-loaded, dendritic cell, *TLR* toll-like receptor, *VEGFR-2* vascular endothelial growth factor receptor-2, *WT-1* Wilms tumor gene-1

METHODOLOGY

We performed a thorough review using the databases PubMed and the American Society of Clinical Oncology (ASCO), both the American

Association of Cancer Research (AARC) meeting abstract databases, and ClinicalTrials.gov updated through October 5, 2017. We narrowed our research with the following keywords and MeSH terms: immunotherapy, tumor directed monoclonal antibodies, antibody drug conjugates, chimeric antigen receptor T-Cells, oncolytic viruses, oncologic vaccines, and adjuvants in immunotherapy. We focused our attention on phase I and phase II clinical trials of new agents in immunotherapy being used with or without other form of immunotherapy. Inclusion criteria included published trials or reported preliminary results during the time of the data collection. Exclusion criteria included phase III or more advanced clinical trials, clinical trials focusing only on immune checkpoint therapy, clinical trials in pediatric population, and non-clinical trials. Finally, we collected 65 phase I and 53 phase II clinical trials for this review.

TUMOR-DIRECTED MONOCLONAL ANTIBODIES

A range of mAbs directed against tumor-specific antigens are currently under development. These mAbs can bind specific tumor antigens, stay on the surface, and activate antibody/complement-dependent cytotoxicity, or affect downstream signals. Monoclonal antibodies promote tumor killing by different mechanisms including direct cell killing by induction of apoptosis, receptor blockade or agonist activity, delivery of cytotoxic agents, radiation, immune-mediated cell killing, or through specific effects on tumor vasculature and stroma [4]. Of particular interest are the recently developed bispecific antibodies that combine antigen-binding specificities on tumor cells and effector immune cells. Among these, bispecific T cell engager (BiTE) and dual-affinity re-targeting (DART) are particularly attractive [5–7]. BiTEs recombinantly link the four variable domains of heavy and light chains with a flexible linker peptide allowing to bypass MHC/peptide recognition and co-stimulation and also to bring effector cells and target cells close together to form cytolytic synapses. This therapy has revealed impressive clinical activity in relapsed non-Hodgkin's lymphoma, chronic lymphocytic leukemia, and acute lymphoblastic leukemia at doses much lower than those administered in conventional monoclonal antibody therapy. DART consists of a diabody that separates variable domains of heavy and light chains of the two antigen-binding specificities on two separate polypeptide chains stabilized through a C-terminal disulfide bridge which acts as a linker [5]. Compared with BiTE, DART has shown a moderately higher association rate constant for CD3 and an ability to cross-link T cells and B cells more efficiently [8, 9]. Ongoing clinical trials will provide more insightful understanding through side-by-side comparison of DART, BiTE, and other bispecific antibody with identical antigen-binding specificities. The quality, stability, and drug distribution of antibodies remain a challenge. Below outlines the ongoing early development of agents within this form of therapy.

Ensituximab (NPC-1C) is a chimeric IgG1 monoclonal antibody that promotes antibody-dependent cellular cytotoxicity after binding its target, a tumor-specific variant of MUC5AC, an antigen that is specifically expressed by colorectal (CRC) and pancreatic cancers. Results from a completed phase I/II clinical trial were recently published and are encouraging [10]. The therapy was well tolerated with < 2% of patients experiencing grade 3 toxicities and there were no grade 4 adverse events. Furthermore, median overall survival (OS) was significantly longer than historical control: 6.8 vs 5.0 months, 21 out of 56 patients survived > 12 months.

BiTEs simultaneously target two different antigens and thus target two different mediators and pathways [6]. CEA CD3 TCB (RG7802, RO6958688) is an IgG1 BiTE that simultaneously binds carcinoembryonic antigen (CEA) on tumor cells and CD3 on T cells to increase tumor-infiltrating lymphocytes (TIL) activation, infiltration, and expression of PD-1/PD-L1 [11]. Two ongoing phase I clinical trials using this new drug are currently recruiting patients (NCT02324257, NCT02650713). Preliminary results reveal that the most common adverse events were mild and 16% of patients develop grade 3 or more adverse events. Five percent and 20% of patients in each study, respectively, showed partial response (PR), and more importantly, activity appeared to be enhanced if it was combined with the anti-PD-L1 antibody, atezolizumab [11].

Blinatumomab is another BiTE that binds CD3 on T cells as well as CD19 on malignant B cells. The antibody is FDA-approved for the use of Philadelphia chromosome-negative B cell acute lymphoblastic leukemia (B-ALL) [12]. A phase II clinical trial in relapsed/refractory diffuse large B cell lymphoma demonstrated a complete response (CR) rate of 19% and progression-free survival (PFS) of up to 20 months [13].

BAY2010112 (AMG212, MT112) and MOR209/ES414 are prostate-specific membrane antigen (PSMA)/CD3 BiTEs that are being investigated in phase I clinical trials in patients with castration-resistant prostate cancer (NCT01723475, NCT02262910).

DARTs differ structurally to BiTEs as stated above [5]. MGD009 is a humanized DART protein that binds both T cells and tumor-associated B7-H3 and is being studied in a phase I clinical study in patients with B7-H3 expressing tumors including melanoma, non-small cell lung carcinoma (NSCLC), mesothelioma, and urothelial cancers [14]. The trial is ongoing and recruiting patients (NCT02628535). Flotetuzumab, another DART that binds CD3 and CD123, is currently being studied in a phase I clinical trial in patients with relapsed or refractory acute myeloid leukemia (AML) and intermediate/high-risk myelodysplastic syndrome (MDS) (NCT02152956).

AFM13 is a tetravalent bispecific antibody that is directed against CD30 and CD16A, this latter found over natural killer (NK) cells. Pharmacokinetics, therapeutic index, and efficacy make this agent a NK cell activator [15]. A phase I clinical trial used AFM13 on patients with relapsed/refractory CD30+ Hodgkin's disease (HD) and concluded that three out of 26 patients achieved PR and 13 obtained a SD with an overall disease control rate (DCR) of 61.5% [16]. Adverse events were mostly mild to moderate and ranged from fever to infusion reactions and pneumonia. A phase II clinical trial using this agent is also being done on patients with HD; however, results have not yet been published (NCT02321592).

ANTIBODY DRUG CONJUGATES

Antibody drug conjugates (ADCs), an emerging therapeutic approach in oncology, combine a monoclonal antibody with a high selectivity for specific targets with a cytotoxic agent. Microtubule inhibitors or DNA-damaging chemotherapeutic agents are the two main cytotoxic agents used in ADCs. One of the most important aspects in ADC therapy consists on the appropriate selection of the target antigen. An ideal antigen is one that is overexpressed by malignant cells with very limited or no expression by normal tissue [17]. For example, nectin-4 is often overexpressed in bladder, breast, lung, and pancreatic cancer, and thus, ACDs against this peptide are indicated in these malignancies. Similarly, folate receptor alpha is more often expressed by ovarian and endometrial carcinomas, and CEA cell adhesion molecule (CEACAM) 5 is commonly found on CRC [17]. Another important factor is the conjugate linker, which largely influences the pharmacokinetics and therapeutic efficacy of ADCs [17]. As with other drugs, resistance can emerge through different mechanisms such as limitation of intracellular concentration of ADC, downregulation of target antigens, reduction of internalization of ADC, increased ADC recycling to mask the antigen epitopes, or activation of alternative signal pathways. Class side effects are often associated with the linked cytotoxic agents. The major challenges of ADCs include target antigen specificity and drug delivery efficiency.

ABBV-399 is an ADC composed of an anti-c-Met antibody (ABT-700) conjugated to a microtubule inhibitor (monomethyl auristatin E). As the c-Met receptor is commonly overexpressed in patients with NSCLC, a phase I clinical trial is using this agent as monotherapy or in combination with erlotinib in this patient population (NCT02099058). Preliminary results demonstrated adverse events in > 10% of patients across all grades, 19% of patients (3 out of 16) had a PR, and, at week 12, 37.5% (6 out of 16) had disease control including stable disease and partial response [18].

Glembatumumab vedotin (GV, CDX-011) is an ADC that contains an antibody that targets glycoprotein non-metastatic b (gpNMB), a transmembrane glycoprotein usually overexpressed in melanoma and other tumors, conjugated to monomethyl auristatin E. A phase II clinical trial using this agent as monotherapy in patients with advanced melanoma is recruiting patients (NCT02302339). Preliminary results show 1 CR, 6 PR, and 33 SD out of 62 patients enrolled [19].

Losatuxizumab vedotin (ABBV-221), an ADC that targets EGFR, is being investigated in a phase I clinical trial as monotherapy on patients with EGFR-dependent tumors (NCT02365662). Preliminary results reveal that the most common adverse events were infusion reactions and fatigue and 16 out of 42 patients (38%) showed SD and 1 patient had an unconfirmed PR [20].

Mirvetuximab soravtansine (IMGN853) is an ADC containing the tubulin inhibitor (maytansinoid) DM4, targeting the folate receptor alpha (FRα). It is being studied in two phase I clinical trials as monotherapy (FORWARD I-NCT01609556) and in combination with bevacizumab, carboplatin, liposomal doxorubicin, or pembrolizumab (FORWARD II-NCT02606305) in patients with ovarian cancer. Preliminary results of FORWARD I reveal an objective response rate (ORR) of 46% with one patient having a CR and 16 out of 37 patients having PR [21]. FORWARD II preliminary results do not mention efficacy but do show a good safety profile with most adverse events being grade 2 or less [21].

Enfortumab vedotin (ASG-22CE; ASG-22ME), an ADC that targets nectin-4, is in a phase I clinical trial that is investigating its potential role as monotherapy in patients with metastatic urothelial tumors (NCT02091999).

Preliminary results show that adverse events occurred in 85% of patients; however, most were grade 2 or less. ORR was 40%, CR was seen in 3 out of 68 patients with a median duration of response of 18 weeks, and the median PFS was 17 weeks [22].

Sacituzumab govitecan (IMMU-132) is an ADC against Trop-2 antigen expressed in many solid tumors and carrying the topoisomerase inhibitor, SN-38. A phase I/II clinical trial as monotherapy in patients with epithelial cell tumors is undergoing (NCT01631552). Preliminary results in triple negative breast cancer show the agent is well tolerated demonstrating an ORR of 30% with two cases of CR and 19 out of 69 patients demonstrating a PR. Median PFS was 6 months and median OS was 16.6 months [23]. In NSCLC, an ORR of 19% was observed among the 47 patients, with a median response duration of 6 months, a median PFS of 5.2 months, and a median OS of 9.5 months [24].

Inotuzumab ozogamicin (InO/CMC-544) is a humanized ADC directed against CD22, coupled to a DNA breaking calicheamicin. The compound was recently approved by the FDA for use in relapsed or refractory B-ALL [25]. This agent was recently evaluated in a phase I clinical trial in conjunction with rituximab, gemcitabine, dexamethasone, and cisplatin in patients with refractory CD22+ non-Hodgkin's lymphoma (NHL) [26]. Results demonstrated an 85% incidence of thrombocytopenia and a 69% of neutropenia with an ORR of 53%.

Labetuzumab govitecan (IMMU-130) is an ADC that targets CEACAM 5 which is expressed by > 80% of CRC [27]. This ADC is being evaluated in a phase II clinical trial in patients with metastatic CRC (NCT01605318). Results reveal that one out of 86 patients enrolled had a PR that extended beyond 2 years, 42 patients had SD, OS was 6.9 months, and PFS was 3.6 months [27].

Lorvotuzumab mertansine (IMGN901), an ADC against CD56 conjugated to the tubulin inhibitor DM1, that was recently studied in a phase I/II clinical trial in combination with carboplatin and etoposide in small cell lung cancer (SCLC) patients with extensive disease [28]. The two-arm cohort of 94 patients (combination ADC with chemotherapy) and 47 patients (no ADC) achieved a PFS of 6.2 and 6.7 months, respectively, with a median OS of 10 months in both cohorts. The ORR was 67% for the combination cohort compared to 59% in the non-combination with no statistical significance. Thus, the authors concluded that the combination of lorvotuzumab mertansine did not improve efficacy over standard therapy [28].

Rovalpituzumab tesirine (Rova-T) is an ADC that targets delta-like protein 3 (DLL3) which has been found to be elevated in patients with SCLC. The antibody is conjugated to a DNA cross-linker tesirine, and results from the first-in-human trial in patients with recurrent SCLC showed that 28 out of 74 patients (38%) developed serious adverse events consisting of pleural and pericardial effusions, 18% demonstrated OR, and 54% had SD [29]. Other phase I/II clinical trials evaluating this agent are currently ongoing (NCT02674568, NCT03026166).

ADCT-301 is the first ADC against CD25, a receptor for IL-2 often found on hematological tumors that has a role in prognosis and oncogenesis in these malignancies [30]. This molecule is being studied in a phase I clinical trial in patients with relapsed or refractory HD and NHL (NCT02432235). Preliminary results on 18 patients show 1 CR, 1 PR, and 6 SD, one of which has remained progression-free for over 30 weeks and four developing adverse events consisting of rash, mucositis, enteritis, and elevated creatine phosphokinase [31]. Another phase I clinical trial is being conducted in patients with refractory or relapsing CD25-positive AML and ALL (NCT02588092).

TAK-264 (MLN0264), a novel ADC that targets guanylyl cyclase C (GCC), has been recently studied in a phase II clinical trial in patients with advanced or metastatic pancreatic adenocarcinoma expressing GCC [32]. The cohort of 43 patients achieved an ORR of only 3%, and all patients experience at least one adverse event; thus, the authors concluded these results did not support further clinical investigation of this molecule.

HER2, a well-studied member of the epidermal growth factor tyrosine kinase receptor family, plays an important role in breast cancer and has been a target for ADCs. T-DM1 (Kadcyla), an ADC consisting of trastuzumab (T) and a microtubule inhibitor (DM1), was the first ADC approved by the FDA to use in solid tumors [33]. Other anti-HER2 ADCs are being studied in multiple phase I/II clinical trials including MEDI-4276 (NCT02576548), XMT-1522 (NCT02952729), ARX-788 (NCT02512237), DS-8201a (NCT02564900), SDY985 (NCT02277717), and ADCT-502 (NCT03125200).

Glycoprotein non-metastatic b (GPNMB), a highly expressed protein in melanoma and breast cancer, plays an important role with modulation, and is an important target in ADC. Glembatumumab vedotin (CDX-011, CR-11-vc-MMAE) targets GPNMB, and a detailed review on this agent has been recently published [34]. A phase II clinical trial on patients with melanoma (NCT02302339) demonstrated that, while using this ADC, 1 out of 62 patients achieved CR, 6 achieved PR, and 33 had SD, and the median OS was 9.8 months. Toxicities were manageable and included alopecia, neutropenia, and rash [19]. Another phase II clinical trial using glembatumumab vedotin on patients with advanced breast cancer demonstrated an ORR of 6% with mild toxicities including rash and pruritus [35].

CA6 is a tumor-associated antigen that can be overexpressed in many solid tumors with a low expression on normal tissues [33]. SAR566658 targets CA6 and delivers maytansinoid DM4. On a first-in-human clinical trial on patients with solid tumors expressing CA6 using this agent, 114 patients were enrolled. Results revealed 1 CR, 8 PR, and 39% SD with overall mild toxicities including fatigue, neuropathy, neutropenia, and gastrointestinal symptoms [36].

LIV-1 is a transmembrane protein that is highly expressed in breast cancer cells. SGN-LIV1A is an anti-LIV-1 ADC that is being tested on a phase I clinical trial on patients with metastatic LIV-1-postive breast cancer (NCT01969643). Preliminary results demonstrate no dose-limiting toxicities, an ORR of 11% and an SD or better achieved in 63% of patients [37].

PF-06647020 is an ADC directed against protein tyrosine kinase 7 (PTK7) which is overexpressed in a variety of tumors including lung, CRC, breast, and ovarian cancers [33]. This molecule is being studied in a phase I clinical trial on patients with advanced solid tumors (NCT02222922). Preliminary results revealed that 1 out of 76 patients had a CR, 5 had PR, 12 had SD, and 4 had PD. Most toxicities were grades 1–2 and included nausea, alopecia, neutropenia, and gastrointestinal complaints [38].

Ephrin-A4 has been found to be elevated in multiple malignancies including lung, pancreas, and breast [33]. PF-06647263 is an anti-Ephrin-A4 ADC that is coupled to calicheamicin and that was studied in a phase I clinical trial on patients with advanced solid tumors [39]. Results revealed that 5 out of 48 patients achieved PR and that dose-limiting toxicities were observed in 6 patients [39].

LAMP-1 is a protein highly expressed by lysosomes that translocates to the surface of tumor cells, and its expression may influence invasiveness and metastatic behavior including CRC, melanoma, and laryngeal cancers [33]. SAR428926 is an anti-LAMP-1 ADC coupled with DM4 that is currently being studied on a phase I clinical trial in patients with advanced solid tumors (NCT02575781). Preliminary results have not been published.

PCA062 is another ADC that is targeted against P-cadherin 3, which is often overexpressed in epithelial tumors [33]. This agent is being studied in a phase I clinical trial on patients with P-cadherin positive tumors (NCT02375958). No preliminary results have been published.

HER-3 overexpression in breast cancer is associated with poor prognosis, and an anti-HER3 ADC therapy, U3-1402, is being studied in a phase I clinical trial on patients with HER-3-positive metastatic breast cancer (NCT02980341) [40]. Preliminary results have not been published. HuMax-Axl is an ADC against an Axl-specific immunoglobulin (IgG1κ) which is present in malignancies like pancreatic, thyroid, and lung cancer and melanoma [41]. A phase I/II clinical trial on patients with solid tumors is studying the use of this agent (NCT02988817); however, preliminary results are not yet available.

PSMA is highly expressed in prostate cancer, and MEDI3726, an anti-PSMA ADC, is being studied on a phase I/Ib clinical trial in patients with metastatic castration-resistant prostate cancer (NCT02991911). Preliminary results are yet to be published.

CHIMERIC ANTIGEN RECEPTOR (CAR) T CELLS

CARs are typically genetically engineered T cell receptors with an antibody-based extracellular domain that specifically recognizes a tumor antigen, a transmembrane portion, and an intracellular domain that activates the T cell. By antigen-specific recognition in a MHC-independent manner, CAR T cells are activated in vivo through phosphorylation of immune receptor tyrosine-based activation motifs (ITAMs) leading to cytokine secretion, T cell proliferation, and antigen-specific cytotoxicity. CAR T cells are produced by inserting specific CAR genes via viral vectors into autologous or allogeneic T cells [42]. New-generation CARs have two or more co-stimulatory domains (e.g., 4-1BB, OX 40) that boost the stimulatory signal [43, 44]. Anti-CD19 CAR T cells were recently FDA-approved for B-ALL in pediatric and young adult population [45]. Other CARs rely on the ligand of the receptor of interest rather than on an antibody [46]. Although impressive clinical activities of CAR T cells in hematological malignancies are reported, several obstacles need to be overcome for a successful application of CAR T cells in solid tumor [47]. Some of these obstacles include the lack of ideal tumor-specific antigens, the inefficient trafficking of CAR T cells to tumor sites, the immune-suppressive tumor microenvironment, and the risk of developing on-target/off-tumor toxicities which results from an attack to host cells that express the targeted tumor antigen. One of the major challenges with this therapy is the cytokine release syndrome (CRS). This potentially fatal toxicity occurs after a massive release of cytokines and causes complications ranging from mild fever and fatigue to severe respiratory distress, cardiac dysfunction, or even disseminated intravascular coagulation [48]. Future investigation to improve the safety, specificity, and efficiency will likely take CAR T cell therapy into the central stage in cancer immunotherapy [49].

T4 Immunotherapy

T4 immunotherapy uses genetically engineered T cells that co-express two CARs, T1E28z that targets ErbB dimers, and 4αβ that binds IL-4 and promotes T cell expansion [50]. These CAR T cells are currently undergoing a phase I clinical trial on patients with head and neck squamous cell carcinoma (HNSCC) (NCT01818323). Preliminary results show limited adverse events and a disease control rate (DCR) of 44% [51].

Anti-CD19 CAR T Cells (CART-19)

19-28z CAR (JCAR015) consists of a single-chain murine antibody against human CD19 (expressed by B cell malignancies) fused with the transmembrane and cytoplasmic domains of the human CD28 co-stimulatory molecule [52]. Currently, these CAR T cells have been studied in a phase I clinical trial in patients with relapsed B-ALL (ROCKET Trial-NCT01044069). In this trial, patients were divided into two cohorts, those with $< 5\%$ of blasts in the bone marrow and those with $> 5\%$. CR rates were 95 and 77%, respectively. The duration of this response, however, was directly proportional to the disease burden, and therefore, authors suggest an early use of this therapy before $> 5\%$ of blasts occupy the bone marrow [53].

The ZUMA-1 trial is studying anti-CD-19 CAR T cells (KTE-C19) in patients with refractory aggressive NHL (NCT02348216). Preliminary results reveal an ORR of 82%, and with a median follow-up of 8 months, 39% remained in CR. Importantly, 43 patients received tocilizumab and 27 out of 111 received steroids; however, ORR did not change significantly with the addition of this therapy [54].

CART-19 are also being studied in other malignancies including Richter syndrome [55] and relapsed chronic lymphocytic leukemia, myeloma, or NHL (NCT01865617). Although in one study 4% died due to CRS [56], efficacy results seem promising with some patients achieving CR or remission at the time of result publication [56, 57].

Anti-GPC3 CAR T Cells

Glypican-3 (GPC3) is a membrane proteoglycan selectively expressed by hepatocellular carcinoma (HCC) cells that is being used as a target CAR therapy [58]. GPC3 CAR T cells were studied in a phase I clinical trial of patients with GPC3+ HCC. Preliminary results showed no dose-limiting toxicities; 1 out of 13 patients had a PR, and 3 showed SD. These responses were seen in patients who received lymphodepletive conditioning [59].

Anti-CD133 CAR T Cells (CART-133)

CD133 is expressed by many tumors of epithelial origin, and therefore, the use of CAR T cells against this molecule (CART-133) is undergoing investigation being in a phase I clinical trial in patients with advanced metastatic malignancies including HCC, pancreatic carcinoma, CRC, and cholangiocarcinoma (NCT02541370). Preliminary results show some grade 3 adverse events consisting of hyperbilirubinemia and CRS. However, 21 out of 23 patients had PFS periods ranging from 8 to 22 weeks. Furthermore, 2 patients maintained greater than 8-month SD at the time of publication [60].

Anti-BCMA CAR T Cells (bb2121)

B cell maturation antigen (BCMA) is expressed by multiple myeloma (MM) cells, and anti-BCMA CARs have been genetically incorporated to T cells [61]. One of these CAR T cells, bb2121, is currently under investigation in a phase I clinical trial on patients with refractory MM who have ≥ 50% BCMA expression on their plasma cells (NCT02658929). Preliminary results show only minor adverse events with grade 1 or 2 CRS, and an ORR was seen in all patients (6 out of 11) who received the higher doses of T cells [62].

LCAR-B38 is another CAR T cell that targets BCMA that was recently studied in a phase I clinical trial of patients with refractory or relapsed MM. Preliminary results were encouraging. Seventy-four percent of patients (14 out 19) developed CRS, but most of these were mild in severity. Importantly,

ORR and CR/near CR rates were 100 and 95%, respectively, at a median follow-up of 6 months [57].

Anti-CD138 CAR T Cells

CD138 is a highly expressed molecule on MM cells and has a role in their development and proliferation [63]. A phase I clinical trial using anti-CD138 CAR T cells was conducted in patients with chemotherapy-refractory MM [63]. Five patients were enrolled, of which 4 achieved SD for more than 3 months and 1 with PD. The main toxicity observed was CTS.

Anti-immunoglobulin Kappa Light Chain CAR T Cells

As a way to spare normal cells from being targeted by this therapy, CAR T cells directed against more tumor-specific proteins like kappa light chains are also being developed [64]. A phase I clinical trial testing this therapy on patients with kappa-positive chronic lymphocytic leukemia (CLL), NHL, or MM is currently undergoing (NCT00881920). Results of this trial showed that 2 out of 9 patients with NHL or CLL had CR and 1 had PR. Four of the 7 patients with MM showed SD that lasted 2–17 months. No toxicities were seen with this therapy [64].

Anti-CD30 CAR T Cells

CD30 is expressed in a limited amount on normal tissues whereas is often overexpressed in patients with HD and NHL [65]. The use of anti-CD30 CAR T cells was recently studied in a phase I clinical trial on 9 patients with refractory or relapsed HD and anaplastic large cell lymphoma (ALCL) [66]. No toxicities were observed with this therapy, 2 patients with HD and 1 with ALCL achieved CR, and 3 patients with HD achieved SD. The use of anti-CD30 CAR T cells in combination with bendamustine is being studied on a phase Ib/II clinical trial on patients with CD30+ HD and NHL (NCT02690545).

Anti-IL13 CAR T Cells

Recently, the use of CAR T cells targeting IL13Rα2 in a patient with recurrent multifocal glioblastoma demonstrated CR sustained for 7.5 months with no associated systemic side effects [67].

CARs on the Horizon: NKG2D, NKR-2, and Mesothelin CAR T Cells

Natural killer group 2D (NKG2D) receptor recognizes specific ligands over tumor cells and promotes T cell activation to eliminate NKG2D ligand-expressing cells [68]. A first-in-human phase I clinical trial is using genetically modified T cells expressing NKG2D in patients with AML, MDS, and MM (NCT02203825).

NKR-2 CAR T cells consist of a fusion of NKG2D receptor with CD3 signaling domain, which are under investigation on a phase I clinical trial in patients with both solid and hematologic malignancies (THINK–Therapeutic immunotherapy with NKR-2–trial) (NCT03018405).

Finally, mesothelin as a target for CAR T therapy in patients with mesothelioma has been recently reviewed [69]. Various phase I clinical trials are exploring treatment approach (NCT02414269, NCT01583686, NCT02580747). Results for these future studies have yet to be reported.

T CELL RECEPTOR (TCR) GENE-MODIFIED T CELL THERAPY

In contrast to CAR T cell therapy, TCR gene-modified T cell therapy functions by targeting the surface antigens of tumor cells to specifically recognize intracellular tumor antigens presented by HLA molecules. By genetic transfer of TCR directed against specific tumor antigens into normal T cells, T cells are able to perform antigen-specific tumor killing. In addition, genetic engineering of T cell also offers the advantage of introducing molecules that can enhance T cell function or overcome tumor escape mechanisms, such as adding genes encoding cytokines, chemokine receptors and costimulatory factors, as well as elements to silence inhibitory molecules. Currently, genes encoding TCRs that are specific for a variety of tumor antigens such as MART-1, gp100, p53, NY-ESO-1, MAGE-A3, and MAGE-A4 have been studied as therapeutic targets for TCR gene-modified T cell therapy in clinical trials for melanoma, lung cancer, and breast cancer patients [70]. Choosing highly specific tumor antigens, maintaining TCR expression over time, and limiting off tumor/on target toxicity are remaining challenges [70].

Anti-NY-ESO-1 TCR T Cells (NY-ESO-1c259t)

NY-ESO-1 is expressed in up to 70% of synovial sarcomas and other mesenchymal tumors and the use of TCR gene-modified cells, specifically, NY-ESO-1c259T cells that recognize an HLA-A2 peptide are being studied. An ongoing phase I/II clinical trial is recruiting patients with unresectable or metastatic synovial sarcoma who express NY-ESO-1 and HLA-A2 (NCT01343043). Preliminary results demonstrated the development of leukopenia (96%), anemia (79%), thrombocytopenia (79%), one fatal bone marrow failure, and 11 out of 34 cases with CRS. However, ORR was 50% and 1 patient had a PR [71].

Anti-E6 TCR T Cells

The E6 oncoprotein is an essential component of HPV-related tumorigenesis, and its expression is maintained in advanced lesion and represents an ideal tumor-specific antigen. TCR gene-modified T cells that express a TCR that recognize an HLA-A*02:01-restricted epitope of E6 has been developed [72]. A phase I/II clinical trial using these T cells in patients with metastatic HPV 16+ carcinomas, including cervical, anal, and pharyngeal, was recently completed. No dose-limiting toxicity was seen, and 2 out of 12 patients demonstrated PRs [73].

Anti-MAGE A10 TCR T Cells

MAGE-A10 peptide is expressed by different malignancies including urothelial, HNSCC, and melanoma [74]. The use of T cells containing the MAGE-A10c796 CAR is being studied in a first-in-human phase I clinical trial on patients with advanced or inoperable urothelial cancer, HNSCC, or melanoma (NCT02989064). Preliminary results are not available yet.

TUMOR-INFILTRATING T CELL THERAPY

Adoptive cell therapy that utilizes endogenous tumor-infiltrating lymphocytes (TIL), which are expanded in vitro from a surgically resected tumor and then re-infused back into the patient, has demonstrated a 20% complete response lasting beyond 3 years in patients with stage IV melanoma [75]. TILs are naturally occurring T cells in the host able to recognize tumor antigens. This likely explains the highly specific anti-tumor responses and the relatively low toxicity of TILs in comparison with TCR gene-modified T cell therapy and

CAR T cell therapy. In addition, TILs are heterogeneous in their specificity which represents an important advantage for impeding immunologic escape. Furthermore, TIL therapy bypasses the limitation identifying specific tumor antigens or the patient's HLA type. Nonetheless, all its clinical advantages are somewhat blunted by the complex process required to generate patient-specific TILs for clinical use. Strategies to improve and simplify the TIL production are being studied.

Infusion of ex vivo expanded TIL is currently being studied in conjunction with total body irradiation (TBI) in a phase II clinical trial in patients with melanoma (NCT01319565). Preliminary results reveal that 13 out of the 48 patients who received TBI developed thrombotic microangiopathy, but this was not seen in the group without TBI. Regardless, a CR rate was seen in 24% of patients in both groups, and only one of these patients had recurrence of the disease [76].

Marrow-infiltrating lymphocytes (MILs) have been used in patients with newly diagnosed or relapsed MM. In a phase I clinical trial, the overall clinical response was of 54%, with 27% CR and 27% PR, 23% of patients had a SD, and only 14% showed a PD. Patients who achieved at least 90% reduction of disease burden had a PFS almost 13 months longer; however, no difference in OS was seen [77].

ONCOLYTIC VIRUSES

Native or genetically modified viruses are a new therapeutic approach within the immunotherapy spectrum. The mechanisms of action of oncolytic viruses are not fully elucidated but likely depend on viral replication within tumor cells, induction of primary cell death, interaction with tumor cell antiviral elements, and initiation of innate and adaptive anti-tumor immunity. A variety of native and genetically modified viruses have been developed as oncolytic agents [78]. Of note, these viruses selectively infect malignant cells due to the lack of adequate function of anti-viral mechanisms. Though many viruses have been considered, the most widely studied to date include herpes simplex virus type 1 (HSV-1), coxsackievirus, reovirus, and adenovirus.

Talimogene laherparepvec (T-VEC; Imlygic) is the first oncolytic virus approved by the FDA for its use in melanoma. It is an attenuated HSV-1 engineered to replicate within tumor cells and enhance immune responses [79]. Treatment has been relatively well tolerated, with the major side effects including fever, chills, nausea, fatigue, and reaction of local injection site.

Coxsackievirus A21 (CVA21–CAVATAK) preferentially infects tumors that express ICAM-1. This virus increases TIL and PD-L1 expression [80]. Therefore, the use of CVA21 is currently being studied in conjunction with pembrolizumab in two phase Ib clinical trials in patients with solid tumors including melanoma, NSCLC, bladder, and prostate cancer (NCT02043665, NCT02565992). Both STORM [81] and CAPRA [82] trials demonstrated that the combination is generally well tolerated and the latter also showed an ORR of 73% and a DCR of 91% of patients with advanced melanoma. CVA21 is also being studied in conjunction with ipilimumab in patients with unresectable melanoma (NCT02307149). Preliminary results demonstrate minimal additional toxicity with an ORR of 38% (3 out of 8 patients) and a DCR of 88% (7 out of 8 patients) [83].

Pelareorep (Reolysin) is a strain of reovirus serotype-3 which has shown in vitro and in vivo activity against many cancers and synergistic activity with concomitant use of microtubule-targeting drugs [84]. This agent has been studied in a phase II clinical trial in combination with carboplatin and paclitaxel for patients with advanced malignant melanoma. Results revealed and ORR of 21%, no CR, and a 1-year survival of 43% with a DCR of 85% [85]. Another phase II clinical trial using pelareorep with paclitaxel versus paclitaxel alone in patients with metastatic breast cancer was also recently published [84]. Median OS was 17.4 months for patients with both agents and 10.4 months for patients with paclitaxel alone; however, PFS was not different between the groups.

The oncolytic adenovirus DNX-2401 was also studied in a phase I clinical trial using temozolamide in patients with first recurrence of glioblastoma [86]. One patient out of 31 was still alive 30 months after the treatment was started, and 2 other patients were still alive 23 months after the agent was given.

Enadenotucirev (EnAd) is an A11/Ad3 chimeric group B oncolytic adenovirus that is currently under investigation in combination with nivolumab for patients with tumors of epithelial origin such as salivary gland, urothelial, HNSCC, and CRC (NCT02636036). No preliminary results have yet been published.

VACCINES

Therapeutic vaccines are designed to increase immune response against malignant cells by enlarging antigen-specific T cell from endogenous T cell repertoire. Its use has the advantage of producing a specific immune response that potentially spares normal cells [87]. Successful cancer vaccines require the selection of appropriate antigens, a platform able to induce robust effector and memory T cell responses, and strategies to overcome immune evasion and suppression. Although cancer vaccines can be effective in settings of early cancer or minimal residual disease, therapeutic cancer vaccines will most likely require co-treatment, such as immune checkpoint inhibitors, to overcome immune suppression and be clinically effective in established cancers.

Depending on its composition, vaccines can be classified into tumor cell vaccines (autologous/allogenic), genetic vaccines (DNA/RNA/viral/bacterial), dendritic cell (DC) vaccines, and protein/peptide vaccines.

TUMOR CELL VACCINES

Tumor cell vaccines are classified as autologous when patient-derived tumor cells are used, or allogenic if established human tumor cell lines are the utilized. Among the allogenic vaccines, HS-110 (viagenpumatucel-L) is derived from lung adenocarcinoma cells. This vaccine is currently being studied in combination with nivolumab on a phase I/IIb clinical trial (DURGA trial) which is recruiting patients with NSCLC (NCT02439450). Preliminary results showed ORR of 50% in patients with IR (immune response, defined by doubling of IFNγ-secreting cells after re-stimulation with total vaccine antigen and individual cancer antigens) in comparison to 0% in patients without immune response among the initial treated 8 patients. Interestingly, patients with objective responses were also found decrease in MDSC and increase in CD8+ T cells in the blood [88].

An autologous tumor-derived gp96 vaccination was studied in a phase II clinical trial in patients with advanced gastric cancer [89]. The trial enrolled 73 patients of which 38 received both vaccination and chemotherapy and the remainder received chemotherapy alone. Overall vaccination was well tolerated with no clinically significant adverse events. The 2-year OS was 81.9% in the vaccination group compared to 67.9% in the chemotherapy-alone arm, though this was not statistically significant [89].

The GM.CD40L is another allogenic vaccine composed of radiated lung adenocarcinoma cells transduced with the GM-CSF and CD40-ligand (CD40L) genes. This vaccine is being studied in patients with lung adenocarcinoma in a phase I/II clinical trial in combination with nivolumab

(NCT02466568) and in combination with CCL21, a chemokine that enhances T cell response (NCT01433172). Preliminary results reveal an acceptable safety profile with no dose-limiting toxicities and a median OS similar to single-agent nivolumab (9.4 months) [90].

Genetic Vaccines

Genetic vaccines use DNA/RNA plasmids, bacteria, or viruses to deliver antigens by transfection of cells that subsequently process them and present them to immune cells [91].

The use of messenger RNA (mRNA) that encodes tumor antigens is currently being studied in clinical trials. RNA-Lipoplex ($RNA_{(LIP)}$) is a mRNA vaccine that encodes melanoma antigens (e.g., NY-ESO-1). It is being studied in a first-in-human phase I/II (Lipo-MERIT) clinical trial (NCT02410733). Preliminary results in 15 patients show no dose-limiting toxicities, but no efficacy results have been published yet [92].

VXM01 is an oral vaccine derived from live, attenuated Salmonella carrying a DNA plasmid that encodes vascular endothelial growth factor receptor-2 (VEGFR-2). A phase I clinical trial using this vaccine in patients with advanced pancreatic carcinoma showed that 12 out of the 18 patients studied had a considerable increase of specific anti-VEGFR2 T cells and the OS was 9.3 months compared to 8.4 months in those who received placebo. Furthermore, patients who showed T cell response had a longer median OS (10.3 months) compared to those without it (5.4 months) [93].

INO-5150 is a plasmid-based DNA vaccine that encodes for highly expressed prostate cancer antigens with amino acid sequence changes to break immune tolerance. This vaccine is being investigated with and without co-administration of IL-12 (INO-9012) in a phase I clinical trial (NCT02514213). Preliminary results revealed no dose-limiting toxicities and PSA was stable in some patients, whereas 10% reported disease progression [94].

INVAC-1 is another plasmid DNA vaccine that encodes an inactive form of human telomerase, which is expressed in over 85% of human tumors [95]. This vaccine is currently being studied in a phase I clinical trial with refractory and progressive solid tumors (NCT02301754). Preliminary results reveal no dose-limiting toxicities with only mild adverse events, 12 out of 20 patients achieved SD, and anti-human telomerase activity was found in 55% of patients [95].

pTVG-HP, also a plasmid DNA vaccine, encodes prostatic acid phosphatase (PAP). It is being studied in a phase II clinical trial in patients with non-castrate, non-metastatic prostate cancer (NCT01341652).

Listeria monocytogenes can survive in the cytosol of host cells and is therefore considered an ideal vector for tumor antigens [96]. ADXS11-001 is a live, attenuated *Listeria monocytogenes*, which is bioengineered to secrete a HPV-E7 antigen, a marker of HPV transformed cells. This vaccine is currently being investigated in a phase I clinical trial in patients with persistent, recurrent, or metastatic cervical carcinoma (NCT02164461). Preliminary results revealed that most adverse events were mild (grades 1 and 2), with no grade 4 or 5 reported. Tumor response analysis is ongoing and not yet published [97].

Adenovirus is also used as a vaccine vector. In a phase I clinical trial, the use of the cancer antigen MAGE-A3 primed to an adenovirus (AdMA3) is being investigated in conjunction with an oncolytic virus (MG-1 Maraba virus) that also expresses the MAGA-A3 antigen (MG1MA3), in patient with MAGE-A3 expressing solid tumors (NCT02285816). Preliminary results reveal a potent induction of pro-inflammatory genes with subsequent anti-tumor activity. However, dose-limiting toxicities also occurred in 4 out of 41 patients manifested by hypoxia, dyspnea, vomiting, and headache [98].

Dendritic Cell Vaccines

DCs play an important role in bridging innate and adaptive immunity, and as such are considered an important target for immunotherapy. The autologous tumor lysate, particle-loaded, dendritic cell (TLPLDC) vaccine consists of DCs that are exposed to autologous tumor antigens, become particle-loaded, and are infused back to the patient. This vaccine is being investigated in patients with stage III or IV of ovarian cancer and preliminary results reveal minimal toxicity, 1 out of 12 patients demonstrated CR, 1 had a SD, and 4 had a progression of their disease [99]. AdHER2ECTM consists of autologous DCs expressing human HER2 extracellular and transmembrane domains and is currently under investigation in a phase I clinical trial on patients with advanced tumors expressing HER-2 (e.g., colon, breast, ovarian) (NCT01730118). Preliminary results show that adverse events were limited to local injection site reactions, 37% had evidence of response, namely, 1 out of the 27 evaluated patients demonstrated CR, 1 had PR, and 5 had SD [100].

CMB305 is another DC vaccine that carries the NY-ESO-1 gene and a boost with G305, a NY-ESO-1 protein vaccine [101]. This vaccine is currently being investigated in a phase I clinical trial in patients with NY-ESO-1-expressing solid tumors. Preliminary results reveal an increase of anti-NY-ESO-1 T cells in up to 65% of patients and anti-NY-ESO-1 antibodies in up to 68% of patients [101].

A Modified Vaccinia Ankara (MVA) vector vaccine expressing genes for brachyury (a transcription factor important in the epithelial-to-mesenchymal transition and in tumor resistance to treatment) and costimulatory molecules (e.g., ICAM-1) designated TRICOM was developed for transducing DCs [102]. This vaccine was studied in a phase I clinical trial in 38 patients with advanced solid malignancies [102]. No dose-limiting toxicities were observed, and 82% of patients developed brachyury-specific immune responses.

BPX101 is another DC-derived vaccine which was recently evaluated in a phase I clinical trial in 18 men with progressive metastatic castrate-resistant prostate cancer [103]. Results revealed no dose-limiting toxicities, one case of CR, and two PR.

Protein/Peptide-based Vaccines

The use of specific tumor-associated antigens has the advantage of inducing specific immune response against defined antigens, sparing healthy tissue. However, because only defined epitopes are used, a specific but sometimes insufficient response may be generated as tumor cells often exhibit mutations of the epitopes used [104].

Wilms tumor gene-1 (WT-1) is overexpressed in many hematological and solid malignancies where it plays an oncogenic role [105]. The use of WT-1 peptide vaccine with gemcitabine was investigated in a phase II clinical trial on patients with pancreatic adenocarcinoma [106]. Results revealed an increased OS from 21.5 to 34.2 months in patients with gemcitabine alone compared to gemcitabine and WT-1 peptide vaccine, respectively. WT4869 is another peptide vaccine derived from WT1 that was recently studied in a phase I/II clinical trial in 26 patients with MDS. Results reveal an ORR of 18.2% and a median OS of 64.71 weeks, with an anti-WT1 lymphocyte induction occurring in 11 patients [107].

Galinpepimut-S is a WT-1-derived peptide vaccine with GM-CSF and Montanide as adjuvants that is being studied with compared to these last two alone, in a phase II clinical trial in 41 patients with pleural mesothelioma

[108]. Results revealed a PFS of 45% in the patients with the vaccine arm compared to 33% in those without the vaccine. Median OS was 22.8 months in the vaccine group compared to 18.3. Most adverse events were not clinically significant. Unfortunately, the trial did not achieve statistical power.

DPX-Survivac is a peptide-based vaccine containing survivin epitopes to elicit cytotoxic T cell response against survivin-expressing tumors. A phase Ib clinical trial on patients with ovarian/fallopian/peritoneal cancer studied this vaccine in combination with low-dose of cyclophosphamide [109]. Besides being safe, it demonstrated achieving sustained immune responses. This vaccine is also being investigated in hematologic malignancies in a phase II clinical trial that is currently recruiting patients (NCT02323230).

AE37 is a vaccine that contains HER-2-derived epitopes that stimulate T cell response. It is currently being studied on a phase II clinical trial on patients in combination with GM-CSF in patients with HER-2-positive breast cancer (NCT00524277). Preliminary results reveal the vaccine is safe and well tolerated, and disease-free survival was improved from 51% with the use of GM-CSF alone compared to 89% with AE37+ GM-CSF [110].

Heat shock protein 70 (HSP70) and a glypican-3 (GPC-3)-derived peptide is undergoing investigation in a phase I clinical trial in solid tumors that express these antigens (UMIN000020440). Preliminary results revealed no severe adverse events, a decrease tumor marker expression in 6 out of 12 patients, and disease control in 5 patients was observed [111].

URLC10-CDCA1-KOC1 multipeptide vaccine uses three HLA-A-24-restricted epitope peptides derived from cancer cells: upregulated lung cancer 10 (URLC10), cell division cycle-associated 1 (CDCA1), and KH domain-containing protein overexpressed in cancer 1 (KOC1). A phase II clinical trial studied this vaccine on patients with esophageal squamous cell carcinoma and found that it is capable of inducing highly specific T cells against these antigens [112].

Other peptide-based vaccines that are undergoing clinical trials but without preliminary results include the mutation-derived tumor antigen (MTA)-based peptide vaccine (NCT02721043); a personalized neoantigen vaccine NEO-PV-01 (NCT02897765); a vaccine composed of GM-CSF and CD40L (GM.CD40l) (NCT02466568); the OCV-C01 composed of peptides derived from KIF20A, VEGFR1, and VEGFR2 (UMIN000007991) [113]; and a Toll-like receptor-2 (TLR2) ligand-synthetic long peptide (SLP) vaccine containing HPV-16 E6 protein long peptides (2014-000658-12) [114].

In Situ Vaccines

Poly-ICLC is a synthetic immune danger signal and is specifically a mimic of viral dsRNA that can ligate TLR-3 and trigger cytokine production by DCs with subsequent immune activation and enhancement of the vaccine-induced anti-tumor responses [115]. A phase I/II clinical trial was recently conducted using this TLR ligand as an in situ vaccine in patients with multiple solid tumors including melanoma, breast, and HNSCC. Most of adverse events were mild and limited to the site of application; only 1 out of 8 patients achieved SD for 41 weeks, the remainder of patients showed PD [116].

BO-112 is a synthetic dsRNA administered intratumorally; it activates pro-apoptotic signals MDA-5 and NOXA and increases IFN response genes leading to the anti-tumor activity [117]. Its use is being studied in a phase I clinical trial in patients with palpable malignant tumors including melanoma and breast cancer (NCT02828098). Preliminary results revealed only one episode of reversible thrombocytopenia, with increase in circulating immune cells [117].

Neoantigen vaccines

Neoantigens are molecules expressed on tumor cell's surface by DNA mutations that present in tumor cells, but not in normal cells, making it an attractive cancer vaccine target [118]. Although neoantigen cancer vaccines have been long envisioned as ideal, its discovery and evaluation only became feasible recently with the development of highly efficient sequencing. Different from other immunotherapy such as checkpoint inhibitors and CAR T cells, vaccines targeting neoantigens are designed to be individual-specific. This personal vaccine induces a focused T cell response to patient's specific tumor neoantigens and avoids toxicities caused by damage to normal cells and tissues [119]. Multiple studies are ongoing to further explore this novel exciting approach.

A first-in-human clinical trial on patients with advanced melanoma identified individual mutations and neoantigens and developed a vaccine unique to each patient (IVAC MUTANOME) (NCT02035956). All 13 patients showed T cell response against neoantigens. Eight patients remained recurrence-free for the entire follow-up period (12–23 months).

Two of 5 patients that relapsed and achieved objective clinical response, and one of them achieved a CR in multiple metastatic lesions that had been unresponsive to radiation therapy and CTLA-4 blockade. A third patient also achieved CR to vaccination in combination with PD-1 blockade [119]. No major adverse events were reported.

The use of specific neoantigens in personalized vaccines is being explored in a phase I clinical trial in patients with melanoma (NCT01970358). The vaccine targets up to 20 personal tumor neoantigens and preliminary results reveal that 4 of the 6 vaccinated patients had no recurrence of the disease at 25 months after vaccination. The other two patients that experienced recurrence were subsequently managed with anti-PD-1 and achieved a complete tumor regression. Adverse events were mild and consisted of flu-like symptoms, injection site reactions, rash, and fatigue [120].

OTHER APPROACHES IN IMMUNOTHERAPY

Targeting Myeloid-derived Suppressor Cells

Myeloid-derived suppressor cells (MDSCs) are immature myeloid cells that promote immunosuppression and favor tumor growth [121]. The TNF-related apoptosis-inducing ligand receptor (TRAIL-R)-2, also known as death receptor (DR)-5, is found on tumor cells and MDSCs, and its activation promotes apoptosis in these populations [122]. The use of the TRAIL-R2 agonist antibody, DS-8273a, is being studied. Results of one trial in patients with solid tumors revealed only mild to moderate adverse events, no dose-limiting toxicities, and a decrease in blood levels of MDSCs [123].

Cytokine Gene Therapy

IL-12 has been considered a good option for immunotherapy given its potent anti-tumor effect [124]. This cytokine promotes the activation of NK and T cells and synergizes other cytokines with anti-tumor effects [124].

Ad-RTS-hIL-12 is a replication-incompetent adenovirus engineered to express IL-12. By default, IL-12 expression by this virus is "off," but with the use of veledimex, gene is activated and IL-12 production is started [125]. The use of Ad-RTS-hIL-12 with veledimex is being studied in patients with advanced gliomas in a phase I clinical trial (NCT02026271). Preliminary results revealed that the most frequent adverse events were mild flu-like symptoms, grade 3 CRS, and grade 3 transaminitis; however, all were reversed upon discontinuation of therapy. A median OS of 12.5 months was also observed [126]. This technology is also being studied in patients with locally advanced or metastatic breast cancer (NCT02423902); however, no preliminary results have been revealed yet.

IL-2 enhances the immune system through the IL-2 receptor (IL-2R) [127]. NKTR-214, an engineered cytokine that specifically stimulates IL-2R, is being investigated on phase I/II clinical trials on solid tumors (NCT02983045, NCT02869295). Preliminary results on the former trial show no dose-limiting toxicities. One patient had a 40% decrease in LDH, and another patient had an unconfirmed CR after only 6 weeks of treatment [128]. The latter trial revealed no dose-limiting toxicities, a tumor size reduction ranging from 10 to 30% in 6 out of 26 patients (23%), and an increase of T cells and NK cells within the tumor microenvironment in 100% of patients [129].

Targeting Tumor Microenvironment

Cancer cells require a milieu, known as tumor microenvironment, which allows their growth. This microenvironment consists of immune and nonimmune cells and non-cellular factors that interact among each other and promote a chronic inflammatory, immunosuppressive, and pro-angiogenic ecosystem that favors tumor survival, growth, and dissemination [130]. Some of these factors that have been identified are being investigated as potential therapeutic targets, often in conjunction with other immunotherapy agents.

Indoleamine 2,3-dioxygenase (IDO) is an enzyme that converts tryptophan to kynurenines. These latter promote the formation of Tregs, increase the number of MDSCs, and decrease the activity of CD8 T cells with a resulting inhibitory environment [130, 131]. BMS-986205 is an IDO1 inhibitor that is being studied on a phase I clinical trial in conjunction with a PD-1 inhibitor in patients with advanced solid tumors (NCT02658890). Preliminary results reported mild toxicities except for three cases of grade 3 hepatitis, rash, and hypophosphatemia. No efficacy was described [132]. Indoximod is another IDO inhibitor undergoing phase II clinical trials on melanoma (NCT02073123) and pancreatic (NCT02077881) and castrate-resistant prostate cancer (NCT01560923). Preliminary results reveal an ORR of 52% in patients with melanoma when used with immune checkpoint inhibitors [133]. Patients with pancreatic cancer had an ORR of

37% when indoximod was used with both gemcitabine and nab-paclitaxel [134]. Median PFS increased from 4.1 to 10.3 months in castrate-resistant metastatic prostate cancer compared to placebo [135]. Epacadostat also blocks IDO pathway and is being evaluated on phase I/II clinical trials with multiple solid malignancies (NCT02327078, NCT02178722). Preliminary results have demonstrated an ORR ranging from 75% in melanoma to 4% in CRC. No dose-limiting toxicities were identified [136, 137].

Toll-like receptors (TLRs) are critical in the identification of pathogens but play a complex role in tumorigenesis. TLRs like TLR4 promote cancer progression by promoting inflammation in the microenvironment. TLRs like TLR7/8 and TLR9 promote anti-tumor responses by inducing a "danger signal" and activating the immune system against malignant cells [138]. MEDI9197, a dual agonist of TLR7/8, is under phase I clinical testing in combination with durvalumab and radiation therapy on metastatic or locally advanced solid malignancies (NCT02556463). Preliminary results show that the agent is overall safe with only mild adverse events. No efficacy data has been yet reported [139]. PG545 (pixatimod, pINN) is an agonist of TLR9/IL-12 tested in a phase I clinical trial in patients with advanced solid tumors [140]. Results show that 3 out of 23 patients developed dose-limiting toxicities and the best response achieved was a 24-week SD and a DCR of 38%. Polyinosinic-polycytidylic acid polylysine carboxymethylcellulose (poly-ICLC) is a potent TLR3 agonist that was studied in combination with radiation in a phase I clinical trial in patients with hepatocellular carcinoma not eligible for surgery [141]. Intratumoral injection was found to be safe with mostly grade I–II adverse events, a PFS of 66% at 6 months and 28% at 24 months and an OS after 1 year was 69% and 38% after 2 years [141].

Arginine is an amino acid required for T cell activation and proliferation. Malignant cells produce high levels of arginase and deplete arginine interfering with immune activation [142]. CB-1158, an arginase inhibitor, is being studied in a phase I clinical trial alone and in combination with a PD-1 inhibitor in patients with advanced solid tumors (NCT02903914). Preliminary results reveal no dose-limiting toxicities, > 90% of arginase inhibition, and up to a 4-fold increase in plasma arginine levels [143].

Oncolytic Peptides

LTX-315 is a cytotoxic peptide that damages the tumor-mitochondrial membranes and triggers caspase-independent necrosis leading to a massive release of tumor antigens and to an increase in TIL activity [144]. A phase

I clinical trial is investigating this agent as monotherapy or in combination with ICIs in patients with metastatic solid tumors, particularly melanoma and breast cancer (NCT01986426). Preliminary results showed that 2 patients achieved a CR, 5 patients had a decrease of > 50% of the tumor size, and 8 patients achieved SD [145].

CONCLUSIONS

Cancer immunotherapy has changed the landscape of modern oncology in varied cancer types. Immunotherapy with checkpoint inhibitors has significantly improved the clinical outcomes in some, but not all patients. This is likely due to individual differences in immunogenicity of tumor and immunosuppressive tumor microenvironments. The understanding of emerging novel immunotherapeutic approaches beyond immune checkpoints discussed above will likely open the opportunities to patients with cancers that have failed to respond to an immune checkpoint inhibitor alone. Furthermore, combination therapies targeting different immune mechanisms will likely to better modulate the immune systems to boost an anti-tumor response.

The development of tumor-directed antibodies, antibody-drug conjugates, CAR T cells, oncolytic viruses, vaccines, and even genetic therapy has allowed for a more targeted and tumor-specific therapy rather than a non-specific cytolytic chemotherapy or radiation therapy. Next wave of clinical trials are already evaluating the combinations of immunotherapy agents from different classes. Immune-related side effects, cost of treatment, lack of response biomarkers, and tumor relapse are remaining challenges. Nevertheless, this rapidly advancing field is becoming the most promising treatment component of current oncologic therapy.

AUTHORS' CONTRIBUTIONS

JMA drafted the manuscript; YL designed and supervised the study, as well as edited the manuscript; AS, BD, and KK reviewed and edited the text. All authors reviewed and approved the final version of the manuscript.

REFERENCES

1. Khalil DN, Smith EL, Brentjens RJ, Wolchok JD. The future of cancer treatment: immunomodulation, CARs and combination immunotherapy. Nat Rev Clin Oncol. 2016;13:273–290. doi: 10.1038/nrclinonc.2016.25.
2. Dholaria B, Hammond W, Shreders A, Lou Y. Emerging therapeutic agents for lung cancer. J Hematol Oncol. 2016;9:138. doi: 10.1186/s13045-016-0365-z.
3. Sharma P, Allison JP. The future of immune checkpoint therapy. Science. 2015;348:56–61. doi: 10.1126/science.aaa8172.
4. Scott AM, Allison JP, Wolchok JD. Monoclonal antibodies in cancer therapy. Cancer Immun. 2012;12:14.
5. Weidle UH, Kontermann RE, Brinkmann U. Tumor-antigen-binding bispecific antibodies for cancer treatment. Semin Oncol. 2014;41:653–660. doi: 10.1053/j.seminoncol.2014.08.004.
6. Zhang X, Yang Y, Fan D, Xiong D. The development of bispecific antibodies and their applications in tumor immune escape. Exp Hematol Oncol. 2017;6:12.
7. Fan G, Wang Z, Hao M, Li J. Bispecific antibodies and their applications. J Hematol Oncol. 2015;8:130.
8. Moore PA, Zhang WJ, Rainey GJ, et al. Application of dual affinity retargeting molecules to achieve optimal redirected T-cell killing of B-cell lymphoma. Blood. 2011;117:4542–4551. doi: 10.1182/blood-2010-09-306449.
9. Rader C. DARTs take aim at BiTEs. Blood. 2011;117:4403–4404. doi: 10.1182/blood-2011-02-337691.
10. Kim RD, Arlen PM, Tsang KY, et al. Ensituximab (E) in patients (pts) with refractory metastatic colorectal cancer (mCRC): results of a phase I/II clinical trial. J Clin Oncol. 2017;35:3081.
11. Tabernero J, Melero I, Ros W, et al. Phase Ia and Ib studies of the novel carcinoembryonic antigen (CEA) T-cell bispecific (CEA CD3 TCB) antibody as a single agent and in combination with atezolizumab: preliminary efficacy and safety in patients with metastatic colorectal cancer (mCRC) J Clin Oncol. 2017;35:3002. doi: 10.1200/JCO.2017.72.8428.
12. Wu J, Fu J, Zhang M, Liu D. Blinatumomab: a bispecific T cell engager (BiTE) antibody against CD19/CD3 for refractory acute lymphoid

leukemia. J Hematol Oncol. 2015;8:104.

13. Viardot A, Goebeler ME, Hess G, et al. Phase 2 study of the bispecific T-cell engager (BiTE) antibody blinatumomab in relapsed/refractory diffuse large B-cell lymphoma. Blood. 2016;127:1410–1416. doi: 10.1182/blood-2015-06-651380.
14. Tolcher AW, Alley EW, Chichili G, et al. Phase 1, first-in-human, open label, dose escalation study of MGD009, a humanized B7-H3 x CD3 dual-affinity re-targeting (DART) protein in patients with B7-H3-expressing neoplasms or B7-H3 expressing tumor vasculature. J Clin Oncol. 2016;34:TPS3105. doi: 10.1200/JCO.2016.67.2162.
15. Wu J, Fu J, Zhang M, Liu D. AFM13: a first-in-class tetravalent bispecific anti-CD30/CD16A antibody for NK cell-mediated immunotherapy. J Hematol Oncol. 2015;8:96. doi: 10.1186/s13045-015-0188-3.
16. Rothe A, Sasse S, Topp MS, et al. A phase 1 study of the bispecific anti-CD30/CD16A antibody construct AFM13 in patients with relapsed or refractory Hodgkin lymphoma. Blood. 2015;125:4024–4031. doi: 10.1182/blood-2014-12-614636.
17. Diamantis N, Banerji U. Antibody-drug conjugates-an emerging class of cancer treatment. Br J Cancer. 2016;114:362–367. doi: 10.1038/bjc.2015.435.
18. Angevin E, Strickler JH, Weekes CD, et al. Phase I study of ABBV-399, a c-Met antibody-drug conjugate (ADC), as monotherapy and in combination with erlotinib in patients (pts) with non-small cell lung cancer (NSCLC) J Clin Oncol. 2017;35:2509.
19. Ott PA, Pavlick AC, Johnson DB, et al. A phase II study of glembatumumab vedotin (GV), an antibody-drug conjugate (ADC) targeting gpNMB, in advanced melanoma. J Clin Oncol. 2017;35:109.
20. Calvo E, Cleary JM, Moreno V, et al. Preliminary results from a phase 1 study of the antibody-drug conjugate ABBV-221 in patients with solid tumors likely to express EGFR. J Clin Oncol. 2017;35:2510.
21. O'Malley DM, Moore KN, Vergote I, et al. Safety findings from FORWARD II: a phase 1b study evaluating the folate receptor alpha (FRα)-targeting antibody-drug conjugate (ADC) mirvetuximab soravtansine (IMGN853) in combination with bevacizumab, carboplatin, pegylated liposomal doxorubicin (PLD), or pembrolizumab in patients (pts) with ovarian cancer. J Clin Oncol. 2017;35:5553. doi: 10.1200/JCO.2017.35.5_suppl.97.

22. Petrylak DP, Perez RP, Zhang J, et al. A phase I study of enfortumab vedotin (ASG-22CE; ASG-22ME): updated analysis of patients with metastatic urothelial cancer. J Clin Oncol. 2017;35:106.
23. Bardia A, Mayer IA, Diamond JR, et al. Efficacy and safety of anti-Trop-2 antibody drug conjugate sacituzumab govitecan (IMMU-132) in heavily pretreated patients with metastatic triple-negative breast cancer. J Clin Oncol. 2017;35:2141–2148. doi: 10.1200/JCO.2016.70.8297.
24. Heist RS, Guarino MJ, Masters G, et al. Therapy of advanced non-small-cell lung cancer with an SN-38-anti-Trop-2 drug conjugate, Sacituzumab Govitecan. J Clin Oncol. 2017;35:2790–2797. doi: 10.1200/JCO.2016.72.1894.
25. Pazdur R. Inotuzumab ozogamicin (Besponsa) approval letter. In: Drug approvals and Databases. U.S. Food and Drug Administration, Center for Drug Evaluation and Research. 2017. https://www.accessdata.fda.gov/drugsatfda_docs/appletter/2017/761040Orig1s000ltr.pdf. Accessed 3 October 2017.
26. Sangha R, Davies A, Dang NH, et al. Phase 1 study of inotuzumab ozogamicin combined with R-GDP for the treatment of patients with relapsed/refractory CD22+ B-cell non-Hodgkin lymphoma. J Drug Assess. 2017;6:10–17. doi: 10.1080/21556660.2017.1315336.
27. Dotan E, Cohen SJ, Starodub AN, et al. Phase I/II trial of labetuzumab govitecan (anti-CEACAM5/SN-38 antibody-drug conjugate) in patients with refractory or relapsing metastatic colorectal cancer. J Clin Oncol. 2017;35:3338–3346. doi: 10.1200/JCO.2017.73.9011.
28. Socinski MA, Kaye FJ, Spigel DR, et al. Phase 1/2 study of the CD56-targeting antibody-drug conjugate lorvotuzumab mertansine (IMGN901) in combination with carboplatin/etoposide in small-cell lung cancer patients with extensive-stage disease. Clin Lung Cancer. 2017;18:68–76. doi: 10.1016/j.cllc.2016.09.002.
29. Rudin CM, Pietanza MC, Bauer TM, et al. Rovalpituzumab tesirine, a DLL3-targeted antibody-drug conjugate, in recurrent small-cell lung cancer: a first-in-human, first-in-class, open-label, phase 1 study. Lancet Oncol. 2017;18:42–51. doi: 10.1016/S1470-2045(16)30565-4.
30. Flynn MJ, Hartley JA. The emerging role of anti-CD25 directed therapies as both immune modulators and targeted agents in cancer. Br J Haematol. 2017;179:20–35. doi: 10.1111/bjh.14770.
31. Horwitz SM, Fanale MA, Spira AI, et al. Interim data from the

first clinical study of ADCT-301, a novel pyrrolobenzodiazapine-based antibody drug conjugate, in relapsed/refractory Hodgkin/non-Hodgkin lymphoma. Hematol Oncol. 2017;35:270–271. doi: 10.1002/hon.2438_143.

32. Almhanna K, Wright D, Mercade TM, et al. A phase II study of antibody-drug conjugate, TAK-264 (MLN0264) in previously treated patients with advanced or metastatic pancreatic adenocarcinoma expressing guanylyl cyclase C. Investig New Drugs. 2017;35:634–641. doi: 10.1007/s10637-017-0473-9.
33. Trail PA, Dubowchik GM, Lowinger TB. Antibody drug conjugates for treatment of breast cancer: novel targets and diverse approaches in ADC design. Pharmacol Ther. 2018;181:126–142. doi: 10.1016/j.pharmthera.2017.07.013.
34. Rose AAN, Biondini M, Curiel R, Siegel PM. Targeting GPNMB with glembatumumab vedotin: current developments and future opportunities for the treatment of cancer. Pharmacol Ther. 2017;179:127–141. doi: 10.1016/j.pharmthera.2017.05.010.
35. Yardley DA, Weaver R, Melisko ME, et al. EMERGE: a randomized phase II study of the antibody-drug conjugate glembatumumab vedotin in advanced glycoprotein NMB-expressing breast cancer. J Clin Oncol. 2015;33:1609–1619. doi: 10.1200/JCO.2014.56.2959.
36. Gomez-Roca CA, Boni V, Moreno V, et al. A phase I study of SAR566658, an anti CA6-antibody drug conjugate (ADC), in patients (Pts) with CA6-positive advanced solid tumors (STs) (NCT01156870) J Clin Oncol. 2016;34:2511.
37. Forero A, Burris H, III, Mita M, et al. Abstract P3-14-05: interim analysis of a phase 1 study of the antibody-drug conjugate SGN-LIV1A in patients with metastatic breast cancer. Cancer Research. 2016;76:P3–14-05-P13–14-05. doi: 10.1158/1538-7445.SABCS15-P3-14-05.
38. Sachdev JC, Maitland M, Sharma M, et al. A phase 1 study of PF-06647020, an antibody-drug conjugate (ADC) targeting protein tyrosine kinase 7 (PTK7), in patients with advanced solid tumors including platinum resistant ovarian cancer (OVCA) Ann Oncol. 2016;27:LBA35.
39. Garrido-Laguna I, Krop IE, Burris H, et al. A phase I study of PF-06647263, a novel EFNA4-ADC, in patients with metastatic triple negative breast cancer. J Clin Oncol. 2017;35:2511.

40. Kogawa T, Yonemori K, Naito Y, et al. Phase 1/2, multicenter, non-randomized, open-label, multiple-dose first-in-human study of U3-1402 (anti-HER3 antibody drug conjugate) in subjects with HER3-positive metastatic breast cancer. J Clin Oncol. 2017;35:TPS1116.
41. Lassen UN, Ramalingam SS, Lopez JS, et al. GCT1021-01, a first-in-human, open-label, dose-escalation trial with expansion cohorts to evaluate safety of Axl-specific antibody-drug conjugate (HuMax-Axl-ADC) in patients with solid tumors (NCT02988817) J Clin Oncol. 2017;35:TPS2605.
42. Zhang C, Liu J, Zhong JF, Zhang X. Engineering CAR-T cells. Biomark Res. 2017;5:22. doi: 10.1186/s40364-017-0102-y.
43. Fesnak AD, June CH, Levine BL. Engineered T cells: the promise and challenges of cancer immunotherapy. Nat Rev Cancer. 2016;16:566–581. doi: 10.1038/nrc.2016.97.
44. Qin L, Zhao R, Li P. Incorporation of functional elements enhances the antitumor capacity of CAR T cells. Exp Hematol Oncol. 2017;6:28. doi: 10.1186/s40164-017-0088-z.
45. Bryan WW. Kymriah (tisagenlecleucel) approval letter. In: Drug approvals and Databases. U.S. Food and Drug Administration, Center for Drug Evaluation and Research. 2017. https://www.fda.gov/downloads/BiologicsBloodVaccines/CellularGeneTherapyProducts/ApprovedProducts/UCM574106.pdf. Accessed 2 September 2017.
46. Zhang E, Xu H. A new insight in chimeric antigen receptor-engineered T cells for cancer immunotherapy. J Hematol Oncol. 2017;10:1. doi: 10.1186/s13045-016-0379-6.
47. Yu S, Li A, Liu Q, et al. Chimeric antigen receptor T cells: a novel therapy for solid tumors. J Hematol Oncol. 2017;10:78. doi: 10.1186/s13045-017-0444-9.
48. Lee DW, Gardner R, Porter DL, et al. Current concepts in the diagnosis and management of cytokine release syndrome. Blood. 2014;124:188–195. doi: 10.1182/blood-2014-05-552729.
49. Liu B, Song Y, Liu D. Clinical trials of CAR-T cells in China. J Hematol Oncol. 2017;10:166. doi: 10.1186/s13045-017-0535-7.
50. van Schalkwyk MCI, Papa SE, Jeannon JP, et al. Design of a phase I clinical trial to evaluate intratumoral delivery of ErbB-targeted chimeric antigen receptor T-cells in locally advanced or recurrent head and neck cancer. Hum Gene Ther Clin Dev. 2013;24:134–142. doi:

10.1089/humc.2013.144.

51. Papa S, Adami A, Metoudi M, et al. Abstract CT118: T4 immunotherapy of head and neck squamous cell carcinoma using pan-ErbB targeted CAR T-cells. Cancer Res. 2017;77:CT118. doi: 10.1158/1538-7445. AM2017-CT118.
52. Stefanski J, Brentjens R, Hollyman D, et al. CD19-targeted normal and CLL patient T cells expanded with beads can eradicate systemic tumors in vivo. Mol Ther. 2006;13:S102. doi: 10.1016/j.ymthe.2006.08.322.
53. Park JH, Rivere I, Wang X, et al. Abstract CT078: impact of disease burden and transplant on long-term survival after CD19 CAR therapy in adults with relapsed B-cell acute lymphoblastic leukemia. Cancer Res. 2017;77:CT078. doi: 10.1158/1538-7445.AM2017-CT078.
54. Locke FL, Neelapu SS, Bartlett NL, et al. Clinical and biologic covariates of outcomes in ZUMA-1: a pivotal trial of axicabtagene ciloleucel (axi-cel; KTE-C19) in patients with refractory aggressive non-Hodgkin lymphoma (r-NHL) J Clin Oncol. 2017;35:7512.
55. Xia L, Chen Q, Li Q, et al. Abstract CT041: the clinical study on CD19-directed chimeric antigen receptor-modified T cells in patient with Richter syndrome. Cancer Res. 2017;77:CT041. doi: 10.1158/1538-7445.AM2017-CT041.
56. Turtle CJ, Hanafi L-A, Berger C, et al. Rate of durable complete response in ALL, NHL, and CLL after immunotherapy with optimized lymphodepletion and defined composition CD19 CAR-T cells. J Clin Oncol. 2016;34:102.
57. Fan F, Zhao W, Liu J, et al. Durable remissions with BCMA-specific chimeric antigen receptor (CAR)-modified T cells in patients with refractory/relapsed multiple myeloma. J Clin Oncol. 2017;35:LBA3001. doi: 10.1200/JCO.2017.35.18_suppl.LBA3001.
58. Li WP, Guo LJ, Ekaterina M, et al. Immunotherapy of hepatocellular carcinoma with T cells engineered to express glypican-3-specific chimeric antigen receptors. Mol Ther. 2015;23:S164–S165. doi: 10.1016/S1525-0016(16)34025-4.
59. Zhai B, Shi D, Gao H, et al. A phase I study of anti-GPC3 chimeric antigen receptor modified T cells (GPC3 CAR-T) in Chinese patients with refractory or relapsed GPC3+ hepatocellular carcinoma (r/r GPC3+ HCC) J Clin Oncol. 2017;35:3049.
60. Wang Y, Chen M, Wu Z, et al. CD133-redirected chimeric antigen

receptor engineered autologous T-cell treatment in patients with advanced and metastatic malignancies. J Clin Oncol. 2017;35:3042.

61. Cohen AD, Garfall AL, Stadtmauer EA, et al. B-cell maturation antigen (BCMA)-specific chimeric antigen receptor T cells (CART-BCMA) for multiple myeloma (MM): initial safety and efficacy from a phase I study. Blood. 2016;128:1147.
62. Berdeja JG, Lin Y, Raje NS, et al. First-in-human multicenter study of bb2121 anti-BCMA CAR T-cell therapy for relapsed/refractory multiple myeloma: updated results. J Clin Oncol. 2017;35:3010. doi: 10.1200/JCO.2017.72.8519.
63. Guo B, Chen M, Han Q, et al. CD138-directed adoptive immunotherapy of chimeric antigen receptor (CAR)-modified T cells for multiple myeloma. J Cellular Immunotherapy. 2016;2:28–35. doi: 10.1016/j.jocit.2014.11.001.
64. Ramos CA, Savoldo B, Torrano V, et al. Clinical responses with T lymphocytes targeting malignancy-associated kappa light chains. J Clin Investig. 2016;126:2588–2596. doi: 10.1172/JCI86000.
65. Falini B, Pileri S, Pizzolo G, et al. CD30 (Ki-1) molecule: a new cytokine receptor of the tumor necrosis factor receptor superfamily as a tool for diagnosis and immunotherapy. Blood. 1995;85:1–14.
66. Ramos CA, Ballard B, Zhang H, et al. Clinical and immunological responses after CD30-specific chimeric antigen receptor-redirected lymphocytes. J Clin Invest. 2017;127:3462–3471. doi: 10.1172/JCI94306.
67. Brown CE, Alizadeh D, Starr R, et al. Regression of Glioblastoma after chimeric antigen receptor T-cell therapy. N Engl J Med. 2016;375:2561–2569. doi: 10.1056/NEJMoa1610497.
68. Soriani A, Fionda C, Ricci B, et al. Chemotherapy-elicited upregulation of NKG2D and DNAM-1 ligands as a therapeutic target in multiple myeloma. Oncoimmunology. 2013;2:e26663. doi: 10.4161/onci.26663.
69. Klampatsa A, Haas AR, Moon EK, Albelda SM. Chimeric antigen receptor (CAR) T cell therapy for malignant pleural mesothelioma (MPM) Cancers. 2017;9:115. doi: 10.3390/cancers9090115.
70. Govers C, Sebestyen Z, Coccoris M, et al. T cell receptor gene therapy: strategies for optimizing transgenic TCR pairing. Trends Mol Med. 2010;16:77–87. doi: 10.1016/j.molmed.2009.12.004.
71. Mackall C, Tap WD, Glod J, et al. Open label, non-randomized,

multi-cohort pilot study of genetically engineered NY-ESO-1 specific NY-ESO-1c259t in HLA-A2+ patients with synovial sarcoma (NCT01343043) J Clin Oncol. 2017;35:3000.

72. Draper LM, Kwong MLM, Gros A, et al. Targeting of HPV-16(+) epithelial cancer cells by TCR gene engineered T cells directed against E6. Clin Cancer Res. 2015;21:4431–4439. doi: 10.1158/1078-0432.CCR-14-3341.

73. Hinrichs CS, Doran SL, Stevanovic S, et al. A phase I/II clinical trial of E6 T-cell receptor gene therapy for human papillomavirus (HPV)-associated epithelial cancers. J Clin Oncol. 2017;35:3009.

74. Hong DS, Butler MO, Sullivan RJ, et al. A phase I single arm, open label clinical trial evaluating safety of MAGE-A10c796T in subjects with advanced or metastatic head and neck, melanoma, or urothelial tumors (NCT02989064) J Clin Oncol. 2017;35:TPS3098.

75. Rosenberg SA, Yang JC, Sherry RM, et al. Durable complete responses in heavily pretreated patients with metastatic melanoma using T-cell transfer immunotherapy. Clin Cancer Res. 2011;17:4550–4557. doi: 10.1158/1078-0432.CCR-11-0116.

76. Goff SL, Dudley M, Citrin DE, et al. A randomized, prospective evaluation comparing intensity of lymphodepletion prior to adoptive transfer of tumor infiltrating lymphocytes for patients with metastatic melanoma. J Clin Oncol. 2016;34:3006. doi: 10.1200/JCO.2016.66.7220.

77. Noonan KA, Huff CA, Davis J, et al. Adoptive transfer of activated marrow-infiltrating lymphocytes induces measurable antitumor immunity in the bone marrow in multiple myeloma. Sci Transl Med. 2015;7:288ra278. doi: 10.1126/scitranslmed.aaa7014.

78. Kaufman HL, Kohlhapp FJ, Zloza A. Oncolytic viruses: a new class of immunotherapy drugs. Nat Rev Drug Discov. 2015;14:642–662. doi: 10.1038/nrd4663.

79. Rehman H, Silk AW, Kane MP, Kaufman HL. Into the clinic: talimogene laherparepvec (T-VEC), a first-in-class intratumoral oncolytic viral therapy. J Immunother Cancer. 2016;4:53. doi: 10.1186/s40425-016-0158-5.

80. Andtbacka RHI, Curti B, Hallmeyer S, et al. Phase II CALM extension study: enhanced immune-cell infiltration within the tumour micro-environment of patients with advanced melanoma

following intralesional delivery of coxsackievirus A21. Eur J Cancer. 2015;51:S677. doi: 10.1016/S0959-8049(16)31854-8.

81. Pandha HS, Ralph C, Harrington K, et al. Keynote-200 phase 1b: a novel combination study of intravenously delivered coxsackievirus A21 and pembrolizumab in advanced cancer patients. J Clin Oncol. 2017;35:TPS3108.

82. Silk AW, Kaufman H, Gabrail N, et al. Abstract CT026: phase 1b study of intratumoral coxsackievirus A21 (CVA21) and systemic pembrolizumab in advanced melanoma patients: interim results of the CAPRA clinical trial. Cancer Res. 2017;77:CT026. doi: 10.1158/1538-7445.AM2017-CT026.

83. Curti BD, Richards JM, Hallmeyer S, et al. Activity of a novel immunotherapy combination of intralesional coxsackievirus A21 and systemic ipilimumab in advanced melanoma patients previously treated with anti-PD1 blockade therapy. J Clin Oncol. 2017;35:3014.

84. Bernstein V, Ellard S, Dent SF, et al. Abstract CT131: a randomized (RCT) phase II study of oncolytic reovirus (pelareorep) plus standard weekly paclitaxel (P) as therapy for metastatic breast cancer (mBC) Cancer Res. 2017;77:CT131. doi: 10.1158/1538-7445.AM2017-CT131.

85. Mahalingam D, Fountzilas C, Moseley J, et al. A phase II study of REOLYSIN(R) (pelareorep) in combination with carboplatin and paclitaxel for patients with advanced malignant melanoma. Cancer Chemother Pharmacol. 2017;79:697–703. doi: 10.1007/s00280-017-3260-6.

86. Alonso MM, García-Moure M, Gonzalez-Huarriz M, et al. Abstract CT027: oncolytic virus DNX-2401 with a short course of temozolomide for glioblastoma at first recurrence: clinical data and prognostic biomarkers. Cancer Res. 2017;77:CT027. doi: 10.1158/1538-7445.AM2017-CT027.

87. van der Burg SH, Arens R, Ossendorp F, et al. Vaccines for established cancer: overcoming the challenges posed by immune evasion. Nat Rev Cancer. 2016;16:219–233. doi: 10.1038/nrc.2016.16.

88. Morgensztern D, Harb W, Schalper K, et al. MA09.06 Viagenpumatucel-L bolsters response to nivolumab therapy in advanced lung adenocarcinoma: preliminary data from the DURGA trial. J Thorac Oncol. 2017;12:S394–S395. doi: 10.1016/j.jtho.2016.11.447.

89. Zhang K, Peng Z, Huang X, et al. Phase II trial of adjuvant immunotherapy with autologous tumor-derived Gp96 vaccination in patients with gastric cancer. J Cancer. 2017;8:1826–1832. doi: 10.7150/jca.18946.
90. Gray JE, Chiappori A, Williams CC, et al. Phase I/II randomized trial of GM.CD40L vaccine plus/minus CCL21 in advanced lung adenocarcinoma: Final results. J Clin Oncol. 2016;34:9037.
91. Guo C, Manjili MH, Subjeck JR, et al. Therapeutic cancer vaccines: past, present, and future. Adv Cancer Res. 2013;119:421–475. doi: 10.1016/B978-0-12-407190-2.00007-1.
92. Jabulowsky RA, Loquai C, Diken M, et al. Abstract CT032: a first-in-human phase I/II clinical trial assessing novel mRNA-lipoplex nanoparticles for potent cancer immunotherapy in patients with malignant melanoma. Cancer Res. 2016;76:CT032. doi: 10.1158/1538-7445.AM2016-CT032.
93. Schmitz-Winnenthal FH, Podola L, Hohmann N, et al. A phase 1 trial extension to assess immunologic efficacy and safety of prime-boost vaccination with VXM01, an oral T cell vaccine against VEGF-receptor 2, in patients with advanced pancreatic cancer. J Clin Oncol. 2016;34:3091.
94. Shore ND, Heath EI, Nordquist LT, et al. A clinical trial for the safety and immunogenicity of a DNA-based immunotherapy in men with biochemically (PSA) relapsed prostate cancer. J Clin Oncol. 2017;35:e14634.
95. Teixeira L, Medioni J, Doucet L, et al. Results of a first-in-human phase I study of INVAC-1, an optimized plasmid DNA encoding an inactive form of human telomerase reverse transcriptase (hTERT), in patients with advanced solid tumors. J Clin Oncol. 2017;35:3087.
96. Singh R, Paterson Y. Listeria monocytogenes as a vector for tumor-associated antigens for cancer immunotherapy. Expert Rev Vaccines. 2006;5:541–552. doi: 10.1586/14760584.5.4.541.
97. Ghamande SA, Platt D, Wheatley D, et al. Phase I study evaluating high-dose treatment with ADXS11-001, a Listeria monocytogenes-listeriolysin O (Lm-LLO) immunotherapy, in women with cervical cancer. J Clin Oncol. 2016;34:e14580.
98. Jonker DJ, Hotte SJ, Abdul Razak AR, et al. Phase I study of oncolytic virus (OV) MG1 maraba/MAGE-A3 (MG1MA3), with and without

transgenic MAGE-A3 adenovirus vaccine (AdMA3) in incurable advanced/metastatic MAGE-A3-expressing solid tumours: CCTG IND.214. J Clin Oncol. 2017;35:e14637.

99. Peace KM, Vreeland TJ, Clifton GT, et al. Abstract CT036: early trial results of an autologous tumor lysate, particle-loaded, dendritic cell (TLPLDC) vaccine in ovarian cancer patients. Cancer Res. 2017;77:CT036. doi: 10.1158/1538-7445.AM2017-CT036.

100. Wood LV, Roberson BD, Agarwal PK, et al. Association of autologous AdHER2 dendritic cell vaccination with antitumor activity and number of circulating tumor cells. J Clin Oncol. 2017;35:3089. doi: 10.1200/JCO.2017.35.8_suppl.87.

101. Pollack S, Lu H, Somaiah N, et al. Association of CMB305 or LV305-induced and baseline anti-NY-ESO-1 immunity with survival in recurrent cancer patients. J Clin Oncol. 2017;35:3090.

102. Heery CR, Palena C, McMahon S, et al. Phase I study of a poxviral TRICOM-based vaccine directed against the transcription factor brachyury. Clin Cancer Res. 2017;23:6833. doi: 10.1158/1078-0432.CCR-17-1087.

103. Sonpavde G, McMannis JD, Bai Y, et al. Phase I trial of antigen-targeted autologous dendritic cell-based vaccine with in vivo activation of inducible CD40 for advanced prostate cancer. Cancer Immunol Immunother. 2017;66:1345–1357. doi: 10.1007/s00262-017-2027-6.

104. Espinoza-Delgado I. Cancer vaccines. Oncologist. 2002;7:20–33. doi: 10.1634/theoncologist.7-suppl_3-20.

105. Oka Y, Tsuboi A, Oji Y, et al. WT1 peptide vaccine for the treatment of cancer. Curr Opin Immunol. 2008;20:211–220. doi: 10.1016/j.coi.2008.04.009.

106. Nishida S, Ishikawa T, Kokura S, et al. Randomized phase II study of WT1 peptide vaccine plus gemcitabine for advanced pancreatic ductal adenocarcinoma (PDAC): clinical efficacy and immune response. J Clin Oncol. 2016;34:3085. doi: 10.1200/jco.2016.34.4_suppl.96.

107. Ueda Y, Ogura M, Miyakoshi S, et al. Phase 1/2 study of the WT1 peptide cancer vaccine WT4869 in patients with myelodysplastic syndrome. Cancer Sci. 2017;108:2445–2453. doi: 10.1111/cas.13409.

108. Zauderer MG, Tsao AS, Dao T, et al. A randomized phase II trial of adjuvant galinpepimut-S, WT-1 analogue peptide vaccine, after multimodality therapy for patients with malignant pleural mesothelioma.

Clin Cancer Res. 2017;23:7483–7489. doi: 10.1158/1078-0432.CCR-17-2169.

109. Villella JA, Wilson MK, Berinstein NL, et al. Determination of optimal dose and treatment schedule of the immunotherapeutic vaccine, DPX-Survivac, for combination immunotherapy treatment of ovarian, fallopian tube or peritoneal cancer (OC): a phase Ib study. J Clin Oncol. 2016;34:e14577.
110. Peace KM, Mittendorf EA, Perez SA, et al. Subgroup efficacy evaluation of the AE37 HER2 vaccine in breast cancer patients in the adjuvant setting. J Clin Oncol. 2017;35:3088.
111. Matsui H, Hazama S, Tamada K, et al. A phase I study of novel multi-HLA-binding peptides and a new combination of immune adjuvants against solid tumors. J Clin Oncol. 2017;35:3086. doi: 10.1200/JCO.2017.74.5273.
112. Yasuda T, Nishiki K, Yoshida K, et al. Cancer peptide vaccine to suppress postoperative recurrence in esophageal SCC patients with induction of antigen-specific CD8+ T cell. J Clin Oncol. 2017;35:e14635.
113. Yamaue H, Miyazawa M, Katsuda M, et al. Phase II clinical trial using novel peptide vaccine cocktail as a postoperative adjuvant treatment for surgically resected pancreatic cancer patients. J Clin Oncol. 2016;34:e14587. doi: 10.1200/jco.2016.34.4_suppl.337.
114. Slingerland M, Speetjens F, Welters M, et al. A phase I study in patients with a human papillomavirus type 16 positive oropharyngeal tumor treated with second generation synthetic long peptide vaccine conjugated to a defined adjuvant. J Clin Oncol. 2016;34:TPS3113. doi: 10.1200/JCO.2016.69.6674.
115. Ammi R, De Waele J, Willemen Y, et al. Poly(I:C) as cancer vaccine adjuvant: knocking on the door of medical breakthroughs. Pharmacol Ther. 2015;146:120–131. doi: 10.1016/j.pharmthera.2014.09.010.
116. Kyi C, Sabado RL, Saenger YM, et al. In situ, therapeutic vaccination against refractory solid cancers with intratumoral poly-ICLC: a phase I study. J Clin Oncol. 2016;34:3086.
117. Marquez Rodas I, Rodriguez-Ruiz ME, Lopez-Tarruella S, et al. First-in-human clinical trial with intratumoral BO-112 in solid malignancies: a novel immunotherapy based in double-stranded RNA (dsRNA) J Clin Oncol. 2017;35:3082.
118. Schumacher TN, Schreiber RD. Neoantigens in cancer immunotherapy.

Science. 2015;348:69–74. doi: 10.1126/science.aaa4971.

119. Sahin U, Derhovanessian E, Miller M, et al. Personalized RNA mutanome vaccines mobilize poly-specific therapeutic immunity against cancer. Nature. 2017;547:222–226. doi: 10.1038/nature23003.

120. Ott PA, Hu Z, Keskin DB, et al. An immunogenic personal neoantigen vaccine for patients with melanoma. Nature. 2017;547:217–221. doi: 10.1038/nature22991.

121. Kumar V, Patel S, Tcyganov E, Gabrilovich DI. The nature of myeloid-derived suppressor cells in the tumor microenvironment. Trends in Immunology. 2016;37:208–220. doi: 10.1016/j.it.2016.01.004.

122. Liguori M, Buracchi C, Pasqualini F, et al. Functional TRAIL receptors in monocytes and tumor-associated macrophages: a possible targeting pathway in the tumor microenvironment. Oncotarget. 2016;7:41662–41676. doi: 10.18632/oncotarget.9340.

123. Forero A, Bendell JC, Kumar P, et al. First-in-human study of the antibody DR5 agonist DS-8273a in patients with advanced solid tumors. Investig New Drugs. 2017;35:298–306. doi: 10.1007/s10637-016-0420-1.

124. Lasek W, Zagozdzon R, Jakobisiak M. Interleukin 12: still a promising candidate for tumor immunotherapy? Cancer Immunol Immunother. 2014;63:419–435. doi: 10.1007/s00262-014-1523-1.

125. Barrett JA, Cai H, Miao J, et al. IMPS-03 intratumoral regulated expression of IL-12 as a gene therapy approach to treatment of glioma. Neuro-Oncology. 2015;17:v113. doi: 10.1093/neuonc/nov217.03.

126. Chiocca EA, Yu J, Phuphanich S, et al. Expanded phase I study of intratumoral Ad-RTS-hIL-12 plus oral veledimex: tolerability and survival in recurrent glioblastoma. J Clin Oncol. 2017;35:2044. doi: 10.1200/JCO.2015.66.0613.

127. Tomala J, Kovar M. IL-2/anti-IL-2 mAb immunocomplexes: a renascence of IL-2 in cancer immunotherapy? Oncoimmunology. 2016;5:e1102829. doi: 10.1080/2162402X.2015.1102829.

128. Diab A, Tannir NM, Bernatchez C, et al. A phase 1/2 study of a novel IL-2 cytokine, NKTR-214, and nivolumab in patients with select locally advanced or metastatic solid tumors. J Clin Oncol. 2017;35:e14040.

129. Bernatchez C, Haymaker CL, Hurwitz ME, et al. Effect of a novel IL-2 cytokine immune agonist (NKTR-214) on proliferating CD8+T cells and PD-1 expression on immune cells in the tumor microenvironment

in patients with prior checkpoint therapy. J Clin Oncol. 2017;35:2545.

130. Pitt JM, Marabelle A, Eggermont A, et al. Targeting the tumor microenvironment: removing obstruction to anticancer immune responses and immunotherapy. Ann Oncol. 2016;27:1482–1492. doi: 10.1093/annonc/mdw168.
131. Moon YW, Hajjar J, Hwu P, Naing A. Targeting the indoleamine 2,3-dioxygenase pathway in cancer. J Immunother Cancer. 2015;3:51. doi: 10.1186/s40425-015-0094-9.
132. Siu LL, Gelmon K, Chu Q, et al. Abstract CT116: BMS-986205, an optimized indoleamine 2,3-dioxygenase 1 (IDO1) inhibitor, is well tolerated with potent pharmacodynamic (PD) activity, alone and in combination with nivolumab (nivo) in advanced cancers in a phase 1/2a trial. Cancer Res. 2017;77:CT116. doi: 10.1158/1538-7445. AM2017-CT116.
133. Zakharia Y, Drabick JJ, Khleif S, et al. Updates on phase1b/2 trial of the indoleamine 2,3-dioxygenase pathway (IDO) inhibitor indoximod plus checkpoint inhibitors for the treatment of unresectable stage 3 or 4 melanoma. J Clin Oncol. 2016;34:3075.
134. Bahary N, Garrido-Laguna I, Cinar P, et al. Phase 2 trial of the indoleamine 2,3-dioxygenase pathway (IDO) inhibitor indoximod plus gemcitabine/nab-paclitaxel for the treatment of metastatic pancreas cancer: interim analysis. J Clin Oncol. 2016;34:3020. doi: 10.1200/jco.2016.34.4_suppl.452.
135. Jha GG, Gupta S, Tagawa ST, et al. A phase II randomized, double-blind study of sipuleucel-T followed by IDO pathway inhibitor, indoximod, or placebo in the treatment of patients with metastatic castration resistant prostate cancer (mCRPC) J Clin Oncol. 2017;35:3066.
136. Hamid O, Bauer TM, Spira AI, et al. Safety of epacadostat 100 mg bid plus pembrolizumab 200 mg Q3W in advanced solid tumors: phase 2 data from ECHO-202/KEYNOTE-037. J Clin Oncol. 2017;35:3012.
137. Perez RP, Riese MJ, Lewis KD, et al. Epacadostat plus nivolumab in patients with advanced solid tumors: preliminary phase I/II results of ECHO-204. J Clin Oncol. 2017;35:3003.
138. Lu H. TLR agonists for cancer immunotherapy: tipping the balance between the immune stimulatory and inhibitory effects. Front Immunol. 2014;5:83.
139. Gupta S, Grilley-Olson J, Hong D, et al. Abstract CT091: safety

and pharmacodynamic activity of MEDI9197, a TLR 7/8 agonist, administered intratumorally in subjects with solid tumors. Cancer Res. 2017;77:CT091. doi: 10.1158/1538-7445.AM2017-CT091.

140. Dredge K, Brennan T, Brown MP, et al. An open-label, multi-center phase I study of the safety and tolerability of the novel immunomodulatory agent PG545 in subjects with advanced solid tumors. J Clin Oncol. 2017;35:3083. doi: 10.1200/JCO.2017.73.9326.

141. de la Torre AN, Contractor S, Castaneda I, et al. A phase I trial using local regional treatment, nonlethal irradiation, intratumoral and systemic polyinosinic-polycytidylic acid polylysine carboxymethylcellulose to treat liver cancer: in search of the abscopal effect. J Hepatocell Carcinoma. 2017;4:111–121. doi: 10.2147/JHC.S136652.

142. Ananieva E. Targeting amino acid metabolism in cancer growth and anti-tumor immune response. World J Biol Chem. 2015;6:281–289. doi: 10.4331/wjbc.v6.i4.281.

143. Papadopoulos KP, Tsai FY-C, Bauer TM, et al. CX-1158-101: a first-in-human phase 1 study of CB-1158, a small molecule inhibitor of arginase, as monotherapy and in combination with an anti-PD-1 checkpoint inhibitor in patients (pts) with solid tumors. J Clin Oncol. 2017;35:3005.

144. Zhou H, Forveille S, Sauvat A, et al. The oncolytic peptide LTX-315 triggers immunogenic cell death. Cell Death Dis. 2016;7:e2134. doi: 10.1038/cddis.2016.47.

145. Spicer JF, Baurain J-F, Awada A, et al. LTX-315, an oncolytic peptide, to convert immunogenically 'cold' tumors to 'hot' in patients with advanced or metastatic tumours: results from an ongoing phase I study. J Clin Oncol. 2017;35:3085. doi: 10.1200/JCO.2017.73.6355.

5

Recent Advances in Cancer Immunotherapy

Weijing Sun

University of Pittsburgh Cancer Institute, 5150 Centre Avenue, Fifth Floor, Pittsburgh, PA 15232, USA

Laboratory and clinical investigations of immunotherapy in cancer treatment have been underway for the past several decades by attempting to stimulate, enhance, and modulate the immune system to detect and destroy cancer cells. Although immunotherapy has shown success in a wide variety of cancers previously: BCG in superficial bladder cancer, interferon-alpha in melanoma, and high-dose interleukin in renal cell carcinoma, allogeneic stem cell transplantation in hematologic malignancies, et al., the benefits are only limited to a small population of patients, and mechanisms are less clear.

Citation: Sun, W. Recent advances in cancer immunotherapy. J Hematol Oncol 10, 96 (2017). https://doi.org/10.1186/s13045-017-0460-9

With the new advances in molecular and tumor biology, new development of technology in "-omics" and data analysis, there have been dramatic successes in the cancer immunotherapy recently. Targeting the immune checkpoint pathways to activate the host immune system against cancer cells is one such novel approach that is rapidly evolving in the recent years. Anti-CTLA4 and anti-PD1 monoclonal antibodies have taken us forward into the realm of longer survival and durable responses with the possibility of cure in a continuously increasing proportion of patients. Combination immunotherapeutic strategies and novel immunotherapeutic agents are being tested at an accelerated pace where the outlook for long-term survival benefits for the majority of patients appears brighter than ever. With the FDA's approval of anti-CTLA-4 and PD-1/PD-L1 agents, and encouraging research data of combination of immune checkpoint inhibitors and success of CAR-T in hematologic malignances, revolutionary changes of cancer therapy will be soon to come.

In this mini-review series, authors have summarized the development of immunotherapy in the field of melanoma, lung cancer, GU cancers, GI cancers, and hematological malignances. Mechanisms of different immunotherapeutic approaches were detailed. Some important issues in immunotherapy were discussed as well, e.g., identification of clinically relevant predictive and prognostic biomarkers to help define subgroups of patients who are most likely to benefit from various immunotherapies; management of immunotherapy-related toxicities and resistance; and future directions and approaches.

6

Advances on Chimeric Antigen Receptor-modified T-cell Therapy for Oncotherapy

Yanyu Pang[1] , Xiaoyang Hou[1], Chunsheng Yang[2] , Yanqun Liu[1] and Guan Jiang[1]

[1] Department of Dermatology, Affiliated Hospital of Xuzhou Medical University, Xuzhou 221002, China
[2] Department of Dermatology, Affiliated Huai'an Hospital of Xuzhou Medical University, the Second People's Hospital of Huai'an, Huai'an 223002, China

ABSTRACT

Tumor treatment is still complicated in the field of medicine. Tumor immunotherapy has been the most interesting research field in cancer therapy. Application of chimeric antigen receptor T (CAR-T) cell therapy has recently achieved excellent clinical outcome in patients, especially

Citation: Pang Y, Hou X, Yang C, Liu Y, Jiang G. Advances on chimeric antigen receptor-modified T-cell therapy for oncotherapy. Mol Cancer. 2018;17(1):91. Published 2018 May 16. doi:10.1186/s12943-018-0840-y.

those with CD19-positive hematologic malignancies. This phenomenon has induced intense interest to develop CAR-T cell therapy for cancer, especially for solid tumors. However, the performance of CAR-T cell treatment in solid tumor is not as satisfactory as that in hematologic disease. Clinical studies on some neoplasms, such as glioblastoma, ovarian cancer, and cholangiocarcinoma, have achieved desirable outcome. This review describes the history and evolution of CAR-T, generalizes the structure and preparation of CAR-T, and summarizes the latest advances on CAR-T cell therapy in different tumor types. The last section presents the current challenges and prospects of CAR-T application to provide guidance for subsequent research.

Keywords: Chimeric antigen receptor T cells, Hematological malignancies, Acute lymphoblastic leukemia, Solid tumors, Cytokine release syndrome

BACKGROUND

Despite the rapid development in medical science and the emergence of new medical technology, tumor therapy is still an intractable problem. Conventional therapies, such as surgery, chemotherapy, and radiotherapy, may provide short-term benefits but have annoying side effects due to their invasiveness and biotoxicity [1, 2]. Furthermore, multidrug resistance for chemotherapy and multiple toxicities of radiotherapy limit their curative effects [3, 4]. Therefore, new and effective treatments must be developed. Typical immunotherapy, including the use of tumor-infiltrating lymphocytes (TILs), T cell receptor (TCR)-engineered T cells, and chimeric antigen receptor (CAR) -modified T cells, has harnessed the immune system against cancer and emerged as a promising treatment modality for human malignancies [5–7]. TILs are cultured from fragments of resected tumors and have produced encouraging results in the therapy of metastatic melanoma [8] but are limited in other solid tumors due to the difficulty in isolation and expansion in vitro [9]. TCR T cell therapy is restricted to major histocompatibility complex (MHC)-expressing antigens [10]. Alternatively, CAR-T cell-based immunotherapy is independent of MHC [11, 12] and has achieved spectacular success in treating cancers, especially B-cell hematologic malignancies [13–15]. CARs are recombinant receptors containing an extracellular antigen recognition domain, a transmembrane domain, and a cytoplasmic signaling domain (such as CD3ζ, CD28, and 4-1BB). Therefore, T cells expressing CAR can recognize a wide range of cell surface antigens, including glycolipids, carbohydrates, and proteins [16], and can attack malignant cells expressing these antigens through the activation of cytoplasmic costimulation [12]. On July 1, 2014, the US Food and Drug Administration (FDA) granted the breakthrough therapy designation to CTL019, which is the CD19-directed CAR-T cell therapy designed by the University of Pennsylvania [17]. In 2017, FDA successively approved two drugs, namely, tisagenlecleucel (CTL019, Novartis) for the treatment of children and young adults with relapsing/refractory acute lymphoblastic leukemia (r/r ALL) and axicabtagene ciloleucel (KTE-C19, Kite Pharma) for the treatment of non-Hodgkin's lymphomas (NHLs) [18, 19]. CAR-T cell therapy also has some effects in other diseases, such as in non-small cell lung cancer (NSCLC), malignant pleural mesothelioma (MPM), metastatic renal cell carcinoma (mRCC), and glioblastoma (GBM) [20–23]. Although the therapeutic efficacy of CAR-T cell in these solid tumors is less effective than that in hematologic diseases, the successful trials achieved by CAR-T cells provide a concrete platform for its further development in solid tumors. In this review, we analyze the reasons why CAR-T cell therapy reaches its limits when targeting solid tumors, conclude the applications of CAR-T cell therapy in different tumors, and discuss the future perspectives on CAR-T cell therapy in cancer treatment.

CAR-T PROFILE

History and Evolution

In 1989, as the beginning of CAR-T cell, Eshhar and colleagues first generated chimeric TCR genes that can be functionally expressed in T cells and endowed the recipient T cell with antibody-type specificity to recognize and respond to the antigen in a non-MHC-restricted manner [12]. In 1993, to achieve the advantages of antibody specificity and T-cell cytotoxic activity, Eshhar combined a single-chain variable region domain (scFv) of an antibody molecule with the constant region domain of the TCR, which is usually the ζ chain of the TCR/CD3 complex [24], to construct a chimeric receptor gene and subsequently induce the T cells to express this gene by generating chimeric scFvRζ T cells [25], which are later called "first-generation CARs" that unfortunately showed limited clinical benefit because of failure in directing T cell expansion upon repeated exposure to the antigen [26] (Table (Table1).1). Hence, a co-stimulatory signaling domain–CD28 or 41BB–was added in between scFv and CD3ζ chain to form the "second-generation CARs." This domain sustainably activates the T cell to augment cytokine secretion and amplify T cell proliferation; thus, the T cells can expand upon repeated antigen exposure and show significant clinical responses [27, 28]. "Third-generation CARs" were formed by incorporating two or more costimulatory domains, usually CD28 and 41BB (CD137) or OX40 (CD134), into the same CAR. However, whether it has better clinical effect than second-generation CAR-T remains unclear [29]. In general, third-generation CARs enhance the expansion and persistence of CAR-T cells after tumor challenge [30, 31]. Notwithstanding, Haso et al. reported that in most vitro cases of anti-CD22 CARs for B-cell ALL, second-generation CAR was superior over third-generation CAR [32]. Moreover, a clinical trial of CEA CAR-T therapy on patients with carcino-embryonic antigen (CEA)-positive colorectal cancer (CRC) carried out by Zhang et al. showed that the third generation of CAR with CD28 and CD137 signaling does not show better performance than the second generation with CD28 signaling [33]. The significant phenotypic heterogeneity of solid tumors makes it difficult for CAR to recognize cancer cells. To circumvent these barriers in solid tumor lesions, Markus Chmielewski et al. developed the "fourth-generation CAR" (TRUCKs, T cells redirected for universal cytokine killing) that include the costimulatory domain and the CAR-inducible interleukin-12 (iIL-12) cassette. When CAR binds to target antigen, it activates T cell signaling; iIL-12 cassette then secretes pro-inflammatory IL-12, which can accumulate in the targeted tissue and thus recruit a second wave of immune cells (NK cells, macrophages) to initiate an attack toward those that would normally escape cancer cells due to the lack of CAR-recognized target and invisibility to CAR-T cells [34, 35].

Table 1: Summary and comparison of four generations of CAR-T therapy

CAR generations	Signal domain	Target antigen	Associated diseases	Profile	References
1st					
	CD3ζ	TAG72	Metastatic colorectal cancer	Limited persistence	[84]
	CD3ζ	FRα	Ovarian cancer	Limited persistence	[26]
	CD3ζ	L1-CAM	Metastatic neuroblastoma	Limited persistence	[85]
2nd					

	CD3ζ + CD28/CD137 (41BB)	CD19	B cell lymphomas	Enhanced expansion, persistence and anti-tumor effect	[28, 40, 86, 87]
	CD3ζ + 41BB(CD137)	IL13Rα2	GBM	Improved anti-tumor activity and T cell persistence	[22]
	CD3ζ + 41BB (CD137)	FRα	Ovarian cancer	Augmented cytokine secretion and proliferation	[88]
3rd					
	CD3ζ + CD28 + 41BB(CD137)	CD19	ALL	Superior activation and proliferation capacity	[89]
	CD3ζ + CD28 + 41BB(CD137)	PMSA	–	Promoted cytokine release, T-cell survival and tumor elimination	[90]
	CD3ζ + CD28 + CD137 (41BB)	Mesothelin	Mesothelioma	Prolonged persistence	[30]
	CD3ζ + CD28 + 41BB(CD137)	CD22	ALL	Inferior antileukemic activity	[32]
4th					
	CD3ζ + iIL-12+ co-stimulator	CEA	CEA^{+} tumors	Improved antitumor efficacy	[35]

TAG72 tumor-associated glycoprotein 72, *CEA* carcinoembryonic antigen, *IL13Rα2* IL-13 receptor α2, *FRα* folate receptor-α, *L1-CAM* L1-cell adhesion molecule, *PSMA* prostate-specific membrane antigen

Structure

CARs are engineered receptors that possess both antigen-binding and T-cell-activating functions. Based on the location of the CAR in the membrane of the T cell, CAR can be divided into three main distinct modules (Fig. 1), that is an extracellular antigen-binding domain, followed by a space region, a transmembrane domain, and the intracellular signaling domain. The antigen-binding moiety, most commonly derived from variable regions of immunoglobulin, is composed of V_H and V_L chains that are joined up by a linker to form the so-called "scFv" [12, 25]. The segment interposing between the scFv and the transmembrane domain is a "spacer domain," that is commonly the constant IgG1 hinge-CH2–CH3 Fc domain [36]. In some cases, the space domain and the transmembrane domain are derived from

CD8 [37]. The intracellular signaling domains mediating T cell activation include a CD3ζ co-receptor signaling domain derived from C-region of the TCR α and β chains [12] and one or more costimulatory domains.

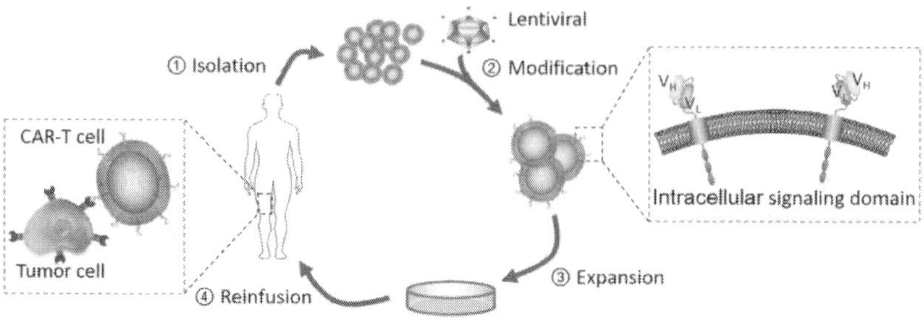

Figure 1: Structure and preparation of CAR-T cells. CARs can be divided into 3 main portions, that is, an extracellular antigen-binding domain followed by a space region, a transmembrane domain, and intracellular signaling domain. The four major steps are as follows: (1) isolation, in which PBMCs is harvested from the patient or donor's peripheral blood; (2) modification, in which the T cells were activated and CARs are transduced into the activated T cells by way of lentiviral; (3) expression, in which the modified T cells expanded ex vivo to obtain clinically relevant cell numbers; and (4) reinfusion, in which the modified T cell that has reached the desired dose were reinfused into the previously lymphocyte-depleted patient.

Preparation

The manufacturing processes of CAR-T cells are complex, and we here briefly summarize their preparation. In general, the process of CAR T-cell manufacturing and delivery involves the following major steps (Fig. (Fig.1):1): (1) Isolation: Peripheral blood mononuclear cells are harvested from the patient or donor's peripheral blood using a standard leukapheresis procedure, a process whereby blood is removed from an individual's antecubital veins, separated into select components, and the remainder of the blood returned to the individual's circulation [38]. (2) Modification: T cells were activated with CD3/CD28 magnetic beads (Dynabead) to be susceptible to viral transduction [39]. Then, CARs with the high affinity to predefined tumor antigens are transduced into these T cells by way of viral (lentiviral or retroviral) or nonviral (transposon) gene transfer systems. Lentiviral vectors and gammaretroviral vectors are currently two standard methods of viral transduction to equip T cells with a CAR [38–40]. The nonviral transduction methods usually used in engineering CAR-T cell are plasmid DNA [41] and RNA electroporation, which are also applied to T cells without pre-activation [42]. In this step, the CARs identifying tumor-associated antigens (TAAs) and, simultaneously, activating T cells were genetically expressed on the collected T cells. (3) Expansion: The CAR-T cells are expanded ex vivo to achieve the desired modified T cell dose. (4) Reinfusion: The modified T cells amplified to clinically relevant cell numbers were subsequently reinfused to the beforehand lymphocyte-depleted patient. Then, a novel CliniMACS Prodigy (Miltenyi Biotec), an automated manufacturing of CAR-T cells, has been adapted for lentiviral transduction of T cells which exhibited enormous potential [43].

THERAPEUTIC EFFECT OF CAR-T IN DIFFERENT SYSTEMS

Clinical trials to date have almost all focused on second- or third-generation CAR constructs. We here concluded the clinical applications of second- or third-generation CAR-T cells in different system tumors and summarized them in Table 2.

Table 2: Clinical trials of CAR-T therapy on different tumors

Tumors	scFv	Single domain	Dose (cells / kg or cells/ m²)	Clinical trials (phage and NCT number) (www.clinicaltrials.gov/)	Number of treated patients	Responses	Persistence	References
ALL	CD19	CD28 + CD3ζ	1.5×10^6 to 3×10^6	Phase I (NCT01044069)	5	5 CR	Uncertain[a]	[87]
ALL	CD19	CD137+ CD3ζ	1.4×10^6 to 1.2×10^7	Phase I (NCT01626495)	2	2 CR	One persisted 11 months, the other relapsed	[45]
ALL	CD19	41BB + CD3ζ	0.76×10^6 to 20.6×10^6	Phase I/ II (NCT01626495) (NCT01029366)	30	27 CR	2 to 3 months	[48]
ALL	CD19	CD28 + CD3ζ	3×10^6	Phase I (NCT01044069)	16	14 CR	2 to 3 months	[52]
ALL	CD19	CD28 + CD3ζ	1×10^6 (maximum)	Phase I (NCT01593696)	21	12 CR	Un stated	[13]
CLL	CD19	CD137+ CD3ζ	1.5×10^5	Phase I (NCT01029366)	3	3CR	10 months	[46]
CLL	CD19	CD28 + CD3ζ	$0.2–1.1 \times 10^7$	Phase I (NCT00466531)	8	1 PR	uncertain	[91]

CLL	CD19	CD28 + CD3ζ	1×10^6, 1.5×10^6, 4×10^6	Phase I (NCT01087294).	10	3 CR	< 1 month	[92]
CLL	CD19	41BB + CD3ζ	0.14×10^8 to 11×10^8	Phase I (NCT01029366)	14	4 CR, 4 PR	14 to 49 months	[49]
CLL	CD19	41BB + CD3ζ	1.6×10^7; 1.0×10^7; 1.46×10^5	Phase I (NCT01029366)	3	2 CR, 1 PR	> 6 months	[93]
CLL/ NHL/ MM	κ light chain	CD28 + CD3ζ	2×10^7, 1×10^8, 2×10^8	NCT00881920	16 (9 CLL/ NHL, 7 MM)	2 CR, 1 PR	6 weeks	[94]
CLL	CD19	CD28 + 41BB+ CD3ζ	2×10^5, 2×10^6, or 2×10^7	unstated	24	4 CR, 10 PR	6 months	[95]
MM	CD19	CD137+ CD3ζ	1×10^7 to 5×10^7	Phase I (NCT02135406)	10	Uncertain	–	[54]
Lymphomas	CD19	41BB+ CD3ζ	3.08×10^6 to 8.87×10^6	NCT02030834	28	16 CR	29.3 months	[47]
NSCLC	EGFR	CD137+ CD3ζ	0.45 to 1.09×10^7	Phase I (NCT01869166)	11	2 PR, 5 SD	2 to 8 months	[20]
CCA	EGFR	CD137+ CD3ζ	$2.2/2.1 \times 10^6$, 1.22×10^6	Phase I (NCT01869166) (NCT02541370)	1	1 PR	13 months	[63]
CRC	CEA	CD28/ CD137+ CD3ζ, CD28+ D137+ CD3ζ	1×10^5 to 1×10^8	Phase I (NCT02349724)	10	7 SD	–	[33]
SVC	MUC1	CD28+ 4-1BB+ CD3ζ	5×10^5	Phase I/II (NCT02587689)	1	Tumor necrosis	Unstated	[68]
GBM	GD2	unstated	2×10^7, 5×10^7, 1×10^8	Phase I (NCT00085930)	19	3CR	> 6 weeks	[69]
GBM	EGFRvIII	41BB + CD3ζ	1×10^7	Phase I (NCT02209376)	10	1SD	–	[70]
GBM	HER2	CD28+ CD3ζ	1×10^6 to 1×10^8	Phase I (NCT01109095)	17	1 PR, 7 SD	> 9 months	[71]
GBM	IL-13Ra2	41BB + CD3ζ	2×10^6, 10×10^6	Phase I (NCT02208362)	1	Tumor necrosis	7.5 months	[72]
Sarcoma	HER2	CD28+ CD3ζ	1×10^4 to 1×10^8	Phase I/II (NCT00902044)	19	4 SD	–	[74]

[a]Four of these patients were treated with subsequent HSCT

CAR-T for Hematological Malignancy Treatment

CAR-T cell therapy is perhaps best known for its role in the treatment of B-cell hematologic malignancies. CD19, a surface protein highly expressed on most B-lineage lymphocytes and not on normal tissue outside the B-lineage [44], is the most thoroughly studied target in all of the hematological malignancy-associated antigens, and CD19-specific CAR-T cell therapy has demonstrated enormous efficiency in inducing endurable remissions of several hematological malignancies, including ALL, chronic lymphocytic leukemia (CLL), and NHL [45–47], with complete remissions (CR) in ALL at 90% and response rates in CLL greater than 50% [48, 49]. In 2008, Till et al. reported that CD20-targeted CAR-T cells have demonstrated potential antitumor activity in treating indolent NHL and mantle cell lymphoma [50]. Later, in 2010, Kochenderfer et al. treated a patient with advanced follicular lymphoma with anti-CD19-CAR-transduced T cells, and the patient underwent a dramatic regression [51]. In 2011, Porter et al. designed a second-generation CAR-T cell in treating a patient with refractory CLL, and all three underwent CR. These findings provoked research exploring the antitumor efficacy of CD19-redirected T cells for B-cell neoplasms [46]. Based on these, in 2013, Grupp et al. extended the application of CAR-T cells to refractory B-cell ALL and established a clinical trial involving two children with ALL treated with CTL019 CAR-T cells. Surprisingly, CR was observed in both patients, demonstrating that CAR-T cells may be favorable for the treatment of patients with refractory ALL. However, cytokine-release syndrome (CRS) was also observed [45]. Subsequently, in 2014, Maude et al. conducted pilot clinical trials of 30 patients (children and adults) with r/r CD19+ ALL, in which infused autologous T cells are transduced with a CTL019. CR was achieved in 27 patients. CTL019 was effective in treating r/r ALL, even in stem cell transplantation-failed patients. Nevertheless, CRS was developed in all the patients [48]. Furthermore, Davila et al. treated 16 adult patients with r/r ALL with 19-28z CAR-T cells specific to the CD19 antigen and achieved a promising outcome, with overall CR rate of 88% (14/16). CRS, which may be related to a systemic inflammatory process induced by the reaction between infused CAR T cells and the targeted CD19 antigen, is almost inevitable. Therefore, the diagnostic criteria for severe CRS were defined, and serum C-reactive protein can serve as a reliable indicator for CRS severity [52]. Although CD19 is not an ideal antigen in multiple myeloma (MM), for its low expression in MM [53], Garfall et al. still reported that CAR-T cell therapy in conjunction with autologous transplantation has achieved durable CR in a patient with advanced MM

[54]. These findings provided a road map for application of CAR-T cells in solid tumors.

CAR-T for Solid Tumor Treatment

The unprecedented success of CAR-T cell therapy in hematological malignancy fostered the enthusiasm to expand this technology to solid tumors. However, this therapy encountered some difficulties in application for solid malignancy. The reasons for this phenomenon are as follows: (1) lack of eligible, effective targets such as CD19 because most target antigens are more or less expressed in normal tissues; (2) hostile immunosuppressive microenvironment of solid tumors that affect the T cells; and (3) heterogeneity of solid tumors. Although the exploration of CAR-T cell treatment in solid tumors is not as definitive as in the research of hematological malignancy, some studies have achieved promising outcomes. Here, we introduced some diseases in which the CAR-T-cell treatment has exhibited benign clinical responses.

NSCLC

Advanced strategies, including surgery, radiotherapy, chemotherapy, and targeted therapy, have improved the survival in patients with NSCLC. Nevertheless, the 5-year survival rate of late-stage NSCLC is still unsatisfactory [55]. The breakthrough treatment of immunotherapy with CAR-T cells in hematology raised the possibility of their use in NSCLC. In 2016, Feng et al. first studied the safety and feasibility of epidermal growth factor receptor (EGFR)-targeted CAR-T cell therapy in treating 11 patients with advanced r/r NSCLC. Two patients obtained partial response (PR), and five had stable diseases (SD) after the infusion of CART-EGFR cells with mild side effects--mild skin toxicity, nausea, vomiting, dyspnea and hypotension [20]. In addition, other TAAs, like erythropoietin-producing hepatocellular carcinoma A2 (EphA2) (Li et al. 2017) [56], prostate stem cell antigen (PSCA), and mucin 1 (MUC1) (Wei et al. 2017) [57], have also been detected in NSCLC and confirmed to be promising targeting antigen for CAR-T cells. These antigen-targeted CAR-T cells have been observed to cause tumor cell lysis in vitro exerting antitumor activity in xenograft mouse models. Furthermore, targeting the combination of PSCA and MUC1 can further enhance the antitumor efficacy of CAR T cells [57].

MPM

MPM is an aggressive malignancy with a median survival of less than one year [58]. Mesothelin plays an important role in screening and detecting the progression of MPM [59]. Basing on this finding, some researchers considered that CAR-T targeting mesothelin can treat MPM. A phase I clinical trial conducted at the University of Pennsylvania was designed to evaluate the manufacturing feasibility and safety of mRNA-transduced CAR T cells that target mesothelin (CART-meso cells) in patients with advanced MPM. In this study, CART-meso cells showed potent antitumor activity with no distinct on target/off-tumor toxicities (pleuritis, pericarditis, or peritonitis) [60].

Digestive System Neoplasm

Cholangiocarcinoma (CCA) is a relatively rare and aggressive malignancy of the biliary tract and is characterized by late diagnosis and poor outcomes [61]. Complete surgical resection can be used as treatment. However, most of the patients will eventually relapse because of the delayed diagnosis and advanced stage of the disease [62]. In 2017, Feng et al. applied EGFR- and CD133-specific CAR-T sequential treatments as CAR-T cocktail immunotherapy for patients with advanced unresectable/metastatic CCA. An 8.5-month PR from the initial CAR-T-EGFR treatment and another 4.5-month PR from the subsequent CD133-specific CAR-T immunotherapy were obtained. However, the epidermal and endothelial damages caused by the infusion of CAR-T cells cannot be disregarded, thereby requiring further investigation [63].

A phase I clinical trial conducted at the University of Pennsylvania was designed to evaluate the manufacturing feasibility and safety of CART-meso cells in patients with advanced MPM and explore the antitumor effect of CART-meso cells in patients with pancreatic cancer. The results showed the antitumor activity. CART-meso cells were also detected in primary and metastatic tumor sites by collecting ascites and conducting a tumor biopsy [60].

Zhang et al. established a clinical trial of CEA CAR-T therapy of 10 patients with CRC by systemic delivery through intravenous (IV) infusion to evaluate its safety and efficacy. Out of the 10 patients, 7 patients who experienced progressive disease in the previous treatments have SD after the CAR-T therapy. Moreover, severe adverse events related to CAR-T therapy are not observed [33].

Genitourinary System Diseases

Epithelial ovarian cancer (EOC) remains to be the most mortal of all gynecological malignancies mainly due to its subtle nature. Despite the fact that most patients with EOC yield a good clinical response following current advanced therapy, almost all patients will ultimately relapse and eventually develop drug resistance [64]. The survival of patients with EOC is positively related to the presence of TILs, which play a significant role in adoptive T-cell therapy [65]. MUC16, a well-known ovarian tumor antigen, is overexpressed by a majority of EOC but at a low level on normal tissues [66]. On the basis of this rationale, Brentjens et al. developed T cells expressing MUC16 to treat EOC. Moreover, to overcome the hostile tumor environment, they co-expressed IL-12 on T cells. Hence, a clinical trial testing the safety of IV and intraperitoneal infusion of genetically modified autologous T cells expressing MUC16 and secreting IL-12 in patients with EOC was conducted. The result demonstrated that the intraperitoneal injections of CAR T cells are superior to that of IV alone [67].

You et al. launched a phase I clinical trial to evaluate the ability of engineered CAR-T cells targeting MUC1 to treat patients with seminal vesicle cancer (SVC). To suppress the unfavorable tumor microenvironment, they induced IL-12 co-expression and constructed two anti-MUC1 CAR-T cell lines, that is, SM3-CAR (co-expressing IL-12) and pSM3-CAR (without IL-12). These two types of CAR-T cells were injected intratumorally into two separate metastatic lesions of the same patient with $MUC1^+$ SVC as part of an interventional treatment strategy. The results showed tumor necrosis induced by pSM3-CAR is more evident than that by SM3-CAR, without significant side effects [68].

GBM

CAR-T cell has also been explored in recent years to treat central nervous system cancers. In 2011, Louis et al. conducted a clinical trial of GD2-specific CAR-T therapy in 19 patients with high-risk neuroblastoma. Three patients had a CR to CAR-T cell infusion, with only slight fever and light-to-moderate local pain being observed [69]. In 2017, a clinical trial of IV administration of EGFRvIII-specific CAR-T cells for the treatment of 10 patients with refractory GBM was established at the University of Pennsylvania. The infusion of CAR-T cells was feasible and safe, without evident off-tumor toxicity or CRS [70]. Furthermore, an open-labeled phase 1 dose-escalation study was conducted at the Baylor College of Medicine,

Houston Methodist Hospital, and Texas Children's Hospital to evaluate the safety and anti-GBM activity of HER2-specific CAR-modified virus-specific T cells in patients with progressive GBM. The results showed that the infusions are well tolerated, with no dose-limiting toxic effects. Moreover, 1 patient showed a PR for more than 9 months, whereas 7 patients had SD for 8 weeks to 29 months [71]. In addition, Brown et al. initiated a clinical trial with one patient with relapsing GBM, who received administration of IL13Rα2 targeted IL13BBζ–CAR-T cells, and regression of tumors was observed and persisted for 7.5 months after the administration of CAR-T cell therapy [72].

Sarcomas

Sarcoma, which can be located anywhere in the body, is usually treated with surgical resection, with or without radiotherapy, and chemotherapy. However, patients with advanced stage sarcomas still have poor prognosis [73]. Hence, several researchers speculated that CAR-T-cell treatment may benefit patients with sarcoma. In 2015, Ahmed et al. designed a phase I/ II clinical study to evaluate the safety and efficacy of HER2-specific CAR-T cells in patients with r/r HER2-positive sarcoma. A total of 19 patients were enrolled in this research, and they received escalating doses of HER2-specific CAR-T cells. Although no CR were observed, 4 out of 17 patients that can be evaluated have SD for 12 weeks to 14 months [74].

CONCLUSION

In this review, we summarized the current clinical studies on CAR-T treatment of hematologic diseases and solid tumors. Clinical outcomes of CAR-T cell therapy in patients with hematologic malignancies have been encouraging. However, in patients with solid tumors, the outcomes have been discouraging, nevertheless, not gloomy. CAR-T cell therapy, as a promising treatment, has the following advantages: (1) binding surface antigen of tumors in non-MHC restriction manner; (2) recognizing multiple antigens simultaneously; and (3) obtaining a large number of CAR-T cells ex vivo in a short term. CRS is an ineluctable complication of CAR-T-cell therapy on the basis of the clinical trials of hematological malignancies mentioned above. The manifestations of CRS include fevers, hypotension, nausea, myalgias, and neurologic dysfunction. When CRS is severe, vasopressors, mechanical ventilation, antiepileptics, and hemodialysis may be required [52]. Fortunately, researchers can now control most cases of CRS with an anti-interleukin 6 antibody, such as tocilizumab, which was approved by the FDA for the treatment of CAR-T-cell therapy-induced CRS in August 2017 [75]. The CRS is not common in solid tumors treated by CAR-T cells; however, on target/off-tumor toxicity has become common due to unavoidable expression, to some extent, of target antigens in normal tissues [60]. This phenomenon could be solved by manufacturing CAR-T cells with dual antigen specificity or switchable dual-receptor [76, 77] or by transfecting T cells with mRNA encoding CAR to reduce their half-life; these CAR-T cells can be repeated administered, and the toxicity to normal tissues can be mitigated [78]. Moreover, the hostile immunosuppressive microenvironment is one of the major challenges in CAR-T-cell treatment of solid tumors. The tumor microenvironment is a complex and dense fibrotic matrix network composed of malignant and nonmalignant cells, in which the infiltrated CAR-T cells can be inhibited by immunosuppressive cells such as regulatory T cells (Tregs) and myeloid-derived suppressive cells [79]. In a recent study conducted by Chen et al., CARs were engineered to target a bunch of soluble ligands, including TGF-β, an otherwise immunosuppressive factor in a variety of solid tumors, and demonstrated the ability to effectively convert TGF-β from a potent immunosuppressive cytokine to a strong stimulant for the primary human T cells [80]. Another research by Batchu et al. also shed light on the immunosuppressive microenvironment in solid tumors. They discovered that suppressing interleukin-10, an immune inhibitory cytokine secreted by Tregs and pancreatic cancer cells, can reverse the negative effect of the tumor microenvironment on mesothelin-CAR-T cells in pancreatic cancer in vitro [81]. In addition, solid tumors have relatively limited body distribution and are concrete compared with hematological malignancies. Hence, we can hypothesize that in some cases, the regional delivery of CAR-T cells may be immensely superior to systemic administration. Several studies reported that local injection, such as intrapleural administration and local intracranial delivery, show greater potential than IV injection [23, 82]. These findings indicated that CAR-T-cell therapy can gain momentum to break through the restriction of tumor microenvironment in treating solid malignancies. In the aggregates, CAR-T-cell therapy is a promising strategy against neoplasms. Several key points must be considered to translate the success of CAR-T cell therapy to extensive solid tumors. These widely acknowledged key points include finding a specific antigen or engineering multiple antigen-targeted CAR-T cells, directly targeting the constituent of immunosuppressive microenvironment, and creating a suitable tumor microenvironment [83]. In addition to these factors, replacing IV infusion with regional delivery, transfecting T cells with mRNA encoding CAR, and combining T cells with oncolytic viruses or immune-checkpoint blockade to bolster the potency of CAR-T cells can also be considered.

ACKNOWLEDGEMENTS

This work was supported by the National Natural Science Foundation of China [No.81572976], the China Postdoctoral Science Foundation [Nos.2016M590505, 2017T100407], the Jiangsu Provincial Medical Talent Foundation, the Postgraduate Research & Practice Innovation Program of Jiangsu Province [No. KYCX17_1715, No. KYCX17_1718].

AUTHORS' CONTRIBUTIONS

YYP and YXH provided direction and guidance throughout the preparation of this manuscript. SCY collected and prepared the related literature. YYP drafted the manuscript. YXH, GJ and QYL reviewed and made significant revisions to the manuscript. All authors have read and approved the final manuscript.

REFERENCES

1. Ruhlmann C, Iversen T, Okera M, Muhic A, Kristensen G, Feyer P, Hansen O, Herrstedt J. Multinational study exploring patients' perceptions of side-effects induced by chemo-radiotherapy. Radiother Oncol. 2015;117:333–337. doi: 10.1016/j.radonc.2015.09.014.
2. Neuman B, Ailon T, Scheer J, Klineberg E, Sciubba D, Jain A, Zebala L, Passias P, Daniels A, Burton D, et al. Development and validation of a novel adult spinal deformity surgical invasiveness score: analysis of 464 patients. Neurosurgery. 2017. 10.1093/neuros/nyx303.
3. Yin Q, Shen J, Zhang Z, Yu H, Li Y. Reversal of multidrug resistance by stimuli-responsive drug delivery systems for therapy of tumor. Adv Drug Deliv Rev. 2013;65:1699–1715. doi: 10.1016/j.addr.2013.04.011.
4. Li R, Pu X, Chang J, Ye Y, Komaki R, Minna J, Roth J, Han B, Wu X. MiRNA-related genetic variations associated with radiotherapy-induced toxicities in patients with locally advanced non-small cell lung Cancer. PLoS One. 2016;11:e0150467. doi: 10.1371/journal.pone.0150467.
5. Crompton J, Klemen N, Kammula U. Metastasectomy for tumor-infiltrating lymphocytes: an emerging operative indication in surgical oncology. Ann Surg Oncol. 2017;
6. Kunert A, Obenaus M, Lamers C, Blankenstein T, Debets R. T-cell receptors for clinical therapy: in vitro assessment of toxicity risk. Clin Cancer Res. 2017;23:6012–6020. doi: 10.1158/1078-0432.CCR-17-1012.
7. Hay K, Turtle C. Chimeric antigen receptor (CAR) T cells: lessons learned from targeting of CD19 in B-cell malignancies. Drugs. 2017;77:237–245. doi: 10.1007/s40265-017-0690-8.
8. Wu R, Forget M, Chacon J, Bernatchez C, Haymaker C, Chen J, Hwu P, Radvanyi L. Adoptive T-cell therapy using autologous tumor-infiltrating lymphocytes for metastatic melanoma: current status and future outlook. Cancer J. 2012;18:160–175. doi: 10.1097/PPO.0b013e31824d4465.
9. Feldman S, Assadipour Y, Kriley I, Goff S, Rosenberg S. Adoptive cell therapy--tumor-infiltrating lymphocytes, T-cell receptors, and chimeric antigen receptors. Semin Oncol. 2015;42:626–639. doi: 10.1053/j.seminoncol.2015.05.005.

10. Mazza C, Malissen B. What guides MHC-restricted TCR recognition? Semin Immunol. 2007;19:225–235. doi: 10.1016/j.smim.2007.03.003.

11. Harris D, Kranz D, Cell Therapies AT, Comparison A. Of T cell receptors and chimeric antigen receptors. Trends Pharmacol Sci. 2016;37:220–230. doi: 10.1016/j.tips.2015.11.004.

12. Gross G, Waks T, Eshhar Z. Expression of immunoglobulin-T-cell receptor chimeric molecules as functional receptors with antibody-type specificity. Proc Natl Acad Sci U S A. 1989;86:10024–10028. doi: 10.1073/pnas.86.24.10024.

13. Lee DW, Kochenderfer JN, Stetler-Stevenson M, Cui YK, Delbrook C, Feldman SA, Fry TJ, Orentas R, Sabatino M, Shah NN, et al. T cells expressing CD19 chimeric antigen receptors for acute lymphoblastic leukaemia in children and young adults: a phase 1 dose-escalation trial. Lancet. 2015;385:517–528. doi: 10.1016/S0140-6736(14)61403-3.

14. Ghorashian S, Pule M, Amrolia P. CD19 chimeric antigen receptor T cell therapy for haematological malignancies. Br J Haematol. 2015;169:463–478. doi: 10.1111/bjh.13340.

15. Park J, Geyer M, Brentjens R. CD19-targeted CAR T-cell therapeutics for hematologic malignancies: interpreting clinical outcomes to date. Blood. 2016;127:3312–3320. doi: 10.1182/blood-2016-02-629063.

16. Morello A, Sadelain M, Adusumilli P. Mesothelin-targeted CARs: driving T cells to solid tumors. Cancer Discov. 2016;6:133–146. doi: 10.1158/2159-8290.CD-15-0583.

17. Gill S, June C. Going viral: chimeric antigen receptor T-cell therapy for hematological malignancies. Immunol Rev. 2015;263:68–89. doi: 10.1111/imr.12243.

18. Ruella M, Kenderian S. Next-generation chimeric antigen receptor T-cell therapy: going off the shelf. BioDrugs. 2017;31:473–481. doi: 10.1007/s40259-017-0247-0.

19. Neelapu S, Locke F, Bartlett N, Lekakis L, Miklos D, Jacobson C, Braunschweig I, Oluwole O, Siddiqi T, Lin Y, et al. Axicabtagene Ciloleucel CAR T-cell therapy in refractory large B-cell lymphoma. N Engl J Med. 2017;377:2531–2544. doi: 10.1056/NEJMoa1707447.

20. Feng K, Guo Y, Dai H, Wang Y, Li X, Jia H, Han W. Chimeric antigen receptor-modified T cells for the immunotherapy of patients with EGFR-expressing advanced relapsed/refractory non-small cell lung cancer. Sci China Life Sci. 2016;59:468–479. doi: 10.1007/s11427-016-5023-8.

21. Lamers C, Klaver Y, Gratama J, Sleijfer S, Debets R. Treatment of metastatic renal cell carcinoma (mRCC) with CAIX CAR-engineered T-cells-a completed study overview. Biochem Soc Trans. 2016;44:951–959. doi: 10.1042/BST20160037.

22. Brown CE, Aguilar B, Starr R, Yang X, Chang WC, Weng L, Chang B, Sarkissian A, Brito A, Sanchez JF, et al. Optimization of IL13Ralpha2-targeted chimeric antigen receptor T cells for improved anti-tumor efficacy against glioblastoma. Mol Ther. 2017;

23. Adusumilli P, Cherkassky L, Villena-Vargas J, Colovos C, Servais E, Plotkin J, Jones D, Sadelain M. Regional delivery of mesothelin-targeted CAR T cell therapy generates potent and long-lasting CD4-dependent tumor immunity. Sci Transl Med. 2014;6:261ra151. doi: 10.1126/scitranslmed.3010162.

24. Letourneur F, Klausner R. T-cell and basophil activation through the cytoplasmic tail of T-cell-receptor zeta family proteins. Proc Natl Acad Sci U S A. 1991;88:8905–8909. doi: 10.1073/pnas.88.20.8905.

25. Eshhar Z, Waks T, Gross G, Schindler D. Specific activation and targeting of cytotoxic lymphocytes through chimeric single chains consisting of antibody-binding domains and the gamma or zeta subunits of the immunoglobulin and T-cell receptors. Proc Natl Acad Sci U S A. 1993;90:720–724. doi: 10.1073/pnas.90.2.720.

26. Kershaw M, Westwood J, Parker L, Wang G, Eshhar Z, Mavroukakis S, White D, Wunderlich J, Canevari S, Rogers-Freezer L, et al. A phase I study on adoptive immunotherapy using gene-modified T cells for ovarian cancer. Clin Cancer Res. 2006;12:6106–6115. doi: 10.1158/1078-0432.CCR-06-1183.

27. van der Stegen S, Hamieh M, Sadelain M. The pharmacology of second-generation chimeric antigen receptors. Nat Rev Drug Discov. 2015;14:499–509. doi: 10.1038/nrd4597.

28. Savoldo B, Ramos C, Liu E, Mims M, Keating M, Carrum G, Kamble R, Bollard C, Gee A, Mei Z, et al. CD28 costimulation improves expansion and persistence of chimeric antigen receptor-modified T cells in lymphoma patients. J Clin Invest. 2011;121:1822–1826. doi: 10.1172/JCI46110.

29. Till B, Jensen M, Wang J, Qian X, Gopal A, Maloney D, Lindgren C, Lin Y, Pagel J, Budde L, et al. CD20-specific adoptive immunotherapy for lymphoma using a chimeric antigen receptor with both CD28 and 4-1BB domains: pilot clinical trial results. Blood. 2012;119:3940–

3950. doi: 10.1182/blood-2011-10-387969.

30. Carpenito C, Milone MC, Hassan R, Simonet JC, Lakhal M, Suhoski MM, Varela-Rohena A, Haines KM, Heitjan DF, Albelda SM, et al. Control of large, established tumor xenografts with genetically retargeted human T cells containing CD28 and CD137 domains. Proc Natl Acad Sci U S A. 2009;106:3360–3365. doi: 10.1073/pnas.0813101106.
31. Wang J, Jensen M, Lin Y, Sui X, Chen E, Lindgren C, Till B, Raubitschek A, Forman S, Qian X, et al. Optimizing adoptive polyclonal T cell immunotherapy of lymphomas, using a chimeric T cell receptor possessing CD28 and CD137 costimulatory domains. Hum Gene Ther. 2007;18:712–725. doi: 10.1089/hum.2007.028.
32. Haso W, Lee D, Shah N, Stetler-Stevenson M, Yuan C, Pastan I, Dimitrov D, Morgan R, FitzGerald D, Barrett D, et al. Anti-CD22-chimeric antigen receptors targeting B-cell precursor acute lymphoblastic leukemia. Blood. 2013;121:1165–1174. doi: 10.1182/blood-2012-06-438002.
33. Zhang C, Wang Z, Yang Z, Wang M, Li S, Li Y, Zhang R, Xiong Z, Wei Z, Shen J, et al. Phase I escalating-dose trial of CAR-T therapy targeting CEA(+) metastatic colorectal cancers. Mol Ther. 2017;25:1248–1258. doi: 10.1016/j.ymthe.2017.03.010.
34. Chmielewski M, Hombach A, Abken H. Of CARs and TRUCKs: chimeric antigen receptor (CAR) T cells engineered with an inducible cytokine to modulate the tumor stroma. Immunol Rev. 2014;257:83–90. doi: 10.1111/imr.12125.
35. Chmielewski M, Kopecky C, Hombach A, Abken H. IL-12 release by engineered T cells expressing chimeric antigen receptors can effectively muster an antigen-independent macrophage response on tumor cells that have shut down tumor antigen expression. Cancer Res. 2011;71:5697–5706. doi: 10.1158/0008-5472.CAN-11-0103.
36. Hombach A, Hombach A, Abken H. Adoptive immunotherapy with genetically engineered T cells: modification of the IgG1 Fc 'spacer' domain in the extracellular moiety of chimeric antigen receptors avoids 'off-target' activation and unintended initiation of an innate immune response. Gene Ther. 2010;17:1206–1213. doi: 10.1038/gt.2010.91.
37. D'Aloia M, Caratelli S, Palumbo C, Battella S, Arriga R, Lauro D, Palmieri G, Sconocchia G, Alimandi M. T lymphocytes engineered to express a CD16-chimeric antigen receptor redirect T-cell immune

responses against immunoglobulin G-opsonized target cells. Cytotherapy. 2016;18:278–290. doi: 10.1016/j.jcyt.2015.10.014.

38. McGuirk J, Waller EK, Qayed M, Abhyankar S, Ericson S, Holman P, Keir C, Myers GD. Building blocks for institutional preparation of CTL019 delivery. Cytotherapy. 2017;19:1015–1024. doi: 10.1016/j.jcyt.2017.06.001.

39. Hollyman D, Stefanski J, Przybylowski M, Bartido S, Borquez-Ojeda O, Taylor C, Yeh R, Capacio V, Olszewska M, Hosey J, et al. Manufacturing validation of biologically functional T cells targeted to CD19 antigen for autologous adoptive cell therapy. J Immunother. 2009;32:169–180. doi: 10.1097/CJI.0b013e318194a6e8.

40. Milone MC, Fish JD, Carpenito C, Carroll RG, Binder GK, Teachey D, Samanta M, Lakhal M, Gloss B, Danet-Desnoyers G, et al. Chimeric receptors containing CD137 signal transduction domains mediate enhanced survival of T cells and increased antileukemic efficacy in vivo. Mol Ther. 2009;17:1453–1464. doi: 10.1038/mt.2009.83.

41. Schmeer M, Buchholz T, Schleef M. Plasmid DNA manufacturing for indirect and direct clinical applications. Hum Gene Ther. 2017;28:856–861. doi: 10.1089/hum.2017.159.

42. Riet T, Holzinger A, Dörrie J, Schaft N, Schuler G, Abken H, Nonviral RNA. Transfection to transiently modify T cells with chimeric antigen receptors for adoptive therapy. Methods Mol Biol. 2013;969:187–201. doi: 10.1007/978-1-62703-260-5_12.

43. Mock U, Nickolay L, Philip B, Cheung GW, Zhan H, Johnston IC, Kaiser AD, Peggs K, Pule M, Thrasher AJ, Qasim W. Automated manufacturing of chimeric antigen receptor T cells for adoptive immunotherapy using CliniMACS prodigy. Cytotherapy. 2016;18:1002–1011. doi: 10.1016/j.jcyt.2016.05.009.

44. Scheuermann R, Racila E. CD19 antigen in leukemia and lymphoma diagnosis and immunotherapy. Leuk Lymphoma. 1995;18:385–397. doi: 10.3109/10428199509059636.

45. Grupp S, Kalos M, Barrett D, Aplenc R, Porter D, Rheingold S, Teachey D, Chew A, Hauck B, Wright J, et al. Chimeric antigen receptor-modified T cells for acute lymphoid leukemia. N Engl J Med. 2013;368:1509–1518. doi: 10.1056/NEJMoa1215134.

46. Porter D, Levine B, Kalos M, Bagg A, June C. Chimeric antigen receptor-modified T cells in chronic lymphoid leukemia. N Engl J

Med. 2011;365:725–733. doi: 10.1056/NEJMoa1103849.

47. Schuster SJ, Svoboda J, Chong EA, Nasta SD, Mato AR, Anak O, Brogdon JL, Pruteanu-Malinici I, Bhoj V, Landsburg D, et al. Chimeric antigen receptor T cells in refractory B-cell lymphomas. N Engl J Med. 2017;

48. Maude S, Frey N, Shaw P, Aplenc R, Barrett D, Bunin N, Chew A, Gonzalez V, Zheng Z, Lacey S, et al. Chimeric antigen receptor T cells for sustained remissions in leukemia. N Engl J Med. 2014;371:1507–1517. doi: 10.1056/NEJMoa1407222.

49. Porter D, Hwang W, Frey N, Lacey S, Shaw P, Loren A, Bagg A, Marcucci K, Shen A, Gonzalez V, et al. Chimeric antigen receptor T cells persist and induce sustained remissions in relapsed refractory chronic lymphocytic leukemia. Sci Transl Med. 2015;7:303ra139. doi: 10.1126/scitranslmed.aac5415.

50. Till B, Jensen M, Wang J, Chen E, Wood B, Greisman H, Qian X, James S, Raubitschek A, Forman S, et al. Adoptive immunotherapy for indolent non-Hodgkin lymphoma and mantle cell lymphoma using genetically modified autologous CD20-specific T cells. Blood. 2008;112:2261–2271. doi: 10.1182/blood-2007-12-128843.

51. Kochenderfer JN, Wilson WH, Janik JE, Dudley ME, Stetler-Stevenson M, Feldman SA, Maric I, Raffeld M, Nathan DA, Lanier BJ, et al. Eradication of B-lineage cells and regression of lymphoma in a patient treated with autologous T cells genetically engineered to recognize CD19. Blood. 2010;116:4099–4102. doi: 10.1182/blood-2010-04-281931.

52. Davila ML, Riviere I, Wang X, Bartido S, Park J, Curran K, Chung SS, Stefanski J, Borquez-Ojeda O, Olszewska M, et al. Efficacy and toxicity management of 19-28z CAR T cell therapy in B cell acute lymphoblastic leukemia. Sci Transl Med. 2014;6:224ra225. doi: 10.1126/scitranslmed.3008226.

53. Tembhare P, Yuan C, Venzon D, Braylan R, Korde N, Manasanch E, Zuchlinsky D, Calvo K, Kurlander R, Bhutani M, et al. Flow cytometric differentiation of abnormal and normal plasma cells in the bone marrow in patients with multiple myeloma and its precursor diseases. Leuk Res. 2014;38:371–376. doi: 10.1016/j.leukres.2013.12.007.

54. Garfall A, Maus M, Hwang W, Lacey S, Mahnke Y, Melenhorst J, Zheng Z, Vogl D, Cohen A, Weiss B, et al. Chimeric antigen receptor T cells

against CD19 for multiple myeloma. N Engl J Med. 2015;373:1040–1047. doi: 10.1056/NEJMoa1504542.

55. Schild SE, Hillman SL, Tan AD, Ross HJ, McGinnis WL, Garces YA, Graham DL, Adjei AA, Jett JR. Long-term results of a trial of concurrent chemotherapy and escalating doses of radiation for Unresectable non-small cell lung Cancer: NCCTG N0028 (alliance) J Thorac Oncol. 2017;12:697–703. doi: 10.1016/j.jtho.2016.12.021.

56. Li N, Liu S, Sun M, Chen W, Xu X, Zeng Z, Tang Y, Dong Y, Chang A, Zhao Q. Chimeric antigen receptor-modified T cells redirected to EphA2 for the immunotherapy of non-small cell lung Cancer. Transl Oncol. 2017;11:11–17. doi: 10.1016/j.tranon.2017.10.009.

57. Wei X, Lai Y, Li J, Qin L, Xu Y, Zhao R, Li B, Lin S, Wang S, Wu Q, et al. PSCA and MUC1 in non-small-cell lung cancer as targets of chimeric antigen receptor T cells. Oncoimmunology. 2017;6:e1284722. doi: 10.1080/2162402X.2017.1284722.

58. Barone E, Gemignani F, Landi S. Overexpressed genes in malignant pleural mesothelioma: implications in clinical management. J Thorac Dis. 2018;10:S369–S382. doi: 10.21037/jtd.2017.10.158.

59. de Fonseka D, Arnold DT, Stadon L, Morley A, Keenan E, Darby M, Armstrong L, Virgo P, Maskell NA: A prospective study to investigate the role of serial serum mesothelin in monitoring mesothelioma. BMC Cancer 2018, 18:199.

60. Beatty GL, Haas AR, Maus MV, Torigian DA, Soulen MC, Plesa G, Chew A, Zhao Y, Levine BL, Albelda SM, et al. Mesothelin-specific chimeric antigen receptor mRNA-engineered T cells induce anti-tumor activity in solid malignancies. Cancer Immunol Res. 2014;2:112–120. doi: 10.1158/2326-6066.CIR-13-0170.

61. Razumilava N, Gores GJ. Cholangiocarcinoma. Lancet. 2014;383:2168–2179. doi: 10.1016/S0140-6736(13)61903-0.

62. Radtke A, Konigsrainer A. Surgical therapy of cholangiocarcinoma. Visc Med. 2016;32:422–426. doi: 10.1159/000452921.

63. Feng KC, Guo YL, Liu Y, Dai HR, Wang Y, Lv HY, Huang JH, Yang QM, Han WD. Cocktail treatment with EGFR-specific and CD133-specific chimeric antigen receptor-modified T cells in a patient with advanced cholangiocarcinoma. J Hematol Oncol. 2017;10:4. doi: 10.1186/s13045-016-0378-7.

64. Padmakumar S, Parayath N, Leslie F, Nair SV, Menon D, Amiji MM.

Intraperitoneal chemotherapy for ovarian cancer using sustained-release implantable devices. Expert Opin Drug Deliv. 2018:1–14.

65. Leffers N, Gooden MJ, de Jong RA, Hoogeboom BN, ten Hoor KA, Hollema H, Boezen HM, van der Zee AG, Daemen T, Nijman HW. Prognostic significance of tumor-infiltrating T-lymphocytes in primary and metastatic lesions of advanced stage ovarian cancer. Cancer Immunol Immunother. 2009;58:449–459. doi: 10.1007/s00262-008-0583-5.

66. Cheon DJ, Wang Y, Deng JM, Lu Z, Xiao L, Chen CM, Bast RC, Behringer RR. CA125/MUC16 is dispensable for mouse development and reproduction. PLoS One. 2009;4:e4675. doi: 10.1371/journal.pone.0004675.

67. Koneru M, O'Cearbhaill R, Pendharkar S, Spriggs DR, Brentjens RJ. A phase I clinical trial of adoptive T cell therapy using IL-12 secreting MUC-16(ecto) directed chimeric antigen receptors for recurrent ovarian cancer. J Transl Med. 2015;13:102. doi: 10.1186/s12967-015-0460-x.

68. You F, Jiang L, Zhang B, Lu Q, Zhou Q, Liao X, Wu H, Du K, Zhu Y, Meng H, et al. Phase 1 clinical trial demonstrated that MUC1 positive metastatic seminal vesicle cancer can be effectively eradicated by modified anti-MUC1 chimeric antigen receptor transduced T cells. Sci China Life Sci. 2016;59:386–397. doi: 10.1007/s11427-016-5024-7.

69. Louis C, Savoldo B, Dotti G, Pule M, Yvon E, Myers G, Rossig C, Russell H, Diouf O, Liu E, et al. Antitumor activity and long-term fate of chimeric antigen receptor-positive T cells in patients with neuroblastoma. Blood. 2011;118:6050–6056. doi: 10.1182/blood-2011-05-354449.

70. O'Rourke DM, Nasrallah MP, Desai A, Melenhorst JJ, Mansfield K, Morrissette JJD, Martinez-Lage M, Brem S, Maloney E, Shen A, et al. A single dose of peripherally infused EGFRvIII-directed CAR T cells mediates antigen loss and induces adaptive resistance in patients with recurrent glioblastoma. Sci Transl Med. 2017;9

71. Ahmed N, Brawley V, Hegde M, Bielamowicz K, Kalra M, Landi D, Robertson C, Gray TL, Diouf O, Wakefield A, et al. HER2-specific chimeric antigen receptor-modified virus-specific T cells for progressive glioblastoma: a phase 1 dose-escalation trial. JAMA Oncol. 2017;3:1094–1101. doi: 10.1001/jamaoncol.2017.0184.

72. Brown C, Alizadeh D, Starr R, Weng L, Wagner J, Naranjo A,

Ostberg J, Blanchard M, Kilpatrick J, Simpson J, et al. Regression of glioblastoma after chimeric antigen receptor T-cell therapy. N Engl J Med. 2016;375:2561–2569. doi: 10.1056/NEJMoa1610497.

73. Linch M, Miah A, Thway K, Judson I, Benson C. Systemic treatment of soft-tissue sarcoma-gold standard and novel therapies. Nat Rev Clin Oncol. 2014;11:187–202. doi: 10.1038/nrclinonc.2014.26.

74. Ahmed N, Brawley VS, Hegde M, Robertson C, Ghazi A, Gerken C, Liu E, Dakhova O, Ashoori A, Corder A, et al. Human epidermal growth factor receptor 2 (HER2) -specific chimeric antigen receptor-modified T cells for the immunotherapy of HER2-positive sarcoma. J Clin Oncol. 2015;33:1688–1696. doi: 10.1200/JCO.2014.58.0225.

75. Le RQ, Li L, Yuan W, Shord SS, Nie L, Habtemariam BA, Przepiorka D, Farrell AT, Pazdur R. FDA approval summary: tocilizumab for treatment of chimeric antigen receptor T cell-induced severe or life-threatening cytokine release syndrome. Oncologist. 2018;

76. Bielamowicz K, Fousek K, Byrd TT, Samaha H, Mukherjee M, Aware N, Wu MF, Orange JS, Sumazin P, Man TK, et al. Trivalent CAR T cells overcome interpatient antigenic variability in glioblastoma. Neuro-Oncology. 2018;20:506–518. doi: 10.1093/neuonc/nox182.

77. Zhang E, Gu J, Xue J, Lin C, Liu C, Li M, Hao J, Setrerrahmane S, Chi X, Qi W, et al. Accurate control of dual-receptor-engineered T cell activity through a bifunctional anti-angiogenic peptide. J Hematol Oncol. 2018;11:44. doi: 10.1186/s13045-018-0591-7.

78. Hung CF, Xu X, Li L, Ma Y, Jin Q, Viley A, Allen C, Natarajan P, Shivakumar R, Peshwa MV, Emens LA. Development of anti-human Mesothelin-targeted chimeric antigen receptor messenger RNA-transfected peripheral blood lymphocytes for ovarian Cancer therapy. Hum Gene Ther. 2018;29(5):614–625. doi: 10.1089/hum.2017.080.

79. Balkwill FR, Capasso M, Hagemann T. The tumor microenvironment at a glance. J Cell Sci. 2012;125:5591–5596. doi: 10.1242/jcs.116392.

80. Chang Z, Lorenzini M, Chen X, Tran U, Bangayan N, Chen Y. Rewiring T-cell responses to soluble factors with chimeric antigen receptors. Nat Chem Biol. 2018;14:317–324. doi: 10.1038/nchembio.2565.

81. Batchu RB, Gruzdyn OV, Mahmud EM, Chukr F, Dachepalli R, Manmari SK, Mostafa G, Weaver DW, Gruber SA. Inhibition of Interleukin-10 in the tumor microenvironment can restore mesothelin chimeric antigen receptor T cell activity in pancreatic cancer in vitro.

Surgery. 2018;163:627–632. doi: 10.1016/j.surg.2017.10.056.

82. Priceman SJ, Tilakawardane D, Jeang B, Aguilar B, Murad JP, Park AK, Chang WC, Ostberg JR, Neman J, Jandial R, et al. Regional delivery of chimeric antigen receptor-engineered T cells effectively targets HER2(+) breast Cancer metastasis to the brain. Clin Cancer Res. 2018;24:95–105. doi: 10.1158/1078-0432.CCR-17-2041.
83. Yu S, Li A, Liu Q, Li T, Yuan X, Han X, Wu K. Chimeric antigen receptor T cells: a novel therapy for solid tumors. J Hematol Oncol. 2017;10:78. doi: 10.1186/s13045-017-0444-9.
84. Hege K, Bergsland E, Fisher G, Nemunaitis J, Warren R, McArthur J, Lin A, Schlom J, June C, Sherwin S. Safety, tumor trafficking and immunogenicity of chimeric antigen receptor (CAR)-T cells specific for TAG-72 in colorectal cancer. J Immunother Cancer. 2017;5:22. doi: 10.1186/s40425-017-0222-9.
85. Park J, Digiusto D, Slovak M, Wright C, Naranjo A, Wagner J, Meechoovet H, Bautista C, Chang W, Ostberg J, Jensen M. Adoptive transfer of chimeric antigen receptor re-directed cytolytic T lymphocyte clones in patients with neuroblastoma. Mol Ther. 2007;15:825–833. doi: 10.1038/sj.mt.6300104.
86. Kowolik C, Topp M, Gonzalez S, Pfeiffer T, Olivares S, Gonzalez N, Smith D, Forman S, Jensen M, Cooper L. CD28 costimulation provided through a CD19-specific chimeric antigen receptor enhances in vivo persistence and antitumor efficacy of adoptively transferred T cells. Cancer Res. 2006;66:10995–11004. doi: 10.1158/0008-5472. CAN-06-0160.
87. Brentjens R, Davila M, Riviere I, Park J, Wang X, Cowell L, Bartido S, Stefanski J, Taylor C, Olszewska M, et al. CD19-targeted T cells rapidly induce molecular remissions in adults with chemotherapy-refractory acute lymphoblastic leukemia. Sci Transl Med. 2013;5:177ra138. doi: 10.1126/scitranslmed.3005930.
88. Kandalaft L, Powell D, Coukos G. A phase I clinical trial of adoptive transfer of folate receptor-alpha redirected autologous T cells for recurrent ovarian cancer. J Transl Med. 2012;10:157. doi: 10.1186/1479-5876-10-157.
89. Tang XY, Sun Y, Zhang A, Hu GL, Cao W, Wang DH, Zhang B, Chen H. Third-generation CD28/4-1BB chimeric antigen receptor T cells for chemotherapy relapsed or refractory acute lymphoblastic leukaemia: a non-randomised, open-label phase I trial protocol. BMJ Open.

2016;6:e013904. doi: 10.1136/bmjopen-2016-013904.

90. Zhong XS, Matsushita M, Plotkin J, Riviere I, Sadelain M. Chimeric antigen receptors combining 4-1BB and CD28 signaling domains augment PI3kinase/AKT/Bcl-XL activation and CD8+ T cell-mediated tumor eradication. Mol Ther. 2010;18:413–420. doi: 10.1038/mt.2009.210.

91. Brentjens R, Rivière I, Park J, Davila M, Wang X, Stefanski J, Taylor C, Yeh R, Bartido S, Borquez-Ojeda O, et al. Safety and persistence of adoptively transferred autologous CD19-targeted T cells in patients with relapsed or chemotherapy refractory B-cell leukemias. Blood. 2011;118:4817–4828. doi: 10.1182/blood-2011-04-348540.

92. Kochenderfer JN, Dudley ME, Carpenter RO, Kassim SH, Rose JJ, Telford WG, Hakim FT, Halverson DC, Fowler DH, Hardy NM, et al. Donor-derived CD19-targeted T cells cause regression of malignancy persisting after allogeneic hematopoietic stem cell transplantation. Blood. 2013;122:4129–4139. doi: 10.1182/blood-2013-08-519413.

93. Kalos M, Levine BL, Porter DL, Katz S, Grupp SA, Bagg A, June CH. T cells with chimeric antigen receptors have potent antitumor effects and can establish memory in patients with advanced leukemia. Sci Transl Med. 2011;3:95ra73. doi: 10.1126/scitranslmed.3002842.

94. Ramos C, Savoldo B, Torrano V, Ballard B, Zhang H, Dakhova O, Liu E, Carrum G, Kamble R, Gee A, et al. Clinical responses with T lymphocytes targeting malignancy-associated κ light chains. J Clin Invest. 2016;126:2588–2596. doi: 10.1172/JCI86000.

95. Turtle C, Hay K, Hanafi L, Li D, Cherian S, Chen X, Wood B, Lozanski A, Byrd J, Heimfeld S, et al. Durable molecular remissions in chronic lymphocytic leukemia treated with CD19-specific chimeric antigen receptor-modified T cells after failure of Ibrutinib. J Clin Oncol. 2017;35:3010–3020. doi: 10.1200/JCO.2017.72.8519.

7

Recent Developments in Immunotherapy of Acute Myeloid Leukemia

Felix S. Lichtenegger[1,2], Christina Krupka[1,2], Sascha Haubner[1,2], Thomas Köhnke[1,2] and Marion Subklewe[1,2,3,4]

[1]Department of Medicine III, University Hospital, LMU Munich, Germany
[2] Laboratory of Translational Cancer Immunology, Gene Center, Munich, Germany
[3] German Cancer Consortium (DKTK), Partner Site, Munich, Germany
[4] German Cancer Research Center (DKFZ), Heidelberg, Germany

ABSTRACT

The advent of new immunotherapeutic agents in clinical practice has revolutionized cancer treatment in the past decade, both in oncology and hematology. The transfer of the immunotherapeutic concepts to the treatment of acute myeloid leukemia (AML) is hampered by various characteristics

Citation: Lichtenegger FS, Krupka C, Haubner S, Köhnke T, Subklewe M. Recent developments in immunotherapy of acute myeloid leukemia. J Hematol Oncol. 2017;10(1):142. Published 2017 Jul 25. doi:10.1186/s13045-017-0505-0.

of the disease, including non-leukemia-restricted target antigen expression profile, low endogenous immune responses, and intrinsic resistance mechanisms of the leukemic blasts against immune responses. However, considerable progress has been made in this field in the past few years.

Within this manuscript, we review the recent developments and the current status of the five currently most prominent immunotherapeutic concepts: (1) antibody-drug conjugates, (2) T cell-recruiting antibody constructs, (3) chimeric antigen receptor (CAR) T cells, (4) checkpoint inhibitors, and (5) dendritic cell vaccination. We focus on the clinical data that has been published so far, both for newly diagnosed and refractory/relapsed AML, but omitting immunotherapeutic concepts in conjunction with hematopoietic stem cell transplantation. Besides, we have included important clinical trials that are currently running or have recently been completed but are still lacking full publication of their results.

While each of the concepts has its particular merits and inherent problems, the field of immunotherapy of AML seems to have taken some significant steps forward. Results of currently running trials will reveal the direction of further development including approaches combining two or more of these concepts.

Keywords: AML, Antibody therapy, Bispecific antibody, CAR T cell, Checkpoint inhibition, Dendritic cell vaccination, Epigenetic therapy, Immunotherapy

BACKGROUND

Advances in immunotherapy have revolutionized cancer therapy in the past few years. Novel immunotherapeutic approaches are entering the mainstream of oncology. In hematology, progress has primarily been made in the field of B-lymphoproliferative diseases including acute lymphoblastic leukemia (ALL). In acute myeloid leukemia (AML), novel strategies utilizing the immune system to eliminate leukemic cells have only recently approached clinical application [1, 2]. This is somewhat surprising, considering that allogeneic hematopoietic stem cell transplantation (HSCT) is one of the oldest immunotherapeutic strategies for postremission therapy in AML. So far, HSCT remains the most successful therapy for prevention of relapse in non-favorable risk patients with AML [3, 4]. However, relapse after allogeneic HSCT does occur, and the vast majority of elderly patients are not eligible for HSCT. Therefore, alternative immunotherapeutic strategies are urgently needed to treat patients not suitable for intensive treatment regimens as well as patients with relapsed or refractory (r/r) disease [5].

In ALL, several antibody-based approaches have already entered standard treatment or are at the verge of approval. Rituximab, an anti-CD20 directed antibody has been shown to be beneficial as an additive to conventional chemotherapeutic agents [6]. Inotuzumab ozogamicin is a toxin-conjugated monoclonal antibody directed against CD22 on the surface of B cells. Approval in r/r ALL is expected in the next year after a phase III trial demonstrated 80.7% overall response rate (ORR) [7]. Moreover, novel T cell-recruiting therapies have opened up an entirely new approach to the treatment of acute leukemias, bypassing typical tumor resistance mechanisms [8]. Blinatumomab, a bispecific molecule connecting CD3 in the T cell receptor complex with CD19 expressed by B cells, was the first T cell-recruiting antibody approved for the treatment of cancer in 2014 [9]. Chimeric antigen receptor (CAR) T cells advance this concept even further by engineering a T cell with the specificity of a monoclonal antibody and a T cell activation domain. The engineered T cells are thus capable of targeting surface molecules of tumor cells in their native conformation independently of MHC [10]. In principle, all of these treatment modalities can be translated to AML.

However, targeted immunotherapy relies on a suitable target antigen to avoid unwanted on-target off-tumor toxicity. In ALL, the restricted expression profile of CD19 and CD20 allows to target these B cell-associated antigens. In AML, it is more difficult to choose an appropriate target antigen due to a more ubiquitous expression pattern overlapping with healthy hematopoiesis. Various potential target antigens are studied for each of the immunotherapeutic strategies [11, 12]. Still, it is to be expected that targeting AML-associated antigens will result in prolonged drug-induced cytopenias. This will require the adjustment of current protocols applied in ALL to the different setting in AML.

Other immunotherapeutic concepts rely on the enhancement of endogenous or the priming of new immune responses. Checkpoint inhibitors have been successfully approved in several solid organ malignancies and are now entering the treatment of hematological diseases [13]. And therapeutic vaccines, particularly those based on dendritic cells (DCs), have been shown to reliably induce anti-leukemic immune responses. Combining these two strategies not only with each other but also with hypomethylating agents

(HMAs), which have been shown to modulate the immune function, seems suitable.

In this review, we will present recent advances made in the aforementioned fields of immunotherapy of AML. HSCT and immunotherapeutic strategies for relapse after HSCT constitute a review topic on their own and have been excluded. As published data from clinical trials is still scarce for the majority of immunotherapeutic approaches, we will integrate currently running clinical trials to point out upcoming directions in this field.

ANTIBODY-DRUG CONJUGATES FOR IMMUNOTHERAPY OF AML

Compared to conventional antibody formats (Fig. 1a), antibody-drug conjugates (ADCs), consisting of monoclonal antibodies conjugated to various toxins, are a tool to bridge conventional chemotherapy and innovative immunotherapy. Upon internalization, the toxin is released in the acidic environment of the lysosomes and reaches the nucleus where it induces cell death through mechanisms like DNA double strand break and cell cycle arrest (Fig. 1b). The prerequisite for successful immunochemotherapy is a rapidly internalizing target antigen, preferably specific to the tumor [14].

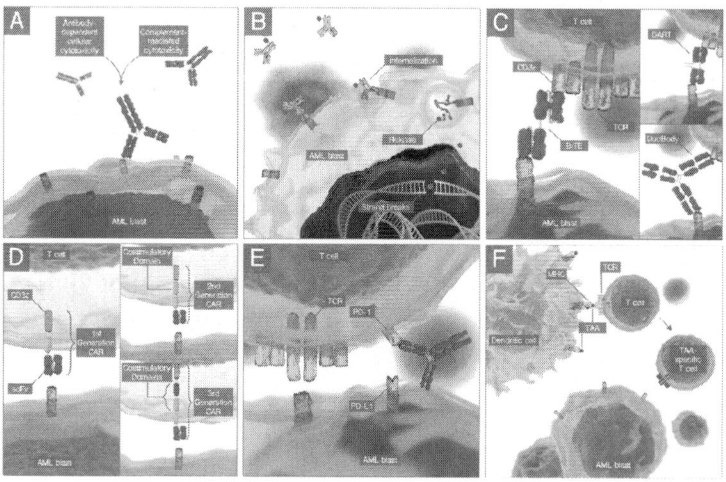

Figure 1: Mechanisms of cancer immunotherapy. Different immunotherapeutic concepts are discussed in the context of AML in this review. a Conventional antibodies directed at AML surface antigens mediate antibody-dependent cellular cytotoxicity as well as complement-mediated cytotoxicity.

b Antibody-drug conjugates consist of monoclonal antibodies conjugated to various toxins, which are released upon internalization and induce cell death through mechanisms like DNA double-strand break and cell cycle arrest. c T cell-recruiting antibody constructs are composed of single-chain variable fragments of two antibodies of different specificity connected by a short peptide linker. Their purpose is to bring malignant cells and T cells in close proximity through simultaneous binding of a tumor-associated antigen and CD3ε in the T cell receptor complex. d Chimeric antigen receptors (CARs) are genetically engineered cell membrane-bound receptors combining extracellular antibody binding and intracellular effector cell signaling. Their structure enables both MHC-independent antigen binding and highly potent cytotoxic effector cell function. Compared to the first generation of CARs, the introduction of various costimulatory domains in later-generation CAR constructs greatly improved their anti-tumor effector function. e Checkpoint inhibitors are monoclonal antibodies binding to inhibitory receptors on T cells or their ligands on antigen-presenting cells or cancer cells, thus boosting the effects of pre-existing T cell responses. f Dendritic cells are professional antigen-presenting cells. Vaccination strategies using in vitro-generated dendritic cells have the purpose to prime new or enhance pre-existing antigen-specific immune responses.

CD33 (SIGLEC-3) is the antigen that has been most commonly targeted so far in AML. The first and most prominent ADC in clinical application was gemtuzumab ozogamicin (GO, Mylotarg, Pfizer), a humanized anti-CD33 IgG4 antibody conjugated to calicheamicin. Promising clinical results lead to an accelerated approval of the antibody by the Food and Drug Administration (FDA) in 2000 [15]. Safety concerns and failure to verify clinical benefit in a confirmatory phase III trial enrolling patients across all cytogenetic risk groups resulted in the voluntary withdrawal of GO from the market in 2010 [16]. In recent years, both retrospective analyses and new clinical trials have been performed to unravel clinical benefits of GO in specific subgroups. A meta-analysis of five randomized controlled trials (RCTs) showed that the addition of GO to conventional chemotherapy significantly reduced the risk of relapse and resulted in an overall survival (OS) benefit mainly for cytogenetically favorable as well as for the intermediate-risk group [17]. Another meta-analysis of 11 RCTs with one arm including GO showed improval in OS only for patients with favorable genetics [18]. A recent clinical trial testing GO vs. best supportive care including hydroxyurea in older patients with newly diagnosed AML unsuitable for intensive chemotherapy confirmed the clinical benefit, particularly in those patients with favorable or intermediate cytogenetic risk profile [19].

In order to further improve the clinical results with GO, several clinical trials have been performed evaluating GO in combination with HMAs. A regimen consisting of hydroxyurea, azacitidine, and GO was tested in a phase II trial for 142 older patients with newly diagnosed AML. The predefined goals concerning efficacy and safety were met for the poor-risk cohort (age ≥70 years and performance status 2 or 3), but not for the good-risk group [20]. GO in combination with both the histone deacetylase inhibitor vorinostat and the DNA methyltransferase I inhibitor azacitidine was studied in a phase I/II trial for older patients with r/r AML. An ORR of 41.9% was seen among the 43 patients that were treated at the maximum tolerated dose, which can be considered rather high in this difficult-to-treat cohort [21]. And finally, 110 patients with newly diagnosed or r/r AML or high-risk myelodysplastic syndrome (MDS) were treated with decitabine and GO within a phase II study. Compared to historical controls, ORR was increased, but not OS [22]. Another combination trial with GO and azacitidine for patients with relapsed AML has not yet been reported (NCT00766116, Table 1).

Table 1: Current clinical trials using antibody-drug conjugates for immunotherapy of AML

Study identifier	Study name	Antigen/ target	Drug name	Combination therapy	Clinical phase	Indication (AML only)	Primary endpoints	(Estimated) Enrollment	Sponsor	Country	Study start	(Estimated) Completion date	Status
NCT00766116	A phase I/II trial of the combination 5-azacitidine and gemtuzumab ozogamicin therapy for treatment of relapsed AML	CD33	Gemtuzumab ozogamicin	Azacitidine	I/II	Relapsed AML	Phase I: MTD; phase II: clinical response (CR rate)	50	University of California, San Diego	USA	2005	2017	Active, not recruiting
NCT01902329	A phase 1 trial of SGN-CD33A in patients with CD33-positive acute myeloid leukemia	CD33	SGN-CD33A	Azacitidine or decitabine	I	Relapsed AML or newly diagnosed AML if not a candidate for intensive chemotherapy; CD33 expression	Toxicity	195	Seattle Genetics	USA	2013	2017	Active, not recruiting
NCT02326584	A phase 1b dose-escalation study of SGN-CD33A in combination with standard-of-care for patients with newly diagnosed acute myeloid	CD33	SGN-CD33A	Standard of care	I	Newly diagnosed AML	Toxicity	144	Seattle Genetics	USA	2014	2017	Active, not recruiting

NCT02674763	A phase 1, multi-center, open-label study of IMGN779 administered intravenously in adult patients with relapsed/refractory CD33-positive	CD33	IMGN779	n.a.	I	r/r AML; CD33 expression	MTD	124	ImmunoGen	USA	2016	2019	Recruiting
NCT02785900	Vadastuximab talirine (SGN-CD33A; 33A) combined with azacitidine or decitabine in older patients with newly diagnosed acute myeloid leukemia (CASCADE)	CD33	SGN-CD33A	azacitidine or decitabine	III	Newly diagnosed AML with non-favorable risk type; not a candidate for allogeneic HSCT	Clinical response (OS)	500	Seattle Genetics	USA, Australia, Korea, Taiwan, various European countries	2016	2021	Recruiting
NCT02848248	A phase 1 study of SGN-CD123A in patients with relapsed or refractory acute myeloid leukemia (AML)	CD123	SGN-CD123A	n.a.	I	r/r AML; CD123 expression	Toxicity	102	Seattle Genetics	USA	2016	2019	Recruiting

As CD33 is expressed on >30% of healthy bone marrow cells, on-target off-leukemia toxicity is inevitable [23–25]. However, a major part of the side effects observed in the clinical trials with GO were attributed to linker instabilities and subsequent off-target toxicities [26, 27]. A lot of effort has therefore been put into the optimization of the ADC technology. An alternative ADC directed against CD33, SGN-CD33A (vadastuximab talirine), has recently entered clinical trials. In this construct, a monoclonal anti-CD33 antibody is conjugated to a highly potent DNA-binding pyrrolobenzodiazepine dimer. The linker technology has been optimized and allows uniform drug loading [28]. Based on promising preclinical data, several clinical trials have been initiated evaluating safety and efficacy of SGN-CD33A alone or in various combinations. Twenty-seven treatment-naive AML patients ineligible for intensive chemotherapy were treated with the recommended monotherapy dose of 40 μg/kg within a phase I study (NCT01902329). The adverse events (AEs) observed were reported to be generally manageable, with a preponderance of myelosuppression. Combined complete remission (CR) and complete remission with incomplete recovery (CRi) rate was 54% [29]. Within another cohort of the same study, 53 patients were treated with a combination of SGN-CD33A and HMAs, resulting in an encouraging CR/CRi rate of 73% [30]. The addition of the ADC to standard 7 + 3 induction chemotherapy is tested within a large phase Ib (NCT02326584) study. Preliminary results have been reported for the first 42 patients of this study. The combination therapy resulted in grade 4 myelosuppression in all patients, but no increase in non-hematological AEs was reported compared to chemotherapy alone. Synergistic effects of HMAs and CD33-directed immunotherapy are supported by a high CR/CRi rate of 78% [31]. This could be due to HMA-induced increase in CD33 expression as well as increased sensitivity to toxin-induced DNA damage [28]. Based on the encouraging response data, a phase III study of SGN-CD33A in combination with azacitidine or decitabine for older patients with newly diagnosed AML (CASCADE study) has recently been initiated (NCT02785900). However, potential hepatotoxicity, including veno-occlusive disease (VOD), is a major concern, particularly in the combination of SGN-CD33A with allogeneic HSCT before or after the treatment. Both phase I studies discussed above have therefore been put on hold by the FDA to explore the incidence of VOD, while the CASCADE trial continues enrollment [32].

SGN-CD123A is a similar ADC with the antibody directed at CD123 instead of CD33. CD123 is more restrictively expressed in the healthy hematopoietic compartment, which might decrease on-target off-leukemia toxicities [24, 33]. This is being tested in the recently initiated phase I trial, which is planned to recruit 102 patients with r/r AML (NCT02848248).

ImmunoGen developed IMGN779, a CD33-directed monoclonal antibody conjugated to the novel DNA-alkylating molecule DGN462. Preclinical data demonstrated highly specific in vitro and in vivo cytotoxicity against primary AML cells, especially in samples with an *FLT-ITD* mutation [34, 35]. The combinatorial approach of IMGN779 with the PARP inhibitor Olaparib resulted in enhanced ex vivo activity and a decreased tumor burden in a xenograft mouse model [36]. A clinical phase I study in r/r AML is currently recruiting patients (124 patients planned, NCT02674763). Results of this study will show if there is any benefit over the usage of SGN-CD33A in terms of the risk-benefit ratio. Apart from the conjugation to toxins, monoclonal anti-CD33 antibodies have also been conjugated to radioisotopes. However, first clinical studies have demonstrated less promising results and most of these strategies are currently not further pursued [37, 38].

Taken together, the field of ADCs finally seems to recover from the huge setback it originally suffered after the voluntary withdrawal of GO in 2010. A lot of effort has been put into the optimization of the ADC technology, and clinical results from early trials demonstrate promising response rates. Results of randomized phase III trials are eagerly awaited in order to estimate the risk-benefit ratio between a potential increase in response rates and the discussed side effects due to on-target off-leukemia toxicities and toxin-induced hepatic toxicity. In order to increase target cell specificity of the therapy, alternative target antigens are being evaluated in preclinical (i.e., CLL-1, SAIL) [39–41] and early clinical studies (i.e., CD25, FLT3) [42, 43].

T CELL-RECRUITING ANTIBODY CONSTRUCTS FOR IMMUNOTHERAPY OF AML

T cell-recruiting antibody constructs are a novel class of molecules composed of the single-chain variable fragments (scFv) of two antibodies of different specificity connected by a short peptide linker (Fig. 1c). Through simultaneous binding of a tumor-associated antigen and CD3ε in the T cell receptor complex, these small adapter molecules bring malignant cells and T cells in close proximity. The binding of CD3ε leads to T cell activation and expansion resulting in Granzyme B/perforin-mediated target cell lysis. The special feature of this strategy is that virtually any memory T cell can be recruited for target cell lysis irrespective of its specificity [44, 45]. Clinical proof of concept has been provided with blinatumomab (BLINCYTO®, AMGEN), a CD19/CD3 T cell-recruiting antibody construct. It was approved as the first in its class by the FDA in 2014 for r/r Ph-negative B-precursor ALL, after a clinical phase II trial demonstrated a CR/CRi rate of 43% after one or two cycles of therapy [9]. Very recently, the superiority of blinatumomab to conventional chemotherapy for patients with r/r B-precusor ALL was proven in a randomized phase III trial [46].

In AML, several T cell-recruiting antibody constructs are under preclinical and early clinical development (Table 2). Similar to the ADCs, the optimal antigen to target is still an open question. The sister molecule of blinatumomab, AMG 330, is a bispecific T cell engager (BiTE) construct targeting CD33 [25, 47]. The high inter- as well as intra-patient variations in CD33 expression levels might influence the success of targeted immunotherapy. Significantly lower expression has been demonstrated for $CD34^{+}/CD38^{-}$ leukemia-initiating cells (LICs) vs. AML bulk cells, but expression was still significantly higher compared to their healthy counterparts ($CD34^{+}/CD38^{-}$ normal hematopoietic stem cells). In preclinical studies, the preincubation of AML cells with AMG 330 and T cells prevented the subsequent engraftment of AML in NOD/SCID gamma null (NSG) mice. This suggests that the CD33 expression level of LICs is sufficient for elimination with T cell-recruiting constructs. Besides, it has been demonstrated in vitro that the CD33 expression level mainly influences kinetics of cytotoxicity, but not necessarily the response rate [25, 48]. Recently, an international, multicenter phase I trial for r/r AML patients (*n* = 50) was initiated (NCT02520427), but data are not yet available. Several other CD33-targeting antibody constructs that differ from AMG 330 in their molecular structure are currently evaluated in preclinical settings [12, 49, 50].

Table 2: Current clinical trials using T cell-recruiting antibody constructs for immunotherapy of AML

Study identifier	Study name	Antigen/ target	Drug name	Combination therapy	Clinical phase	Indication (AML only)	Primary endpoints	(Estimated) Enrollment	Sponsor	Country	Study start	(Estimated) Completion date	Status
NCT02152956	Phase 1, first in human, dose escalation study of MGD006, a CD123 × CD3 dual affinity re-targeting (DART®) bi-specific antibody-based molecule, in patients with relapsed or refractory AML or intermediate-2/high risk MDS	CD123	MGD006	No	I	r/r AML	DLT	124	Macrogenics	USA, France, Germany, Italy, Netherlands	2014	2018	Recruiting
NCT02520427	A phase 1 first-in-human study evaluating the safety, tolerability, pharmacokinetics, pharmacodynamics and efficacy of AMG 330 administered as continuous intravenous infusion in subjects with relapsed/ refractory acute myeloid leukemia	CD33	AMG 330	No	I	r/r AML	DLT, toxicity	50	AMGEN	USA, Germany, Netherlands	2015	2018	Recruiting

NCT02715011	A phase 1, first-in-human, open-label, dose escalation study of JNJ-63709178, a humanized CD123 × CD3 DuoBody in subjects with relapsed or refractory AML	CD123	JNJ-63709178	No	I	r/r AML	DLT, toxicity	60	Janssen Research & Development	USA, Australia, Belgium, Germany	2016	Unknown	Suspended
NCT02730312	A phase 1 multiple dose study to evaluate the safety and tolerability of XmAb®14045 in patients with CD123-expressing hematologic malignancies	CD123	Xmab14045	No	I	Primary or secondary AML	MTD, toxicity	66	Xencor	USA	2016	2018	Recruiting
NCT03038230	A phase 1, multinational study of MCLA-117 in acute myelogenous leukemia	CLL-1	MCLA-117	No	I	r/r AML, newly diagnosed elderly untreated AML patients	DLT, toxicity	50	Merus N.V.	Belgium, France, Italy, Netherlands	2016	2018	Recruiting

To reduce on-target off-leukemia toxicity, alternative AML-associated targets are being explored. CD123 has a lower level of expression on healthy hematopoietic cells compared to CD33 [24, 33]. Therefore, several T cell-recruiting antibody constructs targeting CD123 have been developed and are currently in early clinical studies. One of these constructs is MGD006, developed by MacroGenics. In contrast to the BiTE technology, dual-affinity re-targeting (DART) molecules are composed of heavy and light chain variable domains of two antigen-binding specificities (A + B) on two independent polypeptide chains (VL_A-VH_B-VL_B-VH_A), which are stabilized through an additional C-terminal bridge [51, 52]. Encouraging preclinical data in terms of cytotoxicity against primary AML cells [53] and safe and well-tolerated infusion of MGD006 in cynomolgus monkeys [54] paved the way for the clinical development in a multicenter phase I study of 124 relapsed/refractory AML patients (NCT02152956).

XmAb14045, developed by Xencor, is a structurally distinct anti-CD123 T cell-recruiting antibody construct in early clinical development. The XmAb technology ensures structural stability and an extended serum half-life through the retention of an inactive Fc part. Preclinical studies in cynomolgus monkeys showed rapid clearance of $CD123^+$ cells from the bone marrow as well as from the periphery [55]. These studies formed the basis for the initiation of a clinical phase I study for the evaluation of safety and tolerability of Xmab14045 in 66 patients with CD123-expressing hematological malignancies including primary and secondary AML (NCT02730312).

JNJ-63709178, a CD123/CD3 humanized IgG4 antibody has been developed by Genmab using their DuoBody technology. Preclinical studies in vitro and in vivo showed highly specific T cell activation and targeting of primary AML cells [56, 57], which lead to the initiation of a phase I study in relapsed/refractory AML ($n = 60$, NCT02715011). Currently, the study is on hold because of the occurrence of undisclosed adverse events.

CLL-1 is a novel target antigen in AML characterized by its high expression on AML bulk cells as well as LICs [58, 59]. Recently, a bispecific CLL-1/CD3 antibody construct (MCLA-117) has been developed by Merus B.V. MCLA-117 induced target antigen-specific cytotoxicity against primary AML cells at low E:T ratios using either allogeneic or autologous T cells. This led to the initiation of a clinical phase I trial in r/r or elderly, previously untreated AML patients (NCT03038230, $n = 50$) [60].

Results of the ongoing trials are awaited to see if the success in ALL will translate into the setting of AML. A potential future strategy could be to use the evolving antibody technology to simultaneously target two different AML-associated antigens in order to increase specificity [61]. Apart from that, lots of effort has been put into optimization of the antibody technology to increase safety. The Probody™ technology by CytomX uses antigen-binding site masking peptides attached to antibody constructs by substrate-cleavable linkers. In the tumor microenvironment, linkers are cleaved by highly active proteases generating effective immunotherapeutic agents directly at the tumor site [62]. Recently, an EGFR/CD3 Probody™ has shown promising results in terms of efficacy and increase in therapeutic window in preclinical studies in vitro and in vivo. As the technology relies on tumor site-specific protease activity, it remains to be determined if this approach is also feasible in acute leukemia [63].

Independently of considerations about the optimal target antigen, we are only at the beginning of understanding the exact mechanism of action of those antibody constructs and resistance mechanisms that potentially evolve upon T cell activation. Despite the promising response rate of 43% using blinatumomab in heavily pre-treated ALL patients, reasons for resistance in the remaining patients have not been resolved. Only few biomarkers for response have been determined so far, e.g., in case of the blinatumomab studies, the percentage of blasts in the bone marrow and the degree of T cell expansion [9, 64]. PD-L1 upregulation on AML cells upon T cell activation has been suggested as a potential resistance mechanism in an ex vivo system [48] and in a case report of a blinatumomab-refractory B-precursor ALL patient [65]. Addition of a checkpoint inhibitor to T cell-recruiting antibodies might help to circumvent resistance. A clinical study testing this concept by addition of an anti-PD1 antibody with or without an anti-CTLA4 antibody to blinatumomab for the treatment of r/r ALL patients has been initiated, but is not yet open for patient recruitment (NCT02879695).

CAR T CELLS FOR IMMUNOTHERAPY OF AML

Circumventing T cell exhaustion, anergy and senescence, CAR T cells take the technology of T cell-recruiting antibody constructs one step further and have already shown promising clinical results in various hematologic malignancies. CARs are genetically engineered cell membrane-bound receptors that combine extracellular antibody binding and intracellular effector cell signaling, thereby enabling both MHC-independent antigen binding and highly potent cytotoxic effector cell function (Fig. 1d). Since the first generation of CARs in 1989 [66], the introduction of costimulatory domains (mainly CD28 or 4-1BB) in so-called second-generation CAR constructs greatly improved their anti-tumor effector function and paved their way into clinical trials [67].

To date, the most prominent target antigen for CAR T cell therapy is CD19, due to its restrictive expression pattern and good safety profile. Groundbreaking early clinical trial results could be achieved for various B cell malignancies. In r/r B-ALL, treatment with anti-CD19 4-1BB-costimulatory CAR T cells achieved MRD-negative CR rates of 86% for 29 patients [68]. These are outstanding clinical results, considering the heavily pretreated patient population that was included: in the median, patients had received three prior intensive chemotherapy regimens, and more than one third had relapsed after prior allogeneic HSCT. In another recently published trial, treatment with anti-CD19 CD28-costimulatory CAR T cells showed great clinical efficacy with CR rates of 57% in seven patients with DLBCL refractory to at least three prior lines of therapy [69]. As of November 1, 2016, 1135 patients have been treated with anti-CD19 genetically engineered TCR/CAR T cells [70], leading to high expectations for patients with no therapeutic options until now. Accordingly, there are currently 87 open clinical phase I or II trials involving anti-CD19 CAR T cells in B cell malignancies (ClinicalTrials.gov, last update 03/07/2017).

Despite these promising early results and the rapidly expanding number of anti-CD19 CAR T cell trials, this novel drug format is still incompletely understood and cannot generally be considered safe. In March 2017, Juno announced to shut down development of anti-CD19 CD28-costimulatory JCAR015 CAR T cells and to close their phase II ROCKET trial in r/r adult ALL, after five treatment-related deaths had occurred due to CAR T cell-mediated neurotoxicity [71]. As "living drugs," the in vivo effect of CAR T cells may be dependent on different conditioning chemotherapy regimens, CAR T cell manufacturing protocols and costimulatory domains. Unfortunate combinations of these variables may promote rapid in vivo expansion of CAR T cells with the potential to induce severe systemic and neurological side effects.

Translating CAR T cell therapy to AML is complicated again by the non-restricted expression of AML-associated antigens. Given that current CAR T cell constructs can persist beyond 4 years in the human body [72], several strategies are being explored to circumvent unwanted on-target off-leukemia toxicity, particularly long-term myeloid cell aplasia. Similar to ADCs and T cell-recruiting antibody constructs, the identification of AML-specific target antigens or antigen combinations would be one way to improve safety of future CAR T cell approaches in AML. To date, several target antigens for AML CAR T cell therapy are under preclinical and clinical investigation.

CD33 is the most prominent target antigen for CAR T cells in preclinical trials due to its high and persistent expression in the majority of AML patients [24, 73]. In an in vivo model of AML-xenotransplanted NSG mice, treatment with anti-CD33 CAR T cells resulted in marked reduction of leukemic burden and prolonged survival [74]. However, significant on-target off-leukemia toxicity with reduction of myeloid lineage and hematopoietic stem cells was observed. In another in vivo model of AML-xenotransplantated NSG mice, treatment with only transient CAR expression via electroporation of T cells with anti-CD33 CAR-encoding RNA resulted in similar, but only transient cytotoxicity [75]. Application of CAR T cells directed against CD123 as an alternative target in an in vivo model with AML-xenotransplanted mice resulted in significant reduction of leukemic burden and prolonged survival with only limited on-target off-leukemia toxicity and unaffected healthy hematopoiesis [76–79]. In contrast, eradication of normal human myelopoiesis was demonstrated in another in vivo mouse study with anti-CD123 CAR T cells [80]. Interestingly, modifying the anti-CD123 scFv by utilizing V_H and V_L chains from different monoclonal antibodies could reduce myelotoxicity in an AML mouse model [79]. This conflicting data indicates that variations in antibody clone, costimulatory domain, effector cells, and model system might account for vastly different outcomes. Fine-tuning the development process of CAR T cells might be able to provide differential recognition of target antigens on leukemic vs. healthy cells.

Other potential target antigens identified in preclinical studies include CD44v6 [81], CLL1 [82], FLT3 [83], FRβ [84], LeY [85], NKG2D [86], and PR1/HLA-A2 [87].

To date, only one very small trial evaluating anti-LeY CAR T cells (CTX08-0002) in r/r AML has been completed. None of the four treated patients developed grade 3 or 4 toxicity, and infused CAR T cells persisted for up to 10 months. One patient with active leukemia responded with transient reduction in blast count before progression 1 month later. All patients relapsed 28 days to 23 months after adoptive CAR T cell transfer [88]. Currently, there are four open phase I clinical trials that evaluate the application of CAR constructs in r/r AML (Table 3). One trial recruiting in China is including patients with r/r AML for treatment with anti-CD33 CAR cytokine-induced killer (CIK) cells (NCT01864902). So far, there has only been a report of one patient within this trial who showed a transient decrease in blast count while suffering from cytokine release syndrome and pancytopenia [89]. Trial completion is estimated to be in 2017. Two other trials evaluate lentivirally transduced or mRNA-electroporated anti-CD123 CAR T cells, respectively (NCT02159495, NCT02623582), however, the latter one has been prematurely terminated. Until now, no results have been published. Another phase I trial utilizing allogeneic "off-the-shelf" anti-CD123 CAR T cells (UCART123) was recently opened (NCT03190278 [90]). And finally, a trial applying CAR T cells directed at NKG2D ligands to patients with r/r AML, MDS, and multiple myeloma is estimated to be completed in 2017, but results are still pending (NCT02203825).

Table 3: Current clinical trials using CAR T cells for immunotherapy of AML

Study identifier	Study name	Target	Designation	Generation	Costim. domain	Transduction method	Median dosage	Conditioning chemotherapy	Clinical phase	Indication	Primary endpoints	(Estimated) Enrollment	Sponsor	Country	Study start	(Estimated) Completion date	Status
NCT01864902	Treatment of Relapsed and/or Chemotherapy Refractory CD33 Positive Acute Myeloid Leukemia by CART-33 (CART33)	CD33	CART-33	2nd	4-1BB	Lentiviral	4.26×10^8 CAR T cells	n.a.	I/II	r/r AML or AML in CR2 or later if not a candidate for allo-HSCT; CD33 expression	Toxicity	10 (1 patient reported)	Chinese PLA General Hospital	China	2013	2017	Recruiting
NCT02159495	Genetically Modified T-cell Immunotherapy in Treating Patients With Relapsed/Refractory Acute Myeloid Leukemia and Persistent/Recurrent Blastic Plasmacytoid Dendritic Cell Neoplasm	CD123	CD123R(EQ) 28Z/EGFRt	2nd	CD28	Lentiviral	Variable	Cyclophosphamide +/− fludarabine +/− etoposide	I	r/r AML	DLT, toxicity	30	City of Hope Medical Center	USA	2015	2017	Recruiting

NCT02203825	Safety Study of Chimeric Antigen Receptor Modified T-cells Targeting NKG2D-Ligands	NKG2D-ligands	CM-CS1 T-cells	2nd	DAP10	Retroviral	$1 \times 10^6 - 3 \times 10^9$ CAR T cells/kg	n.a.	I	r/r MDS-RAEB, r/r AML, r/r MM	Toxicity, feasibility	12	Celyad	USA	2015	2017	Active, not recruiting
NCT03190278	Study Evaluating Safety and Efficacy of UCART123 in Patients With Acute Myeloid Leukemia (AML123)	CD123	UCART123	n.a.	n.a.	n.a.	$6.25 \times 10^5 - 6.25 \times 10^6$ CAR T cells/kg	n.a.	I	r/r AML	Safety, efficacy	156	Cellectis S.A.	USA	2017	2021	Recruiting

Novel CAR designs are explored to increase the specificity and to improve safety profiles. In preclinical in vivo models, dual-targeting approaches targeting two independent leukemia-associated antigens were shown to provide increased specificity accompanied by reduced off-leukemia toxicity [91] and to prevent antigen escape mechanisms [92]. In vitro, it was demonstrated that dual targeting of CD33 and CD123 was superior to monospecific approaches in terms of specific cytotoxicity [93]. Further preclinical investigation and translation of dual-targeting strategies into clinics could contribute to efficacy and safety in CAR T cell therapy in AML where target specificity remains a major issue. On-target off-leukemia toxicity could also be further reduced by fine-tuning of CAR density and CAR binding affinity [94]. In light of safety concerns due to unrestricted in vivo CAR T cell expansion and activation, methods of selective CAR T cell depletion are currently being investigated. Integration of so-called suicide gene systems into CAR constructs could act as safety switches enabling rapid on-demand elimination of CAR T cells that would otherwise turn uncontrollable. These suicide gene systems can be based on enzymatic activation of cytotoxic prodrugs, antibody-based targeting of overexpressed surface antigens, or pharmacological induction of apoptosis via inducible caspase 9 which is already tested in clinical phase I CAR T cell trials (NCT03016377 [95]).

CHECKPOINT INHIBITORS FOR IMMUNOTHERAPY OF AML

In contrast to the immunotherapeutic concepts discussed so far, monoclonal antibodies against checkpoint molecules are applied with the idea to unleash pre-existing anti-tumor T cell responses (Fig. 1e). Within recent years, checkpoint inhibition has probably become the single biggest hype in cancer immunotherapy, primarily in solid oncology, but meanwhile, also finding its way into hematology [96]. Most prominently within hematologic diseases, anti-PD-1 antibodies show remarkable success in Hodgkin's lymphoma and are tested in various non-Hodgkin lymphomas. However, there is growing evidence from in vitro experiments and murine models that this strategy could also be applied to AML [96].

Only one clinical study applying a checkpoint antibody as a monotherapy to AML patients has been published so far. Eighteen patients with various hematologic malignancies, including eight patients with AML, were treated with the anti-PD-1 antibody pidilizumab within a phase I study. The antibody was shown to be safe and well tolerable, and one of the AML patients showed a minimal response manifested by a decrease in peripheral blasts from 50 to 5% [97]. A phase I study testing the CTLA-4 antibody ipilimumab in various malignancies including 12 patients with AML has long been completed, but to our knowledge, specific results for AML patients have not been published (NCT00039091, Table 4). Another phase I study, in which ipilimumab was applied to 54 patients with refractory AML, MDS, or chronic myelomonocytic leukemia (CMML), has finished recruiting, but results have not yet been reported (NCT01757639). And three phase II studies (NCT02275533, NCT02532231, NCT02708641) are studying the effect of PD-1 inhibition with either nivolumab or pembrolizumab as a monotherapy on prevention of relapse in remission.

Table 4: Current clinical trials using checkpoint inhibitors for immunotherapy of AML

Study identifier	Study name	Antigen/ target	Drug name	Combination therapy	Clinical phase	Indication (AML only)	Primary endpoints	(Estimated) Enrollment	Sponsor	Country	Study start	(Estimated) Completion date	Status
NCT00039091	Monoclonal antibody therapy in treating patients with ovarian epithelial cancer, melanoma, acute myeloid leukemia, myelodysplastic syndrome, or non-small cell lung cancer	CTLA-4	Ipilimumab	n.a.	I	AML with different recurrent mutations or recurrent AML	Toxicity	12 (AML only)	National Cancer Institute (NCI)	USA	2002	2007	Terminated
NCT01757639	Ipilimumab in treating patients with relapsed or refractory high-risk myelodysplastic syndrome or acute myeloid leukemia	CTLA-4	Ipilimumab	n.a.	I	Refractory AML	Toxicity, regulatory T cells	54 (AML + MDS + CMML)	National Cancer Institute (NCI)	USA	2012	2016	Active, not recruiting
NCT02275533	Nivolumab in eliminating minimal residual disease and preventing relapse in patients with acute myeloid leukemia in remission after chemotherapy	PD-1	nivolumab	n.a.	II	AML in first remission; no eligibility for allo-HSCT	Clinical response (RFS)	80	National Cancer Institute (NCI)	USA	2015	2019	Recruiting

NCT02397720	Study of Nivolumab (BMS-936558) in Combination With 5-azacytidine (Vidaza) for the Treatment of Patients With Refractory/ Relapsed Acute Myeloid Leukemia and Newly Diagnosed Older Acute Myeloid Leukemia (AML) (>65 Years) Patients	PD-1	Nivolumab	Azacitidine	II	r/r AML or newly diagnosed older AML patients	MTD	110	M.D. Anderson Cancer Center	USA	2015	2018	Recruiting
NCT02464657	Study of Idarubicin, Cytarabine, and Nivolumab in Patients With High-Risk Myelodysplastic Syndrome (MDS) and Acute Myeloid Leukemia (AML)	PD-1	Nivolumab	Idarubicin, cytarabine	I/II	De novo AML	MTD	75	M.D. Anderson Cancer Center	USA	2015	2018	Recruiting
NCT02532231	Nivolumab in Acute Myeloid Leukemia (AML) in Remission at High Risk for Relapse	PD-1	Nivolumab	n.a.	II	AML in remission with high risk of relapse	Clinical response (RFS)	30	M.D. Anderson Cancer Center	USA	2015	2018	Recruiting

NCT02708641	A phase II study of pembrolizumab as post-remission treatment of patients ≥60 with AML	PD-1	Pembrolizumab	n.a.	II	AML patients ≥60 years in CR; no eligibility for allo-HSCT	Toxicity, clinical response (time to relapse)	40	Alison Sehgal, MD, MS	USA	2016	2021	Not yet recruiting
NCT02768792	High-dose cytarabine followed by pembrolizumab in relapsed/refractory AML	PD-1	Pembrolizumab	High-dose cytarabine	II	r/r AML	Clinical response (CR rate)	37	UNC Lineberger Comprehensive Cancer Center	USA	2016	2021	Recruiting
NCT02771197	Lymphodepletion and anti-PD-1 blockade to reduce relapse in AML patient not eligible for	PD-1	Pembrolizumab	Fludarabine, melphalane, auto-SCT	II	Non-favorable risk AML in CR	Clinical response (2-y-RR)	20	Northside Hospital, Inc.	USA	2016	2020	Recruiting
NCT02775903	An efficacy and safety study of azacitidine subcutaneous in combination with durvalumab (MEDI4736) in previously untreated subjects with higher-risk myelodysplastic syndromes (MDS) or in elderly subjects with acute myeloid leukemia (AML)	PD-L1	Durvalumab	Azacitidine	II	De novo AML or sAML or tAML in elderly patients	Clinical response (RR)	110 (AML alone)	Celgene Corporation	USA, Canada and various European countries	2016	2019	Recruiting

NCT02845297	Phase 2 study of azacitidine in combination with pembrolizumab in relapsed/refractory acute myeloid leukemia (AML) patients and in newly diagnosed older (≥65 years) AML patients	PD-1	Pembroli-zumab	Azacitidine	II	r/r AML	MTD	40	Sidney Kimmel Compre-hensive Cancer Center	USA	2016	2020	Recruit-ing
NCT02890329	Ipilimumab and decitabine in treating patients with relapsed or refractory myelodysplastic syndrome or acute myeloid	CTLA-4	Ipilimumab	Decitabine	I	r/r AML	MTD	48	National Cancer Institute (NCI)	USA	2017	2019	Not yet recruit-ing
NCT02890329	Ipilimumab and decitabine in treating patients with relapsed or refractory myelodysplastic syndrome or acute myeloid	CTLA-4	Ipilimumab	Decitabine	I	r/r AML or de novo AML in elderly patients	Toxicity, MTD	48 (AML + MDS)	National Cancer Institute (NCI)	USA	2017	2019	Not yet recruit-ing
NCT02892318	A study evaluat-ing the safety and pharmacology of atezolizumab administered in combination with immunomodula-tory agents in participants with acute myeloid leukemia (AML)	PD-L1	Atezolizumab	Gua-decitabine, possibly other im-munomodu-latory agents	I	r/r AML or de novo AML in elderly patients	Toxic-ity, clinical response (CR, CRi, CRp, duration of response)	40	Hoffmann-La Roche	USA	2016	2019	Recruit-ing

NCT02953561	Avelumab (antiPDL1) and azacytidine in acute myeloid leukemia (AML)	PD-L1	Avelumab	Azacitidine	I/II	r/r AML	Toxicity	52	M.D. Anderson Cancer Center	USA	2017	2020	Not yet recruit-ing
NCT02996474	Pembrolizumab and decitabine for refractory or relapsed acute myeloid leukemia	PD-1	Pembroli-zumab	Decitabine	I/II	r/r AML	Feasibility	15	National Heart, Lung, and Blood Institute (NHLBI)	USA	2016	2019	Not yet recruit-ing

While the results of these studies have to be awaited to judge the potential of checkpoint inhibitors as a monotherapy for AML, various combination therapies are already tested in clinical trials. A phase II study is combining lymphodepletion with a fludarabine/melphalane regimen followed by autologous stem cell transplantation with anti-PD-1 therapy with the goal to reduce relapse rates in non-favorable AML patients in remission (NCT02771197). The combination of standard high-dose cytarabine with anti-PD-1 therapy is tested as a salvage therapy in a phase II study planned to recruit 37 patients with r/r AML (NCT02768792). And a phase I/II study analyzes the maximal tolerable dose of an anti-PD-1 antibody in addition to idarubicin and cytarabine for induction of de novo AML (NCT02464657). No results for any of these studies have been reported so far. The combination of a PD-1 antibody with a vaccination strategy based on AML DC hybridoma is described in the DC chapter below (NCT01096602, Table 5).

Table 5: Current clinical trials using dendritic cell vaccination for immunotherapy of AML

Study identifier	Study name	Type of vaccine	Antigen/ target	Antigen source	Combination therapy	Clinical phase	Indication (AML only)	Primary endpoints	(Estimated) Enrollment	Sponsor	Country	Study start	(Estimated) Completion date	Status
NCT00100971	Vaccine therapy in treating patients with acute myeloid leukemia	Fusion of dendritic and leukemic cells	Multiple	Inherent	n.a.	I	De novo AML	MTD, toxicity	9	Boston Medical Center	USA	2004	2007	Terminated early due to slow accrual
NCT00136422	Study of vaccination with autologous acute myeloblastic leukemia cells in patients with advanced myelodysplasia or acute myelogenous leukemia	Lethally irradiated and genetically modified autologous AML cells	Multiple	Inherent	n.a.	I	r/r AML or de novo AML in non-fit patients	Feasibility	30	Dana-Farber Cancer Institute	USA	2000	2006	Completed
NCT00510133	A study of active immunotherapy with GRNVAC1 in patients with acute myelogenous leukemia (AML)	Monocyte-derived dendritic cells	hTERT	mRNA	n.a.	II	AML in CR1 or CR2	Feasibility	21	Asterias Biotherapeutics, Inc.	USA	2007	2014	Completed

NCT00514189	Feasibility study of acute myelogenous leukemia mRNA plus lysate-loaded dendritic cell vaccines	Mono-cyte-derived dendritic cells	Multiple	AML mRNA + lysate	n.a.	I	De novo AML with non-favorable cytoge-netics or AML in first relapse	Feasibil-ity, toxicity, immunoge-nicity	2	M.D. Anderson Cancer Center	USA	2007	2009	Termi-nated early due to slow accrual
NCT00834002	Dendritic cell vaccination for patients with acute myeloid leukemia in remis-sion (CCRG 05–001)	Mono-cyte-derived dendritic cells	WT1	mRNA	n.a.	I/II	AML in CR/ PR with WT1 overex-pression and high risk of relapse	Feasibility, toxicity	10	University Hospital, Antwerp	Belgium	2005	2008	Com-pleted
NCT00963521	Vaccine thera-py in treating patients with acute ,myeloid leukemia in complete	In vitro-differ-entiated leukemic blasts	Multiple	Inherent	n.a.	I	AML in CR (CR2 or later)	Toxicity	10	Institut Paoli-Calmettes	France	2008	2011	Com-pleted
NCT00965224	Efficacy of dendritic cell therapy for myeloid leukemia and myeloma	Mono-cyte-derived dendritic cells	WT1	mRNA	n.a.	II	AML in CR with high risk of relapse	Immuno-genicity, molecular response	50	University Hospital, Antwerp	Belgium	2010	2014	Enroll-ing by invita-tion
NCT01096602	Blockade of PD-1 in con-junction with the dendritic cell/AML vac-cine following chemotherapy induced	Dendritic cell AML fusion vaccine	Multiple	Inherent	PD1 blockade, GM-CSF	II	AML at initial diagnosis or at first relapse	Toxicity	63	Beth Israel Deaconess Medical Center	USA	2010	2017	Active, not recruit-ing

NCT01146262	Vaccination by leukemic apoptotic corpse autologous pulsed dendritic cells for acute myelogenous leukemia (AML) patients in first or second complete remission (CR) (CDlaM)	Monocyte-derived dendritic cells	Multiple	AML apoptotic corpse	n.a.	I/II	AML in CR2 or refractory AML or de novo AML with unfavorable cytogenetics; no eligibility for allo-HSCT	Toxicity	5	Nantes University Hospital	France	2009	2017	Active, not recruiting
NCT01373515	Leukemic dendritic cell vaccination in patients with acute myeloid leukemia	Dendritic-like cells generated from standardized allogeneic AML cells	Multiple	Inherent	n.a.	I/II	AML in CR2 or relapsed AML or de novo AML; no eligibility for intensive therapy	Feasibility, toxicity	12	DCPrime BV	Netherlands	2011	2013	Completed
NCT01686334	Efficacy study of dendritic cell vaccination in patients with acute myeloid leukemia in remission (WIDEA)	Monocyte-derived dendritic cells	WT1	mRNA	n.a.	II	AML in CR or Cri; WT1 overexpression	Clinical response (RR, DFS, OS)	138	University Hospital, Antwerp	Belgium	2012	2020	Recruiting

NCT01734304	DC vaccination for postremission therapy in AML	Monocyte-derived dendritic cells	WT1, PRAME	mRNA	n.a.	I/II	AML in CR or CRi with non-favorable risk profile; no eligibility for allo-	Feasibility, toxicity	20	Ludwig-Maximilians-University of Munich	Germany	2012	2017	Recruiting
NCT02405338	DC vaccination for postremission therapy in AML	Monocyte-derived dendritic cells	WT1, PRAME	mRNA	n.a.	I/II	AML in CR or Cri; WT1 overexpression; no eligibilty for allo-HSCT	Feasibility, toxicity	20	Medigene AG	Norway	2015	2019	Recruiting

A high interest is currently generated by the idea to combine checkpoint inhibition with HMAs. The evaluation of PD-1 as well as PD-L1 expression in patients with MDS or AML receiving HMAs showed upregulation of both markers on mRNA level [98]. Therefore, several trials are evaluating the efficacy of HMAs combined with either CTLA-4, PD-1, or PD-L1 blocking antibodies (Table 4). First results for this strategy within a phase Ib/II study combining the PD-1 blocking antibody nivolumab with azacitidine in patients with r/r AML have recently been presented. Toxicity was comparable with other trials using checkpoint blockade, and outcomes have been encouraging with a median overall survival of 9.3 months in this study with a predominantly poor-risk patient population [99].

Taken together, checkpoint inhibition in AML is still in its infancy, and results of the currently ongoing trials have to be awaited before further conclusions about the applicability of this concept to AML and the existence of any AML-specific side effects of checkpoint inhibition can be drawn. Combination therapies including checkpoint inhibitors, particularly with HMAs, might turn out to be an important step forward.

DENDRITIC CELL VACCINATION FOR IMMUNOTHERAPY OF AML

Vaccination strategies have the purpose to prime new or enhance pre-existing antigen-specific immune responses. DCs are highly eligible for the induction of tailored, strong, and durable responses (Fig. 1f). This is of particular importance for the treatment of tumor entities with low endogenous immune responses, such as AML. In spite of the high costs and efforts accruing for the production of this patient-specific cellular therapy, DC-based vaccination strategies for the treatment of AML are therefore actively pursued. Important variables in these studies are source of DC precursors, DC maturation protocol, target antigen, way of antigen loading route of application, and interval of application [100]. While monocyte-derived DCs are used in the majority of studies and are considered to induce the strongest immune responses, alternative DC-like constructs are also applied [1].

Recently, an interesting clinical trial has been published presenting 17 AML patients that were vaccinated in CR with a hybridoma of AML cells and autologous DCs [101]. The vaccination was well tolerated, and a considerable increase in leukemia-specific T cells was found that persisted for more than 6 months. High relapse-free survival was described, but a strong selection bias for long-term survivors currently impedes further interpretations. This patient cohort is part of a larger study that is designated to analyze the combinatorial effect of PD-1 blockade with the described vaccination strategy (NCT01096602, see Table 5). However, data for the combination therapy has not been released.

DCPrime uses an off-the-shelf product based on a precursor human dendritic cell line. This platform was tested in a phase I/II study for AML patients (NCT01373515), and vaccinations were well tolerated with induction of multi-functional immune responses, resulting in the preparation of a multi-center phase II study. However, there is no full publication of the study results available at present. To our knowledge, no other clinical trial is currently recruiting patients for vaccination concepts with DC-like cells, as a study based on a fusion concept has been terminated early due to slow accrual (NCT00100971), and two studies using modified leukemic blasts (NCT00136422, NCT00963521) have been completed, but their results have not been published (see Table 5).

Monocyte-derived DCs loaded with various antigens are the most commonly used source for DC vaccination trials. Five clinical studies are currently active or recruiting. A small French study ($n = 5$) uses AML apoptotic corpses to load DCs (NCT01146262). A group in Belgium that has already completed a phase I/II study on vaccination with *WT1* mRNA-loaded DCs for 10 AML patients in remission with high risk of relapse demonstrating immunological as well as clinical responses [102] is now conducting a phase II study testing the induction of immune and molecular responses by vaccination with *WT1* mRNA-loaded DCs for AML as well as chronic myeloid leukemia and multiple myeloma patients (NCT 00965224). Besides, the same group also conducts a large (estimated enrollment, 138 patients) randomized phase II study on AML patients in CR/CRi with WT1 overexpression with the goal to determine clinical effects of DC vaccination in terms of relapse rate, disease-free survival, and overall survival (NCT01686334). Results of this study are eagerly awaited, but are not to be expected before 2020.

Our group in Munich has developed a protocol for the generation of DCs by the use of a TLR7/8 agonist [103, 104]. These DCs show improved immunogenicity compared to conventional monocyte-derived DCs [105]. We are currently conducting a phase I/II proof-of-concept study using this type of DCs loaded with mRNA encoding *WT1* and *PRAME* for intradermal vaccination of AML patients in CR with a non-favorable risk profile

(NCT01734304) [106]. Preliminary results for 13 patients have already been reported at ASH, showing that DC generation is feasible, that their application is safe with delayed-type hypersensitivity reactions at the injection sites, but no serious adverse events, and that novel immune responses to both antigens can be induced. Immune responses were markedly increased by combination of DC vaccination with azacitidine within an individual treatment attempt [107]. A very similar study is conducted by our collaborators in Norway (NCT02405338).

Besides current clinical studies, a few interesting new developments in the field of DCs in the context of AML immunotherapy have been described in the past 2 years. In an effort to further optimize the immunostimulatory capacities of monocyte-derived DCs, electroporation of mRNA encoding both for IL-15 and for IL-15 receptor alpha was shown to result in enhanced NK cell activation [108]. Besides, evidence was provided that monocyte-derived DCs express RHAMM independent of RNA electroporation at a level high enough to induce RHAMM-specific T cells [109].

In conclusion, current data suggests that DC vaccination is particularly successful at inducing novel immune responses. Combining this approach with checkpoint inhibition or immunomodulating agents including HMAs in order to further enhance the immune responses seems an interesting way to follow.

CONCLUSIONS

Immunotherapy of cancer has made unprecedented progress in the past few years. While novel immunotherapeutic strategies have already moved into standard clinical practice for various solid cancers as well as selected hematological neoplasms including ALL, a similar development is lagging behind for the treatment of AML. However, different immunotherapeutic concepts are currently being evaluated in clinical trials, with some promising results already published and a lot more of interesting studies expected to be completed within the next couple of years.

The lack of an appropriate target antigen with a restricted expression pattern similar to CD19 or CD20 for B cell neoplasms is a major obstacle for the application of targeted immunotherapy in AML. This problem is shared by ADCs, T cell-recruiting antibody constructs and CAR T cell constructs, where promising leukemia-specific responses seen in early clinical trials are often accompanied by severe on-target off-leukemia toxicity to the myeloid compartment. CD33 and CD123 are the major target antigens of constructs in clinical development so far. Results of the ongoing clinical trials need to be awaited in order to weigh potential benefits vs. side effects. In order to prospectively reduce on-target off-leukemia toxicities, several strategies are followed: The identification of novel leukemia-associated antigens could provide more specific targets. Comprehensive transcriptomic and proteomic analysis is ongoing to fully characterize the AML surfaceome [110]. Alternatively, leukemia-specific neoantigens arising from AML-associated mutations should be further evaluated as source of novel target molecules. Furthermore, dual-targeting approaches could improve treatment specificity while relying on combinations of already known AML-associated antigens.

ADCs have already proven their therapeutic potential in AML. Results of currently running clinical trials will help to identify the optimal clinical setting and to better estimate the risk-benefit ratio. In contrast, T cell-recruiting antibodies and CAR T cell constructs are still in the early phase of clinical development for the therapy of AML, with several currently running phase I trials studying the feasibility and toxicity of their application. Activation of endogenous T cell responses through checkpoint blockade and/or DC vaccines appears to be safe, but has yet to demonstrate its clinical potency when used as a monotherapy for the treatment of AML. Different combinations including HMAs to modulate immune responsiveness appear suitable and are increasingly being tested.

While immunotherapy in AML is complicated by different characteristics including lack of an AML-specific target antigen, low mutational burden resulting in low endogenous immune responses and intrinsic resistance mechanisms of the leukemic blasts against immune responses, remarkable progress has been made with different strategies in the past few years. Hope is high that alternative immunotherapeutic strategies with less treatment-related morbidity and mortality compared to allogeneic HSCT will move into clinical practice within the coming years. Still, many further steps have to be taken before the vision of an individualized immunotherapy for each AML patient based on risk factors and biomarkers can become clinical reality.

AUTHORS' CONTRIBUTIONS

FSL and MS developed the concept for the article and revised the complete manuscript. FSL contributed the sections "Abstract", "Checkpoint inhibition", "Dendritic cell vaccination", and "Conclusions". CK wrote the sections "Antibody-drug conjugates" and "T cell-engaging antibody constructs". SH authored the section about "CAR T cells". TK contributed to the different sections regarding combination strategies and created the figure. MS wrote the section "Background". All authors read and approved the final manuscript.

REFERENCES

1. Lichtenegger FS, Schnorfeil FM, Hiddemann W, Subklewe M. Current strategies in immunotherapy for acute myeloid leukemia. Immunotherapy. 2013;5(1):63–78. doi: 10.2217/imt.12.145.
2. Lichtenegger FS, Krupka C, Köhnke T, Subklewe M. Immunotherapy for acute myeloid leukemia. Semin Hematol. 2015;52(3):207–14. doi: 10.1053/j.seminhematol.2015.03.006.
3. Cornelissen JJ, Gratwohl A, Schlenk RF, Sierra J, Bornhäuser M, Juliusson G, et al. The European LeukemiaNet AML Working Party consensus statement on allogeneic HSCT for patients with AML in remission: an integrated-risk adapted approach. Nat Rev Clin Oncol. 2012;9(10):579–90. doi: 10.1038/nrclinonc.2012.150.
4. Stelljes M, Krug U, Beelen DW, Braess J, Sauerland MC, Heinecke A, et al. Allogeneic transplantation versus chemotherapy as postremission therapy for acute myeloid leukemia: a prospective matched pairs analysis. J Clin Oncol. 2014;32(4):288–96. doi: 10.1200/JCO.2013.50.5768.
5. Ramos NR, Mo CC, Karp JE, Hourigan CS. Current approaches in the treatment of relapsed and refractory acute myeloid leukemia. J Clin Med. 2015;4(4):665–95. doi: 10.3390/jcm4040665.
6. Thomas DA, O'Brien S, Kantarjian HM. Monoclonal antibody therapy with rituximab for acute lymphoblastic leukemia. Hematol Oncol Clin North Am. 2009;23(5):949–71. doi: 10.1016/j.hoc.2009.07.005.
7. Kantarjian HM, DeAngelo DJ, Stelljes M, Martinelli G, Liedtke M, Stock W, et al. Inotuzumab ozogamicin versus standard therapy for acute lymphoblastic leukemia. N Engl J Med. 2016;375(8):740–53. doi: 10.1056/NEJMoa1509277.
8. Aldoss I, Bargou RC, Nagorsen D, Friberg GR, Baeuerle PA, Forman SJ. Redirecting T cells to eradicate B-cell acute lymphoblastic leukemia: bispecific T-cell engagers and chimeric antigen receptors. Leukemia. 2017. [PubMed]
9. Topp MS, Gökbuget N, Stein AS, Zugmaier G, O'Brien S, Bargou RC, et al. Safety and activity of blinatumomab for adult patients with relapsed or refractory B-precursor acute lymphoblastic leukaemia: a multicentre, single-arm, phase 2 study. Lancet Oncol. 2015;16(1):57–66. doi: 10.1016/S1470-2045(14)71170-2.
10. Holzinger A, Barden M, Abken H. The growing world of CAR

T cell trials: a systematic review. Cancer Immunol Immunother. 2016;65(12):1433–50. doi: 10.1007/s00262-016-1895-5.

11. Rotiroti MC, Arcangeli S, Casucci M, Perriello V, Bondanza A, Biondi A, et al. Acute myeloid leukemia targeting by chimeric antigen receptor T cells: bridging the gap from preclinical modeling to human studies. Hum Gene Ther. 2016.
12. Hoseini SS, Cheung NK. Acute myeloid leukemia targets for bispecific antibodies. Blood Cancer J. 2017;7(2) doi: 10.1038/bcj.2017.2.
13. Hude I, Sasse S, Engert A, Bröckelmann PJ. The emerging role of immune checkpoint inhibition in malignant lymphoma. Haematologica. 2017;102(1):30–42. doi: 10.3324/haematol.2016.150656.
14. Peters C, Brown S. Antibody-drug conjugates as novel anti-cancer chemotherapeutics. Biosci Rep. 2015;35(4) doi: 10.1042/BSR20150089.
15. Bross PF, Beitz J, Chen G, Chen XH, Duffy E, Kieffer L, et al. Approval summary: gemtuzumab ozogamicin in relapsed acute myeloid leukemia. Clin Cancer Res. 2001;7(6):1490–6.
16. Petersdorf H, Kopecky J, Slovak M, Willman C, Nevill T, Brandwein J, et al. A phase III study of gemtuzumab ozogamicin during induction and post-consolidation therapy in younger patients with acute myeloid leukemia. Blood. 2013;121(24):4854–60. doi: 10.1182/blood-2013-01-466706.
17. Hills RK, Castaigne S, Appelbaum FR, Delaunay J, Petersdorf S, Othus M, et al. Addition of gemtuzumab ozogamicin to induction chemotherapy in adult patients with acute myeloid leukaemia: a meta-analysis of individual patient data from randomised controlled trials. Lancet Oncol. 2014;15(9):986–96. doi: 10.1016/S1470-2045(14)70281-5.
18. Loke J, Khan JN, Wilson JS, Craddock C, Wheatley K. Mylotarg has potent anti-leukaemic effect: a systematic review and meta-analysis of anti-CD33 antibody treatment in acute myeloid leukaemia. Ann Hematol. 2015;94(3):361–73. doi: 10.1007/s00277-014-2218-6.
19. Amadori S, Suciu S, Selleslag D, Aversa F, Gaidano G, Musso M, et al. Gemtuzumab ozogamicin versus best supportive care in older patients with newly diagnosed acute myeloid leukemia unsuitable for intensive chemotherapy: results of the randomized phase III EORTC-GIMEMA AML-19 trial. J Clin Oncol. 2016;34(9):972–9. doi: 10.1200/

JCO.2015.64.0060.

20. Nand S, Othus M, Godwin JE, Willman CL, Norwood TH, Howard DS, et al. A phase 2 trial of azacitidine and gemtuzumab ozogamicin therapy in older patients with acute myeloid leukemia. Blood. 2013;122(20):3432–9. doi: 10.1182/blood-2013-06-506592.
21. Walter RB, Medeiros BC, Gardner KM, Orlowski KF, Gallegos L, Scott BL, et al. Gemtuzumab ozogamicin in combination with vorinostat and azacitidine in older patients with relapsed or refractory acute myeloid leukemia: a phase I/II study. Haematologica. 2014;99(1):54–9. doi: 10.3324/haematol.2013.096545.
22. Daver N, Kantarjian H, Ravandi F, Estey E, Wang X, Garcia-Manero G, et al. A phase II study of decitabine and gemtuzumab ozogamicin in newly diagnosed and relapsed acute myeloid leukemia and high-risk myelodysplastic syndrome. Leukemia. 2016;30(2):268–73. doi: 10.1038/leu.2015.244.
23. Nguyen DH, Ball ED, Varki A. Myeloid precursors and acute myeloid leukemia cells express multiple CD33-related Siglecs. Exp Hematol. 2006;34(6):728–35. doi: 10.1016/j.exphem.2006.03.003.
24. Ehninger A, Kramer M, Röllig C, Thiede C, Bornhäuser M, von Bonin M, et al. Distribution and levels of cell surface expression of CD33 and CD123 in acute myeloid leukemia. Blood Cancer J. 2014;4 doi: 10.1038/bcj.2014.39.
25. Krupka C, Kufer P, Kischel R, Zugmaier G, Bögeholz J, Köhnke T, et al. CD33 target validation and sustained depletion of AML blasts in long-term cultures by the bispecific T-cell-engaging antibody AMG 330. Blood. 2014;123(3):356–65. doi: 10.1182/blood-2013-08-523548.
26. McCombs JR, Owen SC. Antibody drug conjugates: design and selection of linker, payload and conjugation chemistry. AAPS J. 2015;17(2):339–51. doi: 10.1208/s12248-014-9710-8.
27. Linenberger ML. CD33-directed therapy with gemtuzumab ozogamicin in acute myeloid leukemia: progress in understanding cytotoxicity and potential mechanisms of drug resistance. Leukemia. 2005;19(2):176–82. doi: 10.1038/sj.leu.2403598.
28. Kung Sutherland MS, Walter RB, Jeffrey SC, Burke PJ, Yu C, Kostner H, et al. SGN-CD33A: a novel CD33-targeting antibody-drug conjugate using a pyrrolobenzodiazepine dimer is active in models of drug-resistant AML. Blood. 2013;122(8):1455–63. doi: 10.1182/

blood-2013-03-491506.

29. Bixby DL, Stein AS, Fathi AT, Kovavsovics TJ, Levy MY, Erba HP, et al. Vadastuximab talirine monotherapy in older patients with treatment naive CD33-positive acute myeloid leukemia (AML). Blood. 2016;128:590.
30. Fathi AT, Erba HP, Lancet JE, Stein EM, Ravandi F, Faderl S, et al. Vadastuximab talirine plus hypomethylating agents: a well-tolerated regimen with high remission rate in frontline older patients with acute myeloid leukemia (AML) Blood. 2016;128:591.
31. Erba HP, Levy MY, Vasu S, Stein AS, Fathi AT, Maris MB, et al. A phase 1b study of vadastuximab talirine in combination with 7 + 3 induction therapy for patients with newly diagnosed acute myeloid leukemia (AML). Blood. 2016;128:211.
32. Seattle Genetics, Inc. Seattle Genetics announces clinical hold on several phase 1 trials of vadastuximab talirine (SGN-CD33A). http://www.businesswire.com/news/home/20161227005087/en/Seattle-Genetics-Announces-Clinical-Hold-Phase-1. Accessed 6 Mar 2017.
33. Testa U, Pelosi E, Frankel A. CD 123 is a membrane biomarker and a therapeutic target in hematologic malignancies. Biomark Res. 2014;2(1):4. doi: 10.1186/2050-7771-2-4.
34. Krystal WM, Walker R, Fishkin N, Audette C, Kovtun Y, Romanelli A. IMGN779, a CD33-targeted antibody-drug conjugate (ADC) with a novel DNA-alkylating effector molecule, induces DNA damage, cell cycle arrest, and apoptosis in AML cells. Blood. 2015;126:1366.
35. Whiteman KR, Noordhuis P, Walker R, Watkins K, Kovtun Y, Harvey L, et al. The antibody-drug conjugate (ADC) IMGN779 is highly active in vitro and in vivo against acute myeloid leukemia (AML) with FLT3-ITD mutations. Blood. 2014;124:2321.
36. Portwood S, Puchalski RA, Walker RM, Wang ES. Combining IMGN779, a novel anti-CD33 antibody-drug conjugate (ADC), with the PARP inhibitor, olaparib, results in enhanced anti-tumor activity in preclinical acute myeloid leukemia (AML) models. Blood. 2016;128(22):1645.
37. Laszlo GS, Gudgeon CJ, Harrington KH, Dell'Aringa J, Newhall KJ, Means GD, et al. Cellular determinants for preclinical activity of a novel CD33/CD3 bispecific T-cell engager (BiTE) antibody, AMG 330, against human AML. Blood. 2014;123(4):554–61. doi: 10.1182/

blood-2013-09-527044.

38. Borthakur G, Rosenblum MG, Talpaz M, Daver N, Ravandi F, Faderl S, et al. Phase 1 study of an anti-CD33 immunotoxin, humanized monoclonal antibody M195 conjugated to recombinant gelonin (HUM-195/rGEL), in patients with advanced myeloid malignancies. Haematologica. 2013;98(2):217–21. doi: 10.3324/haematol.2012.071092.
39. Noordhuis P, Terwinjn M, Rutten AP, Smit L, Ossenkoppele GJ, Schuurhuis GJ. Targeting of CLEC12A in acute myeloid leukemia by antibody-drug-conjugates and bispecific CLL-1 × CD3 BiTE antibody. Blood. 2010;116:2890.
40. Kim SY, Theunissen JW, Balibalos J, Liao-Chan S, Babcock MC, Wong T, et al. A novel antibody-drug conjugate targeting SAIL for the treatment of hematologic malignancies. Blood Cancer J. 2015;5 doi: 10.1038/bcj.2015.39.
41. Buckley SA, Walter RB. Antigen-specific immunotherapies for acute myeloid leukemia. Hematology Am Soc Hematol Educ Program. 2015;2015(1):584–95.
42. Flynn MJ, van Berkel PH, Zammarchi F, Tyrer PC, Akarca AU, Janghra N, et al. Mechanistic and pharmacodynamic studies of adct-301, a pyrrolobenzodiazepine (PBD) dimer-containing antibody drug conjugate (ADC) targeting CD25-expressing hematological malignancies. Blood. 2015;126:1559.
43. Rudra-Ganguly N, Lowe C, Virata C, Leavitt M, Jin L, Mendelsohn B, et al. AGS62P1, a novel anti-FLT3 antibody drug conjugate, employing site specific conjugation, demonstrates preclinical anti-tumor efficacy in AML tumor and patient derived xenografts. Blood. 2015;126:3806.
44. Baeuerle PA, Reinhardt C. Bispecific T-cell engaging antibodies for cancer therapy. Cancer Res. 2009;69(12):4941–4. doi: 10.1158/0008-5472.CAN-09-0547.
45. Kontermann RE, Brinkmann U. Bispecific antibodies. Drug Discov Today. 2015;20(7):838–47. doi: 10.1016/j.drudis.2015.02.008.
46. Kantarjian H, Stein A, Gökbuget N, Fielding AK, Schuh AC, Ribera JM, et al. Blinatumomab versus chemotherapy for advanced acute lymphoblastic leukemia. N Engl J Med. 2017;376(9):836–47. doi: 10.1056/NEJMoa1609783.
47. Aigner M, Feulner J, Schaffer S, Kischel R, Kufer P, Schneider K, et al. T lymphocytes can be effectively recruited for ex vivo and in vivo

lysis of AML blasts by a novel CD33/CD3-bispecific BiTE antibody construct. Leukemia. 2013;27(5):1107–15. doi: 10.1038/leu.2012.341.

48. Krupka C, Kufer P, Kischel R, Zugmaier G, Lichtenegger FS, Köhnke T, et al. Blockade of the PD-1/PD-L1 axis augments lysis of AML cells by the CD33/CD3-BiTE® antibody construct AMG 330: reversing a T-cell induced immune escape mechanism. Leukemia. 2015;30(2):484–91. doi: 10.1038/leu.2015.214.

49. Arndt C, Feldmann A, von Bonin M, Cartellieri M, Ewen EM, Koristka S, et al. Costimulation improves the killing capability of T cells redirected to tumor cells expressing low levels of CD33: description of a novel modular targeting system. Leukemia. 2014;28(1):59–69. doi: 10.1038/leu.2013.243.

50. Reusch U, Harrington KH, Gudgeon CJ, Fucek I, Ellwanger K, Weichel M, et al. Characterization of CD33/CD3 tetravalent bispecific tandem diabodies (TandAbs) for the treatment of acute myeloid leukemia. Clin Cancer Res. 2016;22(23):5829–38. doi: 10.1158/1078-0432.CCR-16-0350.

51. Moore PA, Zhang W, Rainey GJ, Burke S, Li H, Huang L, et al. Application of dual affinity retargeting molecules to achieve optimal redirected T-cell killing of B-cell lymphoma. Blood. 2011;117(17):4542–51. doi: 10.1182/blood-2010-09-306449.

52. Rader C. DARTs take aim at BiTEs. Blood. 2011;117(17):4403–4. doi: 10.1182/blood-2011-02-337691.

53. AL Hussaini MH, Ritchey J, Rettig MP, Eissenberg L, Uy GL, Chichili G, et al. Targeting CD123 in leukemic stem cells using dual affinity re-targeting molecules (DARTs®) [abstract] Blood. 2013;122(21):360.

54. Chichili GR, Huang L, Li H, Burke S, He L, Tang Q, et al. A CD3xCD123 bispecific DART for redirecting host T cells to myelogenous leukemia: preclinical activity and safety in nonhuman primates. Sci Transl Med. 2015;7(289):289ra82. doi: 10.1126/scitranslmed.aaa5693.

55. Chu SY, Pong E, Chen H, Phung S, Chan EW, Endo NA, et al. Immunotherapy with long-lived anti-CD123 × anti-CD3 bispecific antibodies stimulates potent T cell-mediated killing of human AML cell lines and of CD123+ cells in monkeys: a potential therapy for acute myelogenous leukemia. Blood. 2014;124:2316. doi: 10.1182/blood-2014-05-576900.

56. Forslund A, Syed K, Axel A, McDaid R, Li Y, Ballesteros J, et al. Ex

vivo activity profile of the CD123xCD3 Duobody® antibody JNJ-63709178 against primary acute myeloid leukemia bone marrow samples. Blood. 2016;128:2875. doi: 10.1182/blood-2016-10-746719.

57. Gaudet F, Nemeth JF, McDaid R, Li Y, Harman B, Millar H, et al. Development of a CD123xCD3 bispecific antibody (JNJ-63709178) for the treatment of acute myeloid leukemia (AML) Blood. 2016;128:2824. doi: 10.1182/blood-2016-05-715987.

58. Bakker ABH, Oudenrijn SVD, Bakker AQ, Feller N, Meijer MV, Bia JA, et al. C-type lectin-like molecule-1: a novel myeloid cell surface marker associated with acute myeloid leukemia. Cancer Res. 2004;64(22):8443–50. doi: 10.1158/0008-5472.CAN-04-1659.

59. Rhenen AV, Dongen GAMSV, Kelder A, Rombouts EJ, Feller N, Moshaver B, et al. The novel AML stem cell–associated antigen CLL-1 aids in discrimination between normal and leukemic stem cells. Blood. 2007;110(7):2659–66. doi: 10.1182/blood-2007-03-083048.

60. Van Loo PF, Doornbos R, Dolstra H, Shamsili S, Bakker L. Preclinical evaluation of MCLA117, a CLEC12AxCD3 bispecific antibody efficiently targeting a novel leukemic stem cell associated antigen in AML. Blood. 2015;126:325.

61. Roskopf CC, Braciak TA, Fenn NC, Kobold S, Fey GH, Hopfner KP, et al. Dual-targeting triplebody 33-3-19 mediates selective lysis of biphenotypic CD19+ CD33+ leukemia cells. Oncotarget. 2016;7(16):22579–89. doi: 10.18632/oncotarget.8022.

62. Polu KR, Lowman HB. Probody therapeutics for targeting antibodies to diseased tissue. Expert Opin Biol Ther. 2014;14(8):1049–53. doi: 10.1517/14712598.2014.920814.

63. LaPorte SL, Hostetter DR, Wong L, Razo J, Diep L, White CW, et al. CD3-EGFR bispecific Probody™ therapeutics induced tumor regressions and increased therapeutic window in preclinical studies. Proceedings of the AACR-NCI-EORTC International Conference: Molecular Targets and Cancer Therapeutics. 2015;14:Abstract nr A203.

64. Zugmaier G, Klinger M, Schmidt M, Subklewe M. Clinical overview of anti-CD19 BiTE(®) and ex vivo data from anti-CD33 BiTE(®) as examples for retargeting T cells in hematologic malignancies. Mol Immunol. 2015.

65. Köhnke T, Krupka C, Tischer J, Knösel T, Subklewe M. Increase of PD-L1 expressing B-precursor ALL cells in a patient resistant to

the CD19/CD3-bispecific T cell engager antibody blinatumomab. J Hematol Oncol. 2015;8:111. doi: 10.1186/s13045-015-0213-6.

66. Gross G, Waks T, Eshhar Z. Expression of immunoglobulin-T-cell receptor chimeric molecules as functional receptors with antibody-type specificity. Proc Natl Acad Sci U S A. 1989;86(24):10024–8. doi: 10.1073/pnas.86.24.10024.
67. Finney HM, Lawson AD, Bebbington CR, Weir AN. Chimeric receptors providing both primary and costimulatory signaling in T cells from a single gene product. J Immunol. 1998;161(6):2791–7.
68. Turtle CJ, Hanafi LA, Berger C, Gooley TA, Cherian S, Hudecek M, et al. CD19 CAR-T cells of defined CD4+:CD8+ composition in adult B cell ALL patients. J Clin Invest. 2016;126(6):2123–38. doi: 10.1172/JCI85309.
69. Locke FL, Neelapu SS, Bartlett NL, Siddiqi T, Chavez JC, Hosing CM, et al. Phase 1 results of ZUMA-1: a multicenter study of KTE-C19 anti-CD19 CAR T cell therapy in refractory aggressive lymphoma. Mol Ther. 2017;25(1):285–95. doi: 10.1016/j.ymthe.2016.10.020.
70. O'Leary M. FDA pilot project to develop a clinical database to examine safety in trials using CAR T-cells. http://www.ema.europa.eu/docs/en_GB/document_library/Presentation/2016/12/WC500217505.pdf. Accessed 6 Mar 2017.
71. Juno Therapeutics, Inc. Juno Therapeutics reports fourth quarter and 2016 financial results. http://ir.junotherapeutics.com/phoenix.zhtml?c=253828&p=irol-newsArticle_Print&ID=2250772. Accessed 6 Mar 2017.
72. Porter DL, Hwang WT, Frey NV, Lacey SF, Shaw PA, Loren AW, et al. Chimeric antigen receptor T cells persist and induce sustained remissions in relapsed refractory chronic lymphocytic leukemia. Sci Transl Med. 2015;7(303):303ra139. doi: 10.1126/scitranslmed.aac5415.
73. Chevallier P, Robillard N, Ayari S, Guillaume T, Delaunay J, Mechinaud F, et al. Persistence of CD33 expression at relapse in CD33(+) acute myeloid leukaemia patients after receiving gemtuzumab in the course of the disease. Br J Haematol. 2008;143(5):744–6. doi: 10.1111/j.1365-2141.2008.07153.x.
74. O'Hear C, Heiber JF, Schubert I, Fey G, Geiger TL. Anti-CD33 chimeric antigen receptor targeting of acute myeloid leukemia. Haematologica.

2015;100(3):336–44. doi: 10.3324/haematol.2014.112748.

75. Kenderian SS, Ruella M, Shestova O, Klichinsky M, Aikawa V, Morrissette JJ, et al. CD33-specific chimeric antigen receptor T cells exhibit potent preclinical activity against human acute myeloid leukemia. Leukemia. 2015;29(8):1637–47. doi: 10.1038/leu.2015.52.

76. Mardiros A, Dos Santos C, McDonald T, Brown CE, Wang X, Budde LE, et al. T cells expressing CD123-specific chimeric antigen receptors exhibit specific cytolytic effector functions and anti-tumor effects against human acute myeloid leukemia. Blood. 2013;122(18):3138–48. doi: 10.1182/blood-2012-12-474056.

77. Tettamanti S, Marin V, Pizzitola I, Magnani CF, Giordano Attianese GM, Cribioli E, et al. Targeting of acute myeloid leukaemia by cytokine-induced killer cells redirected with a novel CD123-specific chimeric antigen receptor. Br J Haematol. 2013;161(3):389–401. doi: 10.1111/bjh.12282.

78. Pizzitola I, Anjos-Afonso F, Rouault-Pierre K, Lassailly F, Tettamanti S, Spinelli O, et al. Chimeric antigen receptors against CD33/CD123 antigens efficiently target primary acute myeloid leukemia cells in vivo. Leukemia. 2014;28(8):1596–605. doi: 10.1038/leu.2014.62.

79. Thokala R, Olivares S, Mi T, Maiti S, Deniger D, Huls H, et al. Redirecting specificity of T cells using the sleeping beauty system to express chimeric antigen receptors by mix-and-matching of VL and VH domains targeting CD123+ tumors. PLoS One. 2016;11(8) doi: 10.1371/journal.pone.0159477.

80. Gill S, Tasian SK, Ruella M, Shestova O, Li Y, Porter DL, et al. Preclinical targeting of human acute myeloid leukemia and myeloablation using chimeric antigen receptor-modified T cells. Blood. 2014;123(15):2343–54. doi: 10.1182/blood-2013-09-529537.

81. Casucci M, di Nicolis Robilant B, Falcone L, Camisa B, Norelli M, Genovese P, et al. CD44v6-targeted T cells mediate potent antitumor effects against acute myeloid leukemia and multiple myeloma. Blood. 2013;122(20):3461. doi: 10.1182/blood-2013-04-493361.

82. Kenderian SS, Ruella M, Shestova O, Klichinsky M, Kim MY, Soderquist C, et al. Leukemia stem cells are characterized by CLEC12A expression and chemotherapy refractoriness that can be overcome by targeting with chimeric antigen receptor T cells. Blood. 2016;128:766.

83. Chien CD, Sauter CT, Ishii K, Nguyen SM, Shen F, Tasian SK, et

al. Preclinical development of FLT3-redirected chimeric antigen receptor T cell immunotherapy for acute myeloid leukemia. Blood. 2016;128:1072. doi: 10.1182/blood-2015-11-683730.

84. Lynn RC, Feng Y, Schutsky K, Poussin M, Kalota A, Dimitrov DS, et al. High-affinity FRβ-specific CAR T cells eradicate AML and normal myeloid lineage without HSC toxicity. Leukemia. 2016;30(6):1355–64. doi: 10.1038/leu.2016.35.

85. Peinert S, Prince HM, Guru PM, Kershaw MH, Smyth MJ, Trapani JA, et al. Gene-modified T cells as immunotherapy for multiple myeloma and acute myeloid leukemia expressing the Lewis Y antigen. Gene Ther. 2010;17(5):678–86. doi: 10.1038/gt.2010.21.

86. Chang YH, Connolly J, Shimasaki N, Mimura K, Kono K, Campana D. A chimeric receptor with NKG2D specificity enhances natural killer cell activation and killing of tumor cells. Cancer Res. 2013;73(6):1777–86. doi: 10.1158/0008-5472.CAN-12-3558.

87. Ma Q, Garber HR, Lu S, He H, Tallis E, Ding X, et al. A novel TCR-like CAR with specificity for PR1/HLA-A2 effectively targets myeloid leukemia in vitro when expressed in human adult peripheral blood and cord blood T cells. Cytotherapy. 2016;18(8):985–94. doi: 10.1016/j.jcyt.2016.05.001.

88. Ritchie DS, Neeson PJ, Khot A, Peinert S, Tai T, Tainton K, et al. Persistence and efficacy of second generation CAR T cell against the LeY antigen in acute myeloid leukemia. Mol Ther. 2013;21(11):2122–9. doi: 10.1038/mt.2013.154.

89. Wang QS, Wang Y, Lv HY, Han QW, Fan H, Guo B, et al. Treatment of CD33-directed chimeric antigen receptor-modified T cells in one patient with relapsed and refractory acute myeloid leukemia. Mol Ther. 2014;23(1):184–91. doi: 10.1038/mt.2014.164.

90. Guzman ML, Sugita M, Zong H, Ewing-Crystal N, Trujillo-Alonso V, Mencia-Trinchant N, et al. Allogeneic Tcrα/β deficient CAR T-cells targeting CD123 prolong overall survival of AML patient-derived xenografts. Blood. 2016;128(22):765.

91. Lanitis E, Poussin M, Klattenhoff AW, Song D, Sandaltzopoulos R, June CH, et al. Chimeric antigen receptor T cells with dissociated signaling domains exhibit focused antitumor activity with reduced potential for toxicity in vivo. Cancer Immunol Res. 2013;1(1):43–53. doi: 10.1158/2326-6066.CIR-13-0008.

92. Zah E, Lin MY, Silva-Benedict A, Jensen MC, Chen YY. T cells expressing CD19/CD20 bispecific chimeric antigen receptors prevent antigen escape by malignant B cells. Cancer Immunol Res. 2016;4(6):498–508. doi: 10.1158/2326-6066.CIR-15-0231.

93. Arndt C, Feldmann A, Koristka S, Cartellieri M, von Bonin M, Ehninger A, et al. Improved killing of AML blasts by dual-targeting of CD123 and CD33 via UniTARG a novel antibody-based modular T cell retargeting system. Blood. 2015;126:2565.

94. Arcangeli S, Bardelli M, Tettamanti S, Rotiroti MC, Simonelli L, Magnani CF, et al. Unraveling the efficacy and safety profiles of anti-CD123 chimeric antigen receptors (CARs) in a model of acute myeloid leukemia immunotherapy by investigating CAR binding affinity and density variables. Blood. 2015;126:1359.

95. Resetca D, Neschadim A, Medin JA. Engineering hematopoietic cells for cancer immunotherapy: strategies to address safety and toxicity concerns. J Immunother. 2016;39(7):249–59. doi: 10.1097/CJI.0000000000000134.

96. Alatrash G, Daver N, Mittendorf EA. Targeting immune checkpoints in hematologic malignancies. Pharmacol Rev. 2016;68(4):1014–25. doi: 10.1124/pr.116.012682.

97. Berger R, Rotem-Yehudar R, Slama G, Landes S, Kneller A, Leiba M, et al. Phase I safety and pharmacokinetic study of CT-011, a humanized antibody interacting with PD-1, in patients with advanced hematologic malignancies. Clin Cancer Res. 2008;14(10):3044–51. doi: 10.1158/1078-0432.CCR-07-4079.

98. Yang H, Bueso-Ramos C, DiNardo C, Estecio MR, Davanlou M, Geng QR, et al. Expression of PD-L1, PD-L2, PD-1 and CTLA4 in myelodysplastic syndromes is enhanced by treatment with hypomethylating agents. Leukemia. 2014;28(6):1280–8. doi: 10.1038/leu.2013.355.

99. Daver N, Basu S, Garcia-Manero G, Cortes JE, Ravandi F, Jabbour EJ, et al. Phase IB/II study of nivolumab in combination with azacytidine (AZA) in patients (pts) with relapsed acute myeloid leukemia (AML) Blood. 2016;128(22):763.

100. Sabado RL, Balan S, Bhardwaj N. Dendritic cell-based immunotherapy. Cell Res. 2017;27(1):74–95. doi: 10.1038/cr.2016.157.

101. Rosenblatt J, Stone RM, Uhl L, Neuberg D, Joyce R, Levine JD, et al.

Individualized vaccination of AML patients in remission is associated with induction of antileukemia immunity and prolonged remissions. Sci Transl Med. 2016;8(368):368ra171. doi: 10.1126/scitranslmed.aag1298.

102. Van Tendeloo VF, Van de Velde A, Van Driessche A, Cools N, Anguille S, Ladell K, et al. Induction of complete and molecular remissions in acute myeloid leukemia by Wilms' tumor 1 antigen-targeted dendritic cell vaccination. Proc Natl Acad Sci U S A. 2010;107(31):13824–9. doi: 10.1073/pnas.1008051107.

103. Spranger S, Javorovic M, Bürdek M, Wilde S, Mosetter B, Tippmer S, et al. Generation of Th1-polarizing dendritic cells using the TLR7/8 agonist CL075. J Immunol. 2010;185(1):738–47. doi: 10.4049/jimmunol.1000060.

104. Beck B, Dörfel D, Lichtenegger FS, Geiger C, Lindner L, Merk M, et al. Effects of TLR agonists on maturation and function of 3-day dendritic cells from AML patients in complete remission. J Transl Med. 2011;9:151. doi: 10.1186/1479-5876-9-151.

105. Lichtenegger FS, Mueller K, Otte B, Beck B, Hiddemann W, Schendel DJ, et al. CD86 and IL-12p70 are key players for T helper 1 polarization and natural killer cell activation by toll-like receptor-induced dendritic cells. PLoS One. 2012;7(9) doi: 10.1371/journal.pone.0044266.

106. Subklewe M, Geiger C, Lichtenegger FS, Javorovic M, Kvalheim G, Schendel DJ, et al. New generation dendritic cell vaccine for immunotherapy of acute myeloid leukemia. Cancer Immunol Immunother. 2014;63(10):1093–103. doi: 10.1007/s00262-014-1600-5.

107. Lichtenegger FS, Deiser K, Rothe M, Schnorfeil FM, Krupka C, Augsberger C, et al. Induction of antigen-specific T-cell responses through dendritic cell vaccination in AML: results of a phase I/II trial and ex vivo enhancement by checkpoint blockade. Blood. 2016;128(22):764.

108. Van den Bergh J, Willemen Y, Lion E, Van Acker H, De Reu H, Anguille S, et al. Transpresentation of interleukin-15 by IL-15/IL-15Rα mRNA-engineered human dendritic cells boosts antitumoral natural killer cell activity. Oncotarget. 2015;6(42):44123–33. doi: 10.18632/oncotarget.6536.

109. Willemen Y, Van den Bergh JM, Bonte SM, Anguille S, Heirman C, Stein BM, et al. The tumor-associated antigen RHAMM (HMMR/

CD168) is expressed by monocyte-derived dendritic cells and presented to T cells. Oncotarget. 2016;7(45):73960–70.

110. Perna F, Berman S, Mansilla-Soto J, Hamieh M, Juthani R, Soni R, et al. Probing the AML surfaceome for chimeric antigen receptor (CAR) targets. Blood. 2016;128(22):526.

8

Assessment by miRNA Microarray of an Autologous Cancer Antigen-pulsed Adoptive Immune Ensemble Cell Therapy (AC-ACT) Approach; Demonstrated Induction of Anti-oncogenic and Anti-PD-L1 miRNAs

Masanobu Chinami[1], Kaoru Iwabuchi[1], Yoshiteru Muto[2], Yasuhiko Uchida[2], Ryu Arita[3], Rana A. Shuraim[1], Chaker N. Adra[1,4]

[1] BFSR Institute, Fukuoka, Japan
[2] The Research Institute of Health Rehabilitation of Tokyo, Tokyo, Japan
[3] Fukuoka MSC Medical Clinics, Fukuoka, Japan
[4] The Adra Institute, Boston, MA, USA

ABSTRACT

A 60-year-old woman with stage IV rectal cancer received adoptive cell therapy with autologous cancer antigen (AC-ACT) causing induction of anti-oncogenic and anti-PD-L1 miRNAs as assessed by miRNA microarray.

Citation: Chinami M, Iwabuchi K, Muto Y, et al. Assessment by miRNA microarray of an autologous cancer antigen-pulsed adoptive immune ensemble cell therapy (AC-ACT) approach; demonstrated induction of anti-oncogenic and anti-PD-L1 miRNAs. Clin Case Rep. 2019;7:2156–2164. https://doi.org/10.1002/ccr3.2343.

More than 1 year after AC-ACT, metastases have been arrested, and the patient reports good quality of life.

Keywords: adoptive cell therapy, cancer, immunotherapy, miRNA analysis, PD-1

INTRODUCTION

A rectal cancer patient (stage IV) was treated with an autologous cancer antigen-pulsed adoptive immune ensemble cell therapy (AC-ACT) and assessed by miRNA microarray. Tumor-suppressive miRNAs for colorectal cancer and the PD-L1 immune checkpoint blocker increased, while tumor-promotive miRNAs decreased.

Three standard cancer therapies, surgery, chemotherapy and radiotherapy, have been established for multiple decades and continued to be developed.1, 2 Recently, a fourth therapy, immunotherapy, has emerged into the therapeutic arena.1, 2 In vivo administration of anticancer agents, even immune checkpoint blockers, give problematic side effects.3 In contrast, ex vivo administration via adoptive cell therapy (ACT) gives reduced adverse effects.4 Adoptive immune cells therapies (ACT) have been performed extensively (>10 years), at many university hospitals and private clinics in Japan, and on more than 5000 patients with cancer including end-stage patients who had exhausted standard therapies. Adoptive immune cell therapies have fairly good performance of therapeutic results in comparison with standard therapies. In this case report, we describe a further improvement of this method by adding autologous antigens to the ACT and using miRNA to assess impact.

Most advanced stage of patients with cancer are suppressed in their antitumor immunity and eventually develop a cachexia, with lymphopenia, fatigue, anorexia, loss of adipose, and muscle tissue. In cancer immunotherapy, T cells, NK cells, NKT cells, DC cells, and others are used for ACT.5 Vaccination with Wilm's tumor peptide 1 (WT1) for patients with cancer is also widely used.6 However, WT1 and other major cancer peptides remain of limited utility for personalized cancer vaccines due to neo-antigens produced by mutations revealed by NGS analysis.7, 8 Thus, highly specific tumor antigens for individual patients are a necessary for therapeutic advance. Autologous cancer antigens from surgically excised tumor material may be ideal. These antigens have been used for in vivo vaccination in clinical cases and have improved outcomes.9

In contrast to in vivo vaccination, adoptive transfer of immune effector cells is able to modulate immunity ex vivo. This method is a highly personalized cancer therapy that involves administration to the cancer-bearing host of immune cells with direct anticancer activity. We adopted ex vivo vaccination to dendritic cells (DC) with autologous cancer antigens and cultured them. These DC were then mixed with separately cultured ensemble immune cells containing T, NK, NKT, and other cells (excepting DC), prior to infusion back into a patient with cancer (we term this novel modification of ACT therapy AC-ACT, autologous cell-adoptive cell therapy). In a further novel modification of the ACT approach, we assessed outcomes using miRNA microarray analysis at pre- and posttherapy. miRNA genes locate intergenic and intragenic noncoding RNA regions in introns or within an exon of the gene. These are suitable markers because they down-regulate target genes in pathways including cell growth, differentiation, metabolism, and the cell cycle.10 Deregulation of miRNAs, both up and down, is found in many cancers. Up-regulated and down-regulated miRNAs targeting oncogene and anti-oncogene mRNAs, respectively, suppress tumor progression.11, 12

Three patients were selected for AC-ACT, representing three distinct groups with interest in the AC-ACT/miRNA approach that we describe here. One is an active patient with cancer (stage IV colorectal cancer). The second and third patients represent emerging classes of patient in Japan, seeking miRNA information on risk or possible early diagnosis while being otherwise healthy, or seeking ACT without autologous cancer antigen as a cancer preventative strategy. For each of these distinct patient types, we present a case report of the protocol used and miRNA outcomes analysis.

MATERIALS AND METHODS

Patients

Subject 1

Enrollment criteria:	Active cancer patient with advanced disease and limited clinical options.
Gender:	Female
Age:	58 y

Disease:	Stage IV rectal cancer patient with multiple metastasis (lung, lumbar vertebra, and peritoneum). Received a surgical operation for rectal cancer with stoma before being diagnosed as stage IV in January 2018. At the time of the stage IV diagnosis, patient was inoperable because of lung and vertebral metastasis. AC-ACT was performed at this stage in disease progression.
Protocol:	Received AC-ACT and miRNA analysis.

Subject 2

Enrollment criteria:	Healthy patient desiring miRNA analysis for risk analysis or early diagnosis.
Gender:	Male
Age:	65 y
Disease:	None.
Protocol:	miRNA analysis only.

Subject 3

Enrollment criteria:	Preliminary diagnosis seeking ACT without autologous cancer antigen as a cancer preventative strategy.
Gender:	Male
Age:	48 y
Disease:	Suspected lung cancer. Tobacco smoker for more than 20 y (20 cigarettes/day) and recently had been coughing for a month. He received medical examinations and showed higher value of a tumor maker, CYFRA (cytokeratin fragment 19, 5.7 ng/mL) and very low values of miR-154-5p, let-7i-3p, miR-3202, and miR-610 by RT-PCR (data not shown).
Protocol:	ACT without autologous cancer antigen pulse and with miRNA analysis.

Ethical Disclosure

Adoptive cell therapy with autologous cancer antigen therapy is approved in Japan. The specific application of this therapy in this study complies with the Declaration of Helsinki and was approved by a Recelling Specific Regeneration Ethics Committee (Osaka, Japan) with the Japan Ministry of Health Protocol Approval Number # PC7180012. In addition, for this therapy involving human subjects, completely informed consent for the three subjects was obtained.

Materials

Antibodies and cytokines, CD3, CD161, IL-2, IL-4, GM-CSF, were purchased from CalBiochem. Ficoll-Paque was from Sigma.

Processing of Cells and Culture of NKT and Other Cells

Ten mL of blood was collected from the subjects. PBMC was separated with Ficoll-Paque, and a buffy coat fraction was obtained, and CD14 + cells were separated by MACSprep CD14 MicroBeads, human (Miltenyi Biotec) in an ice-cold MACS buffer. $CD14^+$ cells were cultured for dendritic cells. $CD14^-$ cells containing T cells, NK cells, NKT cells, and other cells were separately cultured. $CD14^-$ cells were cultured in T-25 flask which had been coated by 1 μg/mL anti-CD3 antibody (BioLegend) in PBS for overnight (16 hours) and further coated by 10 μg/mL anti-CD161 antibody (Abgent) in PBS for another overnight period. All the procedures were done in a cell control center approved by National Health Welfare agency.

DC Cells and Autologous Tumor Antigen Pulse

CD14 + cells containing dendritic cells (DC) were cultured in a complete medium containing 20 ng/mL rhGM-CSF and 20 ng/mL rhIL-4 for 4-5 days in a T-25 flask. For AC-ACT, $CD14+^+$ cells were pulsed for three hours with autologous cancer antigen (0.1 μg/mL protein) extracted by a kit (Formalin Fixed Paraffin Embedded Protein Isolation Kit, ITSI-Bioscience) and were used for mixed culture with $CD4+^-$ cells. For ACT, $CD14+^+$ cells without autologous cancer antigen pulse were used for the mixed culture.

AC-ACT and ACT Therapy

The mixed cells were cultured overnight in the presence of 200 ng/mL α-GalCer at 37°C in 5% CO2, and 100 mL of AC-ACT cells ($2\text{-}3 \times 10^9$) in saline was infused after flowcytometric and aseptic checks. ACT therapy was done in the same manner. Before infusion, the cells were washed three times in saline.

Flow Cytometry

A sample of cell (1×10^5 cells) for the infused cells were characterized flow cytometrically by adding PE anti-Vα24 and APC anti-CD3 antibody for NKT cells; FITC anti-CD57 and PE anti-CD16 antibody for NK cells;

FITC anti-CD4 and PE anti-CD29 antibody for T cells in 500 μL FACS buffer, and measured on an JSAN FACS machine (BayBio Co.).

miRNA Microarray

Paired serums of the two subjects were obtained immediately before and 1 week later of AC-ACT or ACT. Simultaneously, the paired serum of another control subject 2 was obtained. Following procedures were done by Cell Innovator Co. Each sample of total RNA was prepared from 200 μL serum by miRNeasy Serum/Plasma (Qiagen) The samples (100 ng) were labeled with an miRNA Complete Labeling and Hybridization kit (Agilent, 5190-0456) according to the Agilent miRNA microarray protocol. The SurePrint G3 Human miRNA kit 80 × 60 k (Human_miRNA_V21.0) array chip was used.

Statistical Analysis

In order to examine the effects of AC-ACT and ACT, the difference of pre- and postvalues were calculated. From 2550 miRNAs, top 100-200 miRNA of values, either increased or decreased, were used. Shared and specific miRNAs among them were analyzed by Venn Diagram (http://bioinformatics.psb.ugent.be/webtools/Venn/).

RESULTS

Post-AC-ACT Case History

The patient with cancer, subject 1, who received AC-ACT, has been free of progression for 1 year after starting AC-ACT. Metastases have been arrested, and no new metastases have been detected. PET-CT scans showed that lungs and bone metastasis have been stable and some lesions decreased. The subject reports good quality of life (QOL). Biochemical data from 12 months after stage IV diagnosis and AC-ACT are as follows: neutrophil/lymphocyte ratio: 1.76, eosinophil: 6.8%; CRP: 0.46 (lowered); total protein: 6.5 g/dL; and Hb: 14.2 g/dL. In addition, AST and ALT indicate normal liver function. Creatinin, BUN, and eGFR show normal kidney function, and plasma electrolytes are normal. Thus, this patient has no anemia, hypoproteinemia, renal, or liver dysfunction, that would be expected in an end-stage cancer patient in cachexia. We note also that subjects 2 and 3 remain cancer free.

FACS Analysis of the Ensemble Cells

A typical FACS pattern of the infused cells to subject 2 is shown in Figure Figure1.1. NK, NKT, and T-cell fractions comprised 4.96, 0.53 and 30%, respectively.

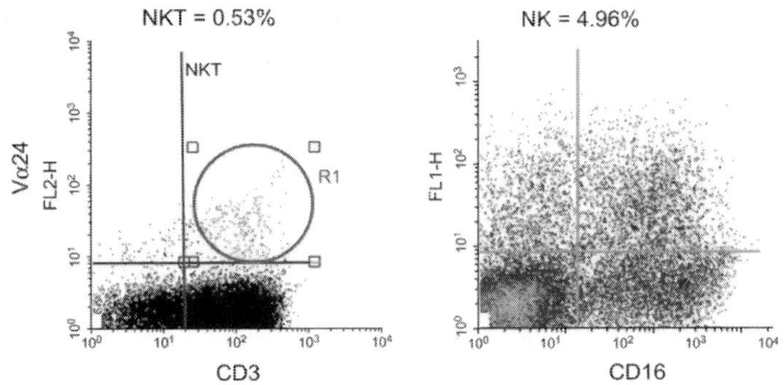

Figure 1: FACS pattern of the infused ACT ensemble cells for subject 3.

Effects of AC-ACT on Oncogenic and Anti-oncogenic miRNA Expression

Paired miRNA microarray examinations were performed for pre- and post-AC-ACT and ACT for subject 1 and subject 3, respectively, and simultaneously for subject 2 without any ACT. Changes in expressed miRNAs of subject 1 were clustered for largely increased and decreased values as shown in Figure 2. The ordered 100-200 miRNAs from each subject were selected with a cutoff value, 0.1, and were analyzed by Venn diagram for shared miRNAs among the subjects as shown in Figure 3. The highly, either positive or negative, ordered relative values of miRNAs which were shared in the subjects received AC-ACT or ACT (areas A and B of subjects 1 and 3, respectively, in Figure 3) are shown in Figure 4. In the same manner, the ordered relative values of specific patient with cancer (areas C and D of subject 1, in Figure 3) are shown in Figure 5. The functions of those miRNAs according to the indicated references are listed in Tables 1, 2.

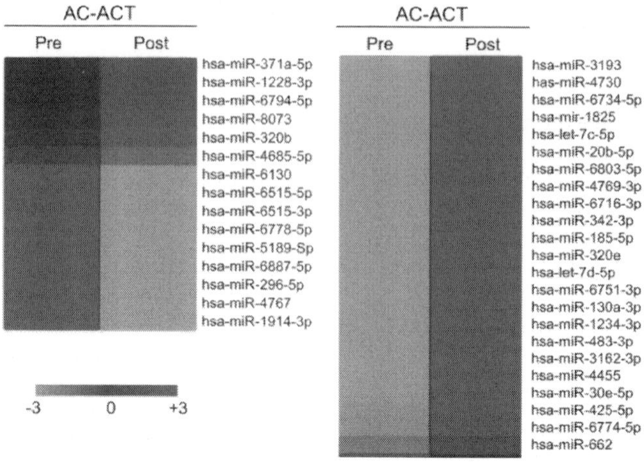

Figure 2: Heatmap of miRNA changes pre- and post-AC-ACT. miRNA in serum was hybridized with probe fixed on solid chip (Agilent) as described in Methods. Intensity differences between pre- and post-AC-ACT were clustered: high (red) to low (green).

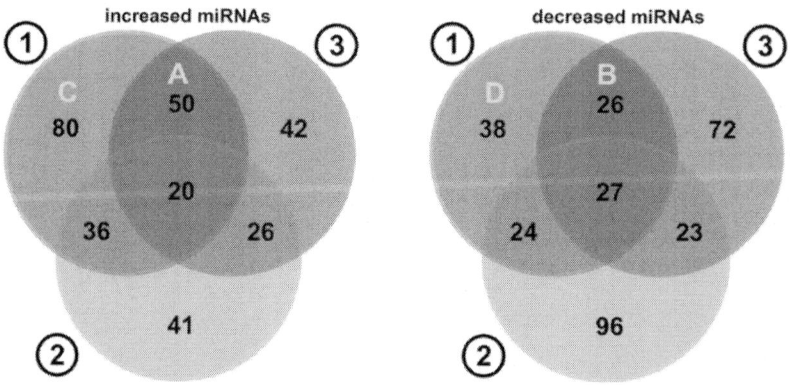

Figure 3: Shared prominently (0.1 cutoff) changed miRNAs by Venn diagram between subjects (1) and (3) after AC-ACT and ACT, respectively, and those of nontherapy subject (2).

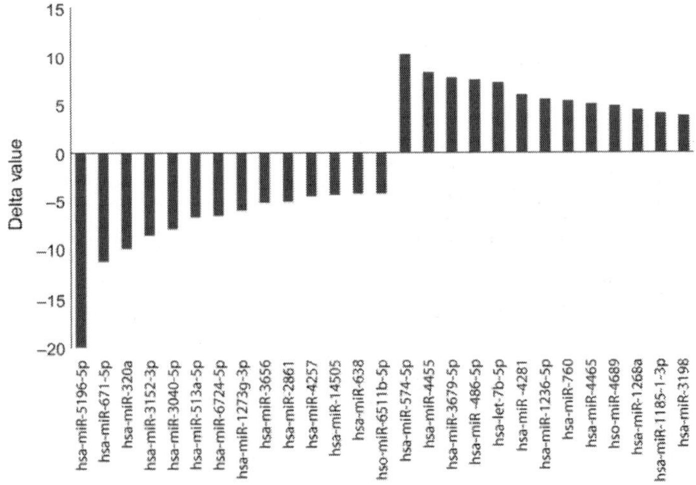

Figure 4: Ordered up/down values shared between subjects 1 and 3 (A and B in Figure 3).

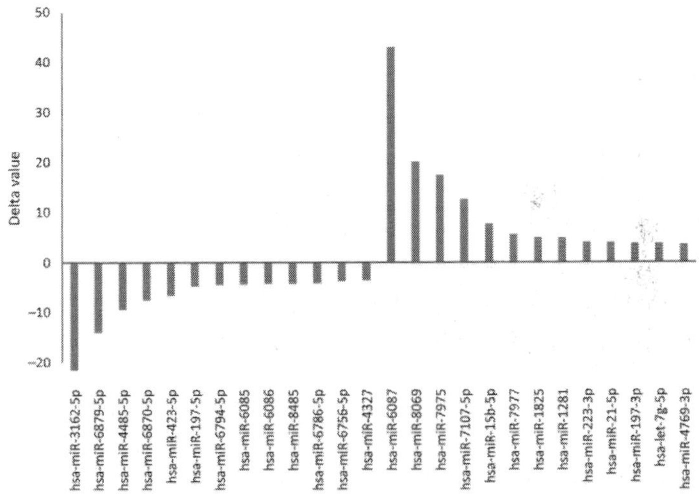

Figure 5: Ordered up/down values specific for subject 1.

Table 1: Reference-based tumor suppressor miRNAs either by increasing (targeting oncogenic genes in CRC) or by decreasing (targeting tumor-suppressive genes in CRC)

AC-ACT or ACT Received Subjects 1, 3)				**Cancer Patient Specific (Subject 1)**			
increased	**Ref.**	**decreased**	**Ref.**	**increased**	**Ref.**	**decreased**	**Ref.**
miR-574-5p	25	miR-671-5p	38	miR-6087	47	miR-3162-5p	54
miR-4455	26	miR-3152-3p	39	miR-8069	48	miR-6879-5p	45
miR-3679-5p	27	miR-940	40	miR-7975	49	miR-6870-5p	55
miR-486-5p	28	miR-6274-5p	41	miR-15b-5p	50	miR-423-5p	56
let-7b-5p	29	miR-1273g-3p	42	miR-7977	51	miR-197-5p	57
miR-4281	30	miR-3656	43	miR-1281	52	miR-6794-5p	59
miR-1236-3p	31	miR-2861	44	miR-223-2p	53	miR-6085	58
miR-760	32	miR-4257	45			miR-8485	59
miR-4465	33	miR-4505	46				
miR-4689	34						
miR-1268a	35						
miR-1185-I-3p	36						
miR-3198	37						

Note

Citations for miRNAs described to have tumor-suppressive effects and that are increased or decreased between pre- and post-AC-ACT or ACT (shared between subjects 1 and 3: A and B and cancer patient specific one subject 1: C, D in Figure 3). There are two modalities of tumor suppressor miRNAs either by increasing (targeting oncogenes) or by decreasing (targeting tumor suppressor gene).58

Table 2: Increased values of miRNAs targeting PD-L1 mRNA after AC-ACT and ACT

miRNA targeting PD-1L	**Relative fold Increase after AC-ACT or ACT**	
	Subject 1 (AC-ACT)	Subject 3 (ACT)
miR-15a-5p	3.06	1.07
miR-16-5p	11.01	56.1
miR-93-5p	2.55	3.23

Note

List of miRNAs that show increases after AC-ACT or ACT, and those target PD-L1 mRNA.

Effects of AC-ACT on miRNAs which target an Immune Checkpoint Blocker

The immunosuppressive protein PD-L1 is up-regulated in many cancers and lead to poor prognosis.13, 14 Targeting PD-L1 is considered as one of therapies for colorectal cancer.15, 16 To assess the effects of AC-ACT on miRNAs which target PD-L1, their levels in pre- and post-AC-ACT samples were examined. As shown in Table Table2,2, increased levels of miR-15a-5p, miR-16-5p, miR-93-5p, and mR-106b-5p, which target PD-L1, are found in both the subject 1 and subject 3, after AC-ACT and ACT, respectively.

DISCUSSION

This paper is a case report on outcomes and a novel miRNA-based assessment system for the outcomes of an approved therapy that is in widespread use in Japan. There is also an emerging patient-led desire to use this therapy for cancer prevention. However, AC-ACT is approved for only patients with cancer by Japanese Health Ministry. Therefore, we sought approval to offer ACT to healthy subjects and assess outcomes by miRNA analysis in order to expand the evidence base for future consideration of its use in monitoring and prevention.

Immunotherapy based on the adoptive transfer of naturally occurring or genetically engineered immune effector cells has therapeutic benefit in clinical trials of advanced cancers. Present AC-ACT is a combination of DC vaccination with other immune cells. Here, DC cells were induced with IL-4 and GM-CSF and were pulsed with autologous cancer antigen. Other mixed immune cells, primarily cultured for NKT cells, a method which was originally developed by Taniguchi,17 were also separately induced with IL-2, αCD3, αCD161, and α-GalCer. AC-ACT has potential advantages over the other ACTs such as NK cells only, which are frequently associated with fever (higher than 38°C) as an adverse effect. In contrasts, AC-ACT is associated with almost no adverse effects after infusion. The patient we presented was largely progression free (assessed by CT examination and quality of life indices) for more than one year, although carcinoembryonic antigen (CEA) gradually elevated from 4.5 to 6.5 ng/mL over this period.

In addition to patient outcomes, the current study has implications for rapid, personalized determination of efficacy of ACT therapies by using miRNA array, as markers of progression or response to therapy. In general, it is difficult in cancer therapy to determine the appropriate treatment for each patient, also the costs of therapy are very expensive and so determining the efficacy rapidly is important. Tumor markers such as CEA and Ca-19-9 are usually used, but the number of them is limited, and the specificity for certain cancers is ambiguous. The type of miRNA analysis performed here may offer new approaches, but there remain ambiguities and challenges in the use of miRNAs for the follow-up of cancer status. Because the seed sequences of miRNAs are only ~7 nucleotides, they can target numerous mRNAs, limiting their diagnostic potential by themselves. However, the scale of array analysis (thousands of miRNAs measured simultaneously) leads to large data sets that may be mined for associations and trends. Our results show most that there may be patterns in changed levels of groups of miRNAs that may have utility in tracking progression after therapy. Most changes in miRNA levels that we observed, either increases or decreases, induced by AC-ACT or ACT, are related to tumor-suppressive functions according to the literature (Tables (Tables1,1, ,2),2), suggesting that they potentiate tumor immunity. These types of personalized, individualized medicine approaches where miRNA profiling is monitored for each patient to develop a unique picture of progression after therapy may have more value than attention to normalized or population-based metrics of therapy effectiveness and progression. Notably, the most significantly decreased miRNA in response to AC-ACT was miR-5196-5p, which targets the Fra2 gene,18 plays a critical role in the progression of human cancers,19 and it is involved in IL-4 production for tumor immunity suppressor.20 Reduction in miR-5196-5p by AC-ACT may suppress cancer progression but also immune potency will be inhibited. miR-320a, another prominently decreased by AC-ACT, suppresses colorectal cancer progression by targeting Rac1,21 although high miR-320a levels appear to induce pro-tumorigenic M2-like macrophages.22 Low miR-320a which we observe with AC-ACT has also been associated with positive efficacy of peptide vaccination for colorectal cancer.23 These counteractive functions, cancer suppressive and immune inhibition, are seen for other miRNAs such as miR-21 and miR-155 24 and underscore the need for further studies.

CONCLUSION

The effectiveness of AC-ACT for colorectal cancer therapy was monitored successfully by miRNA microarray.

AUTHORS' CONTRIBUTIONS

MC and CA: Were responsible for the report conception and design, collection and assembly of data, interpretation, analysis, presentation of data and drafted the manuscript and were responsible for final approval of the manuscript. MC and KI: Performed experiments, such as RNA extraction for microarrays, and were involved in all administrative, experimental, technical, and material support. MC: Was also involved in obtaining the informed consent from all subjects. YM and YU: Provided the medical care to subject 1, the colorectal cancer patient. They were also responsible for obtaining the informed consent from all subjects. RA: Is the owner of the Fukuoka MSC Medical Clinic where AC-ACT was performed, and he was responsible for the oversight processes including accreditation, licensure, permits, informed consent, and certifications. RS: Assisted in the analysis, interpretation, and presentation of data and also participated in manuscript drafting. All authors read and approved the final manuscript.

ACKNOWLEDGMENTS

This work was done in the Fukuoka MSC Medical Clinic. Some of the authors are supported by BFSR Co. Ltd. Dr Chaker N. Adra was supported by The Adra Family.

REFERENCES

1. Chen DS, Mellman I. Elements of cancer immunity and the cancer-immune set point. Nature. 2017;541:321-330.
2. Santegoets SJ, Welters MJ, van derBurg SH Monitoring of the immune dysfunction in cancer patients. Vaccines (Basel). 2016;4:29.
3. Urruticoechea A, Alemany R, Balart J, Villanueva A, Vinals F, Capella G. Recent advances in cancer therapy: an overview. Curr Pharm Des. 2010;16:3-10.
4. Yang JC, Rosenberg SA. Adoptive T-Cell therapy for cancer. Adv Immunol. 2016;130:279-294.
5. Exley MA, Friedlander P, Alatrakchi N, et al. Adoptive transfer of invariant NKT cells as immunotherapy for advanced melanoma: a phase I clinical trial. Clin Cancer Res. 2017;23:3510-3519.
6. Oka Y, Tsuboi A, Nakata J, et al. Wilms' Tumor Gene 1 (WT1) peptide vaccine therapy for hematological malignancies: from CTL epitope identification to recent progress in clinical studies including a cure-oriented strategy. Oncol Res Treat. 2017;40:682-690.
7. Angelova M, Charoentong P, Hackl H, et al. Characterization of the immunophenotypes and antigenomes of colorectal cancers reveals distinct tumor escape mechanisms and novel targets for immunotherapy. Genome Biol. 2015;16:64.
8. Brown SD, Warren RL, Gibb EA, et al. Neo-antigens predicted by tumor genome meta-analysis correlate with increased patient survival. Genome Res. 2014;24:743-750.
9. Imaoka Y, Kuranishi F, Miyazaki T, Yasuda H, Ohno T. Long-lasting complete response status of advanced stage IV gall bladder cancer and colon cancer after combined treatment including autologous formalin-fixed tumor vaccine: two case reports. World J Surg Oncol. 2017;15:170.
10. Hammond SM. An overview of microRNAs. Adv Drug Deliv Rev. 2015;87:3-14.
11. Gambari R, Brognara E, Spandidos DA, Fabbri E. Targeting oncomiRNAs and mimicking tumor suppressor miRNAs: new trends in the development of miRNA therapeutic strategies in oncology (Review). Int J Oncol. 2016;49:5-32.
12. Huang J, Wu J, Li J,et al. Deregulation of serum microRNA expression is associated with cigarette smoking and lung cancer. Biomed Res Int.

2014;2014:364316. doi: 10.1155/2014/364316. Epub 2014 Oct 20.

13. Coelho MA, de Carne TS, Rana S, et al. Oncogenic RAS Signaling promotes tumor immunoresistance by stabilizing PD-L1 mRNA. Immunity. 2017;47(1083–1099):e1086.
14. Roussille P, Tachon G, Villalva C, et al Pathological and molecular characteristics of colorectal cancer with brain metastases. Cancers (Basel). 2018;10(12). pii: E504. doi: 10.3390/cancers10120504.
15. Passardi A, Canale M, Valgiusti M, Ulivi P. Ulivi P: immune checkpoints as a target for colorectal cancer treatment. Int J Mol Sci. 2017;18(6):1324.
16. Restifo NP, Dudley ME, Rosenberg SA. Adoptive immunotherapy for cancer: harnessing the T cell response. Nat Rev Immunol. 2012;12(4):269-281.
17. Imai K, Kanno M, Kimoto H, Shigemoto K, Yamamoto S, Taniguchi M. Sequence and expression of transcripts of the T-cell antigen receptor alpha-chain gene in a functional, antigen-specific suppressor-T-cell hybridoma. Proc Natl Acad Sci USA. 1986;83:8708-8712.
18. Ciechomska M, Zarecki P, Merdas M, et al. The role of microRNA-5196 in the pathogenesis of systemic sclerosis. Eur J Clin Invest. 2017;47(8):555-564. doi: 10.1111/eci.12776. Epub 2017 Jul 18
19. Tkach V, Tulchinsky E, Lukanidin E, Vinson C, Bock E, Berezin V. Role of the Fos family members, c-Fos, Fra-1 and Fra-2, in the regulation of cell motility. Oncogene. 2003;22:5045-5054.
20. Zhang CS, Kim H, Mullins G, et al. Interleukin-4 expressed by neoplastic cells provokes an anti-metastatic myeloid immune response. J Clin Cell Immunol. 2015;6:1-9.
21. Zhao H, Dong T, Zhou H, et al. miR-320a suppresses colorectal cancer progression by targeting Rac1. Carcinogenesis. 2014;35:886-895.
22. Yuan F, Fu X, Shi H, Chen G, Dong P, Zhang W. Induction of murine macrophage M2 polarization by cigarette smoke extract via the JAK2/STAT3 pathway. PLoS ONE. 2014;9:e107063.
23. Shindo Y, Hazama S, Nakamura Y, et al. miR-196b, miR-378a and miR-486 are predictive biomarkers for the efficacy of vaccine treatment in colorectal cancer. Oncol Lett. 2017;14:1355-1362.
24. Li L, Zhang J, Diao W, et al. MicroRNA-155 and MicroRNA-21 promote the expansion of functional myeloid-derived suppressor cells. J Immunol. 2014;192:1034-1043.

25. Cui Z, Tang J, Chen J, Wang Z. Hsa-miR-574-5p negatively regulates MACC-1 expression to suppress colorectal cancer liver metastasis. Cancer Cell Int. 2014;14:47.
26. Chen H, Dai G, Cai Y, et al. Vasodilator-stimulated phosphoprotein (VASP), a novel target of miR-4455, promotes gastric cancer cell proliferation, migration, and invasion, through activating the PI3K/AKT signaling pathway. Cancer Cell Int. 2018;18:97.
27. Xie Z, Yin X, Gong B, et al. Salivary microRNAs show potential as a noninvasive biomarker for detecting resectable pancreatic cancer. Cancer Prev Res (Phila). 2015;8:165-173.
28. Chen H, Ren C, Han C, Wang D, Chen Y, Fu D. Expression and prognostic value of miR-486-5p in patients with gastric adenocarcinoma. PLoS ONE. 2015;10:e0119384.
29. DiFazio P, Maass M, Roth S, et al. Expression of hsa-let-7b-5p, hsa-let-7f-5p, and hsa-miR-222-3p and their putative targets HMGA2 and CDKN1B in typical and atypical carcinoid tumors of the lung. Tumour Biol. 2017;39:1010428317728417.
30. Li BB, Huang GL, Li HH, Kong X, He ZW. Epigallocatechin-3-gallate modulates microRNA expression profiles in human nasopharyngeal carcinoma CNE2 cells. Chin Med J (Engl). 2017;130:93-99.
31. Li C, Ge Q, Liu J, et al. Effects of miR-1236-3p and miR-370-5p on activation of p21 in various tumors and its inhibition on the growth of lung cancer cells. Tumour Biol. 2017;39.
32. Cao L, Liu Y, Wang D, et al. MiR-760 suppresses human colorectal cancer growth by targeting BATF3/AP-1/cyclinD1 signaling. J Exp Clin Cancer Res. 2018;37(1):83. doi: 10.1186/s13046–018–0757–8
33. Kang DW, Lee SW, Hwang WC, et al. Phospholipase D1 Acts through Akt/TopBP1 and RB1 to regulate the E2F1-dependent apoptotic program in cancer cells. Cancer Res. 2017;77:142-152.
34. Hiraki M, Nishimura J, Takahashi H, et al. Concurrent targeting of KRAS and AKT by MiR-4689 is a novel treatment against mutant KRAS colorectal cancer. Mol Ther Nucleic Acids. 2015;4:e231.
35. Lu YL, Yao JG, Huang XY, et al. Prognostic significance of miR-1268a expression and its beneficial effects for post-operative adjuvant transarterial chemoembolization in hepatocellular carcinoma. Sci Rep. 2016;6:36104.

36. Seashols-Williams SJ, Budd W, Clark GC, et al. miR-9 acts as an OncomiR in prostate cancer through multiple pathways that drive tumour progression and metastasis. PLoS ONE. 2016;11:e0159601.
37. Slattery ML, Herrick JS, Pellatt DF, et al. Site-specific associations between miRNA expression and survival in colorectal cancer cases. Oncotarget. 2016;7(37):60193-60205. 10.18632/oncotarget.11173.
38. Zhang J, Luo X, Li H, Deng L, Wang Y. Genome-wide uncovering of STAT3-mediated miRNA expression profiles in colorectal cancer cell lines. Biomed Res Int. 2014;2014:187105.
39. Qiu F, Yang L, Ling X, et al. Sequence variation in mature MicroRNA-499 confers unfavorable prognosis of lung cancer patients treated with platinum-based chemotherapy. Clin Cancer Res. 2015;21:1602-1613.
40. Rashed MH, Kanlikilicer P, Rodriguez-Aguayo C, et al. Exosomal miR-940 maintains SRC-mediated oncogenic activity in cancer cells: a possible role for exosomal disposal of tumor suppressor miRNAs. Oncotarget. 2017;8:20145-20164.
41. Usuba W, Urabe F, Yamamoto Y, et al. Circulating miRNA panels for specific and early detection in bladder cancer. Cancer Sci. 2019;110:408-419.
42. Li M, Qian X, Zhu M, et al. miR1273g3p promotes proliferation, migration and invasion of LoVo cells via cannabinoid receptor 1 through activation of ERBB4/PIK3R3/mTOR/S6K2 signaling pathway. Mol Med Rep. 2018;17:4619-4626.
43. Liu J, Ma L, Wang Z, et al. MicroRNA expression profile of gastric cancer stem cells in the MKN-45 cancer cell line. Acta Biochim Biophys Sin (Shanghai). 2014;46:92-99.
44. Zhao M, Li L, Zhou J, et al. MiR-2861 behaves as a biomarker of lung cancer stem cells and regulates the HDAC5-ERK system genes. Cell Reprogram. 2018;20:99-106.
45. Dejima H, Iinuma H, Kanaoka R, Matsutani N, Kawamura M. Exosomal microRNA in plasma as a non-invasive biomarker for the recurrence of non-small cell lung cancer. Oncol Lett. 2017;13:1256-1263.
46. Zhang X, Chen Y, Wang L, et al. MiR-4505 aggravates lipopolysaccharide-induced vascular endothelial injury by targeting heat shock protein A12B. Mol Med Rep. 2018;17:1389-1395.

47. Dallas NA, Samuel S, Xia L, et al. Endoglin (CD105): a marker of tumor vasculature and potential target for therapy. Clin Cancer Res. 2008;14:1931-1937.
48. Teng Y, Li Y, Lin Z, et al. Analysis of miRNA expression profiling identifies miR-214-3p as a novel biomarker in sinonasal inverted papilloma. Epigenomics. 2018;10:1541-2155.
49. Du B, Wu D, Yang X, et al. The expression and significance of microRNA in different stages of colorectal cancer. Medicine. 2018;97(5):e9635.
50. Zhao C, Zhao Q, Zhang C, et al. miR-15b-5p resensitizes colon cancer cells to 5-fluorouracil by promoting apoptosis via the NF-kappaB/XIAP axis. Sci Rep. 2017;7:4194.
51. Horiguchi H, Kobune M, Kikuchi S, et al. Extracellular vesicle miR-7977 is involved in hematopoietic dysfunction of mesenchymal stromal cells via poly(rC) binding protein 1 reduction in myeloid neoplasms. Haematologica. 2016;101:437-447.
52. Jiang J, Ma B, Li X, et al. MiR-1281, a p53-responsive microRNA, impairs the survival of human osteosarcoma cells upon ER stress via targeting USP39. Am J Cancer Res. 2018;8:1764-1774.
53. Sugawara S, Yamada Y, Arai T, et al. Dual strands of the miR-223 duplex (miR-223-5p and miR-223-3p) inhibit cancer cell aggressiveness: targeted genes are involved in bladder cancer pathogenesis. J Hum Genet. 2018;63:657-668.
54. Stegeman S, Amankwah E, Klein K, et al. A Large-Scale Analysis of Genetic Variants within Putative miRNA Binding Sites in Prostate Cancer. Cancer Discov. 2015;5:368-379. https://doi.org10.1158/2159-8290.CD-14-1057
55. Tűzesi A, Kling T, Wenger A, et al. Pediatric brain tumor cells release exosomes with a miRNA repertoire that differs from exosomes secreted by normal cells. Oncotarget. 2017;8:90164-90175.
56. Li S, Zeng A, Hu Q, Yan W, Liu Y, You Y. miR-423-5p contributes to a malignant phenotype and temozolomide chemoresistance in glioblastomas. Neuro Oncol. 2017;19:55-65.
57. Zhang J, Luo X, Li H, et al. Genome-Wide uncovering of STAT3-mediated miRNA expression profiles in colorectal cancer cell lines. Zhang, BioMed Res In. 2014; 2014:1–11.

58. Ghanbari R, Mosakhani N, Asadi J, et al. Decreased expression of fecal miR-4478 and miR-1295b-3p in early-stage colorectal cancer. Cancer Biomarkers. 2015;15:195-201.

59. Chen X, Fan Z, McGee W, et al. TDP-43 regulates cancer-associated microRNAs. Protein Cell. 2018;9(10):848-866.

9

Investigation of Anti-cancer and Migrastatic Properties of Novel Curcumin Derivatives on Breast and Ovarian Cancer Cell Lines

Jinsha Koroth[1,2], Snehal Nirgude[1,2], Shweta Tiwari[4] , Vidya Gopalakrishnan[1,2,3], Raghunandan Mahadeva[1] , Sujeet Kumar[4] , Subhas S. Karki[4] and Bibha Choudhary[1]

[1] Institute of Bioinformatics and Applied Biotechnology, Electronic City Phase 1, Bangalore, Karnataka 560100, India
[2] JK, SN, and VG are graduate students registered under Manipal Academy of Higher Education, Manipal 576104, India
[3] Department of Biochemistry, Indian Institute of Science, Bangalore 560012, India
[4] Department of Pharmaceutical Chemistry, KLE Academy of Higher Education and Research, KLE College of Pharmacy, Rajajinagar, Bangalore, KN, India.

ABSTRACT

Background

Curcumin is known for its multitude of medicinal properties, including anticancer and migrastatic activity. Efforts to overcome poor bioavailability,

Citation: Koroth, J., Nirgude, S., Tiwari, S. et al. Investigation of anti-cancer and migrastatic properties of novel curcumin derivatives on breast and ovarian cancer cell lines. BMC Complement Altern Med 19, 273 (2019). https://doi.org/10.1186/s12906-019-2685-3.

stability, and side effects associated with the higher dose of curcumin has led to the development of newer derivatives of curcumin. Thus, the focus of this study is to screen novel curcumin derivatives, namely ST03 and ST08, which have not been reported before, for their cytotoxicity and migrastatic property on cancer cells.

Methods

Anti-cancer activity of ST03 and ST08 was carried out using standard cytotoxicity assays viz., LDH, MTT, and Trypan blue on both solid and liquid cancer types. Flow cytometric assays and western blotting was used to investigate the cell death mechanisms. Transwell migration assay was carried out to check for migrastatic properties of the compounds.

Results

Both the compounds, ST03 and ST08, showed ~100 fold higher potency on liquid and solid tumour cell lines compared to its parent compound curcumin. They induced cytotoxicity by activating the intrinsic pathway of apoptosis in the breast (MDA-MB-231) and ovarian cancer cell lines (PA-1) bearing metastatic and stem cell properties, respectively. Moreover, ST08 also showed inhibition on breast cancer cell migration by inhibiting MMP1 (matrix metalloproteinase 1).

Conclusion

Both ST03 and ST08 exhibit anti-cancer activity at nanomolar concentration. They induce cell death by activating the intrinsic pathway of apoptosis. Also, they inhibit migration of the cancer cells by inhibiting MMP1 in breast cancer cells.

BACKGROUND

Cancer is a disease characterized by abnormal proliferation of cells, which can evade anti-growth signals and invade other parts of the body [1]. Cancer cells thus affect the normal functioning of the organ, impairing the homeostasis of the body [2]. Various therapeutic strategies such as chemotherapy, hormone therapy, immunotherapy, radiation therapy, targeted therapy, gene therapy, have been utilized for the effective management of the disease [3, 4]. Despite having very advanced treatment, metastasis, and cancer relapse with resistance remain a significant challenge in cancer treatment. The resistance to treatment is due to a subpopulation of cells, called cancer stem cells [5, 6]. Recent evidence has shown that cancer stem cells undergo continuous mutations and are responsible for drug resistance and relapse. Targeting cancer stem cells is challenging, but identifying drugs which can work on stem cells can solve the relapse and resistance seen in most of the high-grade tumors [7].

Currently, cancer reports show that around 8.2 million people around the globe are suffering from the irreversible metastatic condition of malignant tumors with drug resistance [8]. Malignant tumors, characterized by its invasive and metastatic nature, increases the morbidity and mortality rates seen in cancer patients [9]. Cancer cell produces extracellular matrix-degrading enzymes (MMPs), which help metastasize [10]. Re-localization of malignant cells to distant organs leads to secondary tumor growth even after the complete removal of the tumor from the primary site. Available chemo-therapeutic drugs do not efficiently work on metastasized cancer cells [10]. Therefore, given the above facts, the challenge is to eliminate cancer-promoting cells (cancer stem cells) selectively.

Several natural compounds have shown to have anti-cancer activity. Among them Curcumin (Fig. 1) (1,7-bis (4-hydroxy 3-methoxyphenyl)-1,6-heptadione-3,5-dione or diferuloylmethane), has been shown to have anti-cancer on an array of cancer cells regardless of their origin [11,12,13]. Curcumin is derived from turmeric, and is known for it's anti-inflammatory, anti-oxidant, anti-bacterial and anti-malarial properties [14, 15]. Studies have shown that curcumin can target cancer stem cells [16, 17] as well as inhibit cancer cell migration [8]. Curcumin has been shown to act on cancer stem cells of colorectal cancer, pancreatic cancer, breast cancer, brain cancer, and head and neck cancer [17]. Curcumin exhibits its anti-metastatic property by altering several signaling mechanisms, including inhibition of transcription factors, proteases, protein kinases, inflammatory cytokines, and their signaling pathways [10].

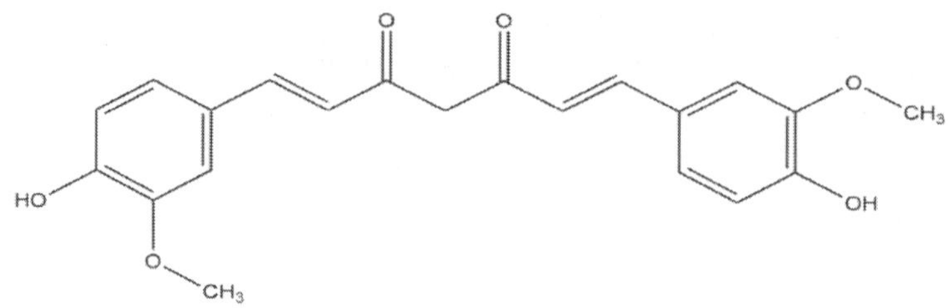

Figure 1: Curcumin chemical structure.

Although curcumin is known for its multitude of activities, it lacks stability and is not bioavailable in the in vivo system [18]. So to enhance the metabolic stability, various modifications were made onto the curcumin structures. DAPs (diarylidenyl-piperidone) is one of the well-studied groups of curcumin derivatives, which exhibited proliferation inhibition on multiple cancer cell lines such as colon, breast, ovarian epithelial cancer, etc. and multidrug resistance reverting property as well [19,20,21,22,23]. The multi-targeted effect of these compounds is documented to have more advantages than the single targeted ligands, as it can interfere with multiple signaling pathways and can have a pleiotropic effect [24,25,26,27]. S. Das et al., have synthesized and demonstrated anti-cancer property of molecular dimers. They have conjugated two moieties of (3E, 5E)-3,5-dibenzylidenepiperidin-4-one pharmacophores via oxamide/propane diamide linkage. Their group has shown the anti-leukemic and anti-lymphoma activity of few 1,2-bis[(3E,5E)-3,5-dibenzylidene-4-oxo-1-piperidyl]ethane-1,2-dione derivatives [28,29,30,31]. The dimers of DAPs or 1,2-bis[(3E,5E)-3,5-dibenzylidene-4-oxo-1-piperidyl]ethane-1,2-dione attracted scientific attention to use as backbone structure due to its anti-cancer effect on various cancer types by activating the apoptotic pathway [29]. 1,2-bis[(3E,5E)-3,5-dibenzylidene-4-oxo-1-piperidyl]ethane-1,2-diones are thus considered as an excellent drug prototype for the development of novel compounds.

The dimers are relatively more stable than curcumin and also known to enhance the anticancer properties. Keeping the backbone of dimer constant, we synthesized two novel compounds, (ST03 (1,2-bis[(3E,5E)-3,5-bis[(2-chlorophenyl)methylene]-4-oxo-1-piperidyl]ethane-1,2-dione) and ST08 ([4-[(E)-[(5E)-1-[2-[(3E,5E)-3,5-bis[(4-hydroxyazonylphenyl)methylene]-4-oxo-1-piperidyl]-2-oxo-acetyl]-5-[(4-hydroxyazonylphenyl)methylene]-4-oxo-3-piperidylidene]methyl]phenyl] azinic acid)). We have checked anti-cancer activities of both the compounds on solid and liquid cancer cells. We have also investigated ST03 and ST08 induced cell death mechanism as well as their migrastatic property. We have carried out these studies on two major gynecological cancer types, breast, and ovarian cancer [32] using breast and ovarian cancer cell lines, respectively.

METHODS

Chemistry

Silica gel plates were used for Thin Layer Chromatography by using toluene and ethyl acetate in 1:1 proportion. The IR spectra were recorded in KBr on a Jasco 430+ (Jasco, Japan); the ^1H NMR spectra were recorded in $CDCl_3$/DMSO on a Bruker (400 MHz), and J values were reported in Hertz (Hz). Mass spectra were recorded in triple quadrupole LCMS-6410 from Agilent technologies.

Procedure for synthesis of ST03 and ST08

ST03

Step 1.

Oxaloyl chloride (0.003 mol, 0.39 g) in DCE (5 mL) was added dropwise to a stirred suspension of a 3,5-bis (2-chlorobenzylidene)piperidin-4-one (0.006 mol) in DCE (20 mL) containing triethylamine (0.006 mol, 0.61 g) at 20 °C for a period of 30 min. The reaction was stirred at room temperature for 12 h. The solvent was removed under reduced pressure at 45 °C. An aqueous solution of potassium carbonate (25 mL, 5% w/v) was added to the crude mass and stirred for 2 h. The solid obtained was fifiltered, dried, and crystallized from 95% ethanol to yield the pure product.

Step 2:

The 2-chlorobenzaldehyde (26.71 mmol) was added dropwise to a suspension of 4-piperidone hydrochloride monohydrate (13.03 mmol) in acetic acid (35 mL). Dry hydrogen chloride gas was passed through this mixture until a clear solution was obtained. After stirring the reaction mixture at room temperature for 24 h, the precipitate was separated through filtration and added to a mixture of a saturated aqueous potassium carbonate solution (25% w/v, 25 mL) and acetone (25 mL); the resultant mixture was stirred for 0.5 h. The free base was collected, washed with water (50 mL), and dried. The compound was recrystallized from 95% ethanol to get the pure compound.

ST08

Step 1:

The 4-nitrobenzaldehyde (26.71 mmol) was added dropwise to a suspension of 4-piperidone hydrochloride monohydrate (13.03 mmol) in acetic acid (35 mL). Dry hydrogen chloride gas was passed through this mixture until a clear solution was obtained. After stirring the reaction mixture at room temperature for 24 h, the precipitate was separated through filtration and added to a mixture of a saturated aqueous potassium carbonate solution (25% w/v, 25 mL) and acetone (25 mL); the resultant mixture was stirred for 0.5 h. The free base was collected, washed with water (50 mL), and dried. The compound was recrystallized from 95% ethanol to get the pure compound.

Step 2:

Oxaloyl chloride (0.003 mol, 0.39 g) in DCE (1,2 Dichloroethane) (5 mL) was added dropwise to a stirred suspension of a 3,5-bis (4-nitrobenzylidene) piperidin-4-one (0.006 mol) in DCE (20 mL) containing triethylamine (0.006 mol, 0.61 g) at 20 °C for a period of 30 min. The reaction was stirred at room temperature for 12 h. The solvent was removed under reduced pressure at 45 °C. An aqueous solution of potassium carbonate (25 mL, 5% w/v) was added to the crude mass and stirred for 2 h. The solid obtained was filtered, dried, and crystallized from 95% ethanol to yield the pure product.

ST-03: Yield 45%, Rf 0.63, MP. 140–145 °C, IR (λ cm^{-1}) 3061, 2975, 1642, 1440, 1260, 1044, 990. ^{1}H NMR (δ): 7.99 (s, 2H), 7.94 (s, 2H), 7.53 (d, 2H, J=9.2), 7.40–7.31 (m, 6H), 7.23 (d, 4H, J8.8), 7.18–7.14 (m, 2H), 7.05 (d, 2H, J=9.2), 4.38 (s, 4H), 4.34 (s, 4H). MS (ESI) m/z: 742.53 (742.47).

ST-08: Yield 40%, Rf 0.55, MP. 180–182 °C, Nitro derivative IR (λ cm^{-1}) 3075, 2932, 2851, 1666, 1599, 1520, 1346, 1263, 987. ^{1}H NMR (δ): 8.33–8.25(m, 8H, ar), 7.77(s, 2H), 7.70 (d, 4H, J=8.8 Hz), 7.63 (d, 4H, J=8.8 Hz), 7.50 (s, 2H), 4.54 (s, 4 HO, 4.48 (s, 4H)). MS (ESI) m/z: 781.58 (784.68).

Cell Lines and Culture

In order to investigate the anti-cancer activity of ST03 and ST08 on solid and liquid type of cancers, they were tested on multiple human cancer cell lines. Human cancer cell lines such as PA1 (ovarian teratocarcinoma cell line), MCF7 (breast adenocarcinoma cell line), MDA-MB-231(breast adenocarcinoma cell line), CEM (T acute lymphoblastic leukemia cell line), K562 (B chronic myelogenous leukemia cell line), A431 (epidermoid carcinoma cell line), HeLa (cervical adenocarcinoma cell line) and 293T (embryonic kidney cell line) cells were purchased from NCCS, Pune, India. A2780 (ovarian endometrioid adenocarcinoma) was purchased from ATCC and Nalm6 (B cell precursor leukaemia cell line) was a kind gift from SCR lab, IISc, Bangalore, India). CEM, K562, Nalm6 cell lines were grown in RPMI-1640 (Lonza). PA1, MCF7, HeLa, A431 were grown in MEM and MDA-MB-231, 293T cell lines were grown in DMEM media (Lonza). All cell lines were supplemented with 10% Fetal bovine serum and 1X antibiotic-antimycotic (GIBCO, Thermo Fisher Scientific, US) and maintained at 37°C in a humidified incubator with 5% CO_2 supply. Cytotoxicity of these compounds on normal cells was examined using peripheral blood mononuclear cells (a kind gift from SCR lab, IISc, Bangalore, India) and 293T cells.

MTT Assay

Cytotoxicity exerted by ST compounds on cell lines were assessed by doing 3-(4,5-dimethylthiazol-2-yl)-2,5-diphenyltetrazolium bromide assay (MTT assay) [33]. Cells were seeded (5000 cells/well) in 96-well plate in triplicates, incubated for 24 h, and treated with respective ST compounds from 1 nM to 1000 nM. Curcumin was used for comparing the potency with its derivatives. Curcumin was added in the range of 1 uM to 100 uM concentrations to compare the cytotoxicity with its derivatives. After 48 and 72 h incubation, 10 µl of MTT (5 mg/mL) reagent was added to each well to a final concentration of 0.25 mg/mL and incubated till the colour developed. Following colour development, the reaction was stopped by adding stopping solution (50% N, N-Dimethylformamide) (Sigma–Aldrich, USA), 10% Sodium dodecyl sulfate, (MP Biomedicals, USA) and kept for 2 h incubation at 37 C for complete solubilization of formazan crystal. Absorbance was measured at 570 nm on a 96 well plate reader (Tecan infinite 200 ELISA plate reader, Tecan Trading AG, Switzerland). Absorbance from culture medium without cells was considered as blank and was subtracted. Cells treated with an appropriate concentration of DMSO was used as vehicle control as the compounds were dissolved in DMSO. The 50% inhibition concentration of the drugs (IC_{50} values) were calculated from the 48 h treatment readings using GraphPad Prism 7 software. Percentage of viable cells in each treatment concentrations were calculated as a ratio of sample OD to the control OD.

LDH Assay

Lactate dehydrogenase (LDH) release is an indication of cell injury. This assay quantitatively measures the stable LDH in the cytosol. To perform this assay, 5000 cells/well were seeded in 96-well plate in triplicates and were treated with ST compounds (1 nM to 1000 nM). After 48 and 72 h of treatment, each well of 96-well plate was washed to remove FBS content and the cells were lysed using 0.5% Triton-X-100 prepared in 1X Phosphate buffer saline. This lysate was mixed with LDH assay reagents, described by OPS Diagnostics LLC, P.O. Box 348, Lebanon, NJ 08833 USA. The absorbance of the orange-red colored formazan product was measured at 490 nm using Tecan infinite 200 ELISA plate reader (Tecan Trading AG, Switzerland).

Trypan Blue Exclusion Assay

The cytotoxic effect of ST compounds on the viability of cancer cell lines was determined by Trypan blue exclusion assay [33]. The cells were seeded in 6 well culture plate at a density of 75,000 cells/mL and incubated for 24 h and treated with different concentrations of ST compounds (1 nM to 1000 nM). Cells were collected at 48 h and 72 h time points and resuspended in 0.4% of trypan blue (Sigma–Aldrich, USA). The number of viable cells was counted using hemocytometer. Percentage of viable cells in each treatment concentrations were calculated as a ratio of sample cell count to the control cell count. The IC_{50} values (50% inhibition concentration of the drugs) were calculated from the 48 h data.

Phosphatidylserine Externalization Assay using AnnexinV-FITC/PI

In order to understand the mode of cell death (apoptosis/necrosis) induced by ST compounds in ovarian and breast cancer cells, annexin V-FITC/PI staining was carried out. Cells grown in 6-well plate with a cell density of 75,000 cells/mL were treated with ST compounds for 48 h. Cells were trypsinized, washed with ice-cold 1X Phosphate buffer saline and resuspended in 1X annexin binding buffer containing annexin V-FITC antibody (Biolegend, San Diego, CA) for 15 min in the dark on ice. PI (propidium iodide) was added (3.3 µg/mL) just before acquiring the samples. Cells incubated in 3% paraformaldehyde was used as a positive control. A total of 10,000 events were acquired for each sample using Beckman coulter Gallios flow cytometer (Beckman Coulter, Miami, FL).

Western Blot Analysis

Western blot analysis was carried out to examine the expression of proteins involved in the apoptotic pathway. To perform this assay, 75,000 cells/mL were seeded and treated with ST compounds (10 nM, 20 nM, 40 nM, 60 nM, 80 nM) for 48 h and whole cell lysate was prepared using RIPA buffer (25 mM Tris-Cl pH 7.6, 150 mM Sodium chloride, 1% NP-40, 1% Sodium deoxycholate, 1% Sodium Dodecyl Sulphate, 1 mM Phenylmethylsulphonyl fluoride, 1 mM Sodium orthovanadate). Crude cell lysates (30–40 µg) were electrophoresed on SDS-PAGE (Sodium dodecyl sulfate-polyacrylamide gel electrophoresis) and were transferred on to Polyvinylidene fluoride membrane (Millipore, USA) to probe with respective antibodies. The primary antibodies against caspase 9, cleaved caspase 9, caspase 3, cleaved caspase 3, caspase 8 and Horseradish peroxidase-labeled secondary anti-rabbit antibodies were purchased from Cell Signalling Technology, Beverly, MA. Anti-tubulin, Anti-MMP1 and its secondary mouse antibody were purchased from Santa-Cruz Biotechnology, Santa Cruz, CA. The membrane was probed with appropriate antibodies and was developed using chemiluminescence reagent (Clarity Western ECL blotting substrate, Biorad). The blot image was captured by using a Syngene G: Box gel doc system. Protein band image quantification was done using GelQuant. Net, Biochem Lab solutions.

Transwell Migration Assay

Transwell assay was performed by seeding 75,000 cells/mL in a 6-well plate and were allowed to grow for 24 h. Forty and eighty nM ST08 treatment was done for 24 h. Permeable migration chambers were purchased from Corning Inc. (24-well insert; pore size, 8 µm) and were coated with 75 µL of matrigel and incubated at 37 °C for 24 h for settling. Fifty thousand cells/ millilitre treated cells were suspended in 200 µL media without FBS and added into the top chamber. Migration was allowed to occur for 5 h in a CO_2 incubator.

Then cells were fixed with 4% paraformaldehyde and stained with 2% crystal violet. Cells that did not migrate to lower compartment were cleaned using a cotton swab. Each insert was imaged in for five random fields at 10X magnification and analysis was done using NIH ImageJ software. Two independent experiments were carried out in duplicates.

Statistical Analysis

Data from 3 different biological replicates were collected and values are expressed as Mean ± SE in bar graphs. One-way or Two-way ANOVA followed by Tukey's multiple comparison test was carried out and significance is represented as **** (p-value ≤ 0.0001), *** (p-value ≤ 0.001), ** (p-value ≤ 0.01), * (p-value ≤ 0.05). Statistical analysis was done using GraphPad Prism 7 tool.

RESULTS

Characterization of ST03 and ST08

ST03 (1,2-bis[(3E,5E)-3,5-bis[(2-chlorophenyl)methylene]-4-oxo-1-piperidyl]ethane-1,2-dione) and ST08 ([4-[(E)-[(5E)-1-[2-[(3E,5E)-3,5-bis[(4-hydroxyazonylphenyl)methylene]-4-oxo-1-piperidyl]-2-oxo-acetyl]-5-[(4-hydroxyazonylphenyl)methylene]-4-oxo-3-piperidylidene]methyl] phenyl] azinic acid) were prepared based on the procedure given above. The structures of the synthesized compounds were confirmed by IR, NMR, and Mass spectrometry. The vibration of C-H bonds was observed between 3075 and 3003 cm^{-1} whereas for aliphatic C-H bonds observed between 2946 and 2840 cm^{-1}, and for C=O bonds observed between 1666 and 1640 cm^{-1}. In ^{1}H NMR, all the synthesized compounds showed prominent signals for aromatic and olefinic protons between δ 7.99–6.64 ppm. The structures of all the compounds were ensured by mass spectrometry. The detailed characterization results of the compounds are provided in the supplementary section (Additional file 1: Figures S1 and S2).

ST03 and ST08 Exert Cytotoxicity on Cancer Cell Lines with Least Effect on Normal Cells

ST03 and ST08 compounds were examined for its cytotoxicity on a) leukemic cell lines: CEM, K562, Nalm6 b) ovarian cancer cell lines: PA1 and A2780 c) Breast cancer cell lines: MCF-7, MDA-MB-231 d) 293T (normal kidney cell line) e) A431 (skin cancer cell line) f) HeLa (cervical cancer cell line) and also on g) PBMC (peripheral blood mononuclear cells) using MTT, LDH, and Trypan blue exclusion assays. In the pilot screening experiment, all cells were treated with five different concentrations (1 nM to1 μM) of compounds (ST03 and ST08) for 48 and 72 h. Both the compound showed cytotoxicity at sub-micromolar concentrations. DMSO was used as the vehicle control. IC_{50} values of these compounds were calculated using all three methods and is listed in (Table 1 and 2).

Table 1: Structure of ST03 and IC_{50} **values. IC_{50} values were calculated from the average IC_{50}**s of all cytotoxicity assays conducted on particular cell line at 48 h, and the results are summarized in micromolar concentrations

Structure of ST03 and IC_{50} values		
ST03		
Cell lines	**IC_{50} (uM)**	
Nalm6	0.033	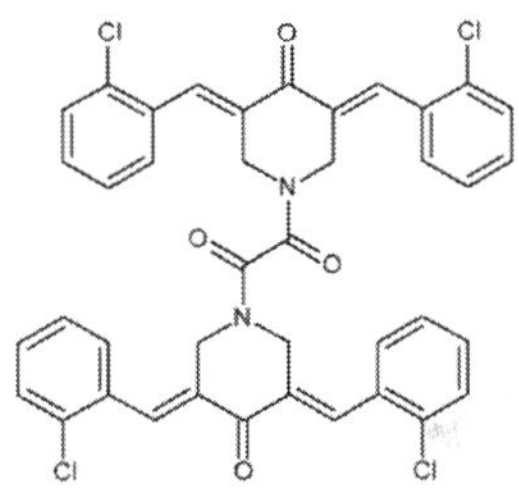
K562	0.044	
CEM	0.029	
PA1	0.040	
A2780	0.051	
HeLa	0.014	
A431	0.044	
293T	1.4	

Table 2: Structure of ST08 and IC_{50} **values. IC_{50} values were calculated from the average IC_{50}s** of all cytotoxicity assays conducted on particular cell line at 48 h, and the results are summarized in micromolar concentrations

Structure of ST08 and IC_{50} values		
ST08		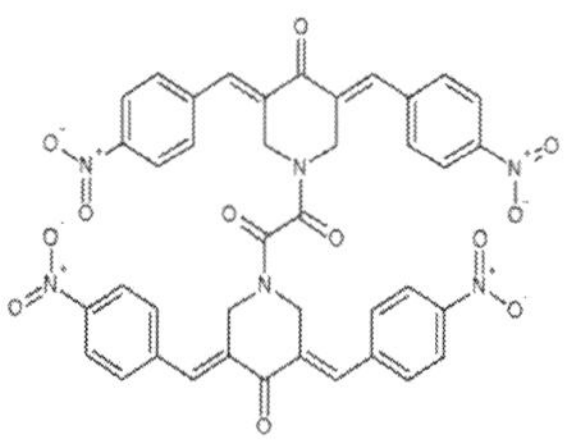
Cell lines	**IC_{50} (uM)**	
Nalm6	0.050	
K562	0.104	
PA1	0.038	
A2780	0.064	
MCF7	0.127	
MDA-MB-231	0.054	
HeLa	0.117	
293T	3.3	

Curcumin showed 50% cytotoxicity in the range of 7–10 μM, ST compounds were in 0.030–0.080 μM range. Both, ST03 and ST08, showed ~100x higher potency than the parent compound curcumin. (Tables 1 and 2, Fig. 3c).

Since ST compounds exhibited cytotoxicity at nanomolar concentrations, the next series of cytotoxicity assays were carried out in a range of 1 nM to 1000 nM to calculate IC_{50}. Initial experiments were carried out on leukemic cell lines (CEM, K562, and Nalm6) using trypan blue assay. Cells were treated with an equivalent amount of DMSO as vehicle control. Dose-Dependent induction of cell death was seen in ST03, and ST08 treated cells. The lowest concentration (1 nM) was found to be least effective on all the cell lines tested (Fig. 2a). Nonetheless, at concentrations such as at ~30–50 nM onwards, we observed effective cytotoxicity, and 1000 nM showed maximum inhibition at 48 h and 72 h (Fig. 2a). No cytotoxicity was observed with the vehicle control, DMSO (0.1%). From these results, it was evident that ST03 and ST08 are toxic to human leukemic cell lines, and ST03 showed relatively higher potency than ST08. Peripheral blood mononuclear cells were used as normal cells to test cytotoxicity. The cells were treated with ST03 and ST08 for 48 h and cell viability assessed. No cytotoxicity was observed at doses three times more (150 nM) than the effective dose on cancer cell lines (~ 30–50 nM).

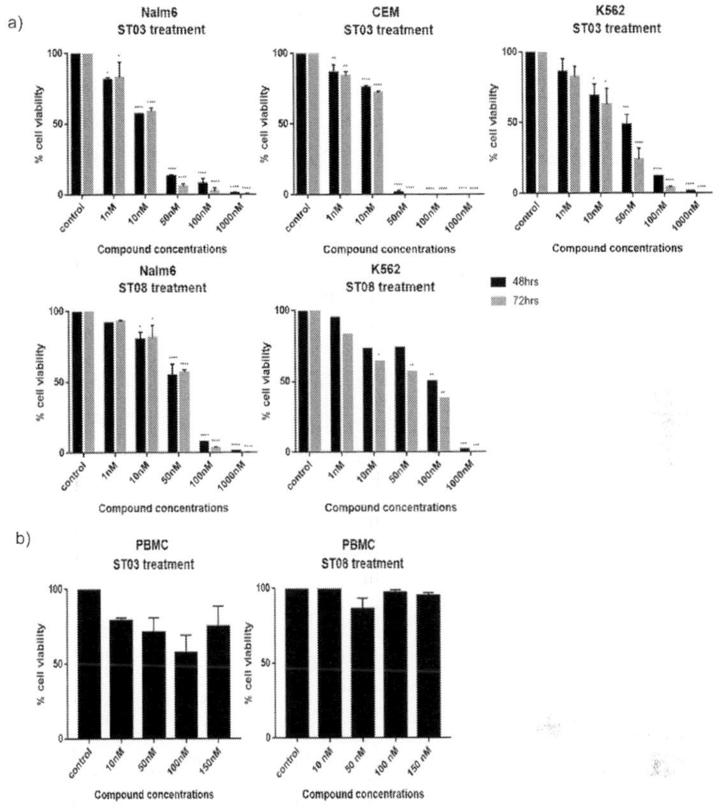

Figure 2: Evaluation of effect of ST03 and ST08 on cancer cell lines and PBMCs by Trypan Blue exclusion assay. a Viable cells after ST03 (Nalm6, CEM, K562) and ST08 (Nalm6, K562) treatment b Viable PBMC cell count after ST03 and ST08 treatments. Each experiment was repeated for a minimum of 3 times and plotted as bar graphs with error bars. Two-way ANOVA was conducted using Graph pad prism 7 tool and the p value was calculated between control and ST compound treated samples, where, *: p value <0.05, **: p value <0.005, ***: p value <0.0001, ****: p value <0.00001.

Additionally, MTT assay was also performed to check cytotoxicity in both liquid (CEM, K562) and solid cancer cell lines (PA1, A2780, A431, MDA-MB-231, MCF7, and 293 T (Fig. 3) following treatment with ST03 and ST08 for 48 h and 72 h. The reduction of MTT by live-cell mitochondria was considered directly proportional to cell proliferation. A dose-dependent

(Fig. 3), decrease in the cell viability was observed. Both the drugs showed IC50 in the range of 30-50 nM in all the cell lines tested. Keeping our interest in developing drugs against metastatic ovarian and breast cancer, we tested, ST03, and ST08 induced cell death in (PA-1, ovarian teratocarcinoma vs. A2780, epithelial cancer). ST03 showed IC50 of 41 nM in PA1 vs. 54 nM in A2780 (Fig. 3a). Whereas, ST08 showed higher inhibition on PA1 (~38 nM) among the other cell lines tested (Fig. 3b). Among the breast cancer cell line MCF-7 (epithelial) and MDA-MB-231 (metastatic), MDA-MB-231(54 nM) showed better cytotoxicity than MCF-7 (127 nM) upon ST08 treatment. ST03, and ST08 showed the least toxic effect on normal kidney cell lines (293 T) as expected. These results again confirmed that both the compounds are cytotoxic to cancer cell lines rather than on normal cells. We also performed MTT with curcumin and found its IC_{50} in most cell lines to be in the range of ~ 10 uM-100 uM which is approximately 100 times more than ST03 and ST08 (Fig. 3c).

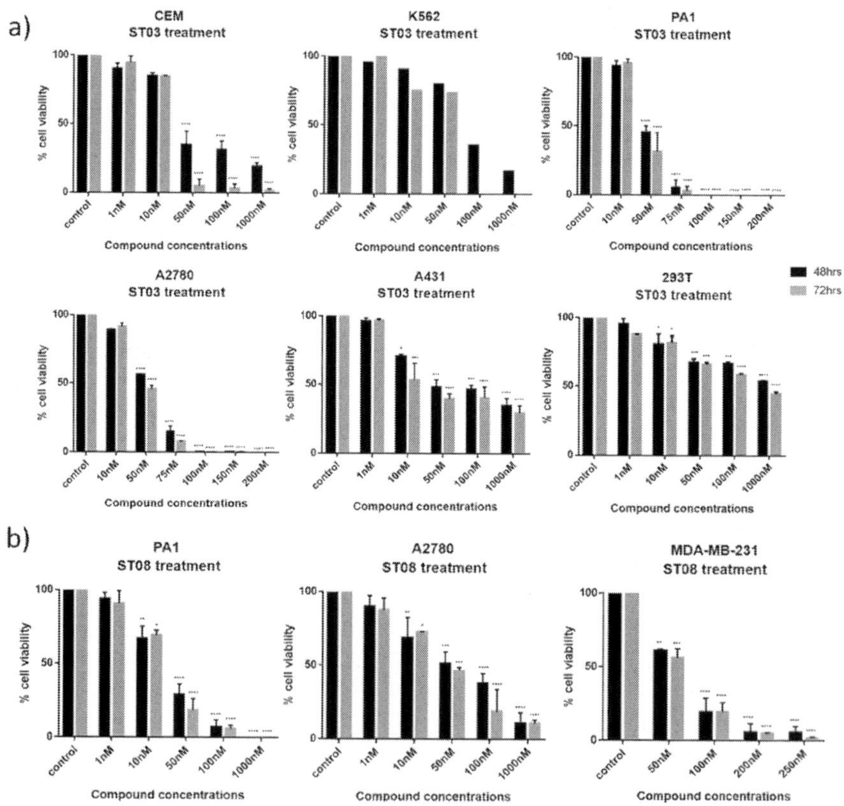

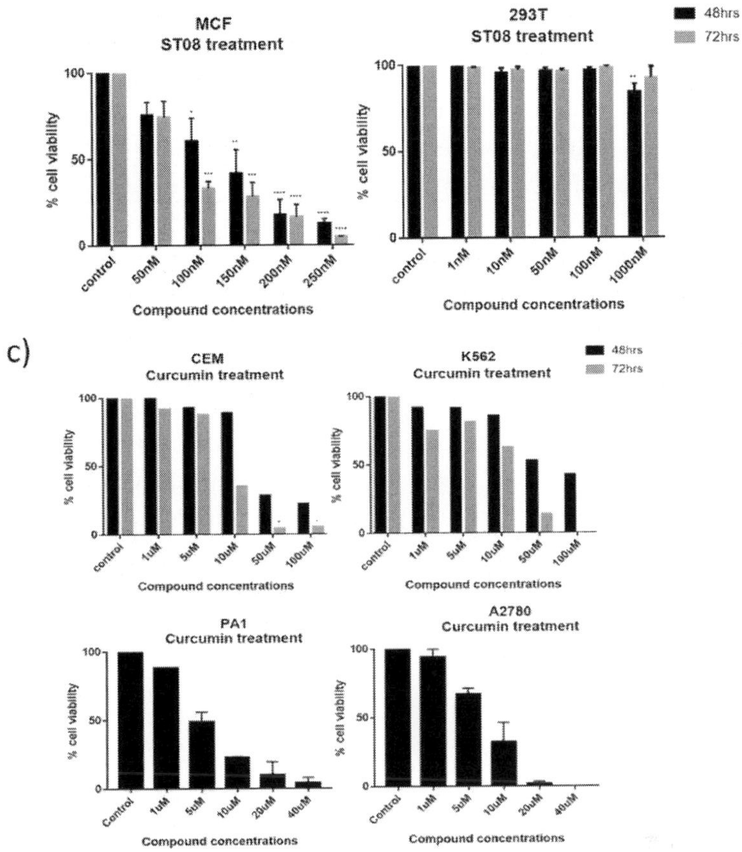

Figure 3: Evaluation of the effect of ST03, ST08 and curcumin compounds on cell lines by MTT assay. a ST03 treatment on PA1, A2780, K562, CEM,A431 and 293T. b ST08 treatment on PA1, A2780, MDA-MB-231, MCF-7 and 293T. c Curcumin treatment on PA1, A2780, CEM and K562. Each experiment was repeated for a minimum of 3 times and plotted as bar graphs with error bars. Two-way ANOVA was conducted using Graph pad prism 7 tool and the p value was calculated between control and compound treated groups, where,*: p value <0.05, **: p value <0.005, ***: p value <0.0001, ****: p value <0.00001.

Also, LDH (lactate dehydrogenase) assay was conducted to confirm the above-observed results by examining the cell damage caused by ST03 and ST08 treatment. LDH is a cytosolic enzyme present in most of the eukaryotic cells. When the plasma membrane integrity is lost during the cell death process, it leaks out of the cell [34]. LDH reduces NAD^+ to NADH with the help of lactate, and reduced NADH catalyses the formation of red formazan from INT with the help of PMS in the assay system. Thus, the presence of cytosolic LDH leads to the formation of red-coloured formazan, which directly correlates to the live cells present in the assay system.

PA1, A2780, A431, HeLa (Fig. 4a) cells were treated with ST03 and MDA-MB-231, MCF7 and HeLa (Fig. 4b) were treated with ST08 for 48 and 72 h and checked for cytoplasmic LDH content in each treatment (Fig. 4). LDH assay showed better sensitivity than MTT upon ST03 and ST08 treatment on cancer cell lines. However, the inhibitory concentration of both the compounds was in the same range. Here, among the ovarian cell lines tested, ST03 showed a better effect on PA1 cells (~39 nM) which is an ovarian teratocarcinoma cell line than A2780 (~49 nM) epithelial cancer of ovary (Fig. 4a). On the other hand, ST08 was more effective on the MDA-MB-231 cell line (~53 nM) than MCF7, the other breast cancer cell line tested. Although MCF7 is also a breast cancer cell type, it showed relatively less effectiveness (~125 nM) when compared to the triple-negative breast cancer cell type, MDA-MB-231 (Fig. 4b).

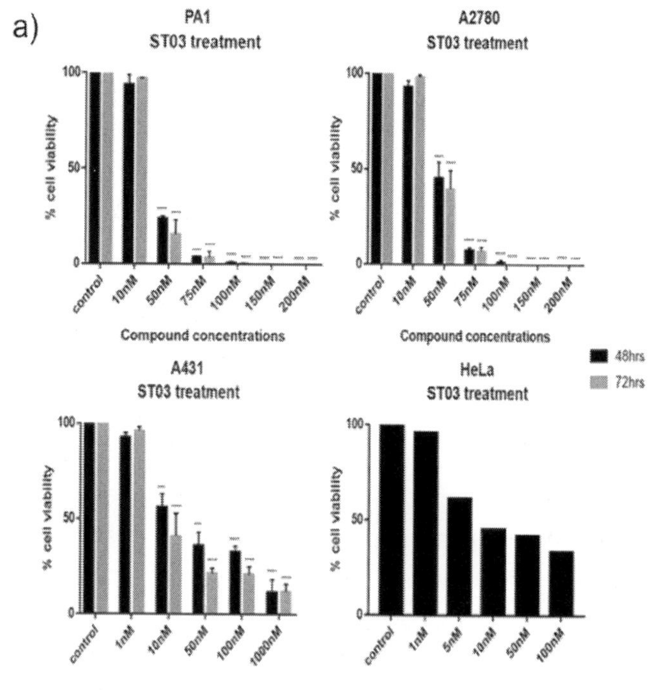

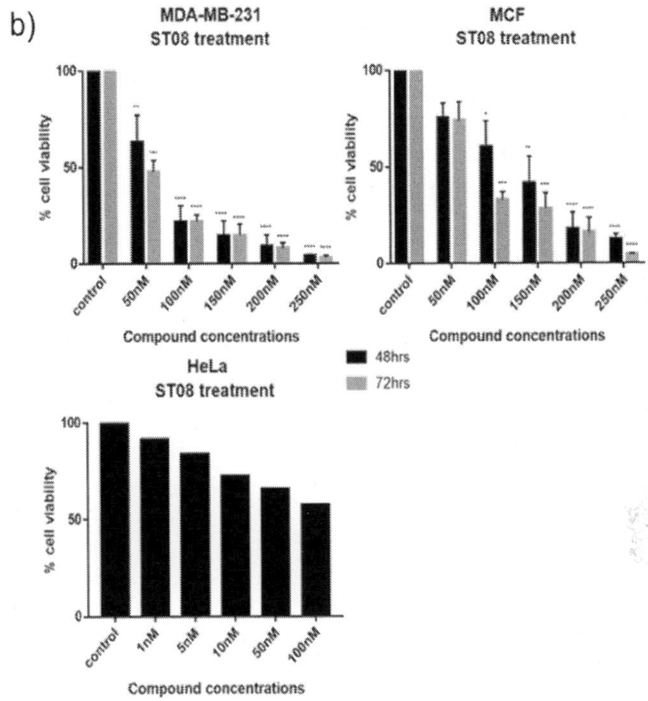

Figure 4: Examination of the effect of ST03 and ST08 compounds on cancer cell lines by LDH assay. **a** ST03 treatment on PA1, A2780, A431, HeLa and **b** ST08 treatment on MDA-MB-231, MCF-7, HeLa. Each experiment was repeated for a minimum of 3 times and plotted as bar graphs with error bars. Two-way ANOVA was conducted using Graph pad prism 7 tool and the p value was calculated between control and ST treated groups, where,*: p value <0.05, **: p value <0.005, ***: p value <0.0001, ****: p value <0.00001.

ST03 and ST08 Induced Apoptosis in Ovarian and Breast Cancer Cell Lines

Since our novel compounds ST03 and ST08 inhibited the cancer cell proliferation, we were interested in analyzing the mechanism by which the compounds induce cytotoxicity on ovarian and breast cancer cell lines. It is well known that anti-cancer compounds can cause different types of cell deaths such as apoptosis, necrosis or autophagy, thereby determining the fate of cell [35]. Thus, to dig into the causative mechanism, ovarian and breast cancer cells were treated with ST03 and ST08, subjected to flow cytometry analysis after staining with annexin V-FITC-PI staining [36]. In live cells, PS (phosphatidylserine) component of the plasma membrane is facing the cytoplasmic side. In a cell undergoing apoptosis, PS flips towards the outer side. This can be recognised by Annexin V. In cells undergoing cell death via necrosis, PS is not flipped, but cells are leaky; therefore, PI enters the nuclei and stains DNA. Annexin V, together with Propidium iodide (PI) is used for the detection of apoptotic cell population based on the integrity of the plasma membrane. The dot blots obtained from flow cytometry analysis shown in Fig. 5 describes the population of cells at different stages of cell death. The lower left quadrant contains cells that are live, negative for both the stains (Annexin $^{(-)}$ PI $^{(-)}$), lower right quadrant is occupied by early apoptotic cells (Annexin $^{(+)}$ PI $^{(-)}$), upper right quadrant contains late apoptotic cells (Annexin $^{(+)}$ PI $^{(+)}$) and the upper left quadrant has necrotic/ dead cells (Annexin $^{(-)}$ PI $^{(+)}$). The results shown in Fig. 5a-d depict that, both ST03 and ST08 induce a greater extent of apoptotic cell death in ovarian and breast cancer cell lines, respectively.

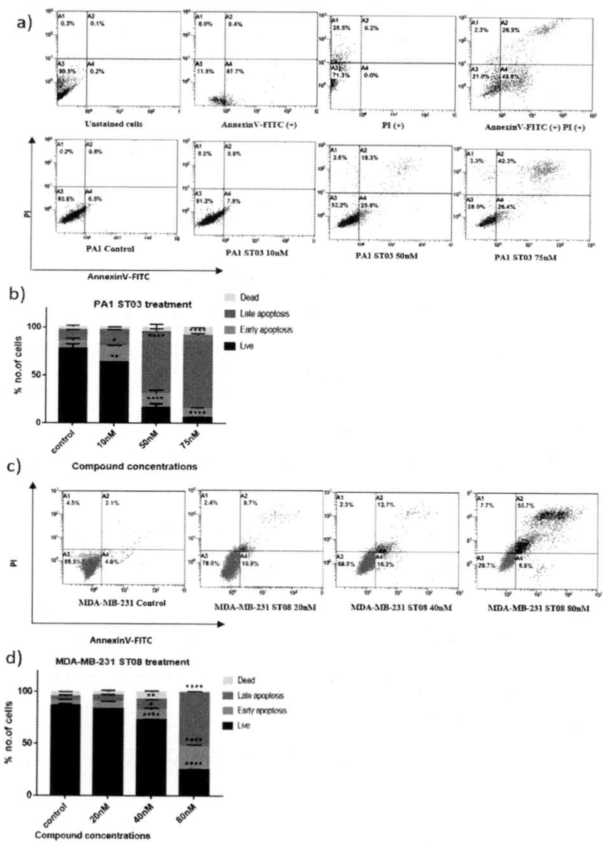

Figure 5: Evaluation of cell death by ST compound treatments using AnnexinV-FITC/PI. a ST03 treated PA1 cells stained with Annexin/FITC-PI

and b Quantification of percentage of cells in each stage c ST08 treated MDA-MB-231 cells stained with Annexin/FITC-PI and d) Quantification of cells in each stage. Each experiment was repeated for a minimum of 3 times and plotted as bar graphs with error bars. Two-way ANOVA was conducted and the p value was calculated between control and ST03 treated groups, where, *: p value <0.05, **: p value <0.005, ***: p value <0.0001, ****: p value <0.00001.

The representative dot plot shows that the 50 nM concentration of ST03 could induce 47.8% of cells death, in which 25.9% were early apoptotic (Annexin $^{(+)}$ PI $^{(-)}$), 19.3% were late apoptotic (Annexin $^{(+)}$ PI $^{(+)}$), 2.6% were necrotic (Annexin $^{(-)}$ PI $^{(+)}$) in PA1 (Fig. 5a, b). Maximal accumulation of early apoptotic population (16.2%) was seen at 40 nM of ST08 on breast cancer cell line MDA-MB-231. And at 80 nM, these early populations have migrated to the third compartment, leaving 29.7% unstained, 6.9% in early, 55.7% in late and 3.2% in necrotic stages (Fig. 5c, d). From these results, it is evident that both compounds exhibited cell death via apoptosis.

ST03 and ST08 Induce the Intrinsic Apoptotic Pathway in Breast and Ovarian Cancer Cell Lines

As we observed apoptotic populations in the treated cells from annexin V-FITC/PI assay results; we were curious to know the pathway of apoptosis. Thus, to further investigate the cell death pathway (intrinsic vs extrinsic pathway of apoptosis) induced by ST03 and ST08, we checked the expression of key proteins involved in apoptotic pathways using western blotting. Caspase-9 is an initiator caspase which is a part of mitochondria-mediated intrinsic apoptotic pathway and is activated by cytochrome-c released from mitochondria. At the same time, caspase-3, being the executioner caspase, is responsible for the proteolysis of α-fodrin, PARP, gelsolin, ICAD and other caspases leading to the effective completion of apoptosis process [37, 38]. In our treatments, both the compounds significantly increased the amount of cleaved/active caspase 9 and cleaved/ active caspase 3 in PA1 and MDA-MB-231 cancer cell lines Fig. 6a-b. On the other hand, the extrinsic apoptotic marker procaspase 8 and its active form were found to be downregulated. Both ST03 and ST08 induce the intrinsic apoptotic pathway in ovarian cancer cell line PA1 and in breast cancer cell line MDA-MB-231.

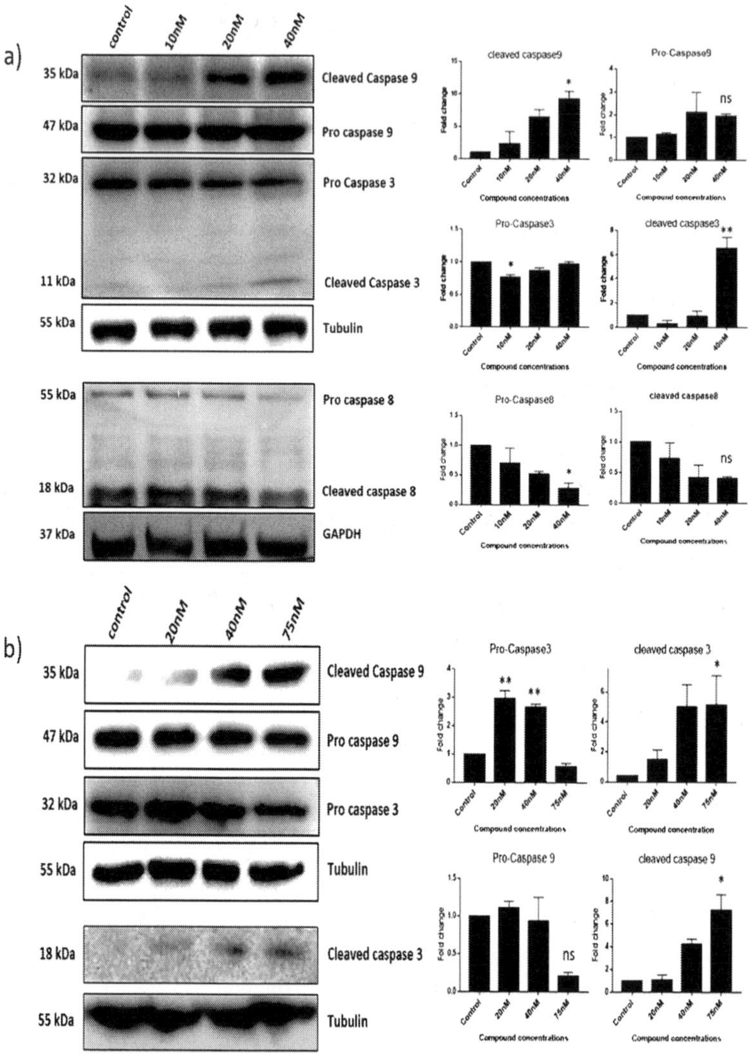

Figure 6: Expression of apoptotic proteins PA1 and MDA-MB-231 cells on compound treatment. a Western blot image of PA1 cell lysate treated with ST03 compound for 48 h b Western blot image of MDA-MB-231 cell lysate treated with ST08 compound for 48 h. Each experiment was repeated for a minimum of 3 times and plotted as bar graphs with error bars. One way ANOVA was conducted and the p value was calculated between control and ST03 treated groups, where, *: p value <0.05, **: p value <0.005.

ST08 Inhibits Breast Cancer Cell Migration

The effect of ST08 on the migratory ability of MDA-MB-231 was evaluated by Transwell migration assay. Compared with untreated control cells, the migration capacity of the 24 h treated cells was significantly diminished (Fig. 7a, b). The migration inhibition rate was 3 fold in the treated cells (Fig. 7b) when compared to the untreated control groups. Besides, we examined the expression of MMP1 (Fig. 7c) in the ST08 treated MDA-MB-231 cells. Interestingly, MMP1 showed reduced expression at higher concentrations (80 nM) of ST08. Cell migration is the most important event that happens as an initial step of metastasis [39]. This result indicates that ST08 can inhibit the cancer cell migration effectively, which can prevent metastasis and progression of cancer.

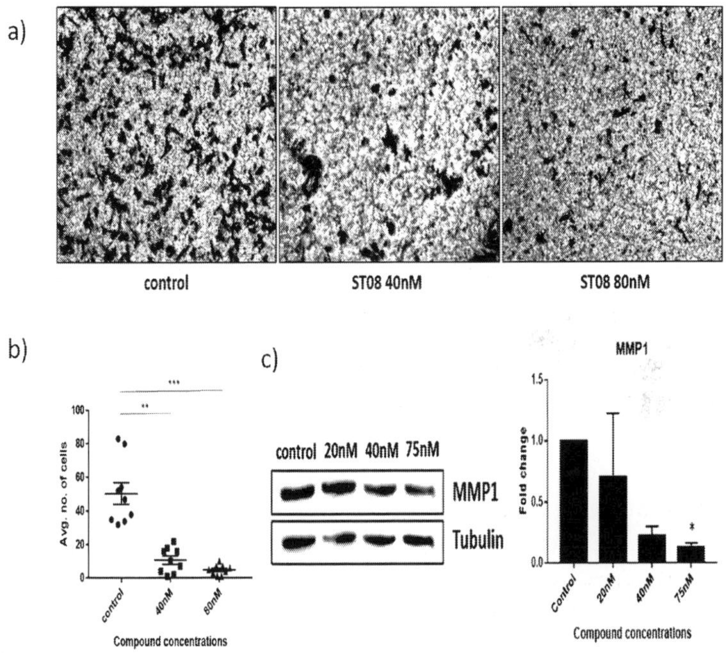

Figure 7: Effect of ST08 on MDA-MB-231 cell migration a ST08 treated MDA-MB-231 migration in vitro and b its quantification c) Expression of MMP1 in ST08 treated MDA-MB-231 cells. Each experiment was repeated for a minimum of 3 times and plotted as bar graphs with error bars. One-way ANOVA was conducted and the p value was calculated between control and ST03 treated groups, where, *: *p* value <0.05, **: *p* value <0.005 *: *p* value <0.05, **: *p* value <0.005, ***: *p* value <0.0001, ****: *p* value <0.00001.

DISCUSSION

Drug toxicity and disease relapse remain as major hurdles in the management of cancer despite having advanced targeted therapies. Plant-derived low toxic phenolic compounds are known to exhibit anti-cancer activity by inducing apoptosis in cancer cells [27, 40, 41]. Curcumin is one such example which exhibits anti-cancer activity against various cancer types [42,43,44,45,46,47]. Curcumin has been replaced by more potent and bioavailable derivatives, which have attracted attention for the development of novel drugs [48]. The novel curcumin derivatives developed in this study has exhibited remarkable cytotoxicity (~ 100X) better than parent compound at nanomolar concentrations on a wide range of cancer cell lines which includes both representative liquid and solid tumour cell lines. Various types of curcumin derivatives have been reported, to overcome the low potency, poor bioavailability and side effects demonstrated by the parent compound curcumin [49]. One among them is 1,2-bis[(3E,5E)-3,5-dibenzylidene-4-oxo-1-piperidyl]ethane-1,2-diones, which exhibited anti-cancer activity on variety of cancer cell lines [27,28,29]. Our novel compounds ST08 and ST03 synthesised from 1,2-bis[(3E,5E)-3,5-dibenzylidene-4-oxo-1-piperidyl]ethane-1,2-dione, by adding strong electron-withdrawing groups such as NO_2 and Cl- respectively to the backbone, has enhanced the anti-proliferative effect on the cancer cells particularly on ovarian cells (PA1) with stem cell-like properties. The cytotoxicity assays performed showed inhibition of cell growth by ST03 and ST08 in the range of 30–125 nM in both adherent and non- adherent cell lines which is ~ 100 fold better than the parent compound curcumin, which was effective in the range of ~ 10–100 μM on most of the cancer cells lines reported till date [43, 44, 50]. ST03 and ST08 are stable and bioavailable better than Curcumin (data not shown).

Further analysis showed that ST03 was consistent in its effect among all cell lines tested with an average IC_{50} value of ~36 nM. Among the ovarian cancer cell lines, PA-1 (teratocarcinoma, undifferentiated) and A2780 (epithelial cancer cell line, differentiated), ST03 showed the better effect on PA1 cells (41 nM) as compared to A2780 (54 nM). Similarly, ST08 was found to show differential cytotoxicity with PA1 (38 nM) better than A2780 (64 nM). Since ovarian cancer cell lines showed promising results, we checked the effect of ST08 on breast cancer cell lines. We took MDA-MB231 (triple negative, basal, metastatic) and MCF-7 (epithelial, luminal) for the comparison. As observed in the ovarian cancer cell line, the highly metastatic MDA-MB231 showed better cytotoxicity (54 nM) than MCF-7 (127 nM). It is to be noted that, ST03 and ST08 showed cytotoxicity on undifferentiated cancer cells populations which have the potency to migrate (metastatic) and differentiate (stem cell characteristics). A comparison of the IC_{50} value of both the ST compounds has revealed the peculiar characteristics of them on the MDA-MB-231 and PA1 cell lines tested. It showed that these novel compounds are highly potent on PA1 and MDA-MB-231 cell lines (Table 1 and 2), which possess some of the properties of undifferentiated cells [51,52,53,54].

PA1 belongs to the category of teratocarcinoma, where the tumour cells have originated from germLine cells which have the characteristics of stem cells [52]. For instance, although we observe reasonable response rates to first-line chemotherapy, recurrence with resistance occurs in most of the cases of ovarian cancer, which is delivered by cancer stem cells that remain even after treatment [55, 56]. It is known that cancer stem cells expansion and characteristics lead to drug resistance [57, 58]. For the improved treatment outcome without relapse, it is necessary to eliminate cancer stem cells. The new treatment strategies of ovarian cancer are thus looking for drugs that can remove cancer stem cells very effectively. Drugs such as AS602801, and 673A has proved to eradicate ovarian cancer stem cells, and it increased the sensitivity of cancer stem cells to standard drugs such as cisplatin and paclitaxel [59, 60]. When compared to the potency of these drugs, ST03 showed cytotoxicity at a nanomolar concentration in PA1 cells (undifferentiated stem cell-like) in vitro. Thus the likelihood of improvement in anti-cancer activity cannot be ruled out if ST03 and ST08 are used in combination with standard of care drugs. Owing to the fact that ovarian cancer stem cells are relatively challenging to eradicate, the observed cytotoxicity of ST03 on PA1 is promising.

In the case of epithelial ovarian cancer cell line A2780, the effect of ST compounds were observed at a little higher concentration than on PA1. This observation is crucial as it indicates the differential cytotoxicity of ST compounds on tumors of different origin. Differential sensitivity of cell lines to ST compounds would help in understanding drug metabolism in cell types and thereby help to combat the acquired resistance. This knowledge would ultimately help in deciding combination therapies [61] .

Lack of multiple receptors is the major characteristics of the triple-negative breast cancer cell line MDA-MB-231 [62] and these cells respond the least to hormone therapies [63]. As a reason, potent inhibitors derived from natural compounds that can kill triple-negative breast cancer cells were developed in our laboratory. The effectiveness was predominant on

MDA-MB-231, the triple-negative breast (highly metastatic) cancer cell line [39, 64,65,66] than on the non-metastatic cell line MCF7. This result is very promising since the triple-negative breast cancer types are highly metastatic and difficult to treat, as there are a fewer treatment strategy for triple-negative breast cancer types [67].

To study the selectivity of ST03 and ST08 between cancer and normal, peripheral blood mononuclear cells and 293T cell lines were used. It was found that ST03 (>150nM) and ST08 (>150nM) are not cytotoxic to them at concentrations, at which it was cytotoxic to the non- adherent cancer cell lines. This points onto a fascinating fact that, these compounds are selective towards cancer cells and are less effective on normal cells which would render less systemic toxicity.

Metastasis is the primary cause of high mortality rate seen in cancer patients [68] which justifies the requirement of drugs with migrastatic potential. Previous reports have established the migrastatic property of curcumin [69] in cancer cells. Metastatic cell types undergo EMT (Epithelial to mesenchymal transition) during metastasis [70]. EMT also occurs not only in metastasis but also during wound healing and helps in tissue regeneration [70]. Recently, reports have linked EMT and dedifferentiated cancer stem cell-like properties of cancer cells. This points out the fact that cancer cells when they undergo EMT transition, it also acquire cancer stem cell-like features which initiates metastasis [71, 72]. The anti-cancer agent which targets the metastatic cancer cells can restrain the cancer stem cell-like cells or vice versa. Our study has demonstrated the cytotoxic effect of ST compounds on both liquid and solid cancer types, including stem cell-like PA1 and highly metastatic MDA-MB-231. This observation made us hypothesise the anti-metastatic potential of ST compounds.

Interestingly, we observed anti-metastatic property of ST08 compound on the highly aggressive, triple-negative, metastatic, MDA-MB-231 cell line. There was 4–5 fold reduction observed in migration of the MDA-MB-231 cells upon ST08 treatment. In addition, MMP1, the matrix metalloprotease reported to be one of the key players in breast cancer cell metastasis [73], was also found to be downregulated on ST08 treatment. Our results are in agreement with the characteristics of parent compound curcumin which also can block metastasis of cancer cells. Hence, our results report for the first time that ST compounds can suppress cancer cells possessing metastatic and cancer stem cell-like properties in vitro. These characteristics of our compounds might, therefore, open up a new strategy for treating recurrent malignant cancer types.

For the elimination of old cells and the growth of new cells, tightly regulated apoptosis process is required, and it plays a vital role in the growth and development of an organism [74]. Deregulation of apoptosis in a group of cells leads to proliferation of cells, cancer. Anti-cancer drugs such as platinum compounds and taxanes usually induce apoptosis in cancer cells [75]. Here in our study, we observed early and late apoptotic events in both the treatments. Further, there are majorly two types of apoptosis, mitochondrial-mediated intrinsic and death receptor-mediated extrinsic [76, 77]. Our results demonstrated that both the compounds are inducing cell death via intrinsic apoptosis as there is an upregulation in the expression of caspases such as caspase 3 and caspase-9. Caspases such as caspase-9 and caspase-3 are predominant proteins involved in the process of the intrinsic apoptotic pathway. A cascade activation of these molecules results in activation of PARP, cleavage of cytoskeleton proteins and chromatin fragmentation leading to apoptosis [78, 79]. We observed an increase in the expression of a cleaved form of caspase-9 in both the cell lines treated with ST03 or ST08. In the PA1 cell line, ST03 treatment has increased the expression of cleaved caspase-3 and caspase-9 indicative of the intrinsic pathway of apoptosis. Similarly increase in cleaved caspase-9 and caspase-3 was indicative of intrinsic pathway of apoptosis in ST08 treated MDA-MB-231.

The in vitro experiments conducted here, demonstrate the effective anti-cancer property of our novel curcumin derivatives ST03 and ST08. The compounds were found to be ~100 fold more potent than its parent compound, curcumin on both solid and liquid tumour types. The fascinating fact here is that they exhibited higher toxicity on cancer cell lines such as PA1 and MDA-MB-231, which possess cancer stem cell-like property to differentiate and migrate. The reduced expression of MMP1 and inhibition of migration on ST08 treatment indicate that this compound can modulate metastasis by inhibiting migration of cancer cells. Eradication of cancer stem cells and inhibition of metastasis could effectively reduce cancer recurrence and drug resistance which is a major hurdle in the cancer treatment. Hence, our results suggest that ST03 and ST08 can be considered as a promising candidate anticancer drug that can target both cancer stem cells and metastatic cells mostly responsible for the failure of cancer therapy. Further studies are required to understand the mechanism of cytotoxicity of these compounds, which is in progress.

CONCLUSION

We report for the first time two novel compounds, ST03 and ST08, which exhibit anti-cancer activity on both liquid and solid cancer types. These are ~ 100 fold better than its parent compound curcumin. We also show that these compounds inhibit cell proliferation by inducing intrinsic apoptosis and also suppressing cell migration in vitro. Due to the anti-cancer and migrastatic effect of ST03 and ST08 on stem cell-like ovary and breast cancer cell lines. These compounds have the potential to be developed as a novel anticancer agent towards the treatment of the metastatic, invasive, and recurrent cancer types.

ACKNOWLEDGEMENTS

We thank Dr. Venketesh from SSIHL, Puttaparthi for help with Mass Spec. We would like to acknowledge Anjana Elizabeth Jose and Hassan. A. Swarup for assistance with NMR analysis, Meghana Manjunath, Dr. Raksha Rao K and Suran R Nambeesan in manuscript reading.

REFERENCES

1. Amin ARMR, Karpowicz PA, Carey TE, Arbiser J, Nahta R, Chen ZG, et al. Evasion of anti-growth signaling: a key step in tumorigenesis and potential target for treatment and prophylaxis by natural compounds. Semin Cancer Biol. 2015;35:S55-S77.
2. Hanahan D, Weinberg RA. The hallmarks of cancer. Cell. 2000;100:57–70.
3. Arruebo M, Vilaboa N, Sáez-Gutierrez B, Lambea J, Tres A, Valladares M, et al. Assessment of the evolution of cancer treatment therapies. Cancers. 2011;3:3279–330.
4. Urruticoechea A, Alemany R, Balart J, Villanueva A, Vinals F, Capella G. Recent advances in cancer therapy: an overview. Curr Pharm Des. 2010;16:3–10.
5. Prieto-Vila M, Takahashi RU, Usuba W, Kohama I, Ochiya T. Drug resistance driven by cancer stem cells and their niche. Int J Mol Sci. 2017;18:2574.
6. Chen K, Huang YH, Chen JL. Understanding and targeting cancer stem cells: therapeutic implications and challenges. Acta Pharmacol Sin. 2013;34:732–40.
7. Torquato HF, Goettert MI, Justo GZ, Paredes-Gamero EJ. Anti-cancer Phytometabolites targeting cancer stem cells. Curr Genomics. 2017;18:156–74.
8. Bachmeier BE, Killian PH, Melchart D. The role of curcumin in prevention and management of metastatic disease. Int J Mol Sci. 2018;19:1716.
9. Gandalovičová A, Rosel D, Fernandes M, Veselý P, Heneberg P, Čermák V, et al. Migrastatics—anti-metastatic and anti-invasion drugs: promises and challenges. Trends Cancer. 2017;3:391–406.
10. Deng Y, Verron E, Rohanizadeh R. Molecular mechanisms of anti-metastatic activity of curcumin. Anticancer Res. 2016;36:5639–47.
11. Gupta AP, Khan S, Manzoor MM, Yadav AK, Sharma G, Anand R, et al. Anticancer curcumin: natural analogues and structure-activity relationship. In: Studies in natural products chemistry, vol. 54; 2017. p. 355–401.**Google Scholar**
12. Fadus MC, Lau C, Bikhchandani J, Lynch HT. Curcumin: an age-old anti-inflammatory and anti-neoplastic agent. J Tradit Complement Med. 2017;7:339–46.

13. Perrone D, Ardito F, Giannatempo G, Dioguardi M, Troiano G, Lo Russo L, et al. Biological and therapeutic activities, and anticancer properties of curcumin (review). Exp Ther Med. 2015;10:1615–23.
14. Reddy RC, Vatsala PG, Keshamouni VG, Padmanaban G, Rangarajan PN. Curcumin for malaria therapy. Biochem Biophys Res Commun. 2005;326:472–4.
15. Xu YX, Pindolia KR, Janakiraman N, Noth CJ, Chapman RA, Gautam SC. Curcumin, a compound with anti-inflammatory and anti-oxidant properties, down-regulates chemokine expression in bone marrow stromal cells. Exp Hematol. 1997;25:413–22.
16. Li Y, Zhang T. Targeting cancer stem cells by curcumin and clinical applications. Cancer Lett. 2014;346:197–205.
17. Zang S, Liu T, Shi J, Qiao L. Curcumin: a promising agent targeting cancer stem cells. Curr Med Chem Anticancer Agents. 2014;14:787–92.
18. Anand P, Kunnumakkara AB, Newman RA, Aggarwal BB. Bioavailability of curcumin : problems and promises. Mol Pharm. 2007;4:807–18.
19. Kálai T, Kuppusamy ML, Balog M, Selvendiran K, Rivera BK, Kuppusamy P, et al. Synthesis of N-substituted 3,5-bis (arylidene)-4-piperidones with high antitumor and antioxidant activity. J Med Chem. 2011;54:5414–21.
20. Selvendiran K, Ahmed S, Dayton A, Kuppusamy ML, Tazi M, Bratasz A, et al. Safe and targeted anticancer efficacy of a novel class of antioxidant-conjugated difluorodiarylidenyl piperidones: differential cytotoxicity in healthy and cancer cells. Free Radic Biol Med. 2010;48:1228–35.
21. Terlikowska KM, Witkowska AM, Zujko ME, Dobrzycka B, Terlikowski SJ. Potential application of curcumin and its analogues in the treatment strategy of patients with primary epithelial ovarian cancer. Int J Mol Sci. 2014;15:21703–22.
22. Adams BK, Ferstl EM, Davis MC, Herold M, Kurtkaya S, Camalier RF, et al. Synthesis and biological evaluation of novel curcumin analogs as anti-cancer and anti-angiogenesis agents. Bioorg Med Chem. 2004;12:3871–83.
23. Adams BK, Cai J, Armstrong J, Herold M, Lu YJ, Sun A, et al. EF24, a novel synthetic curcumin analog, induces apoptosis in cancer cells via

a redox-dependent mechanism. Anticancer Drugs. 2005;16:263–75.

24. Kearns PR, Hall AG. Glutathione and the response of malignant cells to chemotherapy. Drug Discov Today. 1998;3:113–21.
25. Espinoza-Fonseca LM. The benefits of the multi-target approach in drug design and discovery. Bioorg Med Chem. 2006;14:896–7.
26. Frantz S. Playing dirty. Nature. 2005;437:942–3.
27. Block KI, Gyllenhaal C, Lowe L, Amedei A, Ruhul Amin ARM, Amin A, et al. Designing a broad-spectrum integrative approach for cancer prevention and treatment. Semin Cancer Biol. 2015;35(Suppl):S276–304.
28. Das S, Das U, Varela-Ramírez A, Lema C, Aguilera RJ, Balzarini J, et al. Bis [3,5-bis (benzylidene)-4-oxo-1-piperidinyl]amides: a novel class of potent Cytotoxins. ChemMedChem. 2011;6:1892–9.
29. Das S, Das U, Sakagami H, Umemura N, Iwamoto S, Matsuta T, et al. Dimeric 3,5-bis (benzylidene)-4-piperidones: a novel cluster of tumour-selective cytotoxins possessing multidrug-resistant properties. Eur J Med Chem. 2012;51:193–9.
30. Santiago-Vazquez Y, Das S, Das U, Robles-Escajeda E, Ortega NM, Lema C, et al. Novel 3,5-bis (arylidene)-4-oxo-1-piperidinyl dimers: structure-activity relationships and potent antileukemic and antilymphoma cytotoxicity. Eur J Med Chem. 2014;77:315–22.
31. Das S, Das U, Michel D, Gorecki DKJ, Dimmock JR. Novel 3,5-bis (arylidene)-4-piperidone dimers: potent cytotoxins against colon cancer cells. Eur J Med Chem. 2013;64:321–8.
32. Edlich RF, Winters KL, Lin KY. Breast cancer and ovarian cancer genetics. J Long Term Eff Med Implants. 2005;15:533–45.
33. Kavitha CV, Nambiar M, Ananda Kumar CS, Choudhary B, Muniyappa K, Rangappa KS, et al. Novel derivatives of spirohydantoin induce growth inhibition followed by apoptosis in leukemia cells. Biochem Pharmacol. 2009;77:348–63.
34. Allen M, Millett P, Dawes E, Rushton N. Lactate dehydrogenase activity as a rapid and sensitive test for the quantification of cell numbers in vitro. Clin Mater. 1994;16:189–94.
35. Kim R. Recent advances in understanding the cell death pathways activated by anticancer therapy. Cancer. 2005;103:1551–60.
36. Boersma HH, Kietselaer BLJH, Stolk LML, Bennaghmouch A, Hofstra L, Narula J, et al. Past, present, and future of annexin A5: from protein

discovery to clinical applications. J Nucl Med. 2005;46:2035–50.

37. Porter AG, Jänicke RU. Emerging roles of caspase-3 in apoptosis. Cell Death Differ. 1999;6:99–104.
38. McIlwain DR, Berger T, Mak TW. Caspase functions in cell death and disease. Cold Spring Harb Perspect Biol. 2013;5:1–28.
39. Chambers AF, Groom AC, MacDonald IC. Dissemination and growth of cancer cells in metastatic sites. Nat Rev Cancer. 2002;2:563–72.
40. Al-Akhras MAH, Aljarrah K, Al-Khateeb H, Jaradat A, Al-Omari A, Al-Nasser A, et al. Introducing cichorium pumilum as a potential therapeutical agent against drug-induced benign breast tumor in rats. Electromagn Biol Med. 2012;31:299–309.
41. Hamza AA, Heeba GH, Elwy HM, Murali C, El-Awady R, Amin A. Molecular characterization of the grape seeds extract's effect against chemically induced liver cancer: in vivo and in vitro analyses. Sci Rep. 2018;8:1270.
42. Lotempio MM, Veena MS, Steele HL, Ramamurthy B, Ramalingam TS, Cohen AN, et al. Curcumin suppresses growth of head and neck squamous cell carcinoma. Clin Cancer Res. 2005;11:6994–7002.
43. Mukhopadhyay A, Bueso-Ramos C, Chatterjee D, Pantazis P, Aggarwal BB. Curcumin downregulates cell survival mechanisms in human prostate cancer cell lines. Oncogene. 2001;20:7597–609.
44. Siwak DR, Shishodia S, Aggarwal BB, Kurzrock R. Curcumin-induced antiproliferative and proapoptotic effects in melanoma cells are associated with suppression of IκB kinase and nuclear factor κB activity and are independent of the B-Raf/mitogen-activated/extracellular signal-regulated protein kinase pathway and the Akt pathway. Cancer. 2005;104:879–90.
45. Mehta K, Pantazis P, McQueen T, Aggarwal BB. Antiproliferative effect of curcumin (diferuloylmethane) against human breast tumor cell lines. Anticancer Drugs. 1997;8:470–81.
46. Elattar TMA, Virji AS. The inhibitory effect of curcumin, genistein, quercetin and cisplatin on the growth of oral cancer cells in vitro. Anticancer Res. 2000;20:1733–8.
47. Lin YG, Kunnumakkara AB, Nair A, Merritt WM, Han LY, Armaiz-Pena GN, et al. Curcumin inhibits tumor growth and angiogenesis in ovarian carcinoma by targeting the nuclear factor-κB pathway. Clin Cancer Res. 2007;13:3423–30.

48. Ye C, Wang W, Xia G, Yu C, Yi Y, Hua C, et al. A novel curcumin derivative CL-6 exerts antitumor effect in human gastric cancer cells by inducing apoptosis through hippo-YAP signaling pathway. OncoTargets Therapy. 2019;12:2259–69.
49. Vyas A, Dandawate P, Padhye S, Ahmad A, Fazlul S. Perspectives on new synthetic curcumin analogs and their potential anticancer properties. Curr Pharm Des. 2013;19:2047–69.
50. Shi M, Cai Q, Yao L, Mao Y, Ming Y, Ouyang G. Antiproliferation and apoptosis induced by curcumin in human ovarian cancer cells. Cell Biol Int. 2006;30:221–6.
51. Zeuthen J. Human teratocarcinoma cell lines. A review. Int J Androl. 1981;4:61–77.
52. Zeuthen J, Norgaard JO, Avner P, Fellous M, Wartiovaara J, Vaheri A, et al. Characterization of a human ovarian teratocarcinoma-derived cell line. Int J Cancer. 1980;25:19–32.
53. Lawson DA, Bhakta NR, Kessenbrock K, Prummel KD, Yu Y, Takai K, et al. Single-cell analysis reveals a stem-cell program in human metastatic breast cancer cells. Nature. 2015;526:131–5.
54. Morata-Tarifa C, Jiménez G, García MA, Entrena JM, Griñán-Lisón C, Aguilera M, et al. Low adherent cancer cell subpopulations are enriched in tumorigenic and metastatic epithelial-to-mesenchymal transition-induced cancer stem-like cells. Sci Rep. 2016;6:18772.
55. Zhang QH, Dou HT, Xu P, Zhuang SC, Liu PS. Tumor recurrence and drug resistance properties of side population cells in high grade ovary cancer. Drug Res. 2015;65:153–7.
56. Li SS, Ma J, Wong AST. Chemoresistance in ovarian cancer: exploiting cancer stem cell metabolism. J Gynecol Oncol. 2018;29:e32.
57. Dean M, Fojo T, Bates S. Tumour stem cells and drug resistance. Nat Rev Cancer. 2005;5:275–84.
58. Singh A, Settleman J. EMT, cancer stem cells and drug resistance: an emerging axis of evil in the war on cancer. Oncogene. 2010;29:4741–51.
59. Yamamoto M, Suzuki S, Togashi K, Sanomachi T, Seino S, Kitanaka C, et al. AS602801, an anticancer stem cell candidate drug, reduces survivin expression and sensitizes A2780 ovarian cancer stem cells to carboplatin and paclitaxel. Anticancer Res. 2018;38:6699–706.
60. Raghavan S, Mehta P, Ward MR, Bregenzer ME, Fleck EMA, Tan

L, et al. Personalized medicine–based approach to model patterns of chemoresistance and tumor recurrence using ovarian cancer stem cell spheroids. Clin Cancer Res. 2017;23:6934–45.

61. Jensen PB, Christensen IJ, Sehested M, Hansen HH, Vindeløv L. Differential cytotoxicity of 19 anticancer agents in wild type and etoposide resistant small cell lung cancer cell lines. Br J Cancer. 1993;67:311–20.
62. ATCC. Triple negative breast cancer panels- MDA-MB-231 WwwAtccOrg; 2004.
63. Pal SK, Childs BH, Pegram M. Triple negative breast cancer: unmet medical needs. Breast Cancer Res Treat. 2011;125:627–36.
64. Neve RM, Chin K, Fridlyand J, Yeh J, Baehner FL, Fevr T, et al. A collection of breast cancer cell lines for the study of functionally distinct cancer subtypes. Cancer Cell. 2006;10:515–27.
65. Cailleau R, Olivé M, Cruciger QVJ. Long-term human breast carcinoma cell lines of metastatic origin: preliminary characterization. In Vitro. 1978;14:911–5.
66. Holliday DL, Speirs V. Choosing the right cell line for breast cancer research. Breast Cancer Res. 2011;13:215.
67. Ahmed Wahba H, Ahmed E-HH. Current approaches in treatment of triple-negative breast cancer. Cancer Biol Med. 2015;12:106–16.
68. Mehlen P, Puisieux A. Metastasis: a question of life or death. Nat Rev Cancer. 2006;6:449–58.
69. Kuttan G, Hari Kumar KB, Guruvayoorappan C, Kuttan R. Antitumor, anti-invasion, and antimetastatic effects of curcumin. Adv Exp Med Biol. 2007;595:173–84.
70. Thiery JP, Acloque H, Huang RYJ, Nieto MA. Epithelial-mesenchymal transitions in development and disease. Cell. 2009;139:871–90.
71. Polyak K, Weinberg RA. Transitions between epithelial and mesenchymal states: acquisition of malignant and stem cell traits. Nat Rev Cancer. 2009;9:265–73.
72. Herreros-Villanueva M, Zhang JS, Koenig A, Abel EV, Smyrk TC, Bamlet WR, et al. SOX2 promotes dedifferentiation and imparts stem cell-like features to pancreatic cancer cells. Oncogenesis. 2013;2:e61.
73. Liu H, Kato Y, Erzinger SA, Kiriakova GM, Qian Y, Palmieri D, et al. The role of MMP-1 in breast cancer growth and metastasis to the brain in a xenograft model. BMC Cancer. 2012;12:583.

74. Negroni A, Cucchiara S, Stronati L. Apoptosis, necrosis, and necroptosis in the gut and intestinal homeostasis. Mediators Inflamm. 2015;2015:250762.

75. Friesen C, Fulda S, Debatin KM. Cytotoxic drugs and the CD95 pathway. Leukemia. 1999;13:1854–8.

76. Tameire F, Verginadis II, Koumenis C. Cell intrinsic and extrinsic activators of the unfolded protein response in cancer: mechanisms and targets for therapy. Semin Cancer Biol. 2015;33:3–15.

77. Fulda S, Debatin KM. Extrinsic versus intrinsic apoptosis pathways in anticancer chemotherapy. Oncogene. 2006;25:4798–811.

78. COHEN GM. Caspases: the executioners of apoptosis. Biochem J. 1997;326:1–16.

79. Elmore S. Apoptosis: a review of programmed cell death. Toxicol Pathol. 2007;35:495–516.

10

Recent Advances in Cancer Therapy Based on Dual Mode Gold Nanoparticles

Ellas Spyratou [1], Mersini Makropoulou [2], Efstathios P. Efstathopoulos [1], Alexandros G. Georgakilas [2] and Lembit Sihver [3]

[1]2nd Department of Radiology, Medical School, National and Kapodistrian University of Athens, 12462 Athens, Greece
[2]Department of Physics, School of Applied Mathematical and Physical Sciences, National Technical University of Athens, 15780 Athens, Greece
[3]Atominstitut, Technische Universität Wien, Stadionallee 2, 1020 Vienna, Austria

ABSTRACT

Many tumor-targeted strategies have been used worldwide to limit the side effects and improve the effectiveness of therapies, such as chemotherapy, radiotherapy (RT), etc. Biophotonic therapy modalities comprise very promising alternative techniques for cancer treatment with minimal invasiveness and side-effects. These modalities use light e.g., laser irradiation

Citation: Spyratou, E.; Makropoulou, M.; Efstathopoulos, E.P.; Georgakilas, A.G.; Sihver, L. Recent Advances in Cancer Therapy Based on Dual Mode Gold Nanoparticles. Cancers 2017, 9, 173. https://doi.org/10.3390/cancers9120173.

in an extracorporeal or intravenous mode to activate photosensitizer agents with selectivity in the target tissue. Photothermal therapy (PTT) is a minimally invasive technique for cancer treatment which uses laser-activated photoabsorbers to convert photon energy into heat sufficient to induce cells destruction via apoptosis, necroptosis and/or necrosis. During the last decade, PTT has attracted an increased interest since the therapy can be combined with customized functionalized nanoparticles (NPs). Recent advances in nanotechnology have given rise to generation of various types of NPs, like gold NPs (AuNPs), designed to act both as radiosensitizers and photothermal sensitizing agents due to their unique optical and electrical properties i.e., functioning in dual mode. Functionalized AuNPS can be employed in combination with non-ionizing and ionizing radiation to significantly improve the efficacy of cancer treatment while at the same time sparing normal tissues. Here, we first provide an overview of the use of NPs for cancer therapy. Then we review many recent advances on the use of gold NPs in PTT, RT and PTT/RT based on different types of AuNPs, irradiation conditions and protocols. We refer to the interaction mechanisms of AuNPs with cancer cells via the effects of non-ionizing and ionizing radiations and we provide recent existing experimental data as a baseline for the design of optimized protocols in PTT, RT and PTT/RT combined treatment.

Keywords: gold nanoparticles, ionizing radiation, photothermal therapy, radiotherapy, radiosensitizing

INTRODUCTION

With the rapid development of nanotechnology, there has been a revolutionary growth in proposals of smart nanosystems for use in cancer imaging, targeted administration of drugs and post-therapeutical monitoring of cancer regression in oncology. For example, nanoparticles (NPs), such as gold, gadolinium, hafnium, silicon, ferromagnetic and polymer NPs, quantum dots, nanorods, nanotubes, liposomes and dendrimers have been examined and utilized—at least preclinically—as drug delivery systems in personalized nanomedicine. Typically, NPs, with their dimensions of less than 100 nm, exploit the increased vasculature permeability and decreased lymphatic function of tumors. One "competitive edge" of NPs is that they can be attached to specific cancer cell targets with non-invasive implementation, increasing the cellular uptake efficiency, selectivity and localization in tumor cells and tissues. Metallic NPs, semiconductor quantum dots and carbon nanotubes have been used as light-activated heating nanosystems which can be incorporated into tumors allowing high heat administration in the tumor area in hyperthermia treatment, and minimizing the damage in the surrounding healthy tissue [**1**,**2**]. However, one of the main challenges to this approach is the "thermotolerance" of the tumors whereby cells or tissues become resistant to elevated temperatures as a result of prior or continuous exposure to hyperthermia. Metallic NPs can also act as radiosensitizers in radiotherapy (RT) treatment providing "dose enhancement". One challenge in RT is how to deal with the radiation resistance of hypoxic tumor cells. This is especially a problem when using conventional low linear energy transfer (LET) X-rays/γ-rays, and it is not possible to increase the radiation dose to levels high enough to kill the hypoxic tumor cells while sparing normal proximal tissues. The use of radiosensitizers could be a way to enhance the radiotherapy efficacy and make it more specific against the malignant tumor cells while drastically reducing the damage to surrounding normal tissues. Among the various types of NPs, gold NPs AuNPs (or GNPs) have been brought to the forefront of cancer research due to their unique properties, exhibiting a combination of characteristic physical, chemical, optical and electronic properties.

The Basic Physics of Interactions of AuNPs with Ionizing Radiation and Laser Radiation

As the prospective for nanotechnology in personalized medicine is undergoing considerable progress, several studies have aimed at investigations of the actions of both ionizing and non-ionizing radiation on nanosystems, regarding the elucidation of the principal mechanisms of radiation—NP interaction, depending on both the physico-chemical properties of the NPs and the irradiation parameters.

The main physical mechanisms through which ionizing radiation interacts with NPs (in the keV range) are the Compton and photoelectric phenomena, where an incident photon can either be partially or fully absorbed by an atom, resulting in the ejection of an electron. The parameters and the relevant mechanisms of NP and X-ray interactions are the subject of numerous reports and the interested reader should refer to recent reviews of [**3**,**4**,**5**]. Having, as a starting point, the basic physical interactions of X-rays with biological tissues (the photoelectric and Compton effects), it is expected that exposure of high atomic number (Z) NPs to X-rays enhances the photoelectric and Compton effects, when compared to exposure of relatively light elements, such as hydrogen, carbon, and oxygen, in biological tissues. This would lead to an increased radiation therapy efficiency. However, the increase in the interaction cross-sections of AuNPs vs. soft tissue decreases with increasing X-ray energy, and it has indeed been found that the energy of the radiation plays a major role in the radiosensitization effect [3].

Obviously, the energy of the X-rays is important, since the photoelectric effect is dominant at lower energies and prevails until the photon energy reaches a medium energy (e.g., around 500 keV for Au) with a cross section varying with Z^4 or Z^5, depending on the target material, and is enhanced by an increased absorption by electron shells (K, L, M, etc.) at low energies [4]. When using X-rays, mainly the inner electron shells are ionized and this creates cascades of both low and high energy Auger electrons. Increasing the atomic number Z of the NP is enhancing the photoelectric and Compton effects when they are exposed to X-rays. That is why high Z NPs are more radiosensitizing when using X-rays than low-Z NPs. Therefore, an approach to maximize the differential response between the tumor and normal tissue, termed therapeutic ratio, is through the introduction of high-Z material into the target. AuNPs with $Z = 79$ are promising radiosensitizers in this regard due to their high atomic number and mass energy coefficient relative to soft tissue [6]. Haume et al. [3] have presented a comprehensive overview on the physico-chemical properties of NPs (shape, size, surface charge and coating, concentration and NP toxicity) and their role in radiosensitization. The concentration and the physicochemical characteristics of the NPs, as well as their location inside the cell [5,7] influence the radiosensitizing action of NPs.

The interaction of non-ionizing light photons with small particles depends strongly on the dimension, shape, surface treatment and composition of the particles, as well as on the composition of the medium in which the particles are embedded. If we consider the special case of metal NPs, absorption, scattering and extinction mechanisms of laser radiation by a single metal nanoparticle determine the subsequent photophysical and photochemical processes [8]. Solving the problem of absorption and scattering of electromagnetic radiation by a small particle involves solving Maxwell's equations with the correct boundary conditions. To solve Maxwell's equations, various analytical or numerical methods can be used [9]. The most popular and simple analytical method is the one developed in 1908 by Gustav Mie [10], while a complete derivation of Mie's theory is given by Bohren and Huffman [11]. Both these works have even been cited in recent years by [12,13]. The general Mie theory of light scattering by a spherical metal particle states that the scattering cross-section of a nanoparticle of radius R, much smaller than the wavelength of the incident light λ ($R << \lambda$), varies as R^6, while its absorption cross-section varies as R^3 [12]. This implies that absorption is more important than scattering for very small particles.

When a metallic nanoparticle interacts with light (an electromagnetic wave), the oscillating electric field causes the conduction electrons to oscillate coherently. The optical properties of metal NPs, are therefore, as a consequence of their reduced dimensions, dominated by the coherent collective oscillation of their conduction band electrons. Collective excitations of conductive electrons in metals are called "plasmons" [9]. Depending on the boundary conditions, it is commonly accepted to distinguish between bulk plasmons (3D plasma), surface propagating plasmons or surface plasmon polaritons (2D films), and surface localized plasmons (NPs) [14]. Surface plasmon resonance (SPR) is a nanoscale electronic effect that causes metallic NPs to absorb and scatter electromagnetic radiation of wavelengths considerably larger than the particles themselves [15]. Metallic NPs can resonate plasmonically under exposure to light when the laser wavelength corresponds to their SPR. The resonated NPs convert the electromagnetic energy into heat via a non-irradiative process causing hyperthermic damage to cancer cells. The plasmonic oscillations can be tuned in a wide spectral range from visible to near infrared (NIR) when the NPs shape deviates from the highly symmetric nanospherical shape to nanoshells (AuNSs), nanocages (AuNCs) and nanorods (AuNRs) [2]. For example, the SPR band shifts to the NIR region by changing the shell to core ratio, the aspect ratio and the particle size, wall thickness and wall porosity for AuNSs, AuNRs and AuNCs, respectively [2]. The optical properties and efficiencies of the gold NPs depend strongly on their sizes, geometries, surface modifications and the refractive index of the surrounding medium [16] together with laser irradiation parameters such as laser wavelength, pulse duration, irradiation and time.

Recently, a fast developing research field is to combine laser based photothermal therapy with conventional ionizing radiation therapy using NPs. As we mentioned above, besides their photothermal properties, NPs can also act as radiosensitizers in RT providing "dose enhancement", increased cellular uptake efficiency, selectivity and localization in tumor cells and tissues. When internalized into tumor, these NPs increase the cell killing after irradiation due to its properties, including high atomic number

(Z), high absorption coefficient and good biocompatibility. The NPs are straightforward to synthesize, in a wide range of sizes, and their surface functionalization can easily be modified by the attachment of ligands required to target cancer cells, and organelles therein, or increase the life time in the bloodstream [**3**,**17**]. Thus, lower radiation doses can be used to eradicate tumors with the same efficiency, or even better, while decreasing the damage to surrounding normal tissues.

APPLICATION OF INTERACTION OF AUNPS WITH NON-IONIZING RADIATION (NON-IR) IN CANCER THERAPY

Introductory Remarks about Non-IR Cancer Treatment Modalities

Human beings, and other mammals, have a recorded history of cancer incidence throughout history. Cancer is treated with both invasive and non-invasive treatment modalities, such as surgery, radiation therapy, chemotherapy, as well as other therapeutic modalities (e.g., immunotherapy, hormone therapy, biological therapy, thermotherapy, photodynamic therapy), and a combination of these (e.g., radiosurgery). Undoubtedly, in the anti-cancer fighting armamentarium, radiation therapy (RT) is widely used for curative treatment. Ionizing radiation mainly kills the cells either by apoptosis or mitotic cell death (mitotic catastrophe). Following radiation, necrosis can also be detected to some extent in both cancer cell lines and normal tissues, but normally less frequently than apoptosis and mitotic cell death. Mitotic cell death occurs during or after aberrant mitosis (cell division) and is caused by mis-segregation of chromosomes leading to formation of giant cells with aberrant nuclear morphology and multiple nuclei. Normally cell death through apoptosis is considered most preferable death since apoptosis is characterized by cell shrinkage and formation of apoptotic bodies. Necrosis is less orderly than apoptosis, and in contrast with apoptosis, cleanup of cell debris by phagocytes of the immune system is generally more difficult, as the disorderly death generally does not send cell signals which tell nearby phagocytes to engulf the dying cell. However, cell death through necrosis produces an inflammatory response which might be beneficial for the treatment. However, RT based on high-energy ionizing radiation (X-rays, gamma rays, electrons, or ions) is not the only option, as nowadays there are possibilities to develop novel cancer treatment strategies specifically based on novel biophotonic principles, e.g., on the use of non-ionizing photon RT [18]. In addition, the revolution in the fields of physics and technology of monochromatic and coherent laser light, has open up new opportunities in oncology, e.g., for diagnosis, drug delivery and therapy. The possibility to combine different monitoring modalities [19] with therapy in the field of *theranostics*, which is the association between thera(peutics) and diag(nostics) [20,21] has also opened up new opportunities for the treatment of cancer. This interdisciplinary approach is a very promising challenge for both the researchers and the physicians and, with the rapid development of nanomedicine and pharmaceutical technology in the last decades, a revolutionary growth of theranostics methods have been proposed using several different nanosystems which have multiple functions: early cancer detection, targeted drug delivery, and monitoring of the cancer progression [22]. Among the new radiation therapy approaches, PhotoDynamic Therapy (PDT) is a non-invasive or minimally invasive biophotonic modality, based on the simultaneous action of the following three factors: (i) photosensitizing agents, able to photochemically eradicate malignant cells; (ii) generation of highly reactive singlet oxygen; and (iii) non-thermal monochromatic light irradiation. Briefly, PDT's activity is based on the photo-generation of singlet oxygen $^{1}O_2$ or reactive oxygen species (ROS), through the photosensitizer excitation due to photons absorption, which can cause cellular destruction [23]. PTT is an extension of PDT which does not require oxygen, as the stimulated photosensitizing agent release vibrational energy (heat) which kills the targeted cells. It is worthy to mention that ancient Greeks, Romans, and Egyptians used heat to treat breast masses, and this is still a recommended self-care treatment for breast engorgement.

Photothermal Therapy (PTT) and Gold NPs: Advantages and Limitations

PTT is a newly developed and encouraging type of hyperthermia which employs photon energy of laser activated photoabsorbers to generate heat sufficient to cause cells' destruction. The hyperthermia treatment can also be conducted with radiofrequency, microwave, ultrasound and laser which are less invasive treatments. Hyperthermia is connected to cell death with three mechanisms, cell apoptosis and cell necrosis and relatively recently with necroptosis which is a type of programmed necrosis [**24**]. Cell apoptosis takes place when the heating temperatures range from 41 °C to 47 °C. Excessive cell necrosis is occurred by heat shock for temperatures usually higher than

50 °C, producing a much quicker cell death than apoptosis and is based on protein denaturing [**25**]. Hyperthermia is a treatment that can minimize the toxicity of chemical and radioactive agents, destroying diseased cells, without harming the healthy ones, but still requires the invasion with a probe to heat the malignant cells. The recent development of AuNPs, as probes capable of efficient heat generation under laser irradiation renders PTT is a very promising tool. Metallic NPs, semiconductor quantum dots, polymeric NPs, nanographene sheets and, carbon nanotubes have been used as light-activated heating nanosystems which can be incorporated into tumors allowing high heat deposition in the tumor area at low laser intensities and minimizing the damage in the surrounding healthy tissue [1,2,26].

Among the various types of NPs, AuNPs have been brought to the forefront of cancer research mainly due to their strongly enhanced and tunable optical properties. AuNPs can accumulate in the tumor tissues via the passive mechanism referred as "Enhanced Permeability and Retention, (EPR) effect" [27] or by active targeting by conjugation of a chemical moiety that is over expressed in cancer cells (monoclonal antibody, folic acid for cancer treatment). In order for the NPs to find and accumulate at their target cells, through the systemic route, they need to overcome several extra- and intracellular barriers by fulfilling several major prerequisites including: (a) their outer periphery must be inert in order to avoid the activation of the reticuloendothelial system (RES) so that they can circulate in the blood compartment for prolonged periods of time increasing the probability of reaching the pathological cells; (b) their dimensions should be close to 100 nm in order to escape the blood compartment and locate the pathological cells; (c) after internalization in the pathological cells, the NPs should be to overcome multiple obstacles related to: (i) cell surface binding; (ii) cellular uptake; (iii) escape from lysosomes/endosomes; and (iv) association with a particular subcellular location, such as nuclei or mitochondria [28].

The sizes of the AuNPs are critical: as the diameter of the AuNPs e.g., nanoshells, become smaller, the plasmon resonance wavelengths are significantly less optically tunable than those of larger diameter nanoshells. Furthermore, NPs smaller than ~10 nm are easily cleared from circulation via the kidneys and thus do not have adequate residence time to accumulate in target tissue. The concentration, the toxicity, the degree of aggregation, the circulation time and the subcellular location of the NPs affect PTT efficacy. Some NPs are not toxic at low concentrations but when the incubation time or concentration is increased they show a degree of toxicity. In addition, high NP concentrations increase the probability of aggregation, which is a critical point since the optical properties of the NPs are strongly deteriorated by the presence of aggregation [**2**]. The surface treatment of the NP is therefore of great importance. Many studies indicate that AuNPs reveal a more effective PTT when they are located in the cell membrane than when they are located in the cells, while some other studies have focused on the potential benefits of targeting AuNPs to specific cell parts, e.g., the nucleus and the DNA [**28**]. When photothermal treatment is conducted by continuous-wave (CW) laser irradiation, AuNPs usually localized in the cytoplasm are more effective in inducing cell destruction than AuNPs localized at the nucleus [**29**]. On the other hand, high localization of AuNPs at the nucleus is needed when nanosecond pulsed laser is used, suggesting that nonlinear absorption processes enhance the plasmonic field. CW laser induces cell death via apoptosis, while the nanosecond pulsed laser leads to cell necrosis [29]. These findings indicate that the cell death mechanism depends of the different mechanism of laser-cells interactions. A number of research projects have been carried out using different types of AuNP: nanospheres, nanoshells, nanocages and nanorods to overcome the limitations for targeted cancer therapy by photothermal treatment [30,31].

Photothermal Therapy and Gold Nanoshells

Plasmonic photothermal therapy using gold-based nanosphere contrast agents was reported for the first time by Lin and co-workers in 2003 [32] in combination with a nanosecond visible pulsed laser. T lymphocytes were labeled with anti-human CD8 mouse IgG and then incubated with 30 nm gold particles conjugated to anti-mouse IgG. The cells were irradiated with 100 laser pulses (565 nm wavelength, 20 ns pulse duration) at an energy fluence of 0.5 J/cm^2 leading to the destruction of more than 90% of the cells according to cell viability tests. Photothermal destruction has also been performed successfully on squamous carcinoma cells lines (HSC 313 and HOC 3 Clone 8) and a benign epithelial cell line (HaCaT) after incubation with ~40 nm anti-EGFR antibody conjugated Au nanospheres [33] and CW laser irradiation at 514 nm, for 4 min at power densities of 57 W/cm^2 and 25 W/cm^2, respectively. No cell death was observed in the absence of NPs at four times the energy required to kill the malignant cells with bonded anti-EGFR/Au conjugates.

Plasmonic hollow gold nanoshells coated with PEGylated thermosensitive lipids were used as an efficient vehicle to co-deliver two drugs, bortezomib and gemcitabine (GNS-L/GB), for combinational chemotherapy and photothermal therapy of pancreatic cancer. GNS-L/GB showed synergistic cytotoxicity and improved internalization by MIA PaCa-2 and PANC-1 cells. High-powered near-infrared continuous wave laser (λ = 808 nm) at 4 W/cm^2 effectively killed cancer cells via the photothermal effect of GNS-L/GB, irrespective of cell type in a power density, time-, and GNS dose-dependent manner. These results suggest that this method can provide a novel approach to achieve synergistic combinational chemotherapy and photothermal therapy, even with resistant pancreatic cancer [34].

For in vivo applications, PTT using NIR light is favorable as it penetrates deeper into the tissue due to minimal absorption by the major components of water and hemoglobin. Hirsch and coworkers [35] were first to develop NIR PTT using PEG-coated gold nanoshells (AuNSs) for human breast cancer cells morbidity in vitro after exposure to CW NIR light (at λ = 820 nm) at a power density of 35 W/cm^2 for 7 min. In vivo studies in mice show that by injecting the PEG-AuNSs interstitially into the tumor volume, and then irradiating with NIR light (at λ = 820 nm) at 4 W/cm^2, temperatures capable of inducing irreversible tissue damage ($\Delta T = 37.4 \pm 6.6$°C) within 4–6 min were raised. However, the desired biodistribution of AuNSs and the prolonged blood circulation remains a challenge. Recently, macrophage cell membrane (MPCM)-camouflaged gold nanoshells served as new generation agents for in vivo photothermal cancer therapy. The tumor area was irradiated by an 808 nm NIR laser at 1 W/cm^2 for 5 min after 20 min of intravenous injection of the MPCM-AuNSs in mice and subsequent tumor volume and body weight measurements were performed every day until the 25th day. This surface functionalization provided active targeting ability by recognizing tumor endothelium and thus improved tumoritropic accumulation, enhanced in vivo blood circulation time and local accumulation at the tumor when administered systematically and improved the efficacy of PTT [36]. Tumor necrosis factor-alpha coated gold nanospheres (Au-TNF) heated by nanosecond laser pulses demonstrate higher therapeutic efficiency against murine carcinoma [37]. Au-TNF are already in phase I trials in humans as nanodrugs emphasizing the clinical relevance of PT nanodrug.

Recently, hollow gold nanoshells with cores densely packed with small interfering RNAs have been efficiently used in photothermal treatment in a nude mice model. The thermal resistance of cancer cells is closed related with the over expression of the heat shock proteins 70 (HSP70s), which are abnormally upregulated when cells are under lethal stresses. Under NIR light irradiation, efficient down regulation of Hsp70 is achieved since the small interfering RNAs can detach form the gold surface, escape from endosomes and target the mRNA leading to down-regulation of HSP70 gene expression. Meanwhile, the temperature increases for hyperthermia therapy due to the high photothermal efficiency of the nanoshells [38].

Photothermal Therapy and Gold Nanorods

NIR PTT based on gold nanorods (AuNRs) was firstly demonstrated using in vitro studies where AuNRs functionalized with anti-EGFR antibodies were bound to human oral cancer cells [31]. After exposure to NIR light (λ = 800 nm) at 10 W/cm^2 the malignant cells were clearly injured while the threshold for normal cells was found at 15 W/cm^2 [33]. Since then, several in vitro and in vivo studies indicate the potential of AuNRs due to their unique properties [39,40]. AuNRs absorb light in the NIR region that penetrates deeper in tissues with higher spatial precision. The absorption range can be also fine-tuned by adjusting the aspect ratios of AuNRs.

The feasibility of the plasmonic PTT as applied to natural tumors in the mammary glands of dogs and cats were demonstrated after injection of pegylated-AuNPS and laser irradiation after 5 min using an 808 nm diode laser with a power density 5.8 W/cm^2. The results of PTT in animals, histopathology evaluation, blood analysis, and X-ray diffraction show a complete regression in the cancer grade after the third treatment with no obvious changes in liver and kidney functions and no metastasis after 1 year of the treatment [41]. Monosaccharide-imprinted gold nanorods using sialic acid (SA) induced a significant decrease of a tumor examined two weeks after photothermal treatment carried out on HepG-2 tumor-bearing mice [42]. Fluorescence images of a tumor-bearing mouse at 24 h post injection of fluorescence-dye doped SA-imprinted AuNRs show a good targeting selectivity to cancer cells. According to the anti-tumor tests, SA-imprinted AuNRs exhibited good biocompatibility and also high photothermal effect. In vitro photothermal test explored with HepG-2 cells and L-02 cells incubated with SA-imprinted AuNpS show that after irradiation with a 750 nm laser beam for different durations (2, 4, 6, 8 min) the most HepG-2 cells were found dead while the viability of L-02 cells was still higher at 85% as it was given by MTT assay.

Anti-EGFR-conjugated AuNRs (antiEGFR-GN) upon NIR-PTT exerted synergetic anti-proliferative and apoptotic actions through upregulation of HS70 protein and cleaved caspase-3, downregulation of Ki-67 and EGFR, and inhibition of several intracellular signaling molecules (mTOR, AKT, ERK1/2 and FAK) (Table 1) [43]. AuNRs, coated with dipicolyl amine (DPA), which forms stable complexes with Zn^{2+} cations, results to Zn(II)/

DPA-GNRNPs which recognize phosphate-containing molecules including siRNA. Zn(II)/DPA-GNR presents strong complexation with the anti-polo-like kinase 1 siRNA (siPLK), which is used for gene silencing. The SiPLK/Zn(II)DPA-GNRs nano-complex was efficiently delivered siRNA into the cancer cells in vitro and in vivo, in PC-3 tumour mouse model (Table 1) [44].

Table 1: Recent advancements in PTT, RT and PTT/RT and characteristics of testing (cells, AuNP type, concentrations, uptake time, therapy type)

Cellline	**AuNP-Type**	**[CNP]**	**Up Take Time/Irradiation Time**	**Therapy Type**	**Shape**	**Size (nm)**	**Reference**
Murine SCK tested in mice	TNF AuNPs	0.1 mL/20 g bodyweight	1 min	690 nm Pulsed laser 0.1–1 J/cm^2	spheres	33 nm	[37]
B16-F10 in vivo tested in nude mice	Doxorubicin-AuNPs	1 μM	24 h/5 min	660 nm laser 8–20 W/cm^2	sphere	20 nm	[45]
MIA PaCa-2 and PANC-1 cells	GNS-L/GB	50 μg Au_{equiv}	48 h/3 min	808 nm laser 4 W/cm^2	shells	85 nm	[34]
MCF-7 tested female BALB/c mice	PEG-AuNSs-Transferrin	1 mg/mL	6 h/3 min	808 nm laser 2 W/cm^2	shells	30 nm	[46]
SK-BR-3 tested in mice	PEG-AuPs	7.3 nM	30 min/4–6 min	CW 820 nm 4 W/cm^2	shells	110 (core) + 10 nm (shell)	[35]
4T1 cancer cells tested in mice	MPCM-AuNSs	1 mg/mL	20 min/5 min	CW 808 nm 1 W/cm^2	shells	80(core) + 7 nm (shell)	[36]
U87MG tumorstested in nude mice	HGN-siHsp70	1.5×10^{-9} M/Kg	90 min/8 min	CW 808 nm 4 W/cm^2	shells	70 nm	[38]
SK-BR3	Anti-HER2-AuNPs	-	5 min	800 nm Pulsed laser 1.6 W/cm^2	nano-cages	65 nm	[47]
MDA-MB-435 tested in nude mice	PEG-AuNP	14 μg/mL	72 h/5 min	CW 810 nm laser 2 W/cm^2	nanorods	Aspect ratio 3.7	[39]
Hs578T, HCC-38, MDA-MB-468 and MDAMB-231	anti-EGFR-GN	0.22 μg/mL	48 h/3 min	820 nm halogen lamp 1.5 W/cm^2	nanorods	Aspect ratio 4.0	[43]

PC3 cells tested in nude mice	Zn(II)/DPA-AuNR	250 µL	24 h/10 min	808 nm 0.5 W/cm^2	nanorods	Length 51.13 ± 5.2 nm	[44]
MDA-MB-231	Thiol-AuNP	12 µM	24 h	X-radiation 160 kVp 4 Gy	sphere	1.9	[48]
EMT-6	PEG-AuNP	500 µM	48 h	X-radiation 10 Gy	sphere	6.1	[46]
HeLa	folic Acid-AuNP	255 µM	6, 12, 24, 48 h	X-radiation 180 kVp	_	50	[49]
MCF-7	Glu-AuNP	100 µM	2 h	X-radiation 100 kVp 10 Gy	sphere	16	[50]
MDA-MB-361	anti-HER2-PEG-AuNP	4.8 mg/(g tumor)	48 h	X-radiation 100 kVp 11 Gy	sphere	30	[51]
SCCVII tested in mice	PEG-AuNPs	1 µg/mL	24 h/5 min	820 nm halogen lamp 1.5 W/cm^2 15 Gy	spheres	15 nm	[52]
PC-3 cells tested in Foxn1 mice	Goserelin-PEG-AuNRs	0.1–10 µg/g body weight *	72 h/5 min	X-radiation 6 MV 5 Gy	nanorods	_	[53]
4T1 cells	CD44-Antibody-PEG-AuNPs	3 nM	24 h	CW 808 nm 2.5 W/cm^2 X-radiation With 2, 4, 6, or 8 Gy of 6-MV	nanocages	58.4 nm	[54]

* The molar concentration µg/g body weight is referred to mg of Au per gram of body weight.

APPLICATION OF INTERACTION OF AUNPS IN COMBINATION WITH IONIZING RADIATION (IR) IN CANCER THERAPY

The Combination of Metallic NPs with Radiotherapy

Metallic NPs can combine multiple functionalities to target and treat cancer cells. Besides their phototothermal properties, they can also act as radiosensitizers for RT due to their unique electrical properties. IR therapy is currently a major and most applicable modality for cancer treatment following surgery employed in the treatments of more than 50% of all treated cancer patients [55]. However, it is still a great challenge to confine the curative dose of radiation within the tumor tissue while sparing the adjacent normal tissues. An important disadvantage of photon therapy is that cancer tissues can have, or develop, resistance to radiation. Especially hypoxic tumor cells show radiation resistance to photon treatment. This is because in the hypoxic cells there is a decreased production of reactive oxygen species (ROS) in response to irradiation, which leads to decreased levels of oxidative stress and therefore a decreased level of apoptosis compared with irradiated normoxic cells. Moreover, due to the physical extent of this tissue, it can be difficult to effectively irradiate the whole tumor, which may lead to its regeneration. To make predictions of radiation effects, popular simulation models of the effect of RT, microdosimetric kinetic model (MKM) originally developed by Hawkins et al. [56,57] and later modified by different groups [58], has been used also in the case of AuNPs in conjunction with the local effect model (LEM) [59,60,61]. The MKM model gives the dependence of the relative biological effectiveness (RBE) in the limit of zero doses on the linear energy transfer (LET) while the foundation of the LEM is that the local biological response to IR is expected to be equal for equal doses and independent of the type of radiation.

Based on the fact that AuNPs, have large interaction cross sections with X-ray radiation, at least up to about 1 MeV, the dose delivery in the area of gold NPs sometimes increases dramatically [3]. This dose enhancement facilitates a possible strategy to address the issue of radio-toxicity and adverse effects. The use of radiosensitizers in the tumor that confer additive and synergistic advantages to the tumor-killing effect of IR, significantly affects in various cases of tumor treatment [62]. If one can achieve radiation dose enhancement, lower radiation doses can be used to eradicate tumors with the same efficiency, or even better, while decreasing the damage to surrounding normal tissues [63]. A number of reviews have been published to date supporting the use of gold NPs in targeted cancer therapy and the augmentation of the effect of radiation (both IR and NIR) [3,64,65,66].

During the last decade, several in vivo and in vitro preclinical studies have been conducted regarding the applications of metal NPs in radiation therapy [67,68]. McMahon [59] and co-workers irradiated the breast cancer cell line MDA-MB-231 with 6 and 15 MV X-rays produced by a linear accelerator (LINAC) with and without the addition of 1.9 nm AuNPs. At MV energies, a significant radiosensitization is not expected based on the ratio of mass energy absorption coefficients of gold. However, the results showed that the sensitization enhancement ratio (SER) was 1.24 ± 0.05 at 6 MV and 1.18 ± 0.04 at 15 MV. Therefore, the contrary to the initial thought that AuNPs could be used only for physical "dose enhancement", by exploiting the enhanced photoelectric absorption of gold, it seems that both chemical and biological mechanisms contribute to their radiosensitization.

Gold nanorods modified with argine (R)-glycine (G)-aspartate (D) (RGD) peptide demonstrate enhanced radiotherapy of melanoma in cancer cells. The results show an enhancement in the radiosenstitivity of A375 cells with a dose modifying factor of 1.35 at 6 MV X-ray and at dose 4 Gy. RGB is the most effective peptide presented in many extracellular matrix proteins and can interact with integrin receptors $\alpha_v\beta_3$ at focal adhesion points. Integrin $\alpha_v\beta_3$ expression affects tumor radiosensitivity and it is also associated with acquired radioresistance through the upregulation of integrin expression by radiation [69]. Griffin and his group have recently showed that tumor necrosis factor-alpha coated gold nanospheres (Au-TNFα) combined with single or fractionated high-dose radiation therapy reduce effectively the tumor interstitial fluid pressure (IFT) and tumor growth in 4T1 murine breast tumour model, giving hopes for clinical translation [70].

The shape/geometry of the AuNPs is found to be critical in RT, as AuNPs coated with the same PEG molecules and a similar average size (~50 nm) exhibit different cellular uptake and sensitization enhancement ratios in KB cells. The SERs were calculated 1.62, 1.37, and 1.21 corresponding to the treatments upon X-ray irradiation (6 MV) at dose of 4 Gy of gold nanospheres, gold nanospikes and gold nanorods, respectively. The results indicated that gold nanospheres showed a higher anticancer efficiency than both AuNSs and AuNRs [71].

The contribution of NPs to radiosensitization has been demonstrated for a range of different external beam radiation types, including kilovoltage (kV) and megavoltage (MV) photons, MeV electrons and heavy charged particles [53,72,73], suggesting broad in vivo applications. The last mentioned RT approach is the ion-based RT consisting of ions of hydrogen (protons), helium, carbon, or oxygen. In radiation therapy, protons and helium ions are often referred to as "light ions" and the others as "heavy ions" [3]. Radiation therapy with protons and carbon ions is usually also called by the collective names "particle therapy" or "hadron therapy". Protons and light ions have unique physical properties, as they penetrate to the patient body with minimal lateral diffusion, depositing most of their energy at the end of their range (in the so-called Bragg peak), protecting in this way the surrounding healthy tissues. Therefore, particle therapy is particularly useful in pediatric radiation oncology.

For irradiation of biological samples, and especially for the applicability of these experiments to patient proton-based therapy, the positioning of the biological samples is very important. The most usual setup and of clinical relevance is the utilization of three points: entry (or plateau), center of spread-out Bragg peak (SOBP) and towards the end of the track (distal end) [74,75]. Logically, the maximization of LET around the Bragg peak region is expected to induce lesions of high complexity. Indeed and based on DSB measurement using surrogate markers like 53BP1, a high persistence of damage is shown when cells are exposed in this SOBP region compared to the entry and a significant increase in the RBE for cell survival reaching values of about 2 [74,75]. In accordance to these experimental studies, recent work by Nikitaki et al. [76], simulating the complex DNA damage induction along the central axis of a representative 200.6 MeV pencil beam (2 mm spot size, 5 mm FWHM) as calculated using Monte Carlo DNA damage simulation software (MCDS) and based on the actual beam energy-to-depth curves, a variation of clustered DNA damage (DSBs and non-DSB clusters) is predicted. A local maximum of clustered DNA lesions around the SOBP region, with maximum complexity, is found which underlies the importance of the SOBP region high-LET. Monte Carlo simulations used to assess the dose enhancement of AUNPs for proton therapy showed that proton therapy can be enhanced significantly only if the GNPs are in close proximity to the biological target [77,78]. According to Schuemann et al., the dominant process for proton interactions with gold is the production of secondary electrons via small angle scattering due to the higher density of gold as compared to soft tissue (19.3 g/cm^3 vs. ~1 g/cm^3) [79].

The dose enhancement caused by AuNPs in radiation therapy differs for photon and proton radiation. The enhancement effect has been simulated [78] for protons, kVp photons and MV photons, and the simulations conclude that AuNPs have the potential to enhance radiation therapy depending on the type of radiation source. In some cases, kilovoltage energies result in higher dose enhancement factors than megavolt energies [6]. For the same amount of energy absorbed in the AuNP, interactions with protons, kVp photons and MV photons produce similar doses within several nanometers of the AuNP surface, and differences are below 15% for the first 10 nm. However, secondary electrons produced by kilovoltage photons have longer ranges in water as compared to protons and MV photons, e.g., they cause a dose enhancement 20 times higher than the one caused by protons 10 μm away from the AuNP surface [78].

AuNPs' surface, depending on the subcellular localization, can trigger radiochemical sensitization of DNA [80], catalysis [81] and radical generation [82,83] upon IR exposure. Electrons with energies below the ionization threshold (e.g., 12 eV) levels fail to produce considerable secondary electrons through interaction with AuNPs, but they can cause a great deal of DNA damage [80]. For example, and as mentioned above, a DNA damage enhancement was showed using HeLa cancer cells, 50 nm AuNPs and a clinical 6 MV beam for various delivery parameters and depths [73]. In this case, increased γH2AX staining was detected suggesting an increased induction of DSBs. In addition, the role of AuNP charge seems to play a significant role in the overall outcome.

The chemical radiosensitization changes depend on the size and charge of AuNPs [84]. In vitro studies have showed that cell survival decreases with increased cellular internalization of AuNPs and the closer AuNPS are located to the nucleus of target cells [63,85]. Recently, simulations with a model considering the dosimetric characteristics of AuNPs, and the Local Effect Model (LEM) [59], showed very good agreements with experimental survival curves for x-rays assuming in all cases a perinuclear distribution of AuNPs.

In order to further understand the mechanism of AuNP dose enhancement using proton radiation, the group of Lin et al. [78] has recently investigated (using a modified LEM based on Monte Carlo simulations) several potential configurations for AuNP-induced radiosensitization for three particle sources: brachytherapy or superficial RT, AuNPs as tumor vascular-disrupting agents, RT application with in situ dose-painting using AuNPs and customizable RT enhancement of wet age-related macular degeneration using AuNPs. Lin et al. consider these AuNPs plus RT modalities as promising emerging procedures, representing three clinically relevant scenarios [86]. The impact of the beam quality on megavoltage RT treatment techniques, in combination with AuNPs for dose enhancement, as a function of irradiation conditions and potential biological targets was reported recently by Tsiamas et al. [87]. Moreover, several theoretical and experimental and/or preclinical studies are concerned with the interdisciplinary area of "gold and other metallic NPs" in combination with "RT using protons and other ions". The interested reader could study a series of comprehensive relative reviews [3,6,79,88]. Indeed, a PubMed search for scientific information and pre-clinical/clinical applications in this interdisciplinary area reveals an ongoing increased number of publications, reflecting the great interest of the biophysical and medical community in this area.

Definitely, NP assisted radiation therapy opens up new perspectives in the fight against cancer. However, there are some problems and limitations that prevent it from moving from bench to bedside. The mechanism of nanoparticle-mediated radiosensitization seems to be simple, since the main target for ionizing radiation is the nuclear DNA and NPs escalate the attack of radiation on this molecule. However, the biological mechanisms of nanoparticle-mediated radiosensitization remain more obscure [89] since many reports failed to demonstrate augmented nanoparticle-mediated DSB damage though a significant radiosensitizing effect had occurred [89].

Among the most important things to be mentioned is the need for ensuring the biocompatibility and the effective delivery of NPs to the target in the patient's body. It was proposed that the enhancement of the uptake of NPs by cancer cells can be reached via passive targeting, active targeting, or a combination of both strategies and functionalization through coating with biocompatible polymer and conjugation with tumor targeting agents [90]. Some more recent studies turn to other nanostructures called hydrogenated nanodiamonds (H-NDs) which has been proposed as radiosensitizers, as they exhibit excellent biocompatibility, negative electron affinity that confers a high reactivity with oxygen species (ROS) and an increased induction of DNA DSBs. The characteristics of H-NDs allows electron emission from H-NDs following irradiation by photons and in consequence may enhance the effects of radiation on cancer cells [91,92].

This year, in an attempt to identify if physical properties are correlated with the survival fraction of cells exposed to low-energy protons in combination with metallic NPs, an interdisciplinary group in Belgium [92] performed Monte Carlo simulations (based on the latest version of the Geant4 toolkit), assessing also in their study the influence of nanoparticle coating on dose enhancement. They conclude that, although significant efforts will be required to unravel the relevant mechanisms of action, ROS-induced damage and local heating generated from metallic NPs are likely to become fields of importance in radiation therapy research, despite a DNA-centered dogma. Another limitation for nanoparticle's assisted hadron (proton, carbon) RT is related to the cost and huge size of a proton-based cancer therapy facility. However, there are several hard works to "shrink" this problem, by the ongoing research and development efforts on the so called "laser-driven ion beam radiation therapy". According to Ledingham et al. [93], the overall challenge for laser-driven ion beam RT is "*to develop well-controlled, reliable energetic ion beams of very high quality that can meet stringent medical requirements with respect to physical parameters and performance*", representing a viable alternative in the 21th century' RT.

The Combination of Photothermal Therapy with Radiotherapy

The combination of PTT with RT can have a synergistic therapeutic effect which can overcome the limitations of each single treatment [94]. When cells are subjected to temperatures above 41 °C, not only protein denaturation is occurring but also temporary cell inactivation that could last for several hours [25]. Beyond that time, the surviving cells appear resistant to further exposure to such temperatures due to thermotolerance. For instance, two heating treatments with an interval of 12–48 h can cause an increased blood flow in the tumor, compared with heating with no interval that can decrease the tumor perfusion [95]. Whereas, increased thermotolerance may limit the efficiency of thermal treatment, this could be an advantage for radiation therapy. The appearance of cell thermotolerance, in some cases, can be accompanied by a modification of the cellular response to increase their sensitivity to X-ray irradiation. PTT can also enhance the effects of RT by attenuating the repair of DSBs caused by RT [96,97]. Clinical results have shown that the combination of hyperthermia and RT (ionizing radiation) can increase the rates from 16% to 26% [25]. Cells in hypoxic, low pH areas and cells in S-phase are both relatively radioresistant. The increased blood flow, caused by hyperthermia, can improve tissue oxygenation, which temporarily increases radiosensitivity [98]. Studies have shown that the addition of thermal treatment to the exposure of radiation for 7 days, may extent the tumor growth up to 17 days [95].

A basic prerequisite for these highly innovative treatments, before entering clinical applications, is the in-depth knowledge of the biophysical mechanisms involved and the exact dosimetry of the radiation-based treatment method used. In addition, there is accumulating evidence of the strong interaction of AuNPs with the immune system [99] and also that hyperthermia therapy can activate systemic anti-tumor immune responses that can slow growth of untreated tumors [99]. Figure 1 shows a schematic representation of the action mechanisms of PT, RT and PTT/RT in cancer treatment based on functionalized gold NPs.

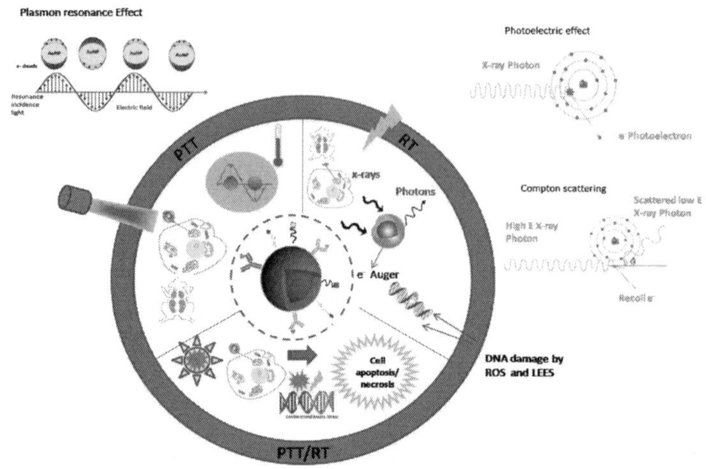

Figure 1: A schematic representation of the basic action mechanisms of photothermal therapy (PTT), ionizing RT and combined PTT/RT in cancer treatment based on functionalized AuNPs.

The co-treatment with PTT and RT based on AuNPs mediated radiosensitization has recently attracted high attention as a very promising dual mode therapy. Zhang and co-workers [100] have demonstrated the synergistic effect of gold nanocages in hyperthermia and RT. They synthesized CD44 Antibody-Conjugated PEG-Modified gold nanocages for 4T1 cells targeting. According to clonogenic survival assays, the combination of CD44-PEG-GNCs, PTT, and RT exhibited a remarkable inhibition rate (96%) compared to the treatment with RT alone (49%). Intratumorally infused small 15 nm gold NPs into radioresistant subcutaneous squamous cell carcinoma (SCCVII) in mice induce a dose decrease when heated by infrared light followed by X-ray treatment. The cells were first irradiated at 1.5 W/cm^2 by a halogen lamp with a 780 nm high pass filter yielding a peak output at $\lambda = 820$ nm and then were irradiated with X-rays generator operating at 100 kVp with a dose rate 7.5 Gy/min at the tumor center. It was found that the dose required to control 50% of the tumors, normally 55 Gy, could be reduced to <15 Gy (a factor of >3.7) [52]. Table 1 provides a list of some indicative publications on different forms of AuNPs that have been used with IR or NIR therapy to treat tumors/cancer cells both in vitro and in vivo.

As can be seen from the published works, the irradiation parameters coupling with AuNPs properties (such as size, morphology, functionalization and concentration) vary significantly and diverse strategies have been developed for PTT and PTT/RT treatments based on the nanosystems, regarding the laser beams and the ionizing radiation beams of each research group. There is still the requirement for optimized protocols in photothermal treatment and photothermal/radiation synergistic treatments for a translation to the clinical practice.

CONCLUSIONS

A large number of in vivo and in vitro studies have been conducted based on AuNP-mediated photon- and radiosensitization in PTT and RT. Different approaches and protocols have been applied regarding the size, shape, concentration, and functionalization of AuNPs and the irradiation conditions. The novel combination of PTT with the common RT, based on AuNP dual-mode agents can have a synergistic therapeutic effect which might overcome the limitations of each single treatment, minimizing the curative doses. However, there is still need for optimized protocols in photothermal treatment, radiotherapy and photothermal/radiation synergetic treatment for a translation to clinical practice. The ideal requirements for designing an effective NP as a diagnostic and therapeutic tool are: controlled particle size, surface modification, functionalization, binding behavior, solubility to aqueous environment, low toxicity, biocompatibility, targeting, permeability, retention and biodegradation. The open challenges for effective cancer treatment coupled with the above requirements are still: the immunoactivation and immunosuppression modulation, the tissue thermotolerance, the radioresistance of cancer cells though proteins upregulation and the fully understanding of the biological mechanisms of nanoparticle-mediated radiosensitization.

ACKNOWLEDGMENTS

Alexandros G. Georgakilas acknowledges support from the DAAD Grant "DNA Damage and Repair and Their Relevance to Carcinogenesis" (No. 57339330). Spyratou Ellas acknowledges support from IKY FELLOWSHIPS OF EXCELLENCE FOR POSTGRADUATE STUDIES IN GREECE-SIEMENS PROGRAM.

REFERENCES

1. Jaque, D.; Martinez Maestro, L.; del Rosal, B.; Haro-Gonzalez, P.; Benayas, A.; Plaza, J.L.; Martin Rodriguez, E.; Garcia Sole, J. Nanoparticles for photothermal therapies. *Nanoscale* 2014, *6*, 9494–9530.
2. Huang, X.; El-Sayed, M.A. Plasmonic photo-thermal therapy (PPTT). *Alex. J. Med.* 2011, *47*, 1–9.
3. Haume, K.; Rosa, S.; Grellet, S.; Śmiałek, M.A.; Butterworth, K.T.; Solovyov, A.; Prise, K.M.; Golding, J.; Mason, N.M. Gold nanoparticles for cancer radiotherapy: A review. *Cancer Nano* 2016, *7*, 1–20.
4. Retif, P.; Pinel, S.; Toussaint, M.; Frochot, C.; Chouikrat, R.; Bastogne, T.; Barberi-Heyob, M. Nanoparticles for radiation therapy enhancement: The key parameters. *Theranostics* 2015, *5*, 1030–1044.
5. Butterworth, K.T.; McMahon, S.J.; Currell, F.J.; Prise, K.M. Physical basis and biological mechanisms of gold nanoparticle radiosensitization. *Nanoscale* 2012, *4*, 4830–4838.
6. Rosa, S.; Connolly, C.; Schettino, G.; Butterworth, K.T.; Prise, K.M. Biological mechanisms of gold nanoparticle radiosensitization. *Cancer Nanotechnol.* 2017, *8*, 2.
7. Hossain, M.; Su, M. Nanoparticle location and material dependent dose enhancement in x-ray radiation therapy. *J. Phys. Chem. C Nanomater. Interfaces* 2012, *116*, 23047–23052.
8. Pustovalov, V.K. Theoretical study of heating of spherical nanoparticle in media by short laser pulses. *Chem. Phys.* 2005, *308*, 103–108.
9. Khlebtsov, N.G.; Dykman, L.A. Optical properties and biomedical applications of plasmonic nanoparticles. *J. Quant. Spectrosc. Radiat. Transf.* 2010, *111*, 1–35.
10. Mie, G. Beiträge zur optik trüber medien, speziell kolloidaler metallösungen. *Ann. Phys.* 1908, *330*, 377–445.
11. Bohren, C.F.; Huffman, D.R. *Absorption and Scattering of Light by Small Particles*; John Wiley & Sons: Weinheim, Germany, 1998.
12. Biswas, A.; Wang, T.; Biris, A.S. Single metal nanoparticle spectroscopy: Optical characterization of individual nanosystems for biomedical applications. *Nanoscale* 2010, *2*, 1560–1572.
13. Sánchez, D.B. The surface plasmon resonance of supported noble metal nanoparticles: Characterization, laser tailoring, and SERS application.

Ph.D. Thesis, University of Kassel, Kassel, Germany, 2007.

14. Khlebtsov, N.; Dykman, L. Biodistribution and toxicity of engineered gold nanoparticles: A review of in vitro and in vivo studies. *Chem. Soc. Rev.* 2011, *40*, 1647–1671.

15. Pissuwan, D.; Valenzuela, S.M.; Cortie, M.B. Therapeutic possibilities of plasmonically heated gold nanoparticles. *Trends Biotechnol.* 2006, *24*, 62–67.

16. Brito-Silva, A.M.; Sobral-Filho, R.G.; Barbosa-Silva, R.; de Araujo, C.B.; Galembeck, A.; Brolo, A.G. Improved synthesis of gold and silver nanoshells. *Langmuir* 2013, *29*, 4366–4372.

17. McMahon, S.J.; Paganetti, H.; Prise, K.M. Optimising element choice for nanoparticle radiosensitisers. *Nanoscale* 2016, *8*, 581–589.

18. Makropoulou, M. Cancer and electromagnetic radiation therapy: Quo Vadis? 2016. Available online: https://pdfs.semanticscholar.org/b6e9/3da094b831761274ab3047683068844fe2838.pdf (accessed on 18 December 2017).

19. Spyratou, E.; Makropoulou, M.; Mourelatou, E.A.; Demetzos, C. Biophotonic techniques for manipulation and characterization of drug delivery nanosystems in cancer therapy. *Cancer Lett.* 2012, *327*, 111–122.

20. Fan, Z.; Fu, P.P.; Yu, H.; Ray, P.C. Theranostic nanomedicine for cancer detection and treatment. *J. Food Drug Anal.* 2014, *22*, 3–17.

21. Josefsen, L.B.; Boyle, R.W. Unique diagnostic and therapeutic roles of porphyrins and phthalocyanines in photodynamic therapy, imaging and theranostics. *Theranostics* 2012, *2*, 916–966.

22. Serafetinides, A.A.; Makropoulou, M.; Kotsifaki, D.G.; Tsigaridas, G. Biophotonics for imaging and cell manipulation: Quo vadis? In Proceedings of the 19th International Conference and School on Quantum Electronics: Laser Physics and Applications, Sozopol, Bulgaria, 26–30 September 2016; Volume 1022613.

23. Doherty, G.J.; McMahon, H.T. Mechanisms of endocytosis. *Ann. Rev. Biochem.* 2009, *78*, 857–902.

24. Mouratidis, P.X.E.; Rivens, I.; Ter Haar, G. A study of thermal dose-induced autophagy, apoptosis and necroptosis in colon cancer cells. *Int. J. Hyperth.* 2015, *31*, 476–488.

25. Cherukuri, P.; Glazer, E.S.; Curley, S.A. Targeted hyperthermia using metal nanoparticles. *Adv. Drug Deliv. Rev.* 2010, *62*, 339–345.

26. Yang, K.; Wan, J.; Zhang, S.; Tian, B.; Zhang, Y.; Liu, Z. The influence of surface chemistry and size of nanoscale graphene oxide on photothermal therapy of cancer using ultra-low laser power. *Biomaterials* 2012, *33*, 2206–2214.

27. Maeda, H. The enhanced permeability and retention (EPR) effect in tumor vasculature: The key role of tumor-selective macromolecular drug targeting. *Adv. Enzyme Regul.* 2001, *41*, 189–207.

28. Kodiha, M.; Wang, Y.M.; Hutter, E.; Maysinger, D.; Stochaj, U. Off to the organelles-killing cancer cells with targeted gold nanoparticles. *Theranostics* 2015, *5*, 357–370.

29. Huang, X.; Kang, B.; Qian, W.; Mackey, M.A.; Chen, P.C.; Oyelere, A.K.; El-Sayed, I.H.; El-Sayed, M.A. Comparative study of photothermolysis of cancer cells with nuclear-targeted or cytoplasm-targeted gold nanospheres: Continuous wave or pulsed lasers. *J. Biomed. Opt.* 2010, *15*, 058002.

30. Rengan, A.K.; Kundu, G.; Banerjee, R.; Srivastava, R. Gold nanocages as effective photothermal transducers in killing highly tumorigenic cancer cells. *Part. Part. Syst. Charact.* 2014, *31*, 398–405.

31. Huang, X.; El-Sayed, I.H.; Qian, W.; El-Sayed, M.A. Cancer cell imaging and photothermal therapy in the near-infrared region by using gold nanorods. *J. Am. Chem. Soc.* 2006, *128*, 2115–2120.

32. Pitsillides, C.M.; Joe, E.K.; Wei, X.; Anderson, R.R.; Lin, C.P. Selective cell targeting with light-absorbing microparticles and nanoparticles. *Biophys. J.* 2003, *84*, 4023–4032.

33. El-Sayed, I.H.; Huang, X.; El-Sayed, M.A. Selective laser photo-thermal therapy of epithelial carcinoma using anti-EGFR antibody conjugated gold nanoparticles. *Cancer Lett.* 2006, *239*, 129–135.

34. Poudel, B.K.; Gupta, B.; Ramasamy, T.; Thapa, R.K.; Pathak, S.; Oh, K.T.; Jeong, J.H.; Choi, H.G.; Yong, C.S.; Kim, J.O. PEGylated thermosensitive lipid-coated hollow gold nanoshells for effective combinational chemo-photothermal therapy of pancreatic cancer. *Colloids Surf. B Biointerfaces* 2017, *160*, 73–83.

35. Hirsch, L.R.; Stafford, R.J.; Bankson, J.A.; Sershen, S.R.; Rivera, B.; Price, R.E.; Hazle, J.D.; Halas, N.J.; West, J.L. Nanoshell-mediated near-infrared thermal therapy of tumors under magnetic resonance guidance. *Proc. Natl. Acad. Sci. USA* 2003, *100*, 13549–13554.

36. Xuan, M.; Shao, J.; Dai, L.; Li, J.; He, Q. Macrophage cell membrane

camouflaged au nanoshells for in vivo prolonged circulation life and enhanced cancer photothermal therapy. *ACS Appl. Mater Interfaces* 2016, *8*, 9610–9618.

37. Shao, J.; Griffin, R.J.; Galanzha, E.I.; Kim, J.W.; Koonce, N.; Webber, J.; Mustafa, T.; Biris, A.S.; Nedosekin, D.A.; Zharov, V.P. Photothermal nanodrugs: Potential of TNF-gold nanospheres for cancer theranostics. *Sci. Rep.* 2013, *3*.
38. Wang, Z.; Li, S.; Zhang, M.; Ma, Y.; Liu, Y.; Gao, W.; Zhang, J.; Gu, Y. Laser-triggered small interfering RNA releasing gold nanoshells against heat shock protein for sensitized photothermal therapy. *Adv. Sci.* 2017, *4*, 1600327.
39. Von Maltzahn, G.; Park, J.H.; Agrawal, A.; Bandaru, N.K.; Das, S.K.; Sailor, M.J.; Bhatia, S.N. Computationally guided photothermal tumor therapy using long-circulating gold nanorod antennas. *Cancer Res.* 2009, *69*, 3892–3900.
40. Tong, L.; Zhao, Y.; Huff, T.B.; Hansen, M.N.; Wei, A.; Cheng, J.X. Gold nanorods mediate tumor cell death by compromising membrane integrity. *Adv. Mat.* 2007, *19*, 3136–3141.
41. Ali, M.R.; Ibrahim, I.M.; Ali, H.R.; Selim, S.A.; El-Sayed, M.A. Treatment of natural mammary gland tumors in canines and felines using gold nanorods-assisted plasmonic photothermal therapy to induce tumor apoptosis. *Int. J. Nanomed.* 2016, *11*, 4849–4863.
42. Yin, D.; Li, X.; Ma, Y.; Liu, Z. Targeted cancer imaging and photothermal therapy via monosaccharide-imprinted gold nanorods. *Chem. Commun.* 2017, *53*, 6716–6719.
43. Zhang, M.; Kim, H.S.; Jin, T.; Woo, J.; Piao, Y.J.; Moon, W.K. Near-infrared photothermal therapy using anti-EGFR-gold nanorod conjugates for triple negative breast cancer. *Oncotarget* 2017, *8*, 86566–86575.
44. Min, K.H.; Kim, Y.H.; Wang, Z.; Kim, J.; Kim, J.S.; Kim, S.H.; Kim, K.; Kwon, I.C.; Kiesewetter, D.O.; Chen, X. Engineered Zn(II)-dipicolylamine-gold nanorod provides effective prostate cancer treatment by combining siRNA delivery and photothermal therapy. *Theranostics* 2017, *7*, 4240–4254.
45. Nam, J.; La, W.G.; Hwang, S.; Ha, Y.S.; Park, N.; Won, N.; Jung, S.; Bhang, S.H.; Ma, Y.J.; Cho, Y.M.; et al. pH-responsive assembly of gold nanoparticles and “spatiotemporally concerted” drug release for

synergistic cancer therapy. *ACS Nano*. 2013, *7*, 3388–3402.

46. Liu, C.J.; Wang, C.H.; Chen, S.T.; Chen, H.H.; Leng, W.H.; Chien, C.C.; Wang, C.L.; Kempson, I.M.; Hwu, Y.; Lai, T.C.; et al. Enhancement of cell radiation sensitivity by pegylated gold nanoparticles. *Phys. Med. Biol.* 2010, *55*, 931–945.
47. Au, L.; Zheng, D.; Zhou, F.; Li, Z.Y.; Li, X.; Xia, Y. A quantitative study on the photothermal effect of immuno gold nanocages targeted to breast cancer cells. *ACS Nano* 2008, *2*, 1645–1652.
48. Jain, S.; Coulter, J.A.; Hounsell, A.R.; Butterworth, K.T.; McMahon, S.J.; Hyland, W.B.; Muir, M.F.; Dickson, G.R.; Prise, K.M.; Currell, F.J.; et al. Cell-specific radiosensitization by gold nanoparticles at megavoltage radiation energies. *Int. J. Radiat. Oncol. Biol. Phys.* 2011, *79*, 531–539.
49. Khoshgard, K.; Hashemi, B.; Arbabi, A.; Rasaee, M.J.; Soleimani, M. Radiosensitization effect of folate-conjugated gold nanoparticles on HeLa cancer cells under orthovoltage superficial radiotherapy techniques. *Phys. Med. Biol.* 2014, *59*, 2249–2263.
50. Rostami, A.; Toossi, M.T.; Sazgarnia, A.; Soleymanifard, S. The effect of glucose-coated gold nanoparticles on radiation bystander effect induced in MCF-7 and QUDB cell lines. *Radiat. Environ. Biophys.* 2016, *55*, 461–466.
51. Chattopadhyay, N.; Cai, Z.; Kwon, Y.L.; Lechtman, E.; Pignol, J.P.; Reilly, R.M. Molecularly targeted gold nanoparticles enhance the radiation response of breast cancer cells and tumor xenografts to x-radiation. *Breast Cancer Res. Treat.* 2013, *137*, 81–91.
52. Hainfeld, J.F.; Lin, L.; Slatkin, D.N.; Avraham Dilmanian, F.; Vadas, T.M.; Smilowitz, H.M. Gold nanoparticle hyperthermia reduces radiotherapy dose. *Nanomedicine* 2014, *10*, 1609–1617.
53. Wolfe, T.; Chatterjee, D.; Lee, J.; Grant, J.D.; Bhatterai, S.; Tailor, R.; Goodrich, G.; Nicolucci, P.; Krishnan, S. Targeted gold nanoparticles enhance sensitization of prostate tumours to megavoltage radiation therapy in vivo. *Nanomedicine* 2015, *11*, 1277–1283.
54. Zhang, X.D.; Chen, J.; Luo, Z.; Wu, D.; Shen, X.; Song, S.S.; Sun, Y.M.; Liu, P.X.; Zhao, J.; Huo, S.; et al. Enhanced tumor accumulation of Sub-2 nm gold nanoclusters for cancer radiation therapy. *Adv. Healthc. Mater* 2014, *3*, 133–141.

55. Atun, R.; Jaffray, D.A.; Barton, M.B.; Bray, F.; Baumann, M.; Vikram, B.; Hanna, T.P.; Knaul, F.M.; Lievens, Y.; Lui, T.Y.; et al. Expanding global access to radiotherapy. *Lancet Oncol.* 2015, *16*, 1153–1186.
56. Hawkins, R.B. A microdosimetric-kinetic theory of the dependence of the RBE for cell death on LET. *Med. Phys.* 1998, *25*, 1157–1170.
57. Hawkins, R.B. A microdosimetric-kinetic model for the effect of non-poisson distribution of lethal lesions on the variation of RBE with LET. *Radiat. Res.* 2003, *160*, 61–69.
58. Inaniwa, T.; Furukawa, T.; Kase, Y.; Matsufuji, N.; Toshito, T.; Matsumoto, Y.; Furusawa, Y.; Noda, K. Treatment planning for a scanned carbon beam with a modified microdosimetric kinetic model. *Phys. Med. Biol.* 2010, *55*, 6721–6737.
59. McMahon, S.J.; Hyland, W.B.; Muir, M.F.; Coulter, J.A.; Jain, S.; Butterworth, K.T.; Schettino, G.; Dickson, G.R.; Hounsell, A.R.; O'Sullivan, J.M.; et al. Biological consequences of nanoscale energy deposition near irradiated heavy atom nanoparticles. *Sci. Rep.* 2011, *1*, 18.
60. Grun, R.; Friedrich, T.; Elsasser, T.; Kramer, M.; Zink, K.; Karger, C.P.; Durante, M.; Engenhart-Cabillic, R.; Scholz, M. Impact of enhancements in the local effect model (LEM) on the predicted RBE-weighted target dose distribution in carbon ion therapy. *Phys. Med. Biol.* 2012, *57*, 7261–7274.
61. Elsasser, T.; Kramer, M.; Scholz, M. Accuracy of the local effect model for the prediction of biologic effects of carbon ion beams in vitro and in vivo. *Int. J. Radiat. Oncol. Biol. Phys.* 2008, *71*, 866–872.
62. Mi, Y.; Shao, Z.; Vang, J.; Kaidar-Person, O.; Wang, A.Z. Application of nanotechnology to cancer radiotherapy. *Cancer Nanotechnol.* 2016, *7*, 11.
63. Hainfeld, J.F.; Smilowitz, H.M.; O'Connor, M.J.; Dilmanian, F.A.; Slatkin, D.N. Gold nanoparticle imaging and radiotherapy of brain tumors in mice. *Nanomedicine* 2013, *8*, 1601–1609.
64. He, C.; Hu, Y.; Yin, L.; Tang, C.; Yin, C. Effects of particle size and surface charge on cellular uptake and biodistribution of polymeric nanoparticles. *Biomaterials* 2010, *31*, 3657–3666.
65. Swain, S.; Sahu, P.K.; Beg, S.; Babu, S.M. Nanoparticles for cancer targeting: Current and future directions. *Curr. Drug Deliv.* 2016, *13*, 1290–1302.

66. Ngwa, W.; Boateng, F.; Kumar, R.; Irvine, D.J.; Formenti, S.; Ngoma, T.; Herskind, C.; Veldwijk, M.R.; Hildenbrand, G.L.; Hausmann, M.; et al. Smart radiation therapy biomaterials. *Int. J. Radiat. Oncol. Biol. Phys.* 2017, *97*, 624–637.

67. Babaei, M.; Ganjalikhani, M. The potential effectiveness of nanoparticles as radio sensitizers for radiotherapy. *BioImpacts* 2014, *4*, 15–20.

68. Su, X.Y.; Liu, P.D.; Wu, H.; Gu, N. Enhancement of radiosensitization by metal-based nanoparticles in cancer radiation therapy. *Cancer Biol. Med.* 2014, *11*, 86–91.

69. Xu, W.; Luo, T.; Li, P.; Zhou, C.; Cui, D.; Pang, B.; Ren, Q.; Fu, S. RGD-conjugated gold nanorods induce radiosensitization in melanoma cancer cells by downregulating αvβ3 expression. *Int. J. Nanomed.* 2012, *7*, 915–924.

70. Koonce, N.A.; Quick, C.M.; Hardee, M.E.; Jamshidi-Parsian, A.; Dent, J.A.; Paciotti, G.F.; Nedosekin, D.; Dings, R.P.; Griffin, R.J. Combination of gold nanoparticle-conjugated tumor necrosis factor-α and radiation therapy results in a synergistic antitumor response in murine carcinoma models. *Int. J. Radiat. Oncol. Biol. Phys.* 2015, *93*, 588–596.

71. Ma, N.; Wu, F.G.; Zhang, X.; Jiang, Y.W.; Jia, H.R.; Wang, H.Y.; Li, Y.H.; Liu, P.; Gu, N.; Chen, Z. Shape-dependent radiosensitization effect of gold nanostructures in cancer radiotherapy: Comparison of gold nanoparticles, nanospikes, and nanorods. *ACS Appl. Mater. Interfaces* 2017, *9*, 13037–13048.

72. Hainfeld, J.F.; Slatkin, D.N.; Smilowitz, H.M. The use of gold nanoparticles to enhance radiotherapy in mice. *Phys. Med. Biol.* 2004, *49*, N309–N315.

73. Berbeco, R.I.; Korideck, H.; Ngwa, W.; Kumar, R.; Patel, J.; Sridhar, S.; Johnson, S.; Price, B.D.; Kimmelman, A.; Makrigiorgos, G.M. DNA damage enhancement from gold nanoparticles for clinical MV photon beams. *Radiat. Res.* 2012, *178*, 604–608.

74. Chaudhary, P.; Marshall, T.I.; Currell, F.J.; Kacperek, A.; Schettino, G.; Prise, K.M. Variations in the processing of DNA double-strand breaks along 60-MeV therapeutic proton beams. *Int. J. Radiat. Oncol. Biol. Phys* 2016, *95*, 86–94.

75. Marshall, T.I.; Chaudhary, P.; Michaelidesova, A.; Vachelova, J.;

Davidkova, M.; Vondracek, V.; Schettino, G.; Prise, K.M. Investigating the implications of a variable RBE on proton dose fractionation across a clinical pencil beam scanned spread-out Bragg peak. *Int. J. Radiat. Oncol. Biol. Phys* 2016, *95*, 70–77.

76. Nikitaki, Z.; Mavragani, I.V.; Laskaratou, D.A.; Gika, V.; Moskvin, V.P.; Theofilatos, K.; Vougas, K.; Stewart, R.D.; Georgakilas, A.G. Systemic mechanisms and effects of ionizing radiation: A new 'old' paradigm of how the bystanders and distant can become the players. *Semin. Cancer Biol.* 2016, *37*, 77–95.
77. Torrisi, L. Gold nanoparticles enhancing protontherapy efficiency. *Recent Pat. Nanotechnol.* 2015, *9*, 51–60.
78. Lin, Y.; McMahon, S.J.; Scarpelli, M.; Paganetti, H.; Schuemann, J. Comparing gold nano-particle enhanced radiotherapy with protons, megavoltage photons and kilovoltage photons: A monte carlo simulation. *Phys. Med. Biol.* 2014, *59*, 7675–7689.
79. Schuemann, J.; Berbeco, R.; Chithrani, D.B.; Cho, S.H.; Kumar, R.; McMahon, S.J.; Sridhar, S.; Krishnan, S. Roadmap to clinical use of gold nanoparticles for radiation sensitization. *Int. J. Radiat. Oncol. Biol. Phys.* 2016, *94*, 189–205.
80. Zheng, Y.; Hunting, D.J.; Ayotte, P.; Sanche, L. Radiosensitization of DNA by gold nanoparticles irradiated with high-energy electrons. *Radiat. Res.* 2008, *169*, 19–27.
81. Ionita, P.; Conte, M.; Gilbert, B.C.; Chechik, V. Gold nanoparticle-initiated free radical oxidations and halogen abstractions. *Org. Biomol. Chem.* 2007, *5*, 3504–3509.
82. Chompoosor, A.; Saha, K.; Ghosh, P.S.; Macarthy, D.J.; Miranda, O.R.; Zhu, Z.J.; Arcaro, K.F.; Rotello, V.M. The role of surface functionality on acute cytotoxicity, ROS generation and DNA damage by cationic gold nanoparticles. *Small* 2010, *6*, 2246–2249.
83. Mateo, D.; Morales, P.; Ávalos, A.; Haza, A.I. Oxidative stress contributes to gold nanoparticle-induced cytotoxicity in human tumor cells. *Toxicol. Mech. Methods* 2014, *24*, 161–172.
84. Yao, X.; Huang, C.; Chen, X.; Yi, Z.; Sanche, L. Chemical radiosensitivity of DNA induced by gold nanoparticles. *J. Biomed. Nanotechnol.* 2015, *11*, 478–485.

85. Hainfeld, J.F.; Dilmanian, F.A.; Slatkin, D.N.; Smilowitz, H.M. Radiotherapy enhancement with gold nanoparticles. *J. Pharm. Pharmacol.* 2008, *60*, 977–985.
86. Lin, Y.; McMahon, S.J.; Paganetti, H.; Schuemann, J. Biological modeling of gold nanoparticle enhanced radiotherapy for proton therapy. *Phys. Med. Biol.* 2015, *60*, 4149–4168.
87. Tsiamas, P.; Liu, B.; Cifter, F.; Ngwa, W.F.; Berbeco, R.I.; Kappas, C.; Theodorou, K.; Marcus, K.; Makrigiorgos, M.G.; Sajo, E.; et al. Impact of beam quality on megavoltage radiotherapy treatment techniques utilizing gold nanoparticles for dose enhancement. *Phys. Med. Biol.* 2013, *58*, 451–464.
88. Dimitriou, N.M.; Tsekenis, G.; Balanikas, E.C.; Pavlopoulou, A.; Mitsiogianni, M.; Mantso, T.; Pashos, G.; Boudouvis, A.G.; Lykakis, I.N.; Tsigaridas, G.; et al. Gold nanoparticles, radiations and the immune system: Current insights into the physical mechanisms and the biological interactions of this new alliance towards cancer therapy. *Pharmacol. Ther.* 2017, *178*, 1–17.
89. Stefancikova, L.; Lacombe, S.; Salado, D.; Porcel, E.; Pagacova, E.; Tillement, O.; Lux, F.; Depes, D.; Kozubek, S.; Falk, M. Effect of gadolinium-based nanoparticles on nuclear DNA damage and repair in glioblastoma tumor cells. *J. Nanobiotechnol.* 2016, *14*, 63.
90. Dreaden, E.C.; Mwakwari, S.C.; Sodji, Q.H.; Oyelère, A.K.; El-Sayed, M.A. Tamoxifen-poly(ethylene glycol)-thiol gold nanoparticle conjugates: Enhanced potency and selective delivery for breast cancer treatment. *Bioconjug. Chem.* 2009, *20*, 2247–2253.
91. Grall, R.; Girard, H.; Saad, L.; Petit, T.; Gesset, C.; Combis-Schlumberger, M.; Paget, V.; Delic, J.; Arnault, J.C.; Chevillard, S. Impairing the radioresistance of cancer cells by hydrogenated nanodiamonds. *Biomaterials* 2015, *61*, 290–298.
92. Heuskin, A.C.; Gallez, B.; Feron, O.; Martinive, P.; Michiels, C.; Lucas, S. Metallic nanoparticles irradiated by low-energy protons for radiation therapy: Are there significant physical effects to enhance the dose delivery? *Med. Phys.* 2017, *44*, 4299–4312.
93. Ledingham, K.; Bolton, P.; Shikazono, N.; Ma, C.M. Towards laser driven hadron cancer radiotherapy: A review of progress. *Appl. Sci.* 2014, *4*, 402–443.

94. Zhang, A.W.; Guo, W.H.; Qi, Y.F.; Wang, J.Z.; Ma, X.X.; Yu, D.X. Synergistic effects of gold nanocages in hyperthermia and radiotherapy treatment. *Nanoscale Res. Lett.* 2016, *11*, 279.
95. Griffin, R.J.; Dings, R.P.M.; Jamshidi-Parsian, A.; Song, C.W. Mild temperature hyperthermia and radiation therapy: Role of tumour vascular thermotolerance and relevant physiological factors. *Int. J. Hyperth.* 2010, *26*, 256–263.
96. Lee, J.; Chatterjee, D.K.; Lee, M.H.; Krishnan, S. Gold nanoparticles in breast cancer treatment: Promise and potential pitfalls. *Cancer Lett.* 2014, *347*, 46–53.
97. Atkinson, R.L.; Zhang, M.; Diagaradjane, P.; Peddibhotla, S.; Contreras, A.; Hilsenbeck, S.G.; Woodward, W.A.; Krishnan, S.; Chang, J.C.; Rosen, J.M. Thermal enhancement with optically activated gold nanoshells sensitizes breast cancer stem cells to radiation therapy. *Sci. Transl. Med.* 2010, *2*, 55ra79.
98. Van der Zee, J. Heating the patient: A promising approach? *Ann. Oncol.* 2002, *13*, 1173–1184.
99. Toraya-Brown, S.; Fiering, S. Local tumour hyperthermia as immunotherapy for metastatic cancer. *Int. J. Hyperth.* 2014, *30*, 531–539.
100. Zhang, L.; Xiao, H.; Li, J.; Cheng, D.; Shuai, X. Co-delivery of doxorubicin and arsenite with reduction and pH dual-sensitive vesicle for synergistic cancer therapy. *Nanoscale* 2016, *8*, 12608–12617.

11

Current Landscape and Future of Dual Anti-CTLA4 and PD-1/PD-L1 Blockade Immunotherapy in Cancer; Lessons Learned from Clinical Trials with Melanoma and Non-small Cell Lung Cancer (NSCLC)

Young Kwang Chae[1,2,3], Ayush Arya[1] , Wade Iams[3] , Marcelo R. Cruz[1] , Sunandana Chandra[1,2,3], Jaehyuk Choi[2,3] and Francis Giles[1,2,3]

[1] Developmental Therapeutics Program of the Division of Hematology Oncology, Early Phase Clinical Trials Unit, 645 N. Michigan Avenue, Suite 1006, Chicago, IL 60611, USA

[2] Robert H. Lurie Comprehensive Cancer Center of Northwestern University, 645 N. Michigan Avenue, Suite 1006, Chicago, IL 60611, USA

[3] Northwestern University Feinberg School of Medicine, 645 N. Michigan Avenue, Suite 1006, Chicago, IL 60611, USA.

Citation: Chae YK, Arya A, Iams W, et al. Current landscape and future of dual anti-CTLA4 and PD-1/PD-L1 blockade immunotherapy in cancer; lessons learned from clinical trials with melanoma and non-small cell lung cancer (NSCLC). J Immunother Cancer. 2018;6(1):39. Published 2018 May 16. doi:10.1186/s40425-018-0349-3.

ABSTRACT

Immunotherapy is among the most rapidly evolving treatment strategies in oncology. The therapeutic potential of immune-checkpoint inhibitors is exemplified by the recent hail of Food and Drug Administration (FDA) approvals for their use in various malignancies. Continued efforts to enhance outcomes with immunotherapy agents have led to the formulation of advanced treatment strategies. Recent evidence from pre-clinical studies evaluating immune-checkpoint inhibitors in various cancer cell-lines has suggested that combinatorial approaches may have superior survival outcomes compared to single-agent immunotherapy regimens. Preliminary trials assessing combination therapy with anti-PD-1/PD-L1 plus anti-CTLA-4 immune-checkpoint inhibitors have documented considerable advantages in survival indices over single-agent immunotherapy. The therapeutic potential of combinatorial approaches is highlighted by the recent FDA approval of nivolumab plus ipilimumab for patients with advanced melanoma. Presently, dual-immune checkpoint inhibition with anti-programmed death receptor-1/ programmed cell death receptor- ligand-1 (anti-PD-1/PD-L1) plus anti-cytotoxic T lymphocyte associated antigen-4 (anti-CTLA-4) monoclonal antibodies (MoAbs) is being evaluated for a wide range of tumor histologies. Furthermore, several ongoing clinical trials are investigating combination checkpoint inhibition in association with traditional treatment modalities such as chemotherapy, surgery, and radiation. In this review, we summarize the current landscape of combination therapy with anti-PD-1/PD-L1 plus anti-CTLA-4 MoAbs for patients with melanoma and non-small cell lung cancer (NSCLC). We present a synopsis of the prospects for expanding the indications of dual immune-checkpoint inhibition therapy to a more diverse set of tumor histologies.

BACKGROUND

The regulation of immune responses through MoAbs is a ground-breaking therapeutic strategy in oncology. Based on substantial pre-clinical and clinical evidence, several immunotherapy agents have received approval by the FDA as standard of care treatment for various malignancies over the past two decades [1, 2]. However, with increasing experience in the use of immunotherapy agents in clinical settings, several limitations, such as treatment resistance and undesired immunogenicity, have been observed [3, 4]. Extensive efforts have been made to meet such challenges, and novel immune checkpoints are being tested and are expected to pave the way for the next generation of immunotherapy agents [5].

The fundamental goal in advancing anti-cancer immunotherapy is to improve clinical outcomes. The use of combination checkpoint inhibition is being applied to meet this goal. This approach intends to exploit the distinct mechanisms of immunomodulation of two MoAbs in a single treatment regimen. Recent evidence suggests that the combined use of an anti-CTLA-4 immune-checkpoint inhibitor with an anti-PD-1/PD-L1 MoAb may have complementary action, thus yielding a higher clinical efficacy than either agent individually [6, 7]. Comprehensive data on the efficacy of MoAb combination therapy in clinical settings is warranted in order to ascertain the true therapeutic value of this treatment strategy.

Presently, combination checkpoint inhibition is being extensively evaluated for potential clinical benefit in a large number of tumor histologies. Due to positive outcomes in preliminary trials, nivolumab (IgG4 anti-PD-1 MoAb) plus ipilimumab (fully humanized IgG1 anti-CTLA-4 MoAb) is one of the most enthusiastically investigated combined immunotherapy regimens, with over 100 clinical trials in various stages [8, 9]. Of note, nivolumab plus ipilimumab received approval for use in BRAF V600 wild-type metastatic/unresectable melanoma, making it the first combination checkpoint inhibition regimen to be approved by the FDA [9]. In addition, other PD-1/PD-L1 inhibitors plus CTLA-4 inhibitor combination checkpoint inhibition regimens that are presently in clinical trials include atezolizumab (anti-PD-L1 MoAb) plus ipilimumab, pembrolizumab (IgG4 anti-PD-1 MoAb) plus ipilimumab, and tremelimumab (IgG2 anti-CTLA-4 MoAb) plus durvalumab (Fc optimized anti-PD-L1 MoAb) [10]. The data published from these trials will be crucial to appraise the efficacy of combination immune checkpoint inhibitor regimens in varying clinical scenarios.

In this review, we describe the rationale for combined immunotherapy with PD-1/PD-L1 plus CTLA-4 checkpoint inhibitors. Building on what we have learned through studies of combination checkpoint inhibition in patients with melanoma and NSCLC, we shall also critically assess the current landscape and future prospects for the development of an ideal combination checkpoint inhibition regimen.

Role of PD-1/PD-L1 and CTLA-4 in Modulation of Anti-tumor T-cell Activity

The process of T cell activation requires two signals. The primary signal comes from the binding of the T cell receptor (TCR) to the major histocompatibility complex (MHC) molecule presented by an antigen presenting cell (APC) [11]. The costimulatory signal may arise from one of several distinct T cell-APC interactions. One such pathway is the engagement of CD28 on T cells with CD80 (B7–1) or CD86 (B7–2) on APCs [11] (Fig. 1). T-cell activity can be modulated by regulating the generation of costimulatory signals through various mechanisms.

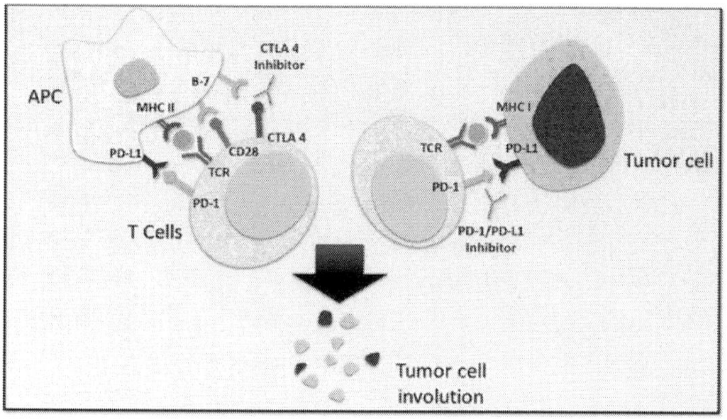

Figure 1: Mechanism of CTLA 4 and PD-1/PD-L1 inhibition. The activation of T cells is mediated by the interaction of T cell receptor and the CD28 receptor with class II major histocompatibility complex and B7 co-stimulatory molecule located on the antigen presenting cells. The interaction of CTLA-4 with the B7 molecule delivers an inhibitory signal, effectively checked by CTLA-4 inhibitors. On the other hand, the negative regulation of T cells resulting from PD-1/PD-L1 interaction between T cells and tumor cells is prevented by PD-1/PD-L1 inhibitors. Abbreviations: APC, antigen presenting cell; PD-1, programmed death receptor-1; PD-L1, programmed cell death receptor ligand-1; TCR, T cell receptor; MHC I, major histocompatibility complex class I; MHC II, major histocompatibility complex class II.

Several signaling pathways have been implicated in the modulation of T cell activity. The CTLA-4 molecule is a homolog of CD28 and is expressed by T cells (Fig. (Fig.1)1) [12, 13]. The influence of CTLA-4 on T cell activity primarily occurs in the priming phase of T cell activation [14]. CTLA-4 competitively binds to B7 on APCs and inhibits the costimulatory signal that arises from the engagement of CD28 with B7, thereby diminishing the T cell immune response [13, 15–18]. The upregulation of CTLA-4 expression on CD8+ and CD4+ T cells precludes stimulatory signaling from CD28-B7 binding and TCR-MHC binding [11, 19]. On the other hand, regulatory T cells (T_{reg} cells) exhibit constitutive expression of CTLA-4 [20]. Murine models bearing T_{reg} cells deficient in *CTLA-4* exhibited attenuated immunosuppressive activity, thus highlighting the significance of CTLA-4 in regulating immunological self-tolerance [20].

The PD-1 molecule is akin to CTLA-4. It is a member of the B7-CD28 family and is expressed by myeloid derived cells, B cells, and T cells [21, 22]. PD-1 has two corresponding ligands, PD-L1 and PD-L2 [14]. PD-L1 is expressed by a diverse set of cells including hematopoietic cells, leucocytes, parenchymal cells, and tumor cells, whereas PD-L2 is expressed by dendritic cells and macrophages [22–24]. The PD-1 receptor on T cells binds PD-L1 expressed by APCs and inhibits pro-inflammatory events such as T cell proliferation and cytokine production (Fig. (Fig.1)1) [14, 25]. Of note, recent evidence has suggested that PD-1/PD-L1 interactions facilitate immune escape by tumor cells [24, 26]. This phenomenon has been attributed to PD-1/PD-L1 mediated induction of anergy and apoptosis of activated T cells, tumor resistance to the cytotoxic T cell response, and differentiation of CD4+ T cells into Fox3p+CD4+ T_{reg} cells [27–29].

An intricate knowledge of various pathways regulating T cell-APC interactions has been central to identifying the points of intervention that allow us to modulate host immune responses. The aforementioned evidence and other studies with similar outcomes prompted the development of PD-1/PD-L1 and CTLA-4 checkpoint inhibitors for potential application in anti-cancer therapy.

Preclinical Rationale for Combination Checkpoint Inhibition

Combination immunotherapy with PD-1/PD-L1 plus CTLA-4 checkpoint inhibitors has been studied in multiple cancer cell lines. In an experiment on murine preclinical models, vaccination with B16-Flt-3 ligand (Fvax) along with the use of CTLA-4 antibody promoted tumor rejection in 10% of mice with pre-implanted B16-BL6 melanoma [7]. Fvax plus PD-1 blockade exhibited tumor rejection in 25% of mice. The combined use of CTLA-4 and PD-1 checkpoint inhibitors resulted in the rejection of B16-BL6 melanoma in 50% of test animals. Upon the addition of a PD-L1 inhibitor to the above, 65% of test animals exhibited rejection of melanoma. The outcomes observed with combined PD-1 and CTLA-4 blockade were found to correlate with an increase in CD4+ effector T cell (CD4+ T_{eff}) to T_{reg} cell ratio and CD8+ T cell to T_{reg} cell ratio in tumor tissue. Another significant observation was that a high percentage of T cells positive for CTLA-4 and PD-1 that would have undergone anergy remained active with combined PD-1 plus CTLA-4 blockade [7].

Similar findings were documented in another study investigating the effects of checkpoint inhibition (PD-1/PD-L1 and CTLA-4) in murine models of ovarian (ID8-VEGF) and colonic carcinoma (CT26) [30]. For ID8-VEGF, tumor regression was observed in 25% of test animals after PD-1 blockade, 25% with CTLA-4 blockade, and 37.5% with PD-L1 blockade, as compared to 50% with combined CTLA-4 plus PD-1 or PD-L1 blockade. Likewise, in mice bearing CT26, tumor regression was observed in 25% of test animals with PD-1 blockade, 50% with CTLA-4 blockade, and 33% with PD-L1 blockade, as compared to 75% ($p<0.01$ for the combination to checkpoint inhibitor monotherapy comparison) with combined CTLA-4 plus PD-1 or PD-L1 blockade. Combined therapy with CTLA-4 and PD-1 blockade was associated with a greater enhancement in tumor induced lymphocyte (TIL) activity and proliferation similar to CTLA-4 or PD-1 blockade alone [30]. Combination therapy also decreased the frequency of T_{reg} cells and functional markers of activated T_{reg} cells such as glucocorticoid induced tumor necrosis factor receptor (GITR) [30]. This suggested that dual checkpoint inhibition simultaneously blunted the function and decreased the number of T_{reg} cells.

In view of these and other preclinical studies, it was established that combined immune checkpoint blockade decreases suppression of the host immune system, while promoting inflammation in the tumor microenvironment. Moreover, the vast amount of preclinical data also suggested that the anti-tumor activity of combination therapy with CTLA-4 plus PD-1/PD-L1 checkpoint inhibitors may have superior outcomes compared to CTLA-4 or PD-1 monotherapy. It was thus prudent to investigate the therapeutic potential of CTLA-4 plus PD-1/PD-L1 checkpoint inhibitors in clinical settings.

Emergence of Combination Checkpoint Inhibition

Cancer immunosurveillance has been considered one of the primary natural mechanisms of defense against aberrant cell populations [31, 32]. Immune cells recognize and eliminate transformed cells through various cellular interactions [33]. In addition, the immune system is also involved in shaping tumor immunogenicity, a phenomenon theorized as immunoediting [34]. However, it was observed that cancer cells gradually undergo immune selection that disrupts the equilibrium with immune cells, consequently generating a tumor cell population that effectively evades immune surveillance [35–38]. Furthermore, alterations in the tumor microenvironment have also been reported to mediate immune escape [39]. It was thus theorized that therapeutic agents with the ability to restore immune surveillance or prevent immune escape of tumor cell populations could potentially have a significant impact in clinical oncology.

The knowledge of various factors that influence tumorigenesis has been exploited for developing a wide range of immunotherapy agents. Immunogenic cell death inducers, immunostimulatory cytokines, pattern recognition receptor agonists (PRR agonists), and tumor targeting antibodies have benefitted cancer patients for nearly two decades [40]. Numerous immunotherapy agents have received FDA approval for use as monotherapy in a variety of tumor histologies. For instance, nivolumab has been approved for patients with melanoma, renal cell carcinoma (RCC), metastatic squamous cell carcinoma of the head and neck (SCCHN), classical Hodgkin's lymphoma (HL) and advanced lung cancer [41–46]. Pembrolizumab, another anti-PD-1 MoAb, is approved for use in patients with SCCHN, NSCLC, and melanoma [47–49]. Likewise, atezolizumab is approved for patients with NSCLC and urothelial carcinoma [50–52]. However, any further enhancements in clinical benefit beyond checkpoint inhibitor monotherapy are still in clinical trials.

Recent clinical trial data comparing combination therapy with nivolumab and ipilimumab versus ipilimumab monotherapy in treatment naïve melanoma patients drew much attention (CheckMate 069, NCT01927419). In patients with *BRAF*-wild type melanoma, the investigators reported an objective response of 61% (95% CI: 49–72) as opposed to 11% (95% CI: 3–25) with combination checkpoint inhibition and ipilimumab monotherapy, respectively [53]. Furthermore, 22% of study participants receiving combination therapy exhibited a complete response, as compared to none among those receiving ipilimumab monotherapy [53]. In view of these findings, combined therapy with nivolumab and ipilimumab in melanoma patients became the first FDA approved (accelerated approval) indication for combination checkpoint inhibition. Furthermore, this study prompted comprehensive efforts to explore the application of combined immunotherapy with anti-PD-1/PD-L1 and anti-CTLA-4 MoAbs for patients with various malignancies.

Currently there are a large number of clinical trials evaluating several PD-1/PD-L1 plus CTLA-4 checkpoint inhibitor combination regimens. Positive outcomes in preliminary trials have made way for more intensive efforts to explore the full potential of combination checkpoint inhibition regimens, particularly nivolumab plus ipilimumab. It is widely anticipated that the ongoing trials will provide formidable evidence to support clinical applications of PD-1/PD-L1 plus CTLA-4 checkpoint inhibitor combination regimens for patients with a substantial variety of tumor histologies.

METHODS

The details of pertinent clinical trials were gathered from clinicaltrials.gov (United States National Institute of Health) from the database available to public domain. The database was narrowed using the search queries "nivolumab" with "ipilimumab", "pembrolizumab" with "ipilimumab", "atezolizumab" with "ipilimumab" and "tremelimumab" with "durvalumab". All prospective clinical trials using aforementioned immunotherapy agents as treatment intervention in NSCLC or melanoma patients were selected from the filtered results. Additionally, all phase 1 solid tumor trials aimed at providing a recommendation for the appropriate phase 2 dose of above combinations were also selected. The NCT number of each trial was used to search for published results on online databases including PubMed, American Association of Cancer Research (AACR), American Society of Clinical Oncology (ASCO) and European Society for Medical Oncology (ESMO).

Clinical Trials: Current Landscape

The United States National Institutes of Health lists a total of 44 ongoing clinical trials (clinicaltrials.gov) evaluating combined immunotherapy with anti-PD-1/PD-L1 plus anti-CTLA-4 MoAbs for patients with NSCLC and melanoma (Tables 1 and 2, Fig. 2a). The designated combination therapy regimens include nivolumab plus ipilimumab, pembrolizumab plus ipilimumab, atezolizumab plus ipilimumab, and tremelimumab plus durvalumab. A majority of these trials focus on survival and other treatment response indices, whereas a limited number of trials are investigating the safety profile and maximum tolerable dose (MTD) of combination therapy protocols. Furthermore, eight phase 1/phase 2 clinical trials are in the process of enrolling participants with solid tumors in order to determine recommended phase 2 doses (RP2D) for combination checkpoint inhibitor therapy with various anti-PD-1/PD-L1 and anti-CTLA-4 MoAbs. Of note, over three quarters (37 of 44 trials) of ongoing combination therapy trials for patients with NSCLC and melanoma are investigating treatment regimens that involve the application of nivolumab and ipilimumab (Fig. 2a and b).

Table 1: Melanoma trials evaluating combination checkpoint inhibition with anti-CTLA-4 plus anti- PD-1/PD-L1 monoclonal antibodies

Cancer type	Phase	Primary outcome	Dosing regimen	Enrollment number	Status	Results	Clinical trials identification number
Unresectable Stage III/stage IV malignant melanoma	Phase 1	Assess the safety of specified doses of NIVO + IPI combination therapy for up to 5.5 years; secondary outcome: assessment of tumor response	Cohort 1: induction with NIVO 0.3 mg/kg (Q3W for 21 weeks)+IPI 3 mg/kg (Q3W for 9 weeks), maintenance with NIVO 0.3 mg/kg+IPI 3 mg/kg (Q12W for a total of 84 weeks); cohort 2: induction with NIVO 1 mg/kg (Q3W for 21 weeks)+IPI 3 mg/kg (Q3W for 9 weeks), maintenance with NIVO 1 mg/kg+IPI 3 mg/kg (Q12W for 84 weeks); cohort 3: induction with NIVO 3 mg/kg (Q3W for 21 weeks)+IPI 3 mg/kg (Q3W for 9 weeks), maintenance with NIVO 3 mg/kg+IPI 3 mg/kg (Q12W for 84 weeks); cohort 4: induction with NIVO 10 mg/kg (Q3W for 21 weeks)+IPI 3 mg/kg (Q3W for 9 weeks), maintenance with NIVO 10 mg/kg+IPI 3 mg/kg (Q12W for a total of 84 weeks); cohort 5: induction with NIVO 10 mg/kg (Q3W for 21 weeks)+IPI 10 mg/kg (Q3W for 9 weeks), maintenance with NIVO 10 mg/kg+IPI 10 mg/kg (Q12W for 84 weeks); cohort 6: NIVO 1 mg/kg (Q2W for 96 weeks); cohort 7: NIVO 3 mg/kg (Q2W for 96 weeks); cohort 8: 4 doses of IPI 3 mg/kg+NIVO 1 mg/kg (Q3W for 12 weeks), then monotherapy with NIVO 3 mg/kg (Q2W for 96 weeks)	136	Completed, results available	53% patients had grade 3–4 AE; NIVO 0.3 mg/kg+IPI 3 mg/kg (n=14): 1 year OS rate=56%, median OS=14.8 months; NIVO 1 mg/kg+IPI 3 mg/kg (n=17): 1 year OS rate=94%, median OS non recordable; NIVO 3 mg/kg+IPI 1 mg/kg (n=16): 1 year OS rate=89%, median OS non recordable; NIVO 3 mg/kg+IPI 3 mg/kg (n=6): 1 year OS rate=100%, median OS non recordable; concurrent NIVO + IPI (n=53): 1 year OS rate=82%, median OS=39.7 months; sequenced regimen (NIVO + IPI followed by monotherapy with NIVO Q2W for 48 doses, n=32): median OS=13.0 months, insufficient follow-up of 1 year OS rate. Subgroup analysis: OR in PD-L1 positive tumors: concurrent therapy: 6 of 13 patients, sequential therapy: 4 of 8 patients; OR in PD-L1 negative tumors: concurrent therapy: 9 of 22 patients; sequential therapy: 1 of 13 patients	NCT01024231 [54]
Advanced melanoma	Phase 1	Incidence of serious AE and treatment emergent AE; secondary outcomes: PFS, OS, ORR and duration of response (assessed up to week 49)	Arm 1: NIVO Q2W+TAK 580 QW; arm 2: plozalizumab 2 mg QW for weeks 1, 3, 5 and 9, then plozalizumab Q4W+NIVO Q2W; arm 3: IPI (once in week 3, 6, 9 and 12)+NIVO (administered in weeks 3, 6, 9, 12 and 15, Q2W thereafter) and vedolizumab (once in weeks 1, 3, 5 and 13)	156	Recruiting	NA	NCT02723006

Stage III unresectable or stage IV melanoma	Phase 1b	To determine MTD of specified treatment regimen (time frame: 12 months); secondary outcome: ORR	NIVO + IPI and ACY 241	36	Recruiting	NA	NCT02935790
BRAF mutant metastatic/ unresectable melanoma	Phase 1 RCT	Incidence of AE (≥ grade 3 NCI CTCAE v4.0) evaluated up to 3 weeks after induction therapy with IPI; secondary outcomes: disease control rate and RR evaluated till 4 weeks after completion of therapy; proportion of patients with ≥ grade 3 AE after progression of disease on IPI	Arm A1: trametinib QID + dabrafenib BID for the first 25 days, then 4 courses of IPI Q3W; arm A2: 25 days of trametinib QID + dabrafenib BID, then 4 courses of IPI + NIVO Q3W, maintenance with NIVO Q2W (42 courses); arm B1: 25 days of trametinib QID, then 4 courses of IPI Q3W; arm B2: 25 days of trametinib QID, then 4 courses of IPI + NIVO Q3W, maintenance with NIVO Q2W (42 courses); arm C1: dabrafenib BID for 25 days, then 4 courses of IPI Q3W; arm C2: 25 days of dabrafenib BID, then 4 courses of IPI + NIVO Q3W, maintenance with NIVO Q2W (42 courses); arm D1: 4 courses of IPI Q3W; arm D2: 4 courses of IPI + NIVO Q3W, maintenance with NIVO Q2W (42 courses)	40	Recruiting	NA	NCT01940809
Stage IIIC/ IV skin melanoma	Phase 1/ Phase 2	Assess the safety of adjuvant NIVO + low-dose IPI	Low fixed dose NIVO in combination with low fixed dose IPI	6	Recruiting	NA	NCT02941744
Uveal melanoma with liver metastases	Phase 1/ Phase 2	Evaluate tolerance and safety profile for treatment regimen (time frame: 3 years); secondary outcomes: PFS and RR (as per RECIST criteria)	Yttrium 90 followed by 4 doses of NIVO 1 mg/ kg + IPI 3 mg/kg Q3W; maintenance with NIVO 3 mg/kg monotherapy Q2W for a maximum of 3 years	18	Not yet recruiting	NA	NCT02913417
Metastatic melanoma	Phase 1/ Phase 2 RCT	Evaluation of grade 3–4 toxicity (treatment associated) event-free survival for up to 6 months	Active comparator: 4 doses of NIVO 1 mg/kg IV + IPI 3 mg/kg IV Q3W, then monotherapy with NIVO 3 mg/ kg Q2W; experimental arm: 4 doses of NIVO 1 mg/kg IV + IPI 0.3 mg/kg IT Q3W, followed by monotherapy with NIVO 3 mg/kg Q2W	65	Recruiting	NA	NCT02857569 (NIVIPIT)

Recurrent/ advanced melanoma	Phase 1/ Phase 2 RCT	Pathological complete RR; secondary outcome: ORR, PFS (time frame: 5 years)	Arm A: 3 doses of NIVO 3 mg/kg Q2W, then surgery at weeks 6 to 8; maintenance with NIVO 3 mg/kg Q3W after surgery; Arm B: 2 doses of IPI 3 mg/kg+NIVO 1 mg/kg Q3W and then surgery; maintenance with 2 doses of NIVO 1 mg/kg+IPI 3 mg/kg Q3W, continued as NIVO 3 mg/kg at Q3W	66	Recruiting	NA	NCT02736123
Advanced melanoma/ renal cell carcinoma	Phase 1/ Phase 2 RCT	PFS, the number of study participants with AE (evaluation for up to 2 years), those discontinuing of study due to AE (evaluation for up to 2 years) and the patients that experienced DLT (assessed up to 6 weeks); secondary outcomes: OS, duration of response and ORR for phase 1b and 2	Arm 1: PEMBRO monotherapy; arm 2: two 6-weeks cycles of PEMBRO + IPI Q3W; arm 3: pegIFN-2b QW for each cycle (cycle duration: 6 weeks)+PEMBRO Q3W	343	Completed	(October 17, 2016 cut-off date): 153 participants received a minimum of 1 dose of PEMBRO + IPI; 72% (110 of 153 patients) received all 4 doses of PMEBRO + IPI; 42% (64 of 153 patients) were on PEMBRO monotherapy; - TRAEs: 45% (69 of 153 patients) developed grade 3–4 TRAEs, 51% (78 of 153 patients) developed grade 1–2 TRAEs; irAEs: seen in 60% (92 of 153) participants, 27% (42 of 153 patients) documented to have grade 3–4 irAEs OR: 61% (93 of 153 patients, 95%CI: 53–69) - CR: 15% (23 of 153 patients) - PR: 46% (70 of 153 patients) - SD: 18% (28 of 153 patients) - PD: 19% (29 of 153 patients) - PFS at 12 months: 69% (95%CI: 60–75)	NCT02089685; KEYNOTE-029 [58]
Melanoma with leptomeningeal metastases	Phase 2	OS rate assessed at 2 years	Pre-determined doses of IPI+NIVO combination therapy, then monotherapy with NIVO, duration of each treatment cycle to be 6 weeks	18	Not yet recruiting	NA	NCT02939300
Resected stage IIIB/ IIIC/IV melanoma	Phase 2	Evaluation of adverse effects with specified treatment regimen; secondary outcomes: time to relapse and immunological response	Cycle 1: 4 doses of IPI 1 mg/kg+NIVO 3 mg/kg Q3W for 12 weeks; cycles 2, 3, 4 and 5: monotherapy with NIVO 480 mg Q4W for 48 weeks	25	Not yet recruiting	NA	NCT02970981
Stage III/ resected stage IV melanoma	Phase 2	Recurrence free survival and OS; evaluate toxicity of adjuvant low-dose IPI and NIVO for up to 7 months	Adjuvant therapy with NIVO 3 mg/kg Q2W+IPI 1 mg/kg Q6W, for a total of 6 months duration	25	Recruiting	NA	NCT02656706; BrUOG 324

Uveal melanoma	Phase 2	OS at 12 months	4 doses of IPI+NIVO Q3W, then NIVO monotherapy Q2W	48	Recruiting	NA	NCT02626962; GEM1402
Uveal melanoma	Phase 2	ORR at 12 weeks	Induction phase: 4 doses of IPI 3 mg/kg+NIVO 1 mg/kg at week 1, 4, 7 and 10, continued through week 12; maintenance phase: monotherapy with NIVO 3 mg/kg Q2W for study participants with unmanageable toxicity or no progression after 12 weeks of induction therapy, to be continued till progression or unacceptable toxicity	52	Recruiting	NA	NCT01585194
Advanced melanoma	Phase 2	RR, evaluated up to 16 weeks; secondary outcomes: PFS assessed up to 24 months and safety	IPI+PEMBRO (all study participants to have received initial PD-1/PD-L1 antibody therapy prior to enrollment as per selection criteria)	70	Recruiting	NA	NCT02743819
Advanced mucosal/acral lentiginous melanoma	Phase 2	OR rate for mucosal melanoma (assessed up to 2 years); secondary outcomes: OR rate for acral lentiginous melanoma, PFS, OS	Induction phase: 4 doses of IPI 3 mg/kg+NIVO 1 mg/kg Q3W; maintenance phase: monotherapy with NIVO 3 mg/kg Q2W for 48 doses; NIVO monotherapy to be continued for another 12 weeks in study participants demonstrating CR	72	Not yet recruiting	NA	NCT02978443
Melanoma	Phase 2	Evaluation of clinical benefit rate up to 6 months	Induction: IPI+NIVO; maintenance phase: NIVO only	110	Recruiting	NA	NCT02320058; CheckMate 204
Advanced melanoma/ bladder cancer	Phase 2	Evaluation of RR up to 12 weeks	Treatment regimen for melanoma patients: 4 doses of NIVO 1 mg/kg+IPI 3 mg/kg Q2W	120	Recruiting	NA	NCT02553642
Stage III/IV melanoma with progression/relapse on PD-1 inhibitor therapy	Phase 2 RCT	Pathological CR with neoadjuvant IPI+NIVO combination therapy and neoadjuvant NIVO monotherapy, assessed at day 57	Group A: Neoadjuvant therapy with NIVO 3 mg/kg on weeks 1, 3, 5 and 7, then surgical excision, later adjuvant phase with NIVO 3 mg/kg Q2W for 6 months; group B: neoadjuvant therapy with IPI 3 mg/kg and NIVO 1 mg/kg on weeks 1, 4 and 7, then surgical excision and later adjuvant therapy same as group A	40	Recruiting	NA	NCT02519322
Stage III/IV melanoma with progression/relapse on PD-1 inhibitor therapy	Phase 2 RCT	OR as per RECIST v1.1 (time frame: 18 weeks)	Experimental arm 1: 4 doses of NIVO 1 mg/kg+IPI 3 mg/kg Q3W; experimental arm 2: 4 doses of IPI 3 mg/kg monotherapy Q3W	70	Recruiting	NA	NCT02731729

Melanoma with brain metastases	Phase 2 RCT	Intracranial response rate assessed for up to 3 years; secondary outcomes: ORR, PFS, OS, extracranial response	Cohort 1: monotherapy with NIVO 3 mg/kg Q2W; cohort 2: monotherapy with NIVO 3 mg/kg Q2W; cohort 3: 4 doses of NIVO 1 mg/kg+IPI 3 mg/kg Q3W, followed by NIVO 3 mg/kg Q2W	76	Recruiting	NA	NCT02374242
Stage III melanoma	Phase 2 RCT	Pathological response rate at 6 weeks; RR (assessed up to 6 weeks); incidence of treatment related AE (time frame: 12 weeks)	Arm A: 2 cycles of NIVO 1 mg/kg+IPI 3 mg/kg before surgery at week 6; arm B: 2 cycles of NIVO 3 mg/kg+IPI 1 mg/kg before surgery at week 6; arm C: 2 cycles of NIVO 3 mg/kg+IPI 3 mg/kg before surgery at week 6	90	Recruiting	NA	NCT02977052
Advanced/ metastatic melanoma	Phase 2 RCT	Determining the percent candidates that develop grade 3–5 AE attributable to induction therapy (time frame: 25 weeks); secondary outcomes: investigator assessed duration of response, RR and rate of progression	Cohort A: NIVO preceding IPI: induction with 6 doses of NIVO 3 mg/kg IV Q2W, to be continued in maintenance phase for up to 2 years; 4 doses of IPI 3 mg/kg Q3W in induction phase only; Cohort B: IPI preceding NIVO: 4 doses of IPI 3 mg/kg Q3W in induction phase only; 6 doses of NIVO 3 mg/kg IV Q2W during induction, continued in maintenance phase for up to 2 years	177	Results available	- Grade 3–5 TRAEs Cohort A: 34 of 68 patients (95%CI: 37.6–62.4); Cohort B: 30 of 70 patients (95%CI: 31.1–55.3) - Response at 25 weeks: Cohort A: 41% (28 patients, 95%CI: 29.4–53.8); Cohort B: 20% (14 patients, 95%CI: 11.4–31.3) - 12 months OS: Cohort A: 76% (95%CI: 64–85) Cohort B: 54% (95%CI: 42–65)	NCT01783938; CheckMate 064 [57]

Previously untreated unresectable/ metastatic melanoma	Phase 2 RCT	Percent study participants (BRAF wild-type) with OR (time frame: 6 months or more)	Experimental regimen 1: 4 doses of NIVO 1 mg/ kg+IPI 3 mg/kg Q3W followed by monotherapy with NIVO 3 mg/kg Q2W; experimental regimen 2: placebo +4 doses of IPI 3 mg/kg Q3W followed by placebo Q2W	179	Completed, results available	Participants with BRAF wild-type tumors: OR and CR with placebo: 11% (4 of 37 patients, 95%CI: 3 to 25) and 0%, respectively; OR and CR with combination therapy: 61% (44 of 72 patients, 95%CI: 49 to 72) and 22% (16 patients), respectively; HR=0.40 (95%CI: 0.23 to 0.68, $p<0.001$) for death or progression of disease when comparing IPI+NIVO combination therapy versus IPI+placebo. Participants with BRAF mutant tumors: OR and CR with placebo: 10% (1 of 10 patients, 95%CI: 0 to 45) and 0%, respectively; OR and CR with combination therapy: 52% (12 of 23 patients, 95%CI: 31 to 73) and 22% (5 patients), respectively; HR=0.38 (95%CI: 0.15 to 1.00); Subgroup analysis: OR rate with combination therapy: PD-L1 positive tumors: 58, 95%CI: 37 to 78, PD-L1 negative tumors: 55, 95%CI: 41 to 69); OR with IPI monotherapy: PD-L1 positive tumors: 18% (95%CI: 2 to 52), PD-L1 negative tumors: 4% (95%CI: 0 to 19)	NCT01927419; CheckMate 069 [53]
Melanoma	Phase 2 RCT	Evaluation of the best ORR at 18 weeks; secondary outcomes: OS and PFS	Experimental arm: induction with cobimetinib + vemurafenib for 6 weeks followed by NIVO + IPI; Control arm: NIVO + IPI only	200	Not yet recruiting	NA	NCT02968303
BRAF mutant metastatic melanoma	Phase 2 RCT	OS evaluated up to 24 months; secondary outcomes: PFS (time frame: 2 years), 3 year PFS rate, duration of response (evaluated for 24 months)	Arm A: binimetinib 45 mg BID + encorafenib 450 mg OD till progression, then 4 doses of NIVO 1 mg/ kg+IPI 3 mg/kg Q3W and later NIVO 3 mg/kg Q2W; arm B: 4 doses of IPI 3 mg/ kg+NIVO 1 mg/kg Q3W, later NIVO 3 mg/kg Q2W till disease progression, binimetinib 45 mg BID + encorafenib 450 mg OD till progression; arm C: binimetinib 45 mg BID + encorafenib 450 mg OD for 8 weeks, then 4 doses of NIVO 1 mg/kg+IPI 3 mg/ kg Q3W and later NIVO 3 mg/kg Q2W till progression. Thereafter, treatment continued with binimetinib 45 mg BID + encorafenib 450 mg OD till progression of disease	230	Recruiting	NA	NCT02631447; SECOMBIT study

Malignant melanoma	Phase 2 RCT	Evaluation of efficacy with adjuvant therapy with NIVO + IPI or NIVO monotherapy (time frame: 24 months)	Active comparator: Placebo + NIVO: IPI substituted with placebo for weeks 1, 4, 7 and 10; NIVO substituted with placebo for weeks 4 and 10; Experimental regimen: 4 doses of IPI 3 mg/kg+NIVO 1 mg/kg Q3W, placebo to replace NIVO on weeks 3, 5, 9 and 11, monotherapy with NIVO 3 mg/kg Q2W to be continued during maintenance phase for 1 year after induction or disease progression	312	Recruiting	NA	NCT02523313
Treatment naïve unresectable/ metastatic advanced melanoma	Phase 3	To determine the incidence of TRAE; Secondary outcomes: ORR and PFS (time frame: 20 weeks)	Arm 1: concomitant NIVO + IPI followed by monotherapy with NIVO; arm 2: sequential administration of NIVO and IPI, followed by monotherapy with NIVO	102	Recruiting	NA	NCT02905266
Stage III unresectable or stage IV melanoma	Phase 3	TRAE for up to 24 months; secondary outcomes: PFS (investigator assessed), OS, ORR	NIVO monotherapy or NIVO + IPI combination therapy	615	Recruiting	NA	NCT02599402; CheckMate 401
Stage III/IV melanoma	Phase 3 RCT	OS rate at the end of 2 years follow-up; secondary outcomes: PFS and RR	Arm A: induction with 2 courses of NIVO + IPI Q6W, maintenance with NIVO monotherapy Q6W for a total of 12 courses, patients cross over to arm C if disease progresses; arm D: induction with 2 courses of NIVO + IPI Q6W, maintenance with NIVO monotherapy Q6W for a total of 12 courses	300	Recruiting	NA	NCT02224781
Treatment naïve unresectable/ metastatic advanced melanoma	Phase 3 RCT	To determine the incidence of TRAE; Secondary outcomes: ORR, PFS, OS	Arm 1: IPI 1 mg/kg+NIVO 3 mg/kg; arm 2: IPI 1 mg/kg+NIVO 6 mg/kg; arm 3: IPI 3 mg/kg+NIVO 1 mg/kg	340	Recruiting	NA	NCT02714218

Treatment naïve unresectable/ metastatic advanced melanoma	Phase 3 RCT	OS (time frame: 44.1 months), PFS (time frame: up to 5 years)	Arm A: NIVO 3 mg/kg Q2W+placebo substituting IPI on weeks 1 and 4, followed by placebo substituting NIVO on week 4 for cycle 1 and 2; Arm B: 4 doses of NIVO 1 mg/kg+IPI 3 mg/kg Q3W followed by NIVO 3 mg/kg Q2W (placebo substituting NIVO on 3rd and 5th weeks for cycle 1 and 2; Arm C: 4 doses of IPI 3 mg/kg Q3W+placebo (replacing NIVO on 3rd and 5th weeks for cycles 1 and 2)	915	Completed, results available	OR: NIVO+IPI 58%, NIVO mono 44%, IPI mono 19%; CR: NIVO+IPI 19%, NIVO mono 16%, IPI mono 5%; Median PFS: NIVO monotherapy: 6.9 months (95%CI: 4.3 to 9.5), IPI monotherapy: 2.9 months (95%CI: 2.8 to 3.4) and IPI+NIVO: 11.5 months (95%CI: 8.9 to 16.7); median PFS for participants with PD-L1 positive malignancy: 14.0 months in NIVO arm and combination therapy arm; median PFS for participants with PD-L1 negative malignancy: 5.3 months (95%CI: 2.8 to 7.1) for NIVO monotherapy and 11.2 months (95%CI: 8.0 - not reached) for IPI+NIVO; OS at 2 years: NIVO+ IPI 64%, NIVO monotherapy 59% and IPI monotherapy 45%; OS at 3 years: NIVO+ IPI 58%, NIVO monotherapy 52% and IPI monotherapy 34%; median OS NIVO+ IPI- not reached (95%CI 38.2 months - not reached), NIVO monotherapy-37.6 months (95%CI: 29.1 - not reached) and IPI monotherapy-19.9 months (95%CI: 16.9–24.9); Grade 3–4 TRAEs: NIVO+IPI 59%, NIVO mono 21%, IPI mono 28%	NCT01844505; CheckMate 067 [56]

Abbreviations: MTD maximum tolerable dose, *DLT* dose limiting toxicity, *RCT* randomized controlled trial, *NA* not available, *CI* confidence interval, *ORR* overall response rate, *OS* overall survival, *PFS* progression free survival, *RR* response rate, *OR* objective response, *CR* complete response, *PR* partial response, *SD* stable disease, *PD* progressive disease, *TRAE* Treatment-related adverse events, *irAE* immune related adverse events, *OD* once daily dosing, *BID* twice daily, *QID* four times a day, *NIVO*, nivolumab, *IPI* ipilimumab, *PEMBRO* pembrolizumab, *RECIST* response evaluation criteria in solid tumors, *NCI CTCAE* National Cancer Institute common terminology criteria for adverse events, Q(x)W, every (x) weeks; D(x), day(x); *TRAE* treatment related adverse events, *AE* adverse events

Table 2: NSCLC trials evaluating combination checkpoint inhibition with anti-CTLA-4 plus anti- PD-1/PD-L1 monoclonal antibodies

Cancer type	Phase	Primary outcome	Dosing regimen	Enroll-ment number	Status	Results	Clinical trials identification number
Stage IV NSCLC	Phase 1	Evaluation of treatment related toxicity of specified treatment regimen in ALK/EGFR mutated NSCLC patients for up to 36 months; secondary outcomes: OS, PFS, RR evaluated up to 36 months	Arm 1 (EGFR mutant NSCLC): 4 doses of IPI 3 mg/kg+erlotinib 150 mg OD (or the tolerable dose); arm 2 (ALK positive NSCLC): 4 doses of IPI 3 mg/kg+crizotinib 250 mg BID (or tolerable dose); arm 3 (EGFR mutant NSCLC): erlotinib 150 mg OD (or tolerable dose)+NIVO 240 mg Q2W; arm D (ALK positive NSCLC): crizotinib 250 mg BID (or tolerable dose)+NIVO 240 mg Q2W	14	Active, not recruiting	NA	NCT01998126
Stage IIIb/IV NSCLC	Phase 1 RCT	Evaluate safety of specified treatment regimens; secondary outcomes: Evaluation of ORR and PFS rate (time frame: up to 24 weeks)	Arm A: gemcitabine + cisplatin and NIVO; arm B: pemetrexed + cisplatin and NIVO; arm C: carboplatin + paclitaxel and NIVO; arm D: NIVO + maintenance with bevacizumab; arm E: erlotinib + NIVO; arm F: NIVO; arm G (squamous NSCLC): IPI+NIVO; arm H (non-squamous NSCLC): IPI+NIVO; arm I (squamous NSCLC): IPI+NIVO; arm J (non-squamous NSCLC): IPI+NIVO; arm K (squamous NSCLC): NIVO; arm L (non-squamous NSCLC): NIVO; arm M (NSCLC with asymptomatic and untreated brain metastases): NIVO; arm N (NSCLC with any histology): IPI+NIVO; arm O: IPI+NIVO; arm P: IPI+NIVO; arm Q: IPI+NIVO; arm R: IPI+NIVO; arm S: IPI+NIVO	412	Completed	NIVO 3 mg/kg Q2W+IPI 1 mg/kg Q12W (n=38): grade 3–4 TRAE: 37%, ORR: 47% (95%CI: 31–64), median PFS: 8.1 months (95%CI: 5.6–13.6), 1 year OS rate: not calculated; NIVO 3 mg/kg Q2W+IPI 1 mg/kg Q6W (n=39): grade 3–4 TRAE: 33%, ORR: 39% (95%CI: 23–55), median PFS: 3.9 months (95%CI: 2.6–13.2), 1 year OS rate: 69% (95% CI: 52–81); NIVO 3 mg/kg Q2W (n=52): grade 3–4 TRAE: 19%, ORR: 23% (95%CI: 13–37), median PFS: 3.6 months (95%CI: 2.3–6.6), 1 year OS rate: 73% (95% CI: 59–83); ORR for tumors having PD-L1 expression of ≥1% was reported as 57% and ORR for tumors having PD-L1 expression of ≥50% was reported as 92% with NIVO + IPI; ORR for tumors having PD-L1 expression of ≥1% was reported as 28% and ORR for tumors having PD-L1 expression of ≥50% was reported as 50% with NIVO monotherapy	NCT01454102; CheckMate 012 [60]

Advanced NSCLC	Phase 1b	OR assessed at 24 weeks, number of study participants experiencing DLT and number of participants that report treatment associated toxicities	Participants enrolled in dose escalation arm and all experimental arms receive MEDI4736+tremelimumab combination therapy	747	Active, not recruiting, preliminary results available	OR in tremelimumab 1 mg/kg combination therapy cohort: Overall: 23% (6 of 26 patients, 95%CI: 9 to 44); PD-L1 positive tumors: 22% (2 of 9 patients, 95%CI: 3 to 60); PD-L1 negative tumors: 29% (4 of 14 patients, 95%CI: 8 to 58); MTD exceeded for therapy with tremelimumab 3 mg/kg+MEDI4736 20 mg/kg with DLT of 30% (2 of 6 patients)	NCT02000947 [62]
NSCLC with brain metastases	Phase 1/ Phase 2	Recommended phase 2 dose of NIVO + stereotactic radiosurgery/ NIVO + whole brain radiation therapy/ NIVO + IPI and stereotactic radiosurgery/ NIVO + IPI and whole brain radiation therapy; the observed MTD from phase 1 to become the starting dose for phase 2	Experimental arm 1: NIVO 3 mg/kg Q2W+stereotactic surgery; experimental arm 2: NIVO 3 mg/kg Q2W+whole brain radiation therapy 30 Gy in 10 fractions; experimental arm 3: NIVO MTD as decided in phase 1+IPI 1 mg/kg Q6W and stereotactic radiosurgery; experimental arm 4: NIVO MTD as decided in phase 1+IPI 1 mg/kg (Q6W) and whole brain radiation therapy (30 Gy in 10 fractions)	80	Not yet recruiting	NA	NCT02696993
Unresectable/ metastatic NSCLC	Phase 1/ Phase 2 RCT	Part I: determine recommended phase 2 dose of PEMBRO; part II: OR rate for cohort G and H; secondary outcomes: PFS, OS and duration of response (assessed up to 2 years)	Cohort H: IPI+PEMBRO (recommended phase 2 dose as per cohort D)	308	Recruiting, preliminary data available	Not available for cohort H.	NCT02039674; KEYNOTE 021 [61]
Advanced NSCLC	Phase 2	Best ORR evaluated Q6W up to 48 weeks	IPI 1 mg/kg Q6W+NIVO 3 mg/kg Q2W	35	Recruiting	NA	NCT02350764

Stage IV NSCLC	Phase 2	OR rate; secondary outcomes: PFS, duration of response (time frame: 6 months)	IPI+NIVO	590	Recruiting	NA	NCT02659059; CheckMate 568
Advanced NSCLC	Phase 2 RCT	PFS rate, ORR and duration of response (assessed up to 24 weeks)	Comparator arm: monotherapy with NIVO; arm 2: BMS-986016+NIVO; arm 3: dasatinib + NIVO; arm 4: IPI+NIVO	504	Recruiting	NA	NCT02750514
Locally advanced/ metastatic NSCLC	Phase 2 RCT	CR rate (assessed up to 2 years); secondary outcomes: ORR and disease control rate (assessed up to 2 years)	Arm 1: gefitinib + MEDI4736; arm 2: AZD9291+MEDI4736; arm 3: docetaxel + selumetinib and MEDI4736; arm 4: tremelimumab + MEDI4736	49	Completed	Not yet available	NCT02179671
Recurrent Stage IV squamous cell lung carcinoma	Phase 2/ Phase 3 RCT	ORR, OS (assessed for 3 years), IA-PFS (evaluated for 18 months), IA-PFS and OS in study participants receiving experimental regimen versus standard of care; secondary outcomes: duration of response, RR, PFS and OS for experimental regimen, frequency of toxicity events (assessed for 3 years)	S1400I arm I: IPI+NIVO; S1400I arm II: NIVO monotherapy	10,000	Recruiting	NA	NCT02154490; lung-MAP trial [79]
Stage IV/ recurrent NSCLC	Phase 3	Percent study participants with high grade toxicity; secondary outcomes: PFS, ORR, duration of response (assessed for 40 months)	NIVO + IPI	1500	Recruiting	NA	NCT02869789

NSCLC	Phase 3 RCT	Major pathological response rate (determined at surgery); secondary outcomes: complete pathological response rate (determined at surgery), OS and event free survival assessed for up to 130 months	Experimental arm: IPI+NIVO	326	Not yet recruit-ing	NA	NCT02998528
Stage IV squa-mous cell car-cinoma	Phase 3 RCT	OS (time frame: 3 years), IA-PFS (time frame: 18 months)	Active comparator: NIVO on D1, repeated every 14 days; experi-mental arm: NIVO + IPI D1 of every 3rd course, course to be repeated every 14 days	350	Recruit-ing	NA	NCT02785952
Chemo-therapy naïve/ recurrent stage IV NSCLC	Phase 3 RCT	OS, assessed up to 48 months; PFS, assessed up to 40 months; Secondary outcome: OR rate, assessed up to 48 months	Arm A: NIVO monotherapy; arm B: IPI+NIVO combina-tion therapy; arm C: platinum doublet chemo (carboplatin/ cisplatin + gemcitabine for squamous histol-ogy and carboplatin/ cisplatin + pemetrexed for non-squamous histology)+NIVO	2220	Recruit-ing	NA	NCT02477826; CheckMate 227
Stage IV/ recurrent NSCLC	Phase 3 RCT	PFS for T790 M negative, EGFR posi-tive NSCLC, evaluated for 33 months; secondary outcomes: PFS rate, ORR and duration of response evaluated for 33 months; OS assessed for 5 years	Arm 1: IPI+NIVO; arm 2: platinum dou-blet therapy (cisplatin/ carboplatin + peme-trexed)+NIVO	465	Recruit-ing	NA	NCT02864251; CheckMate 722
Ad-vanced/ meta-static NSCLC	Phase 3 RCT	OS with MEDI4736 + tremelimumab combination versus standard of care treat-ment (evaluated for 4 years)	Arm 1: MEDI4736+tremeli-mumab	800	Recruit-ing	NA	NCT02542293; NEPTUNE study

Locally advanced/ metastatic NSCLC	Phase 3 RCT	PFS and OS assessed up to 3 years; secondary outcomes: ORR (evaluated for 3 years), proportion of study participants alive at end of 1 year of randomization and duration of response (evaluated for 3 years)	Sub-study-A experimental arm: participants with PD-L1 positive malignancy to receive MEDI4736; sub-study B experimental arm 1: participants with PD-L1 negative malignancy to receive MEDI4736 + tremelimumab; sub-study experimental arm 2: participants with PD-L1 negative malignancy to receive MEDI4736 only; sub-study B experimental arm 3: participants with PD-L1 negative malignancy to receive tremelimumab only	730	Active, not recruiting	NA	NCT02352948; ARCTIC study
Advanced/ metastatic NSCLC	Phase 3 RCT	OS and PFS with MEDI4736 + tremelimumab combination versus standard of care treatment (evaluated for up to 3 years)	Arm 1: MEDI4736 monotherapy; arm 2: MEDI4736 + tremelimumab	1092	Active, not recruiting	NA	NCT02453282; MYSTIC study

Abbreviations: NSCLC non-small cell lung cancer, *EGFR* epithelial growth factor receptor, *ALK* anaplastic lymphoma kinase, *MTD* maximum tolerable dose, *DLT* dose limiting toxicity, *RCT* randomized controlled trial, *NA* not available, *OR* objective response, *CR* complete response, *CI* confidence interval, *TRAE* treatment-related adverse events, *NIVO* nivolumab, *IPI* ipilimumab, *PEMBRO* pembrolizumab, MEDI4736, durvalumab, *ORR* overall response rate, *OS* overall survival, *PFS* progression free survival, *RR* response rate, *OD* once daily dosing, *BID* twice daily, Q(x)W, every (x) weeks; D(x), day(x)

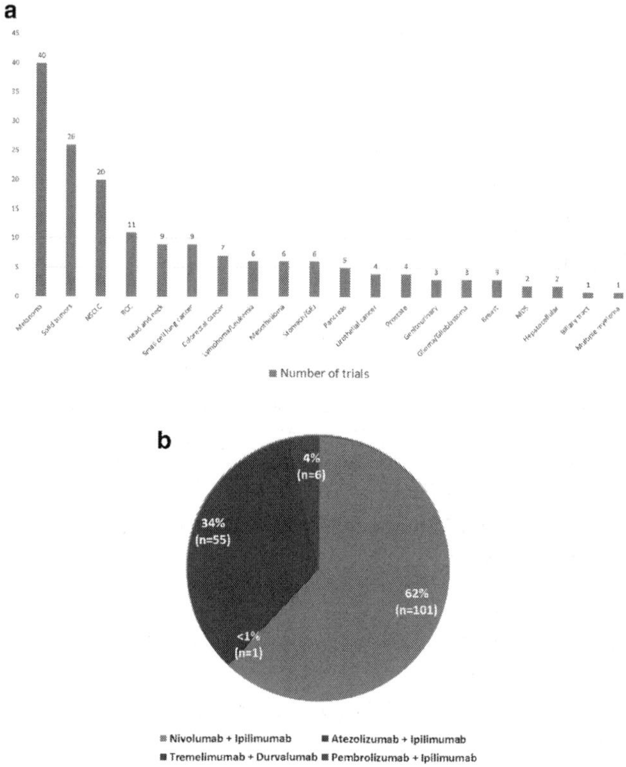

FigURE 2: Current landscape of combination immunotherapy trials for various tumor histologies. a Number of combination checkpoint inhibition trials for various tumor histologies. Abbreviations: NSCLC, non-small cell lung carcinoma; RCC, renal cell carcinoma; GEJ, gastro-esophageal junction; MDS, myelodysplastic syndrome. b Landscape of combination checkpoint inhibition agents. Legend: The figure elaborates relative number of trials for four combinations of immunotherapy agents. Nivolumab plus ipilimumab: 62% (101 trials), pembrolizumab plus ipilimumab 4% (6 trials), tremelimumab plus durvalumab 34% (55 trials) and atezolizumab plus ipilimumab <1% (1 trial).

Combination Checkpoint Inhibition Trials in Melanoma

Massive efforts are testing dual checkpoint inhibition with anti-PD-1/PD-L1 plus anti-CTLA-4 MoAbs in patients with melanoma. Presently, there are 28 clinical trials in various phases committed to this objective (Table 1). Currently, only one trial is in the process of recruiting participants that will receive ipilimumab in combination with pembrolizumab. Approximately two-thirds of all ongoing trials (18 of 28 trials) in melanoma patients are investigating combined therapy with nivolumab plus ipilimumab.

Several distinct methodologies for the combined use of nivolumab and ipilimumab have been identified and are presently being investigated for potential clinical benefit in two trials. One trial will compare the incidence of adverse events, overall response rate (ORR), and progression free survival (PFS) with concurrent versus sequential administration of nivolumab and ipilimumab in the induction phase (NCT02905266). Another trial, CheckMate 064, will evaluate the duration of response, response rate (RR), rate of progression and the proportion of study participants that develop grade 3–5 adverse events when using nivolumab prior to ipilimumab versus ipilimumab prior to nivolumab in the induction phase (NCT01783938).

Treatment protocols in five trials involve the integrated use of nivolumab and/or ipilimumab with additional anti-cancer agents. The assessment of MTD and ORR for combined therapy with nivolumab plus ipilimumab and ACY241 (histone deacetylase {HDAC} inhibitor) will be performed in one trial (NCT02935790) [10]. Another trial will be exploring the clinical efficacy and incidence of adverse events for nivolumab plus TAK580 (pan-Raf kinase inhibitor), nivolumab and plozalizumab (chemokine receptor-2 {anti-CCR-2} MoAb), and the combined use of nivolumab plus ipilimumab and vedolizumab (anti-LPAM-1/α4β7 MoAb), each in separate arms (NCT02723006) [10]. Similarly, a different trial will evaluate the safety profile and clinical efficacy with combination dabrafenib (BRAF protein inhibitor) and/or trametinib (MEK MAPK/ERK kinase {mitogen-activated protein kinase kinase} inhibitor) in the induction phase, followed by ipilimumab and/or nivolumab (NCT01940809) [10]. This trial will also compare the outcomes of the above treatment regimens with nivolumab plus ipilimumab (NCT01940809). The SECOMBIT study will compare survival outcomes among several variations of treatment regimens composed of binimetinib (MEK1/2 inhibitor), encorafenib (Raf kinase inhibitor), nivolumab and ipilimumab (NCT02631447) [10]. Lastly, one trial will analyze response to vemurafenib (BRAF-V600 kinase inhibitor) plus cobimetinib (MAP2K1/MEK1 inhibitor) in the induction phase followed by nivolumab plus ipilimumab, versus nivolumab plus ipilimumab combination therapy alone (NCT02968303) [10].

Several additional treatment modalities with nivolumab plus ipilimumab combination therapy for melanoma is being examined in three trials. Neoadjuvant therapy with nivolumab and combined nivolumab plus ipilimumab is under evaluation in NCT02519322 and NCT02977052, respectively. A different trial will assess the use of Yttrium-90 selective internal hepatic radiation prior to induction with nivolumab plus ipilimumab (NCT02913417).

The safety and survival statistics from five clinical trials that applied combined immunotherapy with various anti-PD-1/PD-L1 plus anti-CTLA-4 MoAbs has been made available. One trial enrolled 136 patients with unresectable stage III/IV malignant melanoma in eight cohorts (NCT01024231). Each cohort received a unique regimen, differing in sequence of administration and dosing for nivolumab and ipilimumab. The doses for nivolumab were 1, 3, or 10 mg/kg whereas ipilimumab was administered at 3 or 10 mg/kg, per predefined criteria for each cohort. The study documented a considerable variation in OS among different cohorts. In one group, among 14 patients that received nivolumab 0.3 mg/kg plus ipilimumab 3 mg/kg, the 1 year OS rate was 56%, whereas among six patients receiving nivolumab 3 mg/kg plus ipilimumab 3 mg/kg the 1 year OS rate was 100% [54]. Another parameter compared in this study was median OS for concurrent and sequential administration of nivolumab and ipilimumab, which was noted to be 39.7 months and 13.0 months, respectively. The study also recorded objective response stratified on the basis of PD-L1 status. In patients with PD-L1 positive melanoma, an objective response was noted in 6 of 13 participants receiving concurrent therapy and 4 of 8 participants receiving sequential therapy. In the PD-L1 negative subgroup 9 of 22 patients responded with concurrent therapy and 1 of 13 patients receiving sequential therapy responded. Treatment related adverse events were noted in 93% (grade 3–4 adverse events: 53%) patients receiving the concurrent regimen and 73% (grade 3–4 adverse events: 18%) in the sequential therapy group [54].

The outcomes for combined therapy with nivolumab plus ipilimumab versus placebo plus ipilimumab in *BRAF*-wild type and *BRAF*-mutant unresectable/metastatic melanoma were recorded in one trial (CheckMate 069, NCT01927419). Combination therapy appeared to be effective in patients with both *BRAF*-wild type (objective response: 61%, 44 of 72 patients, 95%CI: 49 to 72; complete response: 22%, 16 patients) and *BRAF*-mutant tumors (objective response: 52%, 12 of 23 patients, 95%CI: 31 to 73; complete response: 22%, 5 patients) [53]. The placebo group, however, exhibited poor outcomes in both *BRAF*-wild type (objective response: 11%, 4 of 37 patients, 95%CI: 3 to 25; complete response: 0%) and *BRAF* mutant tumors (objective response: 10%, 1 of 10 patients, 95%CI: 0 to 45; complete response: 0%). A subgroup analysis in this study evaluated the influence of tumor PD-L1 status on clinical outcomes. Patients receiving combination therapy demonstrated no significant difference in objective response rates (PD-L1 positive: 58, 95%CI: 37–78; PD-L1 negative: 55, 95%CI: 41–69) [53].

Another trial assessing survival outcomes for combined therapy with nivolumab plus ipilimumab versus placebo in treatment naïve unresectable/metastatic melanoma reported a median PFS for nivolumab monotherapy as 6.9 months (95%CI: 4.3 to 9.5), ipilimumab monotherapy as 2.9 months (95%CI: 2.8 to 3.4), and combined therapy with ipilimumab plus nivolumab as 11.5 months (95%CI: 8.9 to 16.7) (CheckMate 067, NCT01844505) [55, 56]. The study reported an overall survival at 2 years for nivolumab plus ipilimumab combination as 64%, nivolumab monotherapy as 59% and ipilimumab monotherapy as 45%. At 3 years follow-up, these were 58, 52 and 34%, respectively. Median overall survival for ipilimumab monotherapy was recorded as 19.9 months (95%CI: 16.9–24.9), nivolumab monotherapy as 37.6 months (95%CI: 29.1 - not reached) and as not reached (95%CI 38.2 months - not reached) for combination therapy arm. The objective response with combination therapy was noted in 58% patients, 19% with ipilimumab monotherapy and 44% with nivolumab monotherapy whereas complete response was observed in 19, 5 and 16% patients, respectively [56]. This data from CheckMate 067 allowed for confirmatory approval of combined therapy with nivolumab and ipilimumab for melanoma that initially received an accelerated approval by the FDA based upon the findings of CheckMate 069 trial. In addition to the above, the CheckMate 067 study also presented data on the influence of PD-L1 status on clinical outcomes. Patients with PD-L1 positive melanoma that received either combination therapy or nivolumab monotherapy had a median PFS of 14.0 months. However, study participants with PD-L1 negative melanoma exhibited a median PFS of 5.3 months (95%CI: 2.8 to 7.1) with nivolumab monotherapy and 11.2 months (95%CI: 8.0 - not reached) with combination therapy [55]. Lastly, with regards to the safety, this trial noted treatment related grade 3–4 adverse events in 59% patients receiving nivolumab plus ipilimumab therapy, 21% of those receiving nivolumab only and 28% patients treated with ipilimumab monotherapy [56].

A different trial investigating the effects of variation in the sequence of administering different immunotherapy agents observed that the patient cohort receiving nivolumab before ipilimumab exhibited grade 3–5 treatment related adverse events (TRAE) in 50% patients, response at 25 weeks in 41% and 12 months overall survival in 76% patients (CheckMate

064, NCT01783938) [57]. In the patient cohort receiving ipilimumab prior to nivolumab, 42.8% patients had a grade 3–4 TRAE, 20% patients exhibited response at 25 weeks and 12 months overall survival was noted in 54% patients [57].

Finally, one trial investigated the safety and efficacy of combined therapy with reduced dose ipilimumab (1 mg/kg every 3 weeks for 4 doses) plus full-dose pembrolizumab (2 mg/kg every 3 weeks) in patients with advanced melanoma (KEYNOTE-029, NCT02089685) [58]. This trial documented grade 1–2 and grade 3–4 treatment related adverse events in 51% (78 of 153) and 45% (69 of 153) patients, respectively. Immune-related adverse events were reported in 60% (92 of 153) of melanoma patients, of which 27% (42 of 153) patients experienced one or more grade 3–4 immune-related adverse events. A number of study participants discontinued treatment due to treatment related adverse events. This included 14% (22 of 153) patients discontinuing both ipilimumab and pembrolizumab, 9% (14 of 153) patients stopping pembrolizumab only and 8% (12 of 153) patients that discontinued ipilimumab only. With respect to survival outcomes (median 17.0 months follow-up), the study reported a PFS at 12 months of 69% (95%CI: 60–75), objective response of 61% (93 of 153 patients, 95%CI: 53–69), complete response of 15% (23 of 153 patients), partial response of 46% (70 of 153 patients), stable disease in 18% (28 of 153 patients) and progressive disease in 19% (29 of 153 patients) [58]. These results were considerably better in terms of survival benefit and toxicity profile than combination therapy with nivolumab and ipilimumab for melanoma. Combined checkpoint inhibition with nivolumab 1 mg/kg plus ipilimumab 3 mg/kg evaluated in early phase 1 trials reported an objective response rate of 53% (9 of 17 patients, 95%CI: 28 to 77), complete response of 17% (3 of 17 patients), partial response of 35% (6 of 17 patients) and serious treatment related adverse events in 49% patients (NCT01024231) [54, 59].

Combination Checkpoint Inhibition Trials in NSCLC

Combined immunotherapy with anti-PD-1/PD-L1 and anti-CTLA-4 MoAbs in patients with NSCLC is currently being investigated in 16 ongoing trials (Table 2). Over half (10 of 16 trials) of these trials are evaluating combination therapy with nivolumab and ipilimumab with or without other therapeutic modalities. Other combination checkpoint inhibition regimens that are being examined in patients with NSCLC include tremelimumab plus durvalumab (5 of 16 trials) and pembrolizumab plus ipilimumab (1 of 16 trials).

Presently, five ongoing trials are assessing the combined use of anti-PD-1/PD-L1 and/or anti-CTLA-4 MoAbs with various chemotherapy agents. One of the trials has defined treatment cohorts on the basis of patient tumor marker status (NCT01998126). Patients with *EGFR* mutant NSCLC will receive erlotinib with either nivolumab or ipilimumab and those with *ALK* rearranged NSCLC will be administered crizotinib with either nivolumab or ipilimumab. Another trial will compare PFS, ORR, and duration of response in patients with advanced NSCLC after administration of dasatinib (SRC-family protein-tyrosine kinase inhibitor) plus nivolumab, BMS-986016 (anti- LAG-3 {lymphocyte activation gene-3} MoAb) plus nivolumab and ipilimumab plus nivolumab (NCT02750514) [10]. Participants enrolled in CheckMate 227 will be randomized to receive nivolumab plus platinum doublet chemotherapy (carboplatin/cisplatin plus gemcitabine for squamous NSCLC and carboplatin/cisplatin plus pemetrexed for non-squamous NSCLC) or combined therapy with nivolumab plus ipilimumab (NCT02477826). Similarly, CheckMate 722 will evaluate PFS for T790 M negative, *EGFR* mutant NSCLC patients treated with nivolumab plus platinum doublet chemotherapy (carboplatin/cisplatin plus pemetrexed) and ipilimumab plus nivolumab combination therapy (NCT02864251). The KEYNOTE 021 trial will focus on determining the RP2D for pembrolizumab plus ipilimumab, and it has a planned survival assessment to follow (NCT02039674). This trial will also evaluate the combined use of pembrolizumab with one or more standard chemotherapy agents using pre-defined treatment protocols. These include carboplatin, pemetrexed, paclitaxel, bevacizumab, erlotinib, and gefitinib (NCT02039674).

The development of a treatment regimen that integrates immunotherapy with PD-1/PD-L1 and CTLA-4 checkpoint inhibitors with surgery and radiation has been undertaken in one trial (NCT02696993). This trial will determine the RP2D for four combination therapies in NSCLC patients with brain metastases at the time of enrollment. The treatment regimens specified by the study protocol include nivolumab plus stereotactic radiosurgery, nivolumab plus ipilimumab and whole brain radiation therapy, nivolumab plus whole brain radiation therapy, and nivolumab plus ipilimumab and stereotactic radiosurgery.

At present, data on safety and survival benefit from combined immunotherapy with PD-1/PD-L1 and CTLA-4 checkpoint inhibitors in NSCLC is available from three trials. One trial evaluated four experimental dosing schedules of combined therapy with nivolumab and ipilimumab against one monotherapy arm (nivolumab) in order to identify the regimen that delivers maximum clinical benefit with an acceptable adverse-effects profile (CheckMate 012, NCT01454102). In the patient cohort receiving ipilimumab 1 mg/kg every 6 weeks plus nivolumab 3 mg/kg every 2 weeks an ORR of 39% was observed (95%CI: 23–55), median PFS was 3.9 months (95%CI: 2.6–13.2), the 1 year OS rate was 69% (95% CI: 52–81), and grade 3–4 treatment related adverse events (TRAE) occurred in 33% of patients [60]. In another cohort, designated to receive ipilimumab 1 mg/kg every 12 weeks plus nivolumab 3 mg/kg every 2 weeks, reported grade 3–4 TRAE occurred in 37% of patients, the ORR was 47% (95%CI: 31–64), median PFS was 8.1 months (95%CI: 5.6–13.6), and the 1 year OS rate was not calculated. Alternatively, patients treated with nivolumab 3 mg/kg every 2 weeks were noted to have an ORR of 23% (95%CI: 13–37), median PFS of 3.6 months (95%CI: 2.3–6.6), a 1 year OS rate of 73% (95% CI: 59–83), and TRAE occurred in 19% of patients. The study also attempted to correlate the data for treatment response to tumor PD-L1 expression. In patients treated with nivolumab plus ipilimumab combination therapy, the ORR for tumors with PD-L1 expression ≥1% and≥50% were 57 and 92%, respectively. Conversely, patients receiving nivolumab monotherapy experienced an ORR of 28 and 50% for tumors with PD-L1 expression of≥1% and≥50%, respectively. Based on the overall analysis, treatment with ipilimumab 1 mg/kg every 6 weeks plus nivolumab 3 mg/kg every 2 weeks was selected for further investigation [60]. This regimen is presently under evaluation in the CheckMate 227 trial (NCT02477826).

The KEYNOTE 021 trial evaluated the combined use of pembrolizumab and traditional chemotherapy; it noted an ORR of 52% (13 of 25 patients, 95%CI: 31 to 72) and PFS of 10 months (95%CI: 4 - not reached) in patients treated with carboplatin 6 mg/mL/min plus pembrolizumab 2 or 10 mg/kg and paclitaxel 200 mg/m^2 (NCT02039674) [61]. Likewise, the ORR was 48% (12 of 25 patients, 95%CI: 28 to 69) and PFS was "not reached" (95%CI: 4 - not reached) in patients that received paclitaxel 200 mg/m^2 plus pembrolizumab 2 or 10 mg/kg, carboplatin 6 mg/mL/min and bevacizumab 15 mg/kg. Another cohort of patients that were given pemetrexed 500 mg/m^2 plus carboplatin 5 mg/mL/min and pembrolizumab 2 or 10 mg/kg were noted to have an ORR of 71% (17 of 24 patients, 95%CI: 49 to 87) and PFS of 10 months (95%CI: 6 to 15) [61].

A different trial assessing combination checkpoint inhibition in patients with NSCLC involved the use of various doses of tremelimumab (1, 3 or 10 mg/kg) in combination with durvalumab (3, 10, 15 or 20 mg/kg) during the dose escalation phase (NCT02000947). The objective response (investigator assessed) in the tremelimumab 1 mg/kg treatment cohort was 23% (6 of 26 patients, 95%CI: 9 to 44) [62]. Also, an objective response was noted in 22% (2 of 9 patients, 95%CI: 3 to 60) of cases with PD-L1 positive tumors and 29% (4 of 14 patients, 95%CI: 8 to 58) of cases with PD-L1 negative tumors. MTD was exceeded for treatment with tremelimumab 3 mg/kg plus durvalumab 20 mg/kg, with almost 30% (2 of 6 patients) of patients reporting DLTs [62].

DISCUSSION

There is great enthusiasm surrounding combination immunotherapy with PD-1/PD-L1 and CTLA-4 checkpoint inhibitors. The superior outcomes with combined immunotherapy over single-agent regimens in preclinical studies, together with the approval of nivolumab plus ipilimumab combination therapy for patients with melanoma have shed light on the therapeutic potential of this concept. The possibility of expanding the spectrum of indications for combination checkpoint inhibition to a wide range of tumor histologies is being explored in several trials. Simultaneously, extensive efforts have been undertaken to optimize clinical benefit to adverse effects ratios with combination checkpoint inhibition.

Combined therapy with anti-PD-1/PD-L1 and anti-CTLA-4 MoAbs in advanced melanoma has exhibited better survival outcomes in comparison with single-agent immunotherapy (Fig. 3). CheckMate 067 reported a survival benefit with nivolumab plus ipilimumab combination therapy (median PFS: 11.5 months, objective response 58%) over monotherapy with ipilimumab (median PFS: 2.9 months, objective response 19%) and nivolumab (median PFS: 6.9 months, objective response 44%) [55, 56, 63]. It is, however, important to note that though this study met the co-primary end-point of exhibiting improved overall survival with combination therapy versus ipilimumab, it was underpowered to reflect upon the use of nivolumab and ipilimumab combination therapy over anti-PD-1 monotherapy followed by subsequent ipilimumab rescue. Another trial, CheckMate 069, recorded the hazard ratio for death or progression of disease for combined versus single-agent immunotherapy for *BRAF* wild-type as 0.40 (95% CI: 0.23 to 0.68, $p<0.001$) and *BRAF*-mutant as 0.38 (95% CI: 0.15 to 1.00) melanoma patients, underscoring the survival benefit with combination checkpoint inhibition [53]. Akin to the aforementioned trials, several ongoing trials will compare single-agent immunotherapy with combination checkpoint inhibition with anti-PD-1/PD-L1 plus anti-CTLA-4 checkpoint inhibitors (NCT01940809, NCT02736123, NCT02519322, NCT02731729, NCT02374242, NCT02523313, NCT02599402, NCT02460068, NCT02750514, NCT02154490, NCT02785952, NCT02477826, NCT02352948, NCT02453282, NCT01928394, NCT02537418).

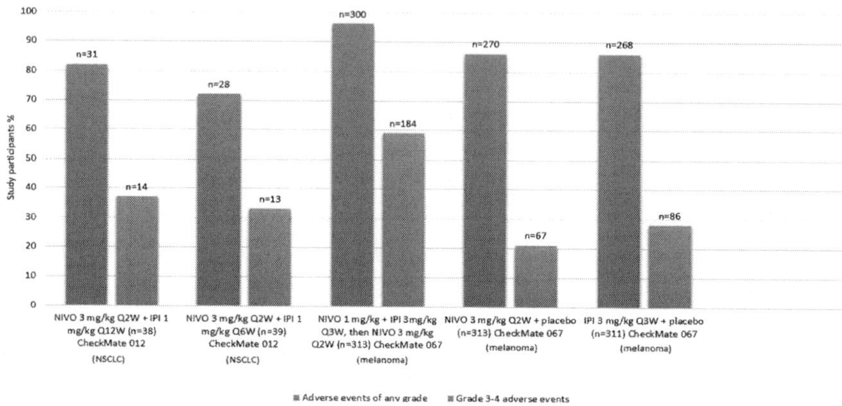

Figre 3: Comparison of objective response in select trials. Legend: Figure presenting a comparison of objective response in study participants, expressed in percent individuals, with different treatment regimens in CheckMate 012, NCT02000947 and CheckMate 067 trials. Abbreviations: NIVO, nivolumab; IPI, ipilimumab; DURVA, durvalumab; TREME, tremelimumab; Q(x)W, every (x) weeks; NSCLC, non-small cell lung cancer.

Several studies have attempted to correlate the expression of PD-L1 and response to combination checkpoint inhibition with anti-PD-1/PD-L1 plus anti-CTLA-4 MoAbs. PD-L1 positivity has classically been defined as visualization of at least 5% of tumor cells with PD-L1 staining in a section containing a minimum of one hundred cells suitable for evaluation [53, 54, 64, 65]. NCT01024231 documented response with combination therapy regardless of tumor PD-L1 status [54]. Likewise, CheckMate 069 reported no significant variation in response to combination checkpoint inhibition based on PD-L1 status [53]. Interestingly, patients with PD-L1 positive melanoma had better objective response rates as compared to those who tested negative for PD-L1 with ipilimumab monotherapy [53]. Another trial, CheckMate 067, noted a considerable advantage in median PFS with combined ipilimumab plus nivolumab (11.2 months) as opposed to nivolumab monotherapy (5.3 months) in patients with PD-L1 negative melanoma. The median PFS for patients with PD-L1 positive tumors, however, was the same for both treatment groups (14 months) [65]. Similarly, a different trial evaluating low-dose ipilimumab plus nivolumab in NSCLC patients reported an ORR of 57% in patients with tumor PD-L1 expression ≥1 and 92% in patients with tumor PD-L1 expression ≥50% (CheckMate 012) [60]. Considering the above, it appears that a pragmatic approach for selecting the most appropriate immunotherapy regimen (monotherapy versus combination therapy) may be through a novel treatment algorithm including tumor PD-L1 status.

The incidence of treatment-related adverse events with combination checkpoint inhibition has been a matter of concern in pivotal trials. The CheckMate 069 trial, which evaluated combination therapy with nivolumab and ipilimumab (nivolumab 1 mg/kg plus ipilimumab 3 mg/kg every 3 weeks for 4 doses followed by monotherapy with nivolumab 3 mg/kg every 2 weeks) versus ipilimumab monotherapy in melanoma patients, documented treatment-associated grade 3–4 adverse events in 54% of patients receiving combination therapy as compared to 24% receiving ipilimumab monotherapy [53]. These results suggest that melanoma patients receiving combination checkpoint inhibition were much more likely to develop severe drug-related adverse events as compared to treatment with ipilimumab monotherapy. Although 68% of patients that discontinued combination therapy due to toxicity exhibited an objective response, concerns over treatment-related toxicity with combination checkpoint inhibition regimens persuade some to favor immune-checkpoint inhibitor monotherapy [53]. In order to validate these findings, the same combination therapy regimen was evaluated in the CheckMate 067 trial and compared ipilimumab monotherapy and nivolumab monotherapy in parallel arms [56]. The frequency of grade 3–4 adverse events in patients treated with combination therapy (59% patients) was higher than that recorded for patients receiving monotherapy with ipilimumab (28% patients) or nivolumab (21% patients) [56]. However, treatment related adverse events with combination therapy were described as manageable and the study concluded that this regimen was suitable for further investigation. Considering the above, it may be stated that one should be cautious in selection of combination immune checkpoint inhibition over monotherapy in elderly patients with high frailty index.

A retrospective pooled analysis conducted to study efficacy and safety of combined therapy with nivolumab and ipilimumab in patients that discontinued therapy due to adverse events presented interesting findings. Data from phase 2 and phase 3 trials for advanced melanoma evaluating nivolumab 1 mg/kg plus ipilimumab 3 mg/kg every 3 weeks followed by monotherapy with nivolumab 3 mg/kg every 2 weeks, was pooled to compare outcomes in participants who discontinued therapy due to adverse events versus those who did not [66]. After a minimum of 18 months follow-up, median PFS and objective response rate for patients that discontinued therapy during induction were found to be 8.4 months and 58.3% respectively. On the other hand, patients that did not discontinue therapy were found to have a median PFS and objective response rate of 10.8 months and 50.2%, respectively. The nearly similar efficacy outcomes in the two groups indicated that despite discontinuing the therapy, patients continued to derive benefits from the treatment [66]. In other words, shorter course of treatment exhibited outcomes comparable to those achieved with full-course combination therapy. Also, it may be argued that immune related adverse events could possibly serve as surrogate markers for patients that may benefit from immunotherapy regimens.

The use of low-dose combination checkpoint inhibition appears to be a promising approach for improving clinical benefit without significantly increasing adverse events. The CheckMate 012 trial assessed several nivolumab plus ipilimumab combination regimens as first-line therapy in NSCLC patients. The frequency of treatment-associated grade 3–4 adverse events with nivolumab 3 mg/kg every 2 weeks plus ipilimumab 1 mg/kg every 12 weeks was 37% and with nivolumab 3 mg/kg every 2 weeks plus ipilimumab 1 mg/kg every 6 weeks was 33% [60]. The frequency of treatment-related grade 3–4 adverse events with nivolumab monotherapy was 19% [60]. Low-dose combination checkpoint inhibition with nivolumab and ipilimumab had a more acceptable adverse-effects profile compared to the combination regimen evaluated in CheckMate 067 (Fig. 4). However, it may be argued that the variation in toxicity profile was secondary to, or at least in part influenced by a difference in tumor histologies. One example may be the FDA approved regimen for melanoma (4 cycles of ipilimumab 3 mg/kg plus nivolumab 1 mg/kg every 3 weeks followed by nivolumab 240 mg every 2 weeks), that exhibited an acceptable safety profile and response in recurrent small cell lung cancer but at the same time demonstrated a poor safety and efficacy profile in NSCLC, leading to abandonment of this strategy in the latter [60, 67]. Data from ongoing studies investigating low-dose combination checkpoint inhibition in a number of tumor histologies will be crucial to the validation of these findings.

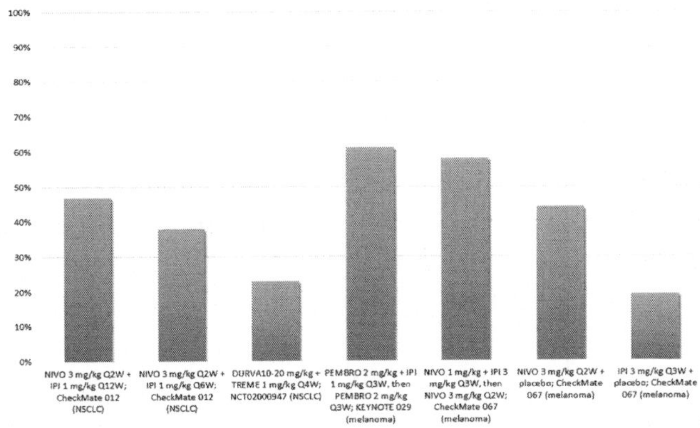

Figure 4: Comparison of treatment related adverse events for different regimens in CheckMate 012 and CheckMate 067. Legend: The figure elaborates percent patients documented to have adverse events with different treatment regimens assessed in the CheckMate 012 and CheckMate 067 trials. Please note that some participants had more than one adverse event. No treatment associated deaths were reported till point of assessment in CheckMate 012 while 2 treatment related deaths in CheckMate 067 nivolumab group were reported >100 days after discontinuation of therapy. Abbreviations: NIVO, nivolumab; IPI, ipilimumab; n, number of patients; X axis: treatment regimens, Y axis: percent patients relative to cohort size.

Another step towards achieving the right balance of efficacy and incidence of adverse events may be to critically assess the use of common terminology criteria for adverse events (CTCAE). CTCAE are used for documenting chemotherapy associated adverse events [68]. They are instrumental to determine the appropriate dose limiting toxicity for the experimental regimen in a trial. This in-turn has a significant bearing on the recommended phase 2 dose for the novel therapeutic agent. CTCAE are, however, now being applied to immune related adverse events for novel immunotherapy regimens [69]. A phase 1 trial investigating nivolumab plus ipilimumab in melanoma patients used an asymptomatic rise in lipase as the primary dose limiting toxicity, central to informing the recommended phase 2 dose in this trial. A retrospective study analyzed the association between asymptomatic rise in lipase and amylase (grade 3 and above) with pancreatitis in 119 participants and found only 2 patients to have pancreatitis. This represented 6.3% of all patients with grade 3 and above lipase and 20% of those with grade 3 or above increase in both amylase and lipase. Thus, in simple terms, lipase did not appear to be a relevant marker for pancreatitis.

This observation thus signifies the need to exercise appropriate caution when grading independent lab values using CTCAE in immunotherapy trials [69].

Existing evidence suggests that single-agent immunotherapy for tumors with high PD-L1 expression (≥50% cells positive for PD-L1 staining) can achieve far superior outcomes than chemotherapy in similar settings. In patients with NSCLC with high PD-L1 expression, pembrolizumab exhibited a response rate of 45%, PFS of 10.3 months, and a 1-year survival rate of 70% [70]. In comparison, treatment with standard of care chemotherapy had a response rate of 28%, PFS of 6 months, and 1-year survival rate of 54% (KEYNOTE 026) [70]. On the other hand, trials involving a lower cut-off value for PD-L1 positivity (≥5% cells positive for PD-L1 staining) failed to demonstrate any advantage in clinical efficacy with single-agent immunotherapy over standard chemotherapy (CheckMate 026) [71, 72]. Reflecting on the data on efficacy and overall toxicity profile of single-agent immunotherapy regimens in tumors with high PD-L1 expression, outperforming these regimens may prove to be a challenge for combination checkpoint inhibition regimens. Further data from ongoing trials will be vital to conclusively determine if single-agent immunotherapy with pembrolizumab can be replaced with a combined immunotherapy regimen.

Immunotherapy with PD-1/PD-L1 plus CTLA-4 checkpoint inhibitors for a diverse set of solid tumors is currently being investigated in 8 trials (Table 3). Half of all phase 1/phase 2 solid tumor trials are evaluating combined therapy with nivolumab and ipilimumab. Others include three trials with combined therapy with tremelimumab plus durvalumab and one trial with atezolizumab plus ipilimumab. Each of these eight trials will evaluate combination checkpoint inhibition regimens in a large number of malignancies. The data gathered from these studies will be crucial to identifying tumor histologies that would benefit most from combination checkpoint inhibition.

Table 3: Phase 1 solid tumor trials investigating combined immunotherapy with anti-CTLA-4 plus anti- PD-1/PD-L1 monoclonal antibodies

Cancer type	Phase	Primary outcome	Dosing regimen	Enrollment number	Status	Results	Reference/ Clinical trials identification number
HIV associated unresectable metastatic solid tumors	Phase 1	MTD of NIVO (time frame: 56 days)	NIVO on D1; study participants in dose level 2 receive IPI on 1st day of every 3rd course of NIVO while those in dose level − 2 receive IPI on 1st day of every 6th course of NIVO; treatment repeated every 14 days for 46 cycles of NIVO	42	Recruiting	NA	NCT02408861
Locally advanced/ metastatic solid tumors	Phase 1	Incidence of TRAE (evaluated up to 30 days after completion of therapy) and the incidence of DLT (assessed for 21 days from initiation of treatment); secondary outcomes: OS, PFS, duration of response, OR and best overall response assessed for 3 years	Arm A: atezolizumab + IPI Q3W for 4 cycles; arm B: Interferon alfa-2b (3 doses/ week) + atezolizumab Q3W	200	Recruiting	NA	NCT02174172
Advanced incurable solid malignancies	Phase 1b	Determine RP2D tremelimumab with/ without MEDI4736 in patients on treatment with standard of care chemotherapy (assessed up to 2 years)	Tremelimumab (D1 of cycles 1, 3 and 5, or, D1 of cycle 1) with/without MEDI4736 (Q3W)	150	Recruiting	NA	NCT02537418

Advanced solid tumors/ relapsed metastatic SCCHN	Phase 1/ Phase 2	Determine the MTD and RP2D; assessment of safety and efficacy of specified treatment regimen; evaluation of ORR for up to 12 months	Experimental part A1: MEDI4736+AZD9150; A2: MEDI4736+AZD5069; B1: (patients pre-treated with PD-L1 inhibitor) MEDI4736+AZD9150; B2: (patients pre-treated with PD-L1 inhibitor) MEDI4736+AZD5069; B3: (treatment naïve patients) MEDI4736+AZD9150; B4: (treatment naïve patients) MEDI4736+AZD5069; B5: AZD9150 until progression, followed by MEDI4736; B6: AZD5069 until progression, followed by MEDI4736; A3: MEDI4736+AZD5069; A4: MEDI4736+AZD9150+tremelimumab; A5: MEDI4736+AZD5069+tremelimumab; A6: MEDI4736+AZD9150; A7: MEDI4736+AZD5069	147	Recruiting	NA	NCT02499328
Advanced solid tumors	Phase 1	Determine the number of patients with DLT, AE and serious AE; secondary outcomes: OS and ORR, assessed up to 2 years or until death	Arm 1: MEDI4736 Q2W; arm 2: MEDI4736 Q3W; arm 3 (dose expansion): MEDI4736 Q2W; arm 4: MEDI4736 Q4W; arm 5: MEDI4736+tremelimumab Q4W	264	Recruiting	NA	NCT01938612
Solid tumors	Phase 1/ Phase 2	ORR assessed up to 10 years; secondary outcomes: clinical benefit rate assessed for 6 months, PFS and OS evaluated for up to 10 years	IPI on D1+NIVO on D1, 15 and 29, course to be repeated every 42 days until unacceptable treatment related toxicity or progression of disease	334	Recruiting	NA	NCT02834013 (S1609 trial/DART trial)
Refractory/recurrent solid tumors	Phase 1/ Phase 2	RR with IPI+NIVO combination therapy, RR with NIVO, MTD of NIVO, phase 2 dose of IPI+NIVO	IPI+NIVO	352	Recruiting	NA	NCT02304458

Metastatic/ advanced solid tumors	Phase 1/ Phase 2 RCT	OR rate; secondary outcomes: PFS, OS (time frame: 5 years)	Arm N: NIVO 3 mg/kg Q2W; arm N-I level 1: 4 doses of NIVO 1 mg/ kg+IPI 1 mg/kg Q3W, later continued on monotherapy with NIVO 3 mg/kg Q2W; arm N-I level 2: 4 doses of IPI 3 mg/kg+NIVO 1 mg/kg Q3W, later on monotherapy with NIVO 3 mg/kg Q2W; arm N-I level 2b: 4 doses of IPI 1 mg/ kg+NIVO 3 mg/kg Q3W, later on monotherapy with NIVO 3 mg/kg Q2W; arm N-I level 2c: IPI 1 mg/ kg Q6W+NIVO 3 mg/kg Q3W; arm N-I level 2d: cobimetinib 60 mg/day for 21 days followed by 7 days off + IPI 1 mg/kg Q6W and NIVO 3 mg/kg Q3W	1150	Recruiting, results available for recurrent small cell lung cancer	Recurrent small cell lung cancer: ORR, disease control rate, SD, PR and progressive disease with NIVO (n=40) noted in 18, 38, 20, 18 and 53% patients who received the drug, respectively; NIVO + IPI combination (n=50) exhibited ORR of 17%, SD in 37%, PR in 15%, disease control rate of 54% and progressive disease in 37% patients	NCT01928394 [80]

Abbreviations: MTD maximum tolerable dose, *DLT* dose limiting toxicity, *SCCHN* squamous cell carcinoma of the head and neck, *RCT* randomized controlled trial, *NA* not available, *CI* confidence interval, *ORR* overall response rate, *OR* objective response, *OS* overall survival, *PFS* progression free survival, *RR* response rate, *RP2D* recommended phase 2 dose, *TRAE* treatment related adverse events, *NIVO* nivolumab, *IPI* ipilimumab, *PEMBRO* pembrolizumab, *MEDI4736* durvalumab, *OD* once daily dosing, *BID* twice daily, *QID* four times a day; Q(x)W, every (x) weeks, *SD* stable disease, *PR* partial response, D(x), day(x), *AE* adverse events

Recently, the FDA approved the use of pembrolizumab for unresectable/ metastatic mismatch repair deficient (dMMR) or microsatellite instability-high (MSI-H) solid tumors and colorectal cancer with progression on prior therapy. This was based upon data from KEYNOTE 012, KEYNOTE 028, KEYNOTE 164, KEYNOTE 016 and KEYNOTE 158 [73–76]. The advised regimen is pembrolizumab 10 mg/kg every 2 weeks or 200 mg every 3 weeks for up to 24 months, unacceptable toxicity or progression of disease. It is of note that this is the first time when a drug has been approved not on the basis of tumor location but a tumor biomarker.

The Dual Anti-CTLA-4 and Anti-PD-1 Blockade in Rare Tumors (DART) trial will evaluate response to combination checkpoint inhibition for a large number of rare tumors in a basket fashion (NCT02834013). In this trial, minimizing toxicity profiles without compromising clinical efficacy is the primary goal. We nominated the low-dose combination therapy with fixed-dose nivolumab and wider interval ipilimumab (nivolumab 240 mg every 2 weeks plus ipilimumab 1 mg/kg every 6 weeks) for assessment in this trial. Our treatment regimen is based on the superior toxicity profile for this regimen observed in CheckMate 012 as compared to the FDA approved combination therapy regimen (nivolumab 1 mg/kg plus ipilimumab 3 mg/kg every 3 weeks for 4 doses followed by monotherapy with nivolumab 3 mg/kg every 2 weeks) in CheckMate 069 [53, 60]. Based on the fact that no significant difference in response was recorded between PD-L1 positive and negative patients receiving combination therapy, we decided to recruit study participants irrespective of tumor PD-L1 status. Through this trial, we expect to provide critical data for expanding the application of low-dose combination therapy in rare tumors.

Recent studies have suggested that sequential administration of immune-checkpoint inhibitors targeting various pathways may benefit cancer patients exhibiting treatment resistance. A multi-center retrospective study evaluated outcomes with ipilimumab and combination therapy with nivolumab and ipilimumab in advanced melanoma patients that previously failed treatment with anti-PD-1 MoAbs [77]. Patients receiving ipilimumab monotherapy were observed to have superior disease control as compared to those receiving combination checkpoint inhibition (42% versus 33%) [77]. Of note, though this was a retrospective study, it was insufficiently powered to detect the difference. A different retrospective analysis of 10 melanoma patients that received ipilimumab after progression on anti-PD-1 therapy found that 1 of 10 patients exhibited partial remission and an additional four patients had stable disease [78]. Similar to the findings of the above and other retrospective studies, the CheckMate 064 documented a higher treatment efficacy in melanoma patients receiving nivolumab prior to ipilimumab versus those receiving ipilimumab prior to nivolumab [57]. However, superior clinical efficacy with the former regimen was associated with an inferior toxicity profile to the latter [57]. Of note, the trial involved a planned switch to reverse sequence at 12 weeks and not at progression. Therefore, it remains unclear if switching to combinatorial regimen at the time of progression is a feasible approach. The outcomes from NCT02731729 trial should perhaps be able to provide some direction on this matter.

CONCLUSION

Combination immunotherapy is evolving at a phenomenal pace. In light of initial success in patients with melanoma, efforts to explore the indications for combination checkpoint inhibition with anti-PD-1/PD-L1 and anti-CTLA-4 MoAbs have diversified to a large number of tumor histologies. Several treatment strategies intended for achieving better clinical efficacy and to overcome challenges such as treatment resistance and toxicity associated with the use of immunotherapy agents, are presently under investigation. Of note, the use of low-dose combination checkpoint inhibition with nivolumab and ipilimumab in NSCLC appears to be a promising approach. Alternatively, the use of nivolumab prior to ipilimumab in the induction phase for melanoma patients may be a simple but effective strategy to achieve superior outcomes. Data from ongoing trials is expected to provide vital evidence for validation of the above preliminary findings and facilitate the application of combination checkpoint inhibition on a larger scale.

AUTHORS' CONTRIBUTIONS

YKC and AA gathered, analyzed and interpreted the data from various published articles. All authors read and approved the final manuscript.

REFERENCES

1. Boyiadzis M, Foon KA. Approved monoclonal antibodies for cancer therapy. Expert Opin Biol Ther. 2008;8(8):1151–1158. doi: 10.1517/14712598.8.8.1151.
2. Scott AM, Wolchok JD, Old LJ. Antibody therapy of cancer. Nat Rev Cancer. 2012;12(4):278–287. doi: 10.1038/nrc3236.
3. Chames P, Van Regenmortel M, Weiss E, Baty D. Therapeutic antibodies: successes, limitations and hopes for the future. Br J Pharmacol. 2009;157(2):220–233. doi: 10.1111/j.1476-5381.2009.00190.x.
4. Reslan L, Dalle S, Dumontet C. Understanding and circumventing resistance to anticancer monoclonal antibodies. Taylor & Francis. MAbs. 2009;1(3):222-29. 10.4161/mabs.1.3.8292.
5. Pillay V, Gan HK, Scott AM. Antibodies in oncology. New Biotechnol. 2011;28(5):518–529. doi: 10.1016/j.nbt.2011.03.021.
6. McKay RR, Choueiri TK: Emerging role of combination immunotherapy regimens in metastatic renal cell carcinoma. In: ASCO Annual Meeting: 2016.
7. Curran MA, Montalvo W, Yagita H, Allison JP. PD-1 and CTLA-4 combination blockade expands infiltrating T cells and reduces regulatory T and myeloid cells within B16 melanoma tumors. Proc Natl Acad Sci. 2010;107(9):4275–4280. doi: 10.1073/pnas.0915174107.
8. Chen L, Han X. Anti–PD-1/PD-L1 therapy of human cancer: past, present, and future. J Clin Invest. 2015;125(9):3384–3391. doi: 10.1172/JCI80011.
9. DiGiulio S. FDA approves Opdivo-Yervoy combo for melanoma, first combo immunotherapy regimen for Cancer. Oncology Times. 2015. https://journals.lww.com/oncology-times/blog/fdaactionsandupdates/pages/post.aspx?PostID=116.
10. Dictionary ND. National Cancer Institute at the National Institutes of Health. https://www.cancer.gov/publications/dictionaries.
11. Buchbinder EI, Desai A. CTLA-4 and PD-1 pathways: similarities, differences, and implications of their inhibition. Am J Clin Oncol. 2016;39(1):98–106. doi: 10.1097/COC.0000000000000239.
12. Collins AV, Brodie DW, Gilbert RJ, Iaboni A, Manso-Sancho R, Walse B, Stuart DI, van der Merwe PA, Davis SJ. The interaction properties of costimulatory molecules revisited. Immunity. 2002;17(2):201–210. doi: 10.1016/S1074-7613(02)00362-X.

13. Chambers CA, Kuhns MS, Egen JG, Allison JP. CTLA-4-mediated inhibition in regulation of T cell responses: mechanisms and manipulation in tumor immunotherapy. Annu Rev Immunol. 2001;19(1):565–594. doi: 10.1146/annurev.immunol.19.1.565.
14. Keir ME, Butte MJ, Freeman GJ, Sharpe AH. PD-1 and its ligands in tolerance and immunity. Annu Rev Immunol. 2008;26:677–704. doi: 10.1146/annurev.immunol.26.021607.090331.
15. Egen JG, Kuhns MS, Allison JP. CTLA-4: new insights into its biological function and use in tumor immunotherapy. Nat Immunol. 2002;3(7):611–618. doi: 10.1038/ni0702-611.
16. Parry RV, Chemnitz JM, Frauwirth KA, Lanfranco AR, Braunstein I, Kobayashi SV, Linsley PS, Thompson CB, Riley JL. CTLA-4 and PD-1 receptors inhibit T-cell activation by distinct mechanisms. Mol Cell Biol. 2005;25(21):9543–9553. doi: 10.1128/MCB.25.21.9543-9553.2005.
17. Masteller EL, Chuang E, Mullen AC, Reiner SL, Thompson CB. Structural analysis of CTLA-4 function in vivo. J Immunol. 2000;164(10):5319–5327. doi: 10.4049/jimmunol.164.10.5319.
18. Krummel MF, Allison JP. CD28 and CTLA-4 have opposing effects on the response of T cells to stimulation. J Exp Med. 1995;182(2):459–465. doi: 10.1084/jem.182.2.459.
19. Linsley PS, Bradshaw J, Greene J, Peach R, Bennett KL, Mittler RS. Intracellular trafficking of CTLA-4 and focal localization towards sites of TCR engagement. Immunity. 1996;4(6):535–543. doi: 10.1016/S1074-7613(00)80480-X.
20. Takahashi T, Tagami T, Yamazaki S, Uede T, Shimizu J, Sakaguchi N, Mak TW, Sakaguchi S. Immunologic self-tolerance maintained by CD25+ CD4+ regulatory T cells constitutively expressing cytotoxic T lymphocyte–associated antigen 4. J Exp Med. 2000;192(2):303–310. doi: 10.1084/jem.192.2.303.
21. Ishida Y, Agata Y, Shibahara K, Honjo T. Induced expression of PD-1, a novel member of the immunoglobulin gene superfamily, upon programmed cell death. EMBO J. 1992;11(11):3887.
22. Chen L. Co-inhibitory molecules of the B7–CD28 family in the control of T-cell immunity. Nat Rev Immunol. 2004;4(5):336–347. doi: 10.1038/nri1349.
23. Taube JM, Klein A, Brahmer JR, Xu H, Pan X, Kim JH, Chen L, Pardoll

DM, Topalian SL, Anders RA. Association of PD-1, PD-1 ligands, and other features of the tumor immune microenvironment with response to anti–PD-1 therapy. Clin Cancer Res. 2014;20(19):5064–5074. doi: 10.1158/1078-0432.CCR-13-3271.

24. Keir ME, Liang SC, Guleria I, Latchman YE, Qipo A, Albacker LA, Koulmanda M, Freeman GJ, Sayegh MH, Sharpe AH. Tissue expression of PD-L1 mediates peripheral T cell tolerance. J Exp Med. 2006;203(4):883–895. doi: 10.1084/jem.20051776.
25. Tseng S-Y, Otsuji M, Gorski K, Huang X, Slansky JE, Pai SI, Shalabi A, Shin T, Pardoll DM, Tsuchiya H. B7-DC, a new dendritic cell molecule with potent costimulatory properties for T cells. J Exp Med. 2001;193(7):839–846. doi: 10.1084/jem.193.7.839.
26. Zitvogel L, Kroemer G. Targeting PD-1/PD-L1 interactions for cancer immunotherapy. Oncoimmunol. 2012;1(8):1223–1225. doi: 10.4161/onci.21335.
27. Dong H, Strome SE, Salomao DR, Tamura H, Hirano F, Flies DB, Roche PC, Lu J, Zhu G, Tamada K. Tumor-associated B7-H1 promotes T-cell apoptosis: a potential mechanism of immune evasion. Nat Med. 2002;8(8):793–800. doi: 10.1038/nm730.
28. Iwai Y, Ishida M, Tanaka Y, Okazaki T, Honjo T, Minato N. Involvement of PD-L1 on tumor cells in the escape from host immune system and tumor immunotherapy by PD-L1 blockade. Proc Natl Acad Sci. 2002;99(19):12293–12297. doi: 10.1073/pnas.192461099.
29. Tsushima F, Yao S, Shin T, Flies A, Flies S, Xu H, Tamada K, Pardoll DM, Chen L. Interaction between B7-H1 and PD-1 determines initiation and reversal of T-cell anergy. Blood. 2007;110(1):180–185. doi: 10.1182/blood-2006-11-060087.
30. Duraiswamy J, Kaluza KM, Freeman GJ, Coukos G. Dual blockade of PD-1 and CTLA-4 combined with tumor vaccine effectively restores T-cell rejection function in tumors. Cancer Res. 2013;73(12):3591–3603. doi: 10.1158/0008-5472.CAN-12-4100.
31. Reiman JM, Kmieciak M, Manjili MH, Knutson KL. Tumor immunoediting and immunosculpting pathways to cancer progression. In: Seminars in Cancer Biology. Elsevier; 2007;17(4):275-87. https://www.sciencedirect.com/science/article/pii/S1044579X0700034X.
32. Dunn GP, Bruce AT, Ikeda H, Old LJ, Schreiber RD. Cancer immunoediting: from immunosurveillance to tumor escape. Nat

Immunol. 2002;3(11):991–998. doi: 10.1038/ni1102-991.

33. Houghton AN, Guevara-Patiño JA. Immune recognition of self in immunity against cancer. J Clin Invest. 2004;114(4):468–471. doi: 10.1172/JCI22685.
34. Schreiber RD, Old LJ, Smyth MJ. Cancer immunoediting: integrating immunity's roles in cancer suppression and promotion. Science. 2011;331(6024):1565–1570. doi: 10.1126/science.1203486.
35. Kim R, Emi M, Tanabe K. Cancer immunoediting from immune surveillance to immune escape. Immunology. 2007;121(1):1–14. doi: 10.1111/j.1365-2567.2007.02587.x.
36. Beatty GL, Gladney WL. Immune escape mechanisms as a guide for cancer immunotherapy. Clin Cancer Res. 2015;21(4):687–692. doi: 10.1158/1078-0432.CCR-14-1860.
37. Drake C. Combination immunotherapy approaches. Ann Oncol. 2012;23(suppl 8):viii41–viii46. doi: 10.1093/annonc/mds262.
38. Poschke I, Mougiakakos D, Kiessling R. Camouflage and sabotage: tumor escape from the immune system. Cancer Immunol Immunother. 2011;60(8):1161–1171. doi: 10.1007/s00262-011-1012-8.
39. Swann JB, Smyth MJ. Immune surveillance of tumors. J Clin Invest. 2007;117(5):1137–1146. doi: 10.1172/JCI31405.
40. Galluzzi L, Vacchelli E, Pedro B-S, Buqué A, Senovilla L, Baracco EE, Bloy N, Castoldi F, Abastado J-P, Agostinis P. Classification of current anticancer immunotherapies. Oncotarget. 2015;5:12472–12508.
41. Ferris RL, Blumenschein G, Jr, Fayette J, Guigay J, Colevas AD, Licitra L, Harrington K, Kasper S, Vokes EE, Even C. Nivolumab for recurrent squamous-cell carcinoma of the head and neck. N Engl J Med. 2016;2016(375):1856–1867. doi: 10.1056/NEJMoa1602252.
42. Motzer RJ, Rini BI, McDermott DF, Redman BG, Kuzel TM, Harrison MR, Vaishampayan UN, Drabkin HA, George S, Logan TF. Nivolumab for metastatic renal cell carcinoma: results of a randomized phase II trial. J Clin Oncol. 2015;33(13):1430–1437. doi: 10.1200/JCO.2014.59.0703.
43. Borghaei H, Paz-Ares L, Horn L, Spigel DR, Steins M, Ready NE, Chow LQ, Vokes EE, Felip E, Holgado E. Nivolumab versus docetaxel in advanced nonsquamous non–small-cell lung cancer. N Engl J Med. 2015;373(17):1627–1639. doi: 10.1056/NEJMoa1507643.
44. Brahmer J, Reckamp KL, Baas P, Crinò L, Eberhardt WE, Poddubskaya

E, Antonia S, Pluzanski A, Vokes EE, Holgado E. Nivolumab versus docetaxel in advanced squamous-cell non–small-cell lung cancer. N Engl J Med. 2015;373(2):123–135. doi: 10.1056/NEJMoa1504627.

45. Weber JS, D'Angelo SP, Minor D, Hodi FS, Gutzmer R, Neyns B, Hoeller C, Khushalani NI, Miller WH, Lao CD. Nivolumab versus chemotherapy in patients with advanced melanoma who progressed after anti-CTLA-4 treatment (CheckMate 037): a randomised, controlled, open-label, phase 3 trial. Lancet Oncol. 2015;16(4):375–384. doi: 10.1016/S1470-2045(15)70076-8.

46. Ansell SM, Lesokhin AM, Borrello I, Halwani A, Scott EC, Gutierrez M, Schuster SJ, Millenson MM, Cattry D, Freeman GJ. PD-1 blockade with nivolumab in relapsed or refractory Hodgkin's lymphoma. N Engl J Med. 2015;372(4):311–319. doi: 10.1056/NEJMoa1411087.

47. Garon EB, Rizvi NA, Hui R, Leighl N, Balmanoukian AS, Eder JP, Patnaik A, Aggarwal C, Gubens M, Horn L, et al. Pembrolizumab for the treatment of non–small-cell lung Cancer. N Engl J Med. 2015;372(21):2018–2028. doi: 10.1056/NEJMoa1501824.

48. Robert C, Schachter J, Long GV, Arance A, Grob JJ, Mortier L, Daud A, Carlino MS, McNeil C, Lotem M, et al. Pembrolizumab versus Ipilimumab in Advanced Melanoma. N Engl J Med. 2015;372(26):2521–2532. doi: 10.1056/NEJMoa1503093.

49. Seiwert TY, Haddad RI, Gupta S, Mehra R, Tahara M, Berger R, Lee S-H, Burtness B, Le DT, Heath K. ASCO annual meeting proceedings. 2015. Antitumor activity and safety of pembrolizumab in patients (pts) with advanced squamous cell carcinoma of the head and neck (SCCHN): preliminary results from KEYNOTE-012 expansion cohort; p. LBA6008.

50. Rosenberg JE, Hoffman-Censits J, Powles T, van der Heijden MS, Balar AV, Necchi A, Dawson N, O'Donnell PH, Balmanoukian A, Loriot Y. Atezolizumab in patients with locally advanced and metastatic urothelial carcinoma who have progressed following treatment with platinum-based chemotherapy: a single-arm, multicentre, phase 2 trial. Lancet. 2016;387(10031):1909–1920. doi: 10.1016/S0140-6736(16)00561-4.

51. Fehrenbacher L, Spira A, Ballinger M, Kowanetz M, Vansteenkiste J, Mazieres J, Park K, Smith D, Artal-Cortes A, Lewanski C. Atezolizumab versus docetaxel for patients with previously treated non-small-cell lung cancer (POPLAR): a multicentre, open-label, phase 2 randomised

controlled trial. Lancet. 2016;387(10030):1837–1846. doi: 10.1016/S0140-6736(16)00587-0.

52. Rittmeyer A, Barlesi F, Waterkamp D, Park K, Ciardiello F, von Pawel J, Gadgeel SM, Hida T, Kowalski DM, Dols MC. Atezolizumab versus docetaxel in patients with previously treated non-small-cell lung cancer (OAK): a phase 3, open-label, multicentre randomised controlled trial. Lancet. 2017;389(10066):255-65.
53. Postow MA, Chesney J, Pavlick AC, Robert C, Grossmann K, McDermott D, Linette GP, Meyer N, Giguere JK, Agarwala SS, et al. Nivolumab and Ipilimumab versus Ipilimumab in Untreated Melanoma. New England J Med. 2015;372(21):2006-17.
54. Wolchok JD, Kluger H, Callahan MK, Postow MA, Rizvi NA, Lesokhin AM, Segal NH, Ariyan CE, Gordon R-A, Reed K. Nivolumab plus ipilimumab in advanced melanoma. N Engl J Med. 2013;369(2):122–133. doi: 10.1056/NEJMoa1302369.
55. Larkin J, Chiarion-Sileni V, Gonzalez R, Grob JJ, Cowey CL, Lao CD, Schadendorf D, Dummer R, Smylie M, Rutkowski P. Combined nivolumab and ipilimumab or monotherapy in untreated melanoma. N Engl J Med. 2015;2015(373):23–34. doi: 10.1056/NEJMoa1504030.
56. Wolchok JD, Chiarion-Sileni V, Gonzalez R, Rutkowski P, Grob J-J, Cowey CL, Lao CD, Wagstaff J, Schadendorf D, Ferrucci PF. Overall survival with combined nivolumab and ipilimumab in advanced melanoma. N Engl J Med. 2017;377(14):1345–1356. doi: 10.1056/NEJMoa1709684.
57. Weber JS, Gibney G, Sullivan RJ, Sosman JA, Slingluff Jr CL, Lawrence DP, Logan TF, Schuchter LM, Nair S, Fecher L. Sequential administration of nivolumab and ipilimumab with a planned switch in patients with advanced melanoma (CheckMate 064): an open-label, randomised, phase 2 trial. Lancet Oncol. 2016;17(7):943–955. doi: 10.1016/S1470-2045(16)30126-7.
58. Long GV, Atkinson V, Cebon JS, Jameson MB, Fitzharris BM, McNeil CM, Hill AG, Ribas A, Atkins MB, Thompson JA. Standard-dose pembrolizumab in combination with reduced-dose ipilimumab for patients with advanced melanoma (KEYNOTE-029): an open-label, phase 1b trial. Lancet Oncol. 2017;18(9):1202–1210. doi: 10.1016/S1470-2045(17)30428-X.
59. Atkins MB, Choueiri TK, Hodi FS, Thompson JA, Hwu W-J, McDermott DF, Brookes M, Tosolini A, Ebbinghaus S, Yang Z.

Pembrolizumab (MK-3475) plus low-dose ipilimumab (IPI) in patients (pts) with advanced melanoma (MEL) or renal cell carcinoma (RCC): data from the KEYNOTE-029 phase 1 study. J Clin Oncol. 2015;33(15_suppl):3009.

60. Hellmann M, Gettinger S, Goldman J, Brahmer J, Borghaei H, Chow L, Ready N, Gerber D, Juergens R, Shepherd F. CheckMate 012: safety and efficacy of first-line (1L) nivolumab (nivo; N) and ipilimumab (ipi; I) in advanced (adv) NSCLC. J Clin Oncol. 2016;34(Suppl):abstr 3001.

61. Gadgeel S, Stevenson J, Langer C, Gandhi L, Borghaei H, Patnaik A. Pembrolizumab (pembro) plus chemotherapy as front-line therapy for advanced NSCLC: KEYNOTE-021 cohorts AC. Proc Am Soc Clin Oncol. 2016;

62. Antonia S, Goldberg SB, Balmanoukian A, Chaft JE, Sanborn RE, Gupta A, Narwal R, Steele K, Gu Y, Karakunnel JJ. Safety and antitumour activity of durvalumab plus tremelimumab in non-small cell lung cancer: a multicentre, phase 1b study. Lancet Oncol. 2016;17(3):299–308. doi: 10.1016/S1470-2045(15)00544-6.

63. Doyle C. Combination immunotherapy superior to monotherapy in patients with melanoma. Am Health Drug Benefits. 2015;8(Spec Issue):41.

64. Topalian SL, Hodi FS, Brahmer JR, Gettinger SN, Smith DC, McDermott DF, Powderly JD, Carvajal RD, Sosman JA, Atkins MB. Safety, activity, and immune correlates of anti–PD-1 antibody in cancer. N Engl J Med. 2012;366(26):2443–2454. doi: 10.1056/NEJMoa1200690.

65. Larkin J, Chiarion-Sileni V, Gonzalez R, Grob JJ, Cowey CL, Lao CD, Schadendorf D, Dummer R, Smylie M, Rutkowski P, et al. Combined Nivolumab and Ipilimumab or monotherapy in untreated melanoma. N Engl J Med. 2015;373(1):23–34. doi: 10.1056/NEJMoa1504030.

66. Schadendorf D, Wolchok JD, Hodi FS, Chiarion-Sileni V, Gonzalez R, Rutkowski P, Grob J-J, Cowey CL, Lao CD, Chesney J. Efficacy and safety outcomes in patients with advanced melanoma who discontinued treatment with nivolumab and ipilimumab because of adverse events: a pooled analysis of randomized phase II and III trials. J Clin Oncol. 2017;35(34):3807–3814. doi: 10.1200/JCO.2017.73.2289.

67. Antonia SJ, López-Martin JA, Bendell J, Ott PA, Taylor M, Eder JP, Jäger D, Pietanza MC, Le DT, de Braud F. Nivolumab alone and nivolumab plus ipilimumab in recurrent small-cell lung cancer

(CheckMate 032): a multicentre, open-label, phase 1/2 trial. Lancet Oncol. 2016;17(7):883–895. doi: 10.1016/S1470-2045(16)30098-5.

68. Basch E, Reeve BB, Mitchell SA, Clauser SB, Minasian LM, Dueck AC, Mendoza TR, Hay J, Atkinson TM, Abernethy AP. Development of the National Cancer Institute's patient-reported outcomes version of the common terminology criteria for adverse events (PRO-CTCAE) J Natl Cancer Inst. 2014;106(9):dju244. doi: 10.1093/jnci/dju244.

69. Friedman CF, Clark V, Raikhel AV, Barz T, Shoushtari AN, Momtaz P, Callahan MK, Wolchok JD, Chapman PB, Hellmann MD. Thinking critically about classifying adverse events: incidence of pancreatitis in patients treated with nivolumab+ ipilimumab. JNCI. 2017;109(4):djw260.

70. Reck M, Rodríguez-Abreu D, Robinson AG, Hui R, Csőszi T, Fülöp A, Gottfried M, Peled N, Tafreshi A, Cuffe S. Pembrolizumab versus chemotherapy for PD-L1–positive non–small-cell lung cancer. N Engl J Med. 2016;375(19):1823–1833. doi: 10.1056/NEJMoa1606774.

71. Top-Line B-MSA: Results from CheckMate− 026, a Phase 3 Study of Opdivo (nivolumab) in Treatment-Naïve Patients with Advanced Non-Small Cell Lung Cancer. *Dostupné na:*<. 2016. p. 8.https://news.bms.com/press-release/bristolmyers/bristol-myers-squibb-announces-top-line-resultscheckmate-026-phase-3-stu.

72. Socinski M, Creelan B, Horn L, Reck M, Paz-Ares L, Steins M, Felip E, van den Heuvel M, Ciuleanu T, Badin F. NSCLC, metastaticCheckMate 026: a phase 3 trial of nivolumab vs investigator's choice (IC) of platinum-based doublet chemotherapy (PT-DC) as first-line therapy for stage iv/recurrent programmed death ligand 1 (PD-L1)− positive NSCLC. Ann Oncol. 2016;27(suppl 6):LBA7_PR.

73. Muro K, Chung HC, Shankaran V, Geva R, Catenacci D, Gupta S, Eder JP, Golan T, Le DT, Burtness B. Pembrolizumab for patients with PD-L1-positive advanced gastric cancer (KEYNOTE-012): a multicentre, open-label, phase 1b trial. Lancet Oncol. 2016;17(6):717–726. doi: 10.1016/S1470-2045(16)00175-3.

74. Le DT, Yoshino T, Jäger D, Andre T, Bendell JC, Wang R, Kang SP, Koshiji M, Diaz LA. KEYNOTE-164: phase II study of pembrolizumab (MK-3475) for patients with previously treated, microsatellite instability-high advanced colorectal carcinoma. Proc Am Soc Clin Oncol. 2016.

75. O'Neil BH, Wallmark JM, Lorente D, Elez E, Raimbourg J, Gomez-Roca

C, Ejadi S, Piha-Paul SA, Stein MN, Razak ARA. Safety and antitumor activity of the anti–PD-1 antibody pembrolizumab in patients with advanced colorectal carcinoma. PLoS One. 2017;12(12):e0189848. doi: 10.1371/journal.pone.0189848.

76. Diaz LA, Marabelle A, Delord J-P, Shapira-Frommer R, Geva R, Peled N, Kim TW, Andre T, Van Cutsem E, Guimbaud R. Pembrolizumab therapy for microsatellite instability high (MSI-H) colorectal cancer (CRC) and non-CRC. Proc Am Soc Clin Oncol. 2017;

77. Zimmer L, Apuri S, Eroglu Z, Kottschade LA, Forschner A, Gutzmer R, Schlaak M, Heinzerling L, Krackhardt AM, Loquai C. Ipilimumab alone or in combination with nivolumab after progression on anti-PD-1 therapy in advanced melanoma. Eur J Cancer. 2017;75:47–55. doi: 10.1016/j.ejca.2017.01.009.

78. Prithviraj P, McArthur GA, Atkinson V, Parente P, Andrews MC, Parakh S, Cebon JS, Klein O. Efficacy and toxicity of treatment with the anti-CTLA-4 antibody Ipilimumab in patients with metastatic melanoma who have progressed on anti-PD-1 therapy. Proc Am Soc Clin Oncol. 2015;33(15_suppl):9059.

79. Steuer C, Papadimitrakopoulou V, Herbst R, Redman M, Hirsch F, Mack P, Ramalingam S, Gandara D. Innovative clinical trials: the LUNG-MAP study. Clin Pharmacol Therapeutics. 2015;97(5):488–491. doi: 10.1002/cpt.88.

80. Taylor M, Antonia S, Bendell J, Calvo E, Jäger D, de Braud F, Ott PA, Pietanza MC, Horn L, Le DT: Phase I/II study of nivolumab with or without ipilimumab for treatment of recurrent small cell lung cancer (SCLC): CA209-032. J Immunotherapy Cancer 2015, 3(2):P376.

12

PD-1: Its Discovery, Involvement in Cancer Immunotherapy, and Beyond

Yasumasa Ishida

Division of Biological Science, Nara Institute of Science and Technology, 8916-5 Takayama-cho, Ikoma-shi, Nara 630-0192, Japan

ABSTRACT

On December 10, 2018, I was sitting among the big crowd of audience, as one of the invited guests to the ceremony, in the Stockholm Concert Hall. When King of Sweden Carl XVI Gustaf bestowed the diploma and medal of Nobel Prize of Physiology or Medicine 2018 on Dr. Tasuku Honjo and shook his hand for a while, surrounded by the thunderous applause and energetically blessing orchestral music, I thought that it had been a long journey for the molecule that we had first isolated in the early 1990s. Although it was truly a commemorable moment in the history of the programmed death-1 (PD-1) research, I believe we still have a long way to go. In this review article, I will explain why I think so, particularly by focusing on the potential role(s) that PD-1 appears to play in self-nonself discrimination by the immune system.

Citation: Ishida, Y. PD-1: Its Discovery, Involvement in Cancer Immunotherapy, and Beyond. Cells 2020, 9, 1376. https://doi.org/10.3390/cells9061376.

Keywords: PD-1; T cell; subtractive hybridization; self-nonself discrimination; cancer; immunotherapy

INTRODUCTION

In the early 1990s, a novel gene was discovered in Kyoto University, Japan, in search for the molecular mechanisms involved in self-nonself discrimination by the immune system [1]. In 1989, a UK team showed that self-reactive (potentially harmful) immature T lymphocytes (T cells) in the thymus undergo programmed cell death (apoptosis) [2]. This observation led the Kyoto University researches to assume that, if they are able to discover the genes strongly associated with the deaths of immature T cells, some of them would become good candidates for the key molecules playing pivotal roles in self-nonself discrimination. Only one gene was discovered at that time in a challenging screening experiment in molecular biology, and the gene (or its product) was named programmed death-1 (PD-1), with a hope that it would be somehow involved in the apoptosis-inducing processes of self-reactive immature T cells [1]. However, it turned out, several years later, that the novel molecule had nothing to do with the induction of programmed cell death/apoptosis [3,4,5,6]. Very interestingly, the wonderful developments in cancer immunotherapy in the recent years strongly suggest that, while PD-1, induced to be expressed on the surface of activated T cells, forces T cells to ignore cancer cells as one of the authentic 'self' components (i.e., prevents T cells from attacking cancer cells), the antibody-mediated blockade of PD-1's function makes activated T cells aware of the 'nonself' nature of cancer cells and unleashes their cytotoxicity. Therefore, although PD-1 was not directly involved in the cell death-inducing processes (despite its ominous name), it could still be playing crucial roles in self-nonself discrimination, as initially expected in the early 1990s.

HISTORICAL BACKGROUND

By the end of the 1970s, critical parts of the human and mouse immunoglobulin (Ig) genes had already been identified, and the genetic basis for the enormous diversification of the variable regions of antibody molecules had been elucidated [7,8]. In the early 1980s, on the other hand, there was a fierce competition among immunologists for the discovery of the molecular nature of T-cell antigen receptors (TCRs). Three groups, led by Dr. Ellis L. Reinherz of Dana-Farber Cancer Institute in Boston, Drs. John Kappler and Philippa Marrack of National Jewish Hospital and Research Center in Denver, and Dr. James P. Allison of the University of Texas System Cancer Center Science Park in Texas, took a similar strategy of affinity purification of the TCR proteins by using anti-clonotypic monoclonal antibodies (that recognized the hypervariable regions of the TCR molecules) and showed remarkable successes [9,10,11]. Many thought at that time (in the middle of 1983) that either of the above three groups in the United States would win the TCR-cloning race.

In March 1984, three Nature Articles were published by the groups of Dr. Tak W. Mak of University of Toronto, Canada, and Dr. Mark M. Davis of Stanford University, USA [12,13,14]. There, they reported the discovery of the TCR genes by using the extremely elegant subtractive (or differential) hybridization technique in molecular biology. Both of the groups were not professionals of protein purification but were able to discover the long-sought molecules only with their well-controlled experiments at the DNA-RNA level. Many immunologists, especially the young ones all over the world, were deeply inspired by the great achievements and strongly wished to perform similar subtractive hybridization experiments in their own careers in the future.

One of the desperate losers in this TCR-cloning race was Dr. James P. Allison, who 34 years later shared the Nobel Prize in Physiology or Medicine 2018 with Dr. Tasuku Honjo for their discovery of the novel type of cancer immunotherapy.

SELF-NONSELF DISCRIMINATION AND T-CELL DEATHS

In the late 1980s, bioscience researchers were beginning to understand an attractive and novel notion of apoptosis and/or programmed cell death. In 1989, a UK team published a Nature paper [2], showing that, when strongly stimulated through their TCRs, developing immature T cells in the thymus immediately realize that they are harmfully reacting against self-components and commit suicide by undergoing apoptosis/programmed cell death.

Dr. Jonathan D. Ashwell and colleagues at National Institutes of Health (NIH), USA, published an interesting paper indicating that mouse T-cell hybridomas produced by the cell fusion between a thymoma cell line BW5147 and a normal mouse peripheral T cell can be induced to die in vitro via strong stimulation towards their TCRs [15]. Their behavior after the TCR stimulation appeared to be very similar to that of self-reactive immature T cells in the thymus [2]. More interestingly, they also showed that such activation-induced death of mouse T-cell hybridomas depended upon de novo gene expression [16].

When the researchers in the Honjo laboratory in Kyoto University, Japan, noticed these interesting papers in 1989, they realized that they might be able to achieve their target by performing the state-of-the-art molecular-biology experiments: if they are able to isolate the genes strongly associated with the apoptosis/programmed death of self-reactive (strongly stimulated) immature T cells (or T-cell hybridomas) by using the subtractive-hybridization technique, such genes must become the very important clues for them to initiate their investigation on the molecular mechanisms involved in self-nonself discrimination.

At that time, however, the members in the Honjo laboratory were afraid that subtraction in just one direction (i.e., 'stimulated T-cell hybridoma' minus 'unstimulated T-cell hybridoma') would not be able to reduce the number of isolated cDNA clones to the practical ones because such simple subtraction should also enrich for the genes that are involved in typical T-cell activation events (like the cDNA for interleukin-2 (IL-2)). Therefore, they decided to incorporate into their subtractive hybridization experiments another type of cell-death induction system in vitro: apoptosis was induced in a mouse hematopoietic progenitor cell line by the deprivation of its growth factor interleukin-3 (IL-3), and they confirmed that the death induction of that cell line was also dependent upon de novo gene expression. If it is possible for them to isolate the gene(s) in which expression is commonly upregulated in two entirely distinct cell lines that are stimulated to undergo apoptosis in different ways, they thought that the chance of discovering the gene(s) deeply involved in the apoptosis-inducing processes would be reasonably high.

CELL-DEATH RESEARCH IN THE EARLY 1990S

Here, I need to briefly review the situation around the cell-death research at that time. The importance of the genetically regulated cell deaths was beginning to be appreciated among bioscience researchers, and Dr. H. Robert Horvitz's team in Massachusetts Institute of Technology in Cambridge, USA, was leading the field. By 1986, they had already identified two genes, ced-3 and ced-4, that are required for the physiological cell deaths taking place during the development of a nematode Caenorhabditis elegans (C. elegans) [17].

In the research field of humans or mice, on the other hand, an Australian group led by Drs. Suzanne Cory and Jerry M. Adams of the Walter and Eliza Hall Institute of Medical Research discovered that Bcl-2, which was first identified from a chromosomal breakpoint in human follicular B lymphoma cells by Drs. Yoshihide Tsujimoto and Carlo M. Croce of the Wistar Institute in Philadelphia, USA [18], is able to suppress cell deaths without causing cell proliferation [19]. It was in 1992 that the ced-9 gene product that negatively regulates programmed cell deaths during the development of C. elegans was shown to possess functional and structural similarities to mammalian Bcl-2 [20].

In 1993, the first member of a group of cysteine proteases (that were later given with the name 'caspases') was reported as a functional and structural mammalian homolog of the nematode cell-death gene ced-3 product [21]. For the discovery of Apaf-1, the mammalian homolog of the nematode ced-4 gene product, it took several more years [22]. In summary, it would be important here to note that nothing had been elucidated as crucial molecular mechanisms involved in mammalian programmed cell deaths before 1990–1991, when the researches initiated their subtractive hybridization experiments in Dr. Honjo's laboratory in Kyoto University.

DISCOVERY OF PD-1

In 1991, Dr. Honjo's group decided to initiate subtractive hybridization in which two independent apoptosis-induction systems (i.e., the strong TCR stimulation of mouse T-cell hybridomas and the growth-factor deprivation from mouse hematopoietic progenitor cells) were combined. They thought that it might be possible for them to discover the gene(s) deeply involved in induction of programmed deaths of mammalian cells, as well as those playing crucial roles in self-nonself discrimination by the immune system [23].

Dr. Yusuke Yanagi, who discovered a gene for human TCR in Dr. Tak W. Mak's laboratory in 1984 [12], provided Dr. Honjo's team with his detailed protocol of subtractive hybridization. During the summer of 1991, the researchers in the Honjo group performed many preliminary control experiments in subtractive hybridization. They initiated the 'two-dimensional' (or combined) subtractive hybridization experiments in the middle of August and, by the end of September, were able to isolate four independent cDNA clones that they initially believed to have been derived from four different candidate genes. Very interestingly, however, the reality was that all of the four cDNA clones were originated from the same single gene, which they later named programmed death-1 (PD-1) in the middle of October 1991, hoping that the novel gene could have something to do with apoptosis/programmed death of immature T cells. Next year (in 1992), they published their first PD-1 paper [1].

THE PD-1 RESEARCH IN THE HONJO LABORATORY IN THE LATE 1990S

In the late 1990s, Dr. Honjo's group at Kyoto University elucidated the basic function(s) of the newly identified molecule. PD-1, one of the type I transmembrane proteins, transiently expressed on the cell surface of activated T cells (and also B cells), was a receptor that receives the negative regulatory signals from the surrounding cells and sends such signals into the cytoplasm of activated T cells [3,4,5,6]. In other words, Dr. Honjo's group showed that: (1) From the functional point of view, PD-1 is not directly involved in the induction of apoptosis/programmed death of immature T cells. (2) Instead, PD-1 is a negative regulator of the immune responses of T cells.

Because of this, many people (including Dr. Honjo) say that PD-1 was discovered just coincidentally. Exactly, PD-1 was not a molecule directly involved in the apoptosis-inducing processes nor an ultimate negative regulator of T-cell survival. Instead, PD-1 was a novel type of the negative regulator on T-cell responses. Therefore, we might be able to say that PD-1 was not so distantly located from the original mark. In addition, the goal of Dr. Honjo's group at that time was to elucidate the molecular mechanisms involved in self-nonself discrimination by the immune system. Today, no one believes that PD-1 is an apoptosis-inducing gene. At the same time, however, no one doubts that PD-1 plays some kind of pivotal roles in self-nonself discrimination by T cells (as described below).

Some of the graduate students in the laboratories of Drs. Nagahiro Minato and Tasuku Honjo also discovered the potential usefulness of the blockade of the PD-1 signaling pathway in cancer therapy [24]. An ally of a group in the Ono pharmaceutical company (Osaka, Japan) led by Dr. Shiro Shibayama, who also had a training in Dr. Honjo's laboratory in the early 1990s, and another led by Drs. Alan J. Korman and Nils Lonberg in

Medarex/Bristol-Myers Squibb (USA) succeeded in the development of a fully human anti-human PD-1 monoclonal blocking antibody of the IgG4 subclass (nivolumab/Opdivo).

RECENT DEVELOPMENTS IN CANCER IMMUNOTHERAPY

Recently, the immunotherapies using anti-CTLA-4 and/or anti-PD-1 monoclonal blocking antibodies have revolutionized the field of human cancer treatment [25,26,27]. For instance, several anti-PD-1 and anti-PD-L1 monoclonal blocking antibodies, including nivolumab, pembrolizumab, and atezolizumab, have been approved by Food and Drug Administration (FDA) of the United States as an effective drug for the treatment of malignant melanoma, non-small cell lung cancer (NSCLC), renal cell carcinoma (RCC), Hodgkin lymphoma, squamous cell carcinoma of the head and neck, hepatocellular carcinoma, and others.

Consequently, Drs. James P. Allison and Tasuku Honjo have been frequently honored with distinguished prizes and awards, including the Tang Prize (Taiwan), the William B. Coley Award (USA), the Lasker Award (USA), and the Nobel Prize (Sweden), for their significant contribution to cancer immunotherapy.

Both CTLA-4 and PD-1 are the negative regulators of immune responses, belonging to the same CD28 family. In addition, both CTLA-4 and PD-1 are currently being used as the targets of antibody-mediated functional blockade in human cancer immunotherapy. Historically speaking, Dr. James P. Allison first showed that the CTLA-4 blockade is effective in the murine model of cancer immunotherapy [28]. Drs. Tasuku Honjo and Nagahiro Minato were able to prove the same thing for PD-1 about six years later [24]. This is why many people think that Dr. James P. Allison was the person who first opened the door of current cancer immunotherapy and why the Lasker award 2015 in the field of clinical medicine went to him alone.

However, the reality of cancer immunotherapy at present in clinics all over the world is another story. Side effects of cancer immunotherapy using the CTLA-4 antibody (ipilimumab) are much more serious and severe than those using the PD-1 antibodies. Although the FDA approval was earlier for CTLA-4 than for PD-1, the majority of clinicians right now believe that the CTLA-4 blockade would not be able to survive as a mono-therapy: it would be utilized only in combination with some other things, including the PD-1 blockade. On the other hand, the list of cancer types that can be treated through the PD-1 blockade is still constantly expanding. Basically, the PD-1 blockade is safer and more effective than the CTLA-4 blockade.

To understand the differential effects between the CTLA-4 blockade and the PD-1 blockade in cancer immunotherapy, we first need to think about the physiological functions of CTLA-4 and PD-1. First of all, the phenotypes of the knockout mice for these two molecules are dramatically different. All of the CTLA-4 knockout mice die of the extremely severe autoimmune diseases within three weeks after birth [29,30]. In contrast, the PD-1 knockout mice (especially on the C57BL/6 genetic background) are perfectly healthy for about one year after birth [4]. Sporadic and mild autoimmune diseases gradually begin to arise in the aged PD-1 knockouts, only about one year after birth.

This is because CTLA-4 is negatively regulating virtually all immune reactions, while PD-1 is suppressing only a limited type of immune responses, like those against cancer cells. Consequently, in the CTLA-4 knockout mice, all immune reactions become unleashed and uncontrolled, while, in the PD-1 knockout mice, only a certain type of immune responses are gradually re-activated [4,29,30]. After blocking the functions of CTLA-4 or PD-1 by using antibodies, the same patterns of things happen in cancer patients: serious and grave side effects with the CTLA-4 antibody; and only mild adverse events with the PD-1 antibody [25,26,27].

Recent findings also suggest that many of the anti-CTLA-4 antibodies with the anti-tumor activity, including ipilimumab (a fully human anti-human CTLA-4 monoclonal antibody of the IgG1 subclass), might not be acting simply as the checkpoint blocker. Instead, such antibodies could also be unleashing a broad spectrum of immune reactions by depleting regulatory T cells (Tregs), which are well-known to express the CTLA-4 molecules on their cell surface constitutively, through their antibody-dependent cellular cytotoxicity (ADCC) activity [31,32,33]. In the experimental mouse studies in which different anti-mouse CTLA-4 monoclonal IgG antibodies (either natural or recombinant) with distinct Fc structures were compared, those with higher affinity to Fc-gamma receptors (FcγRs) showed the stronger Treg-depleting and anti-tumor activities [31,32]. In addition, the anti-tumor activity was completely abolished in the anti-CTLA-4 antibodies without the FcγR-binding capability [32]. In the human immune system that matters for the clinical medicine, the ADCC activity of IgG1 (e.g., ipilimumab) is much higher than that of IgG4 (e.g., nivolumab). Taken together, these numerous findings in recent years strongly suggest that the main effect of anti-CTLA-4 monoclonal antibodies must be mediated by the depletion of Tregs inside tumors, not by the checkpoint blockade.

REMAINING QUESTIONS AND A HYPOTHESIS ABOUT THE 'REAL' PHYSIOLOGICAL FUNCTION(S) OF PD-1

As described above, the PD-1 knockout mice (especially on the C57BL/6 genetic background) are perfectly healthy, at least for about one year after birth [4]. In some of the aged PD-1 knockout mice (that are older than one year after birth), however, we can see the signs of mild glomerulonephritis and arthritis [4]. Then, what is PD-1 doing in young mice? In addition, why does such mild and sporadic autoimmunity arise only in aged mice?

Upon blockade of the PD-1's function(s) in the body of some cancer-bearing animals or patients by using the inhibitory antibodies, the killer T cells begin to attack cancer cells. Dr. Robert D. Schreiber's group in Washington University in St. Louis, MO, USA, showed in their chemically induced sarcoma model in mice that such killer T cells activated after the PD-1 blockade recognized mutated peptide antigens (so called 'neoantigens') derived from the chemically mutagenized genome of the sarcoma cell line in the context of the major histocompatibility complex (MHC) molecules [34]. This result is particularly interesting because it strongly suggests that, although killer T cells are capable of recognizing and responding to the neoantigens expressed on cancer cells, PD-1 usually halts such beneficial immune reactions. Then, what is PD-1 protecting there? In the normal condition (i.e., without injection of the inhibitory anti-PD-1 antibodies), PD-1 appears to be protecting cancer cells from the attack by the immune system. Is PD-1 on our side? Or on the side of cancer cells?

Immunologists usually say that PD-1 is involved in the maintenance of peripheral tolerance. These words give us some kind of impression about the immune reactions, and many of us might be satisfied with it. Unfortunately, however, the image here still looks very vague and ambiguous. We have to precisely understand what kind of immune reactions can be, or cannot be, suppressed by PD-1 because PD-1 is not an all-round player in the negative regulation of immune reactions [35].

We might be able to hypothesize as follows. When we are very young, cells in our body are pretty intact, and the immune system can easily distinguish 'self' from 'nonself' (Figure 1A). As we age, however, tiny alterations are gradually accumulated in a variety of components of our cells, and our highly evolved immune system with superb sensitivity and specificity, equipped with Variable (Diversity) Joining (V(D)J) recombination, somatic hypermutation, and so on, could detect and misrecognize such small alterations ('slightly altered self') as 'nonself', initiating the battles against them. In my personal view, we appear to have acquired PD-1 during evolution so that we can avoid (or suppress) autoimmunity against the gradually accumulated small changes in our cellular components (Figure 1B). In other words, we can be tolerant to the 'slightly altered self', especially in aged individuals, due to the presence of PD-1.

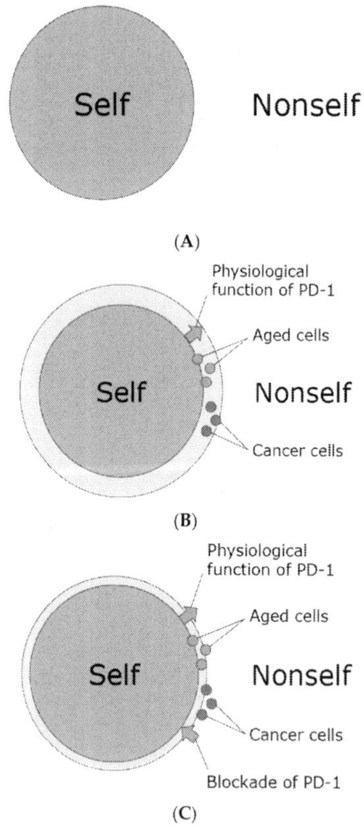

Figure 1: A hypothesis about the physiological function of programmed death-1 (PD-1): Demarcation of the border between 'slightly altered self' and 'nonself'. (A) Originally, when we are young, the border between 'self' and 'nonself' is obvious. (B) However, subtle changes gradually accumulate as cells age, and

the self-nonself border becomes ambiguous. In order to avoid the attacking against 'slightly altered' aged somatic cells by the exquisitely evolved immune system of higher vertebrates, we obtained PD-1. Here, PD-1 functions to widen and re-define the border between 'slightly altered self' in aged individuals and 'nonself'. Cancer cells are clever enough to sneak into this chink between the original self and the slightly altered self, so that they can be protected by PD-1 and avoid the immune-cell attack. (C) When we block the activity of PD-1, the self-nonself border moves back (closer) to the original one, and a fraction of cancer cells (and also aged somatic cells) get re-defined as nonself, allowing the highly evolved immune system to attack them.

As described before, young PD-1 knockout mice do not show any particular immunological abnormalities. This is probably because they still have not accumulated small alterations in their cellular components. Aged PD-1 knockout mice, on the other hand, have accumulated a significant amount of such changes in their cells, and in the absence of PD-1, our immune system misrecognizes them as 'nonself' and initiates the attacks against them.

In the genome of cancer cells (particularly in those of aged individuals), numerous somatic mutations are usually accumulated. While only a limited fraction of them are classified as the driver mutation that could provide a selective growth advantage and thus promote cancer development, the vast majority are the neutral passenger mutations that have nothing to do with the cancer initiation and/or promotion [**36**,**37**]. Some of such mutations in the genome, either the driver ones or passenger ones, could result in the generation of neoantigens, a portion of which might be able to be presented on the cell surface and recognized by T cells as the slightly altered self components [34,38]. However, when PD-1 halts harmful immune reactions against slightly altered normal somatic cells especially in aged individuals, the beneficial responses toward neoantigens generated in cancer cells also appear to be suppressed by PD-1 (Figure 1B).

In a good contrast, PD-1 is not capable of negatively regulating (nor suppressing) the strong immune reactions against the apparent 'nonself' (e.g., microorganisms that cause acute infections in animals like the influenza or measles viruses). Immunity against such apparent 'nonself' can easily and perfectly be provoked even in the presence of PD-1 in part because one of the ligands of PD-1, PD-L1, is sequestered and functionally inactivated in cis by the CD80 (B7-1) molecules the expression of which is strongly upregulated on antigen-presenting cells upon inflammation [35].

Upon cancer immunotherapy with anti-PD-1 or PD-L1 monoclonal antibodies, immune reactions against the slightly altered self-components could be generally re-activated or unleashed. This is probably why cancer patients with checkpoint-blockade immunotherapy sometimes experience so-called immune-related adverse events (irAE) [25,26,27]. Importantly, however, the number of somatic mutations in the genome is usually much larger in cancer cells than in normal somatic cells, and, from the immunological point of view, a fraction of cancer cells are more distantly located than normal somatic cells of aged individuals from the original 'self'. Therefore, we should usually be able to observe some degree of selectivity in the PD-1-mediated cancer immunotherapy, having the stronger immune reaction against cancer cells than toward normal somatic cells (Figure 1C).

With the final proof of the above hypothesis, we would be more firmly convinced that PD-1 plays the extremely important role(s) in self-nonself discrimination.

ACKNOWLEDGMENTS

The author thanks Tasuku Honjo, Philip Leder, Masashi Kawaichi, Nagahiro Minato, Taku Okazaki, Yusuke Yanagi, Kayo Inaba, Osamu Taguchi, Ryuzo Ueda, and Toshitada Takahashi for their invaluable mentorship, suggestions, and helpful comments.

REFERENCES

1. Ishida, Y.; Agata, Y.; Shibahara, K.; Honjo, T. Induced expression of PD-1, a novel member of the immunoglobulin gene superfamily, upon programmed cell death. *EMBO J.* 1992, *11*, 3887–3895.
2. Smith, C.A.; Williams, G.T.; Kingston, R.; Jenkinson, E.J.; Owen, J.J.T. Antibodies to CD3/T-cell receptor complex induce death by apoptosis in immature T cells in thymic cultures. *Nature* 1989, *337*, 181–184.
3. Agata, Y.; Kawasaki, A.; Nishimura, H.; Ishida, Y.; Tsubat, T.; Yagita, H.; Honjo, T. Expression of the PD-1 antigen on the surface of stimulated mouse T and B lymphocytes. *Int. Immunol.* 1996, *8*, 765–772.
4. Nishimura, H.; Nose, M.; Hiai, H.; Minato, N.; Honjo, T. Development of Lupus-like Autoimmune Diseases by Disruption of the PD-1 Gene Encoding an ITIM Motif-Carrying Immunoreceptor. *Immunity* 1999, *11*, 141–151.
5. Nishimura, H.; Okazaki, T.; Tanaka, Y.; Nakatani, K.; Hara, M.; Matsumori, A.; Sasayama, S.; Mizoguchi, A.; Hiai, H.; Minato, N.; et al. Autoimmune Dilated Cardiomyopathy in PD-1 Receptor-Deficient Mice. *Science* 2001, *291*, 319–322.
6. Okazaki, T.; Maeda, A.; Nishimura, H.; Kurosaki, T.; Honjo, T. PD-1 immunoreceptor inhibits B cell receptor-mediated signaling by recruiting src homology 2-domain-containing tyrosine phosphatase 2 to phosphotyrosine. *Proc. Natl. Acad. Sci. USA* 2001, *98*, 13866–13871.
7. Honjo, T. Immunoglobulin Genes. *Annu. Rev. Immunol.* 1983, *1*, 499–528.
8. Tonegawa, S. Somatic generation of antibody diversity. *Nature* 1983, *302*, 575–581.
9. Acuto, O.; Hussey, R.E.; Fitzgerald, K.A.; Protentis, J.P.; Meuer, S.C.; Schlossman, S.F.; Reinherz, E.L. The human T cell receptor: Appearance in ontogeny and biochemical relationship of α and β subunits on IL-2 dependent clones and T cell tumors. *Cell* 1983, *34*, 717–726.
10. Kappler, J.; Kubo, R.; Haskins, K.; White, J.; Marrack, P. The mouse T cell receptor: Comparison of MHC-restricted receptors on two T cell hybridomas. *Cell* 1983, *34*, 727–737.
11. McIntyre, B.W.; Allison, J.P. The mouse T cell receptor: Structural heterogeneity of molecules of normal T cells defined by Xenoantiserum.

Cell 1983, *34*, 739–746.

12. Yanagi, Y.; Yoshikai, Y.; Leggett, K.; Clark, S.P.; Aleksander, I.; Mak, T.W. A human T cell-specific cDNA clone encodes a protein having extensive homology to immunoglobulin chains. *Nature* 1984, *308*, 145–149.
13. Hedrick, S.M.; Cohen, D.I.; Nielsen, E.A.; Davis, M.M. Isolation of cDNA clones encoding T cell-specific membrane-associated proteins. *Nature* 1984, *308*, 149–153.
14. Hedrick, S.M.; Nielsen, E.A.; Kavaler, J.; Cohen, D.I.; Davis, M.M. Sequence relationships between putative T-cell receptor polypeptides and immunoglobulins. *Nature* 1984, *308*, 153–158.
15. Ashwell, J.D.; E Cunningham, R.; Noguchi, P.D.; Hernandez, D. Cell growth cycle block of T cell hybridomas upon activation with antigen. *J. Exp. Med.* 1987, *165*, 173–194.
16. Ucker, D.S.; Ashwell, J.D.; Nickas, G. Activation-driven T cell death. I. Requirements for de novo transcription and translation and association with genome fragmentation. *J. Immunol.* 1989, *143*, 3461–3469.
17. Ellis, H. Genetic control of programmed cell death in the nematode C. elegans. *Cell* 1986, *44*, 817–829.
18. Tsujimoto, Y.; Finger, L.; Yunis, J.; Nowell, P.; Croce, C. Cloning of the chromosome breakpoint of neoplastic B cells with the t(14;18) chromosome translocation. *Science* 1984, *226*, 1097–1099.
19. Vaux, D.L.; Cory, S.; Adams, J.M. Bcl-2 gene promotes haemopoietic cell survival and cooperates with c-myc to immortalize pre-B cells. *Nature* 1988, *335*, 440–442.
20. Hengartner, M.; Ellis, R.; Horvitz, R. Caenorhabditis elegans gene ced-9 protects cells from programmed cell death. *Nature* 1992, *356*, 494–499.
21. Yuan, J. The C. elegans cell death gene ced-3 encodes a protein similar to mammalian interleukin-1β-converting enzyme. *Cell* 1993, *75*, 641–652.
22. Zou, H.; Henzel, W.; Liu, X.; Lutschg, A.; Wang, X. Apaf-1, a Human Protein Homologous to C. elegans CED-4, Participates in Cytochrome c–Dependent Activation of Caspase-3. *Cell* 1997, *90*, 405–413.
23. Cornwall, C.M. *Battling Melanoma: One Couple's Struggle from Diagnosis to Cure*; Rowman & Littlefield: Lanham, MD, USA, 2016; pp. 147–157.

24. Iwai, Y.; Ishida, M.; Tanaka, Y.; Okazaki, T.; Honjo, T.; Minato, N. Involvement of PD-L1 on tumor cells in the escape from host immune system and tumor immunotherapy by PD-L1 blockade. *Proc. Natl. Acad. Sci. USA* 2002, *99*, 12293–12297.
25. Baumeister, S.; Freeman, G.J.; Dranoff, G.; Sharpe, A.H. Coinhibitory Pathways in Immunotherapy for Cancer. *Annu. Rev. Immunol.* 2016, *34*, 539–573.
26. Ribas, A.; Wolchok, J.D. Cancer immunotherapy using checkpoint blockade. *Science* 2018, *359*, 1350–1355.
27. Chamoto, K.; Hatae, R.; Honjo, T. Current issues and perspectives in PD-1 blockade cancer immunotherapy. *Int. J. Clin. Oncol.* 2020, *25*, 790–800.
28. Leach, D.R.; Krummel, M.F.; Allison, J.P. Enhancement of Antitumor Immunity by CTLA-4 Blockade. *Science* 1996, *271*, 1734–1736.
29. Tivol, E.A.; Borriello, F.; Schweitzer, A.; Lynch, W.P.; Bluestone, J.A.; Sharpe, A.H. Loss of CTLA-4 leads to massive lymphoproliferation and fatal multiorgan tissue destruction, revealing a critical negative regulatory role of CTLA-4. *Immunity* 1995, *3*, 541–547.
30. Waterhouse, P.; Penninger, J.M.; Timms, E.; Wakeham, A.; Shahinian, A.; Lee, K.P.; Thompson, C.B.; Griesser, H.; Mak, T.W. Lymphoproliferative Disorders with Early Lethality in Mice Deficient in Ctla-4. *Science* 1995, *270*, 985–988.
31. Simpson, T.R.; Li, F.; Montalvo-Ortiz, W.; Sepulveda, M.A.; Bergerhoff, K.; Arce, F.; Roddie, C.; Henry, J.Y.; Yagita, H.; Wolchok, J.D.; et al. Fc-dependent depletion of tumor-infiltrating regulatory T cells co-defines the efficacy of anti–CTLA-4 therapy against melanoma. *J. Exp. Med.* 2013, *210*, 1695–1710.
32. Selby, M.J.; Engelhardt, J.J.; Quigley, M.; Henning, K.A.; Chen, T.; Srinivasan, M.; Korman, A.J. Anti-CTLA-4 Antibodies of IgG2a Isotype Enhance Antitumor Activity through Reduction of Intratumoral Regulatory T Cells. *Cancer Immunol. Res.* 2013, *1*, 32–42.
33. Vargas, F.A.; Furness, A.J.; Litchfield, K.; Joshi, K.; Rosenthal, R.; Ghorani, E.; Solomon, I.; Lesko, M.H.; Ruef, N.; Roddie, C.; et al. Fc Effector Function Contributes to the Activity of Human Anti-CTLA-4 Antibodies. *Cancer Cell* 2018, *33*, 649–663.e4.
34. Gubin, M.M.; Zhang, X.; Schuster, H.; Caron, E.; Ward, J.; Noguchi, T.; Ivanova, Y.; Hundal, J.; Arthur, C.D.; Krebber, W.-J.; et al.

Checkpoint blockade cancer immunotherapy targets tumour-specific mutant antigens. *Nature* 2014, *515*, 577–581.

35. Sugiura, D.; Maruhashi, T.; Okazaki, I.; Shimizu, K.; Maeda, T.K.; Takemoto, T.; Okazaki, T. Restriction of PD-1 function by cis-PD-L1/CD80 interactions is required for optimal T cell responses. *Science* 2019, *364*, 558–566.
36. Stratton, M.R.; Campbell, P.J.; Futreal, P.A. The cancer genome. *Nature* 2009, *458*, 719–724.
37. Pon, J.R.; Marra, M.A. Driver and Passenger Mutations in Cancer. *Annu. Rev. Pathol. Mech. Dis.* 2015, *10*, 25–50.
38. Alspach, E.; Lussier, D.M.; Miceli, A.P.; Kizhvatov, I.; DuPage, M.; Luoma, A.M.; Meng, W.; Lichti, C.F.; Esaulova, E.; Vomund, A.N.; et al. MHC-II neoantigens shape tumour immunity and response to immunotherapy. *Nature* 2019, *574*, 696–701.

13

A Current Perspective on Cancer Immune Therapy: Step-by-step Approach to Constructing the Magic Bullet

Gabriele D'Errico[1], Heather L. Machado[2] and Bruno Sainz Jr.[1,3,4]

[1]Department of Biochemistry, School of Medicine, Autónoma University of Madrid, Calle del Arzobispo Morcillo 4, 28029 Madrid, Spain
[2] Department of Biochemistry and Molecular Biology, Tulane University School of Medicine, 1430 Tulane Ave, #8543, New Orleans, LA 70112, USA
[3] Department of Cancer Biology, Instituto de Investigaciones Biomédicas "Alberto Sols" CSIC-UAM, Madrid, Spain
[4] Enfermedades Crónicas y Cáncer Area, Instituto Ramón y Cajal de Investigación Sanitaria (IRYCIS), Madrid, Spain.

ABSTRACT

Immunotherapy is the new trend in cancer treatment due to the selectivity, long lasting effects, and demonstrated improved overall survival and tolerance, when compared to patients treated with conventional chemotherapy. Despite these positive results, immunotherapy is still far from becoming the perfect

Citation: D'Errico G, Machado HL, Sainz B Jr. A current perspective on cancer immune therapy: step-by-step approach to constructing the magic bullet. Clin Transl Med. 2017;6(1):3. doi:10.1186/s40169-016-0130-5.

magic bullet to fight cancer, largely due to the facts that immunotherapy is not effective in all patients nor in all cancer types. How and when will immunotherapy overcome these hurdles? In this review we take a step back to walk side by side with the pioneers of immunotherapy in order to understand what steps need to be taken today to make immunotherapy effective across all cancers. While early scientists, such as Coley, elicited an unselective but effective response against cancer, the search for selectivity pushed immunotherapy to the side in favor of drugs focused on targeting cancer cells. Fortunately, the modern era would revive the importance of the immune system in battling cancer by releasing the brakes or checkpoints (anti-CTLA-4 and anti-PD-1/PD-L1) that have been holding the immune system at bay. However, there are still many hurdles to overcome before immunotherapy becomes a universal cancer therapy. For example, we discuss how the redundant and complex nature of the immune system can impede tumor elimination by teeter tottering between different polarization states: one eliciting anti-cancer effects while the other promoting cancer growth and invasion. In addition, we highlight the incapacity of the immune system to choose between a fight or repair action with respect to tumor growth. Finally we combine these concepts to present a new way to think about the immune system and immune tolerance, by introducing two new metaphors, the "push the accelerator" and "repair the car" metaphors, to explain the current limitations associated with cancer immunotherapy.

Keywords: CTLA-4, PD-1, PD-L1, Immunotherapy, Coley's toxin

INTRODUCTION

Since war was declared on cancer in 1971, our arsenal of drugs against this enemy has steadily increased. The first class of drugs developed, conventional chemotherapy [1], provided significant benefits for the treatment and management of different cancers; however, conventional chemotherapy has two intrinsic defects: lack of selectivity [2] and long-term resistance [3]. The subsequent development of targeted therapies, including monoclonal antibody-based therapies, such as rituximab, overcame the lack of selectivity associated with conventional chemotherapy by targeting specific proteins involved in cancer cell stimulation, proliferation or apoptosis evasion [4]. While monoclonal antibody-based strategies have significantly increased in clinical practice over the past decade, mechanisms of acquired resistance remains a hurtle, likely due to genetic and epigenetic instability of cancer cells [5]. Thus, despite significant advances made in cancer treatment over the past 45 years, we are still far from developing therapies capable of effectively ablating cancer while avoiding adverse effects on healthy cells and loss of efficacy over time.

Starting with Coley's toxin, followed by Paul Ehrlich's hypothesis of tumor surveillance and contrasted by Burnet's immunological tolerance theory [6], the idea that the immune system can play a pro- and/or anti-tumor role has been recognized and debated for years. As the immune system is armed to protect against pathogens, it has been long postulated that immune cells should recognize tumor cells as foreign, and effectively eliminate them before spreading to distant organs. This concept, today referred to as cancer immunotherapy, has the potential to be the magic bullet that investigators have been desperately searching for since the early 1900s [7].

The last century was a pendulum, swinging between big hope and deep disappointment in the immunotherapy field. Why has our view of immunotherapy shifted from promising to disappointing? Can immunotherapy be made more effective? In this review we look to the past to understand the current problems associated with immunotherapy. We present a new way to think about the immune system and immune tolerance. In addition to the standard "fuel the engine, release the brake" rules of immunotherapy, we introduce the "push the accelerator" and "repair the car" metaphors to explain part of the current limitations associated with cancer immunotherapy.

Immunotherapy: A Revolutionary View of Cancer Treatment

Immunotherapy was born in 1890. Its father, William Coley [8], observed that a patient with an inoperable sarcoma that suffered a *Streptococcus pyogenes* infection twice obtained complete remission. Based on this observation, Coley treated approximately 1000 patients with inoperable cancers (specially sarcomas) with a mixture consisting of killed *S. pyogenes* and *Serratia marcescens*, achieving a complete remission in 10% of treated patients. Compared with actual 5-year survival rates for metastatic sarcoma (20%) [9], the effects of Coley's toxin were promising and demonstrated that our immune system could effectively eradicate tumors with a low rate of adverse effects in a subset of patients. While the mechanism of action of Coley's toxin was unknown at the time, Coley's toxin essentially activated tumor-infiltrating leukocytes [heterogeneous populations of cells, including varying proportions of neutrophils, macrophages, T and B cells, and natural killer cells (NK)]. Following administration of Coley's toxin, dendritic cells (DCs), professional antigen presenting cells (APCs), initiate an immune response by presenting the captured bacterial antigen to naïve $CD4^+$ or $CD8^+$ T cells [10] in lymphoid tissues, which results in T cell priming [11] and inflammatory interleukin production [IL-1, IL-2, tumor necrosis factor alpha (TNFα,) IL-12] [12, 13]. Clonal T cell expansion triggers a humoral immune response by activating B cells, or cellular immunity by activating Th1 effector cells or NK cells [14]. Thus, as tumors often display a high degree of leukocyte infiltration, it is reasonable that stimulation of tumor-infiltrating leukocytes can result in tumor cell targeting and elimination.

The First Gauntlet: Release the Brake

In 1949, some years after Coley's first experiments, Macfarlane Burnet stated, "if in embryonic life expendable cells from a genetically distinct race are implanted and established, no antibody response should develop against the foreign cell antigen when the animal takes on independent existence" [6, 15]. Peter Medawar would go one step further and propose that the immune system becomes tolerant to cancer cells due to the similarities that exist with normal healthy cells. In his early experiments, Medawar injected embryonic mouse donor cells into mice of a different strain, rendering the recipient mice tolerant to future grafts from the donor and not third-party strains [16]. These experiments set the initial groundwork for what would become the concept of acquired immunological tolerance. Even if the molecular mechanisms explaining Medawar and Burnet's observations would not be discovered until 25 years later, Medawar and Burnet had already thrown down the first gauntlet to modern immunotherapy [16].

Immunological tolerance is a fundamental process, the lack of which would result in numerous pathologies including autoimmune illnesses. Since Burnet and Medawar won the Nobel Prize for their pioneering work in this field more than 50 years ago, an enormous amount of progress has been made to better understand immune tolerance and the numerous redundant mechanisms involved in this biological process. In 1959, Joshua Lederberg published nine propositions on immunity and tolerance [17]. The sixth proposition stated that "the immature antibody-forming cell is hypersensitive to an antigen–antibody combination: it will be suppressed if it encounters the homologous antigen at this time", highlighting that each antibody-producing cell has a single specificity [17]. In 1978, Nossal and Pike experimentally demonstrated that bone marrow-derived cells cultured with an antigen became tolerant to the antigen in a time-dependent manner, with maximal tolerance being achieved only when the antigen was present continuously as the cultured bone marrow cells matured [18]. In the late 1980's, Kappler demonstrated central tolerance in mice and indicated that tolerance induction may occur in the thymus [19]. Goodnow later showed that B cells that reactive to "self" antigens are eliminated or silenced in order to avoid autoimmunity [20]. And finally, Le Douarin demonstrated that the thymus is not only important for central tolerance but it produces cells [i.e. regulatory T (T_{reg}) cells] that strongly regulate effector cells, discovering a third dominant form of immune tolerance [21].

Taken together these results clearly indicated that self-immune tolerance is maintained and regulated by multiple mechanisms, including similarity with self-antigens, regulatory immune cells, a suppressive versus activating cytokine balance and immune checkpoints. Thus, if this process is to be effectively manipulated in order to release the brake, we have to stimulate APCs with an antigen significantly different from "self" to recruit effector cells, balance cytokines to obtain a pro-inflammatory milieu, and finally, inhibit immune checkpoints to avoid tolerance.

Hope Dies Last

The idea of immune tolerance put forth by Medawar and Burnet dampened the hope generated by Coley's experiments that activating the immune system could treat cancer. However, in 1953, Foley demonstrated that methylcholanthrene-induced tumor cells could produce immunogenic antigens in mice, although strain-dependent differences were observed [22]. Nathrath would add additional fuel to the fire by showing that the host immune system is capable of recognizing new molecular properties (antigens) displayed by tumor cell as a result of the changes accumulating during the transition from normal to malignant cells [23]. Interestingly, however, the greatest hope for immunotherapy would come from another illness described in the 1950s, autoimmune diseases such as lupus, which clearly confirmed that auto-reactive cells can elude self tolerance [24]. Finally, towards the later end of the 20th century, immunogenic tumor associated antigens were discovered in mice [25] and humans [26], resulting from cancer cell genomic instability. Unsettling still was the fact that while cancer cells could express a multitude of new and unknown antigens as a result of malignant transformation, the immune system was not stimulated to target these antigens/cells. The problem therefore was no longer the absence of immunogenic antigens on cancer cells, but rather understanding why our immune system "ignores/tolerates" cells harboring these antigens.

In the absence of inflammation, naïve T-cells circulate preferentially to secondary lymphoid tissue [27]. During an infection, APCs respond to inflammatory cytokines (IL-1, TNF-α) [28] and migrate via afferent lymph vessels to lymph nodes where they can interact with naïve T-cells. In a similar manner, macrophage or B cells can take up and process free antigen in the blood or spleen [29]. The interaction of T cell CD28 and DC CD80 (B7-1) or CD86 (B7-2) allows for specific T cells to proliferate in the paracortex [30] and become competent to receive further activation signals from antigen-bearing macrophages and B cells. This process results in the production of cytokines, loss of L-selectin (which is involved in lymph node entry) and increased expression of adhesion molecules like VLA 4 (which facilitates extravasation into non-lymphoid tissue) [31]. Macrophages and parenchymal cells produce inflammatory cytokines (IL-1, TNF-α) that increase expression of selectins and integrin ligands. Ultimately, activated T-cells express adhesion molecules that allow them to selectively enter inflamed tissues expressing counterpart adhesion molecules. In summary, immune cells can present antigens (dendritic cells, macrophages), produce cytokines (macrophage) and interact with cells of the adaptive immune system (B-cells, T-cells) in the lymph nodes, resulting in their subsequent activation via receptor/ligand interactions. Once activated, adaptive immune cells proliferate, alter their receptors and adhesion molecules, and finally migrate to initiate the destruction of foreign pathogens.

To avoid indiscriminate activation and self-destruction, the immune system has developed redundant mechanisms to tolerate self or non-dangerous antigens. Since central tolerance is not the underlying mechanisms by which cancer cells escape immune targeting, we refer the reader to several published reviews detailing the biology of central tolerance (Ref. [32, 33]). Peripheral tolerance, on the other hand, is the primary mechanism utilized by cancer cells to avoid the immune system. In the early 1990s, Jenkins and Schwartz demonstrated that T-cells need a co-stimulatory signal to fully activate, and if T-cells receive only TCR signals they become anergic [34]. The co-stimulatory signals must be received from APCs [35]: B7-1 (CD80) or B7-2 (CD86) on APCs binding to CD28 on T-cells is necessary to fully stimulate T-cells. While an important step forward in our understanding of T-cell regulation, the big discovery was not B7-1 or B7-2, but rather the receptor that inhibits the second co-stimulatory signal. James Allison, director of the UC Berkeley Cancer Research Laboratory, was intrigued with a molecule called cytotoxic T-lymphocyte antigen-4 (CTLA-4) [36], originally discovered in a cDNA library derived from activated T-cells. In the late 1990s, Allison and his group began to study how CTLA-4 inhibits T-cells and if this inhibition could explain why T-cells do not attack cancer cells. They demonstrated that CTLA-4, a homologue of CD28, bound with higher affinity (at least 10-fold) to both B7-1 and B7-2 [37] and inhibited $CD4^+$ T-cell activation. Under certain conditions, T-cells up-regulate CTLA-4, which binds to B7-1 and B7-2 with a higher affinity than CD28, effectively hijacking the second co-stimulatory signal that T-cells require for full T-cell activation, proliferation, and effector function [38]. As a result, T-cells cannot be fully activated and thus undergo anergy. Since its discovery, several groups have worked diligently towards dissecting the role of CTLA-4. Studies with CTLA-$4^{-/-}$ mice confirmed its inhibitory function in vivo. Waterhouse et al. showed that mice lacking CTLA-4 died early on of fatal lymphoproliferative disorders [39], demonstrating that CTLA-4 acts as a negative regulator of T cell activation and is vital for the control of lymphocyte homeostasis. Based on ever increasing data demonstrating a role for CTLA-4 as a negative regulator of T-cell activation, Allison and colleagues went on to show that in vivo administration of antibodies to CTLA-4 promoted the rejection of tumors, including pre-established tumors, confirming that CTLA-4 blockage can allow for, and potentiate, an effective immune responses against tumors [40].

The First Victory in the New Era

These studies and others [41–43] suggested that monoclonal antibody-mediated CTLA-4 blockage could represent an effective anti-cancer therapy. To translate these findings to the clinical setting, the Medarex Corporation generated a series of monoclonal antibodies using a unique transgenic mouse (HuMAb), in which the endogenous murine immunoglobulin genes had been knocked out and replaced with the human loci [44]. Ipilimumab showed safety in a phase I study and efficacy in a phase III study (ClinicalTrials.gov Identifier: NCT00094653) with primary overall survival endpoints. Patients allocated to receive ipilimumab had a median overall survival of 10.1 versus 6.4 months for the control group [hazard ratio (HR), 0.68; $P \leq 0.003$]. Based on the promising results, Bristol-Meyer's Yervoy® (ipilimumab) was approved by the US Food and Drug Administration (FDA) and the European Medicines Agency (EMA) in 2011 for the treatment of metastatic melanoma [38].

While CTLA-4 was the first "checkpoint" inhibitor identified, the list of checkpoint immunomodulators continues to grow, with inhibitors of the programmed cell death protein 1/ligand pathway leading the way. Programmed cell death protein 1, also known as PD-1 and CD279 (cluster of differentiation 279) is a member of the CD28 superfamily expressed on activated $CD4^+$ and $CD8^+$ T-cells as well as NK and B-cells while programmed death-ligand 1 (PD-L1), also known as cluster of differentiation 274 (CD274) or B7 homolog 1 (B7-H1), is expressed predominantly on APCs. The main role of PD-1 is to act like a stopwatch to limit the activity of T-cells in the "battle field" during the effector phase of T-cell activation in peripheral tissues and the tumor microenvironment via the delivery of negative signals upon interaction with its two ligands (PD-L1 or PD-L2). PD-L1 and PD-L2 compete for PD-1 [45] and upon binding both inhibit T cell proliferation, cytokine production and cell adhesion [46], although some contradictory data have suggested a costimulatory function [47]. T-cells begin to express PD-1 when activated and its expression increase over time [48, 49], while PD-L1 is expressed on APCs present in the inflamed tissue. PD-1 is also highly expressed on T_{reg} cells, where it may enhance their proliferation in the presence of ligand [50]. The expression on T_{reg} cells also highlights the role PD-1 plays in regulating the induction and maintenance of peripheral tolerance and protection from autoimmune attack [51]. Our current understanding of the PD-1/PD-L1 pathway shows that engagement of PD-1 and PD-L1 leads to inactivation of effector T-cell molecules such as Zap70 to inhibit T-cell proliferation, thus limiting the inflammatory damage in inflamed tissues [52]. For example, during chronic infections the PD-1/PD-L1 pathway leads to anergy [53]. It appears as though several cancers, including lung, ovarian and colon carcinomas as well as melanomas, have evolved to over express PD-L1 [54]. Thus, chronic tumor associated antigen exposure in cancer can lead to high levels of PD-1 expression on T-cells, which can interact with PD-L1 expressed on cancer cells, inhibiting T-cell activation and perhaps inducing a state of anergy in immune cells present in the tumor. Blocking the PD-1/PD-L1 pathway can revert this condition, promoting cancer cell elimination by activated T-cells. In the past 5 years, Bristol-Myers Squibb has produced a fully humanized antibody against PD-1 named Opdivo (Nivolumab), which obtained FDA accelerated approval in 2014 [55] based on the "Study of Nivolumab (BMS-936558) Compared With Dacarbazine in Untreated, Unresectable, or Metastatic Melanoma".

Today many check-point inhibitors are used by oncologist to achieve significant increases in survival rates (e.g. 1- and 2-year survival rates of 62 and 43%, respectively for melanoma or 1- and 2-year survival rates of 42 and 23%, respectively for lung cancer) and/or to achieve a durable partial or complete response in cancer patients (e.g. 31% for melanoma patients) [56, 57]. Table 1 summarizes the current FDA- and EMA-approved immune checkpoint inhibitors. Clinical studies have also investigated the efficacy of combination therapies using anti-PD-1/PD-L1 therapies together with other checkpoint inhibitors, such as the anti-CTLA4 treatment ipilimumab. The combination of nivolumab and ipilimumab increased overall survival in patients with untreated melanoma. The median progression-free survival was 11.5 months (95% confidence interval [CI], 8.9–16.7) with nivolumab plus ipilimumab, as compared with 2.9 months (95% CI, 2.8–3.4) with ipilimumab (hazard ratio for death or disease progression, 0.42; 99.5% CI, 0.31–0.57; $P < 0.001$), and 6.9 months (95% CI, 4.3–9.5) with nivolumab (hazard ratio for the comparison with ipilimumab, 0.57; 99.5% CI, 0.43–0.76; $P < 0.001$) [58]. It is important to stress that even if combination therapy allows us to obtain better results and improved medium overall survival, it is still far from becoming the perfect therapy. The current underlying problem with immune checkpoint inhibitors is that if there are no T-cells in the tumor border then there are no effector cells capable of eliminating the tumor cells. Therefore, even if we release the brake we cannot obtain clinically relevant results. Thus, the next strategy lies in "fueling the engine", that is allowing T-cells to reach the tumor border.

Table 1: Summary of immune therapies in clinical use

Immune therapy	Target	Stage	Cancer type	Ref.
Ipilimumab	CTLA-4	Clinical use	Advanced melanoma	107
Nivolumab	PD-1	Clinical use	Melanoma Renal cancer NSLC	108 109 110
Pembrolizumab	PD-1	Clinical use	Melanoma	111
Atezolizumab	PD-L1	Clinical use	NSLC Clear renal cancer bladder cancer	112 113
Sipileucel	Peripheral blood mononuclear cells	Clinical use	Prostate cancer	114

Fueling the Engine

As stated above, immune checkpoints inhibitors are only effective if tumors are infiltrated with T-cells. Therefore, if a tumor lacks infiltrated T-cells, immune checkpoint inhibitors are essentially ineffective. Cancer vaccines have been extensively investigated as a strategy to induce T-cell infiltration in the tumor—"fuel the engine". Currently, however, Sipileucel is the only "cellular immunotherapy" (i.e. vaccine therapy) approved by the FDA. Sipileucel consists of autologous peripheral blood mononuclear cells (PBMCs), obtained by leukapheresis and cultured with a prostatic acid phosphatase linked to granulocyte–macrophage colony-stimulating factor (GM-CSF) [59]. Results from the 9902B study in patients with prostate cancer demonstrated an overall survival of 25.8 months for patients receiving Sipileucel compared to 21.7 months for patients who received the control treatment [59].

While Sipileucel demonstrates the clear benefit of "fueling the engine", the use of cancer vaccines in other tumor types is still in the experimental stages. For example, Kleponis et al. developed a GM-CSF-secreting pancreatic cancer vaccine (GVAX) that provided maturation signals to APCs at the local vaccine site. Stimulated APCs processed tumor antigens and presented them to T-effector cells, allowing T-cells to infiltrate the tumor [60]. The authors went on to show that PD-L1 expression was induced in the infiltrating cells (T-cells). This study demonstrated that vaccine-based therapies may have adequately primed pancreatic cancer for anti-PD-1/PD-L1 treatments, highlighting that this typically T-cell-effector poor/T_{reg}-rich tumor [61] could be potentially treated with PD-1/PD-L1 inhibitors. Thus, cancer vaccine-based immunotherapy may overcome the resistance of certain cancers to immune checkpoint inhibitors, while immune checkpoint inhibitors may enhance the efficacy of the cancer-vaccine therapies. The goal of a combination strategy is to combine the strength of each immunotherapy approach, with cancer vaccines functioning to "fuel the engine" and immune checkpoint inhibitors working to "release the brake".

Perplexing is the fact that PD-1 and CTLA-4 checkpoint inhibitors, even when helped by cancer vaccines, are not effective against all cancer types, nor do they work in every patient with the same cancer. Perhaps other immune cell types are negatively affecting cancer immunotherapy? The explanation we put forward to explain this dilemma is that in some cancers we have the machinery (the car) on a downhill slope, so if we release the brake (immune checkpoint inhibitors) the car can move. In contrast, when we are on an uphill slope or on a plain field, releasing the brake simply does not move the car. For such scenarios, we have to release the brake and push the accelerator.

Looking deep inside the immune system we can find a dynamic and complex environment of cells that are different in type, size, complexity, markers and function. Even the same cell can exist in two (or more) different polarized states. For example, macrophages can switch between a pro-inflammatory (classically activated) and a reparative (alternatively activated) state [62]. T-cells can be stimulated into T effector cells or T_{regs} cells, each of which can have a very distinct role within a tumor [63]. Thus, we need to take a step back and understand why, when and how a cell (i.e. macrophage, T-cell, etc.) can switch from a classically activated (inflammatory macrophage or T effector cell) to an alternatively activated (pro-tumorigenic macrophage or T_{reg}) state and what mediates this change. Answering these questions may provide the key to push the accelerator.

Immune Cells: Dr. Jekyll and Mr. Hide

The immune system is comprised of many cells, including but not limited to dendritic cells, mast cells, macrophages, neutrophils and lymphocytes. In addition, and to complicate the matter even more, immune cells are extremely plastic and each cell type can differentiate into at least two forms depending on their environment and the paracrine signals they receive (reviewed in [64–66]). Lymphocytes can differentiate into many subsets. Apart from B and T cells, T lymphocytes can differentiate into $CD8^+$ or $CD4^+$ T cells, the latter of which can in turn differentiate into T_{reg} and T_{helper} cells [67]. DCs that present captured antigens to naïve T-cells, have two major subsets: myeloid (i.e. conventional DC or immunogenic DC) and plasmacytoid form (or tolerogenic DC) [68]. As described above, macrophages can exist in at least two forms [69], and in the context of a tumor, macrophages and neutrophils can differentiate into tumor-associated macrophages (TAMs) or neutrophils (TANs) [70], respectively.

The interest in the different states of immune cells, particularly within the tumor, stems from the different biological affects these cells produce depending on their state, polarization or differentiation (Fig. 1). For example, human DCs can exist as immunogenic or tolerogenic DCs [71], with immunogenic DCs functioning primarily to stimulate a T_{helper} [72] response while tolerogenic DCs function primarily to stimulate a T_{reg} response [73]. Moreover, the ratio of different DCs depends on the cytokine milieu. In vitro studies clearly demonstrate that GM-CSF, interferon alpha (IFNα), or IL-15 can induce the differentiation of inflammatory DCs while IL-10, vitamin A or D3, or immunosuppressive drugs such as cyclosporine A induce tolerogenic DCs through E-cadherin mediated signaling [74, 75]. Thus, depending on the cytokine milieu, DCs can elicit a strong immune response or a tolerogenic state.

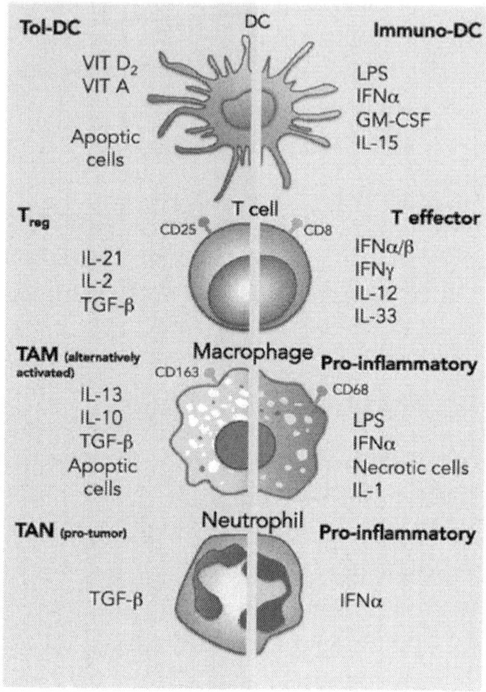

Figure 1: The two faces of immune cells: The Dr. Jekyll and Mr. Hide concept. Immune cells, including dendritic cells (DCs), T-cells, macrophages and neutrophils, are extremely plastic and can assume different roles/functions depending on factors encountered at the site of infection or within the tumor microenvironment. When stimulated by factors such as TGF-ß or when in contact with apoptotic cells, immune cells become pro-tumorigenic (*left side*) and differentiate/polarize towards tolerogenic DCs (Tol-DC), T_{reg} cells, tumor-associated macrophages (TAMs) or tumor-associated neutrophils (TANs). In contrast, when stimulated by pro-inflammatory cytokines, such as IFNs, IL-1 or IL-12, immune cells become anti-tumorigenic (*right side*) and differentiate/polarize towards immunogenic DCs (immuno-DC), T effector cells, activated pro-inflammatory macrophages or neutrophils.

Macrophages represent another important cell type that play a pivotal role in activating and shaping the immune response, and similar to DCs, a dichotomy has been proposed for macrophage activation: classically or alternatively activated [76, 77]. IFNα, LPS or inflammatory cytokines can induce classical activation while IL-4, IL-13, TGF-β and reparatory signals can induce alternative activation of macrophages. In the last decade, if macrophages have gone from a negligible player in tumor progression to a pivotal modulator of tumor growth and metastasis, it is due to the identification of TAMs. There is now solid and continuously growing evidence to show that TAMs actively promote all aspects of tumor growth and development including promotion of angiogenesis, matrix remodeling and suppression of adaptive immunity [78]. Recent studies also show that TAMs share many characteristics with alternatively activated macrophages such as (1) activation of the arginase pathway, implicated in arginine metabolism; (2) promotion of cell repair and proliferation in strong opposition with the NOS pathway that promote cells killing [79]; (3) production of IL-10 and vascular endothelial growth factor (VEGF) over other factors promoting cell survival [80]; (4) production of matrix metalloproteinases (MMPs), implicated in cancer initiation and metastasis [81, 82]; (5) activation of NFκB and STAT3 signaling, enhancing tumor progression by directly communicating with cancer stem cells (CSCs) [62], and (6) activation of STAT6, which possess potent inhibitory activity of T-cell activity [83].

T-cells, the soldiers of the immune system, have a fundamental role in immune surveillance, and even for these cells a dichotomy exists: a T-cell can defend the host from cancer while aiding tumor growth. Zhang et al. showed that ovarian carcinoma patients with high tumor infiltrating lymphocytes had improved 5-year survival rates compared to patients with low tumor infiltrating lymphocytes [84]; however, in other cancers, such as renal cancer, high tumor infiltrating lymphocytes translated into worse prognosis [85]. These clear opposing observations are now explained by differences in the type of infiltrating T-cells: T_{reg} versus T effectors cells ($TCD8^+$). T_{reg} cells, a subgroup of T-cells expressing CD4 and CD25, regulate activation of other T-cells and are necessary to maintain peripheral tolerance to self-antigens. We have now come to understand that increased numbers of T_{reg} cells present in the tumor can have a negative prognostic impact. For example, Sato et al. showed that a high $TCD8^+/T_{reg}$ cell ratio translated into better overall survival while the opposite was seen for a high $T_{reg}/TCD8^+$ ratio [86]. Likewise, Li et al. demonstrated that the efficiency and percent depletion of T_{reg} cells from a tumor can improve cancer outcome [67, 87]. Specifically, they show that in Foxp3.LuciDTR-4 mice, which show 90–95% T_{reg} depletion, large established tumors completely regressed, unlike anti-CD25 antibody-mediated T_{reg} elimination, which is less efficient (approximately 70%). Thus, high-level depletion of T_{reg} cells is necessary for tumor regression.

Like a coin, immune cells have two faces, one with a strong potential to fight cancer and the other (the opposite one) strongly promoting cancer development and immune system escape. This Dr. Jekyll and Mr. Hide concept has been re-coined the "corrupted policemen" concept by Bonavita et al. in order to stress the fact that those cells born to protect the host can become corrupt and turn against it to favor cancer growth [88]. This switch is complex and partly mediated by cytokines: inflammatory cytokines promoting immune system cells to show their anti-cancer face while anti-inflammatory cytokines promote the pro-cancer side to dominate. But cytokines are only mediators. More important are the stimuli that induce immune host cells to convert into either cancer allies (pro-tumor macrophages, T_{reg} cells, or tolerogenic DC) or cancer enemies (inflammatory macrophage, $CD8^{+}$ T-cells, T_{helper} cells or immunogenic DC) (Fig. 1). If we discover which mechanism(s) induce the inflammatory anti-cancer response and which ones induce the pro-cancer response, in theory we could pharmacologically change the face of the coin, favoring an anti-cancer response.

To fight or to repair? That is the Question

The first problem that oncoimmunotherapists faced was immune tolerance. Currently, a more challenging dilemma lies in the ability of the immune system to balance itself between two opposing actions: "fight" the enemy or "repair the damage". A successful immune response can be accompanied by extensive tissue damage [89]. Fortunately, the immune system has the capacity to repair the resulting damage via T_{reg} cells, macrophages and anti-inflammatory cytokines that send a repair message to the site of damage. In the context of cancer, however, the wrong choice can have detrimental effects on tumor eradication, as alluded to above. Numerous studies have shown that tumor growth and even post treatment tumor cell death can be perceived by the immune system as a repair signal, initiating a wound healing response that can favor sustained tumor growth and even tumor chemoresistance via immune cell secreted factors (reviewed in [90, 91]). Dissecting how the immune system senses and responds to damage will improve our efforts of inhibiting the immune system from favoring tumor growth over tumor destruction.

After immune system activation, the battlefield is replete with damaged cells, the majority being apoptotic cells (e.g. bacteria, neutrophils, epithelial cells). It is therefore logical to think that these cells could regulate the immune system or represent the signal that promotes a repair response. Although conventional and targeted therapies often aim to induce apoptosis, these strategies may themselves be carcinogenic [92].

The relation between cancer, immune cells and dead/apoptotic cells has gained increased attention over the past decade. Apoptosis obtained through activation of caspases or mitochondrial chain dysfunction was historically declared as non-immunogenic, while necrotic death accompanied by the release of proteins, lipids and other cellular debris from cells has been widely considered as strongly immunogenic [93, 94]. Chemotherapy-mediated cell death was accepted as an apoptosis-mediated process, inducing an immunosuppressive milieu of cytokines [95], but studies show that depending on the agent used and the degree of cell death induced, chemotherapy can also induce necrotic cells [96]. Moreover, other studies have shown that classically-induced apoptotic cells can profoundly affect the immune system [97], and the idea that chemotherapy-induced apoptosis is non-immunogenic may be overstated [98–100].

In cancer, there appears to be a ying-yang scenario with respect to the presence of dead cells and method of cell death induction. The presence of apoptotic cells can be sensed and translated into a "need to repair" action, allowing T_{reg}/Th2 stimulation through cytokine expression (IL-10, IL-13) [100, 101]. On the other hand, necrotic cells or highly variable tumor associated antigens can be sensed and translated into a "need to fight" action, allowing T-$CD4^+$ stimulation thought inflammatory cytokine expression (TNF-α, IL-1, etc.) [93]. Williams et al. showed that DCs exposed to apoptotic Jurkat cells or apoptotic primary T-cells failed to maturate and were unable to support $CD4^+$ allogeneic T-cell proliferation, as compared to DCs exposed to lipopolysaccharide (LPS) or necrotic cells [102]. Conversely, phosphatidylserine exposed on apoptotic epithelial cells suppressed IFN-β production by dendritic cells via inhibitory signalling mediated by the cell-surface glycoprotein CD300a and thus suppressed T_{reg} cell proliferation [103]. Moreover, Kleinclauss et al. in 2006 demonstrated that apoptotic cells induce $CD4^+$ T-cells to express $CD25^+$ (a marker of T_{reg} cells) inducing a state of tolerance [104]. Taken together, these studies demonstrate that the immune system can sense the type of danger and activate a specific action. In our laboratory we have discovered that apoptotic pancreatic tumor cells, as opposed to necrotic or live cells, can strongly induce immune system suppression or a pro-Th2 state, promoting repair signals that favor pancreatic tumor growth and chemoresistance (unpublished data). Our observations are in line with data published by Wu et al. where they show that intravenous administration of donor apoptotic splenocytes promotes the generation of tolerogenic DCs and the expansion of T_{reg} cells in the pancreas; in vivo clearance of either DCs or T_{reg} cells abrogated immune tolerance induction [105]. Thus, in pancreatic cancer where chemotherapies such as gemcitabine induce tumor cell apoptosis, an immune response that favors tumor growth may be activated.

CONCLUSION

The field of cancer immunotherapy has been driven by scientists such as William Coley whose important results set the foundation for modern day immunotherapy, as well as by scientists who while opposed the concept of immunotherapy made pinnacle discoveries that favored the evolution of immunotherapy into one of the most promising techniques in place today to battle cancer. The scientific community is convinced that immunotherapy is only a step away from becoming the magic bullet to defeat cancer. While the future looks promising, there are still many hurdles that need to be overcome.

We now have the capacity to "fuel the engine" by using vaccine strategies to accumulate effector T-cells at the tumor border, and "release the brake" to allow T-cells to fight the cancer by inhibiting immune system checkpoints. This strategy allows us to fight cancer when the battlefield is downhill or when the tumor entity itself is "immunogenic" due to (1) profound differences in tumor associated antigens between normal cells and tumor cells or (2) the presence of immunogenic necrotic cells. This strategy, however, is ineffective when the battlefield is plain or uphill, or when the tumor itself promotes a "need to repair" response due to chemotherapy-induced apoptosis or other unknown factors. Thus, when faced with this scenario, an approach to ensure that the immune system senses the cancer as danger in order to promote a "need to fight" over a "need to repair" response is essential. Consequently, the "push the accelerator" component of the universal magic bullet is still lacking. This action may possibly be achieved by educating APCs (DCs and macrophages) to sense every cancer as "dangerous", ensuring an anti-tumor immune response in all cases. Moreover, understanding that certain immune cells (e.g. macrophages) can shift the immune response in one direction over another, combination therapies may be necessary to transiently eliminate other immune cells. Ultimately, the goal is to unmask those factors that promote a "need to repair" response while at the same time enhancing those factors that are sensed as a danger signal by the immune system. Only then will immunotherapy truly become the magic bullet for cancer treatment.

AUTHORS' CONTRIBUTIONS

GD prepared, designed and wrote the manuscript. HLM and BSJ assisted in the manuscript design and edited the manuscript. All authors read and approved the final manuscript.

ACKNOWLEDGEMENTS

The authors would like to thank their respective laboratories and clinical collaborators for helpful discussions.

REFERENCES

1. DeVita VT, Jr, Chu E. A history of cancer chemotherapy. Cancer Res. 2008;68:8643–8653. doi: 10.1158/0008-5472.CAN-07-6611.
2. Liang XJ, Chen C, Zhao Y, et al. Circumventing tumor resistance to chemotherapy by nanotechnology. Methods Mol Biol. 2010;596:467–488. doi: 10.1007/978-1-60761-416-6_21.
3. Luqmani YA. Mechanisms of drug resistance in cancer chemotherapy. Med Princ Pract. 2005;14(Suppl 1):35–48.
4. Attarwala H. Role of antibodies in cancer targeting. J Nat Sci Biol Med. 2010;1:53–56. doi: 10.4103/0976-9668.71675.
5. Reslan L, Dalle S, Dumontet C. Understanding and circumventing resistance to anticancer monoclonal antibodies. MAbs. 2009;1:222–229. doi: 10.4161/mabs.1.3.8292.
6. Burnet F, Fenner F. The production of antibodies. 2. Melbourne: Macmillan; 1949.
7. Coussens LM, Zitvogel L, Palucka AK. Neutralizing tumor-promoting chronic inflammation: a magic bullet? Science. 2013;339:286–291. doi: 10.1126/science.1232227.
8. McCarthy EF. The toxins of William B. Coley and the treatment of bone and soft-tissue sarcomas. Iowa Orthop J. 2006;26:154–158.
9. Steen S, Stephenson G. Current treatment of soft tissue sarcoma. Proc (Bayl Univ Med Cent) 2008;21:392–396.
10. Alloatti A, Kotsias F, Magalhaes JG, et al. Dendritic cell maturation and cross-presentation: timing matters! Immunol Rev. 2016;272:97–108. doi: 10.1111/imr.12432.
11. Morelli AE, O'Connell PJ, Khanna A, et al. Preferential induction of Th1 responses by functionally mature hepatic (CD8alpha− and CD8alpha+) dendritic cells: association with conversion from liver transplant tolerance to acute rejection. Transplantation. 2000;69:2647–2657. doi: 10.1097/00007890-200006270-00027.
12. Agrawal S, Agrawal A, Said HM. Biotin deficiency enhances the inflammatory response of human dendritic cells. Am J Physiol Cell Physiol. 2016;00141:02016.
13. Romagnani S. T-cell subsets (Th1 versus Th2) Ann Allergy Asthma Immunol. 2000;85:9–18. doi: 10.1016/S1081-1206(10)62426-X.
14. Chen J, Zurawski G, Zurawski S, et al. A novel vaccine for mantle cell

lymphoma based on targeting cyclin D1 to dendritic cells via CD40. J Hematol Oncol. 2015;8:35. doi: 10.1186/s13045-015-0131-7.

15. Liston A. Immunological tolerance 50 years after the Burnet Nobel Prize. Immunol Cell Biol. 2011;89:14–15. doi: 10.1038/icb.2010.138.
16. Billingham RE, Brent L, Medawar PB. Actively acquired tolerance of foreign cells. Nature. 1953;172:603–606. doi: 10.1038/172603a0.
17. Lederberg J. Genes and antibodies. Science. 1959;129:1649–1653. doi: 10.1126/science.129.3364.1649.
18. Nossal GJ, Pike BL. Mechanisms of clonal abortion tolerogenesis. I. Response of immature hapten-specific B lymphocytes. J Exp Med. 1978;148:1161–1170. doi: 10.1084/jem.148.5.1161.
19. Marrack P, Kappler J. T cell tolerance. Semin Immunol. 1990;2:45–49.
20. Goodnow CC, Adelstein S, Basten A. The need for central and peripheral tolerance in the B cell repertoire. Science. 1990;248:1373–1379. doi: 10.1126/science.2356469.
21. Le Douarin N, Corbel C, Bandeira A, et al. Evidence for a thymus-dependent form of tolerance that is not based on elimination or anergy of reactive T cells. Immunol Rev. 1996;149:35–53. doi: 10.1111/j.1600-065X.1996.tb00898.x.
22. Foley EJ. Antigenic properties of methylcholanthrene-induced tumors in mice of the strain of origin. Cancer Res. 1953;13:835–837.
23. Nathrath WB. Organ and tumour antigens in malignant disease: a review. J R Soc Med. 1978;71:755–761.
24. Hargraves MM. Discovery of the LE cell and its morphology. Mayo Clin Proc. 1969;44:579–599.
25. DeLeo AB, Jay G, Appella E, et al. Detection of a transformation-related antigen in chemically induced sarcomas and other transformed cells of the mouse. Proc Natl Acad Sci USA. 1979;76:2420–2424. doi: 10.1073/pnas.76.5.2420.
26. Boon T, Cerottini JC, Van den Eynde B, et al. Tumor antigens recognized by T lymphocytes. Annu Rev Immunol. 1994;12:337–365. doi: 10.1146/annurev.iy.12.040194.002005.
27. Salmi M, Hellman J, Jalkanen S. The role of two distinct endothelial molecules, vascular adhesion protein-1 and peripheral lymph node addressin, in the binding of lymphocyte subsets to human lymph nodes. J Immunol. 1998;160:5629–5636.

28. Steinman RM. Lasker basic medical research award. Dendritic cells: versatile controllers of the immune system. Nat Med. 2007;13:1155–1159. doi: 10.1038/nm1643.
29. Bronte V, Pittet MJ. The spleen in local and systemic regulation of immunity. Immunity. 2013;39:806–818. doi: 10.1016/j.immuni.2013.10.010.
30. Vasilevko V, Ghochikyan A, Holterman MJ, et al. CD80 (B7-1) and CD86 (B7-2) are functionally equivalent in the initiation and maintenance of CD4+ T-cell proliferation after activation with suboptimal doses of PHA. DNA Cell Biol. 2002;21:137–149. doi: 10.1089/10445490252925404.
31. Sprent J. T and B memory cells. Cell. 1994;76:315–322. doi: 10.1016/0092-8674(94)90338-7.
32. Schietinger A, Greenberg PD. Tolerance and exhaustion: defining mechanisms of T cell dysfunction. Trends Immunol. 2014;35:51–60. doi: 10.1016/j.it.2013.10.001.
33. Sprent J, Kishimoto H. The thymus and central tolerance. Philos Trans R Soc Lond B Biol Sci. 2001;356:609–616. doi: 10.1098/rstb.2001.0846.
34. Schwartz RH. A cell culture model for T lymphocyte clonal anergy. Science. 1990;248:1349–1356. doi: 10.1126/science.2113314.
35. Mueller DL, Jenkins MK, Schwartz RH. Clonal expansion versus functional clonal inactivation: a costimulatory signalling pathway determines the outcome of T cell antigen receptor occupancy. Annu Rev Immunol. 1989;7:445–480. doi: 10.1146/annurev.iy.07.040189.002305.
36. Brunet JF, Denizot F, Luciani MF, et al. A new member of the immunoglobulin superfamily–CTLA-4. Nature. 1987;328:267–270. doi: 10.1038/328267a0.
37. van der Merwe PA, Bodian DL, Daenke S, et al. CD80 (B7-1) binds both CD28 and CTLA-4 with a low affinity and very fast kinetics. J Exp Med. 1997;185:393–403. doi: 10.1084/jem.185.3.393.
38. Lipson EJ, Drake CG. Ipilimumab: an anti-CTLA-4 antibody for metastatic melanoma. Clin Cancer Res. 2011;17:6958–6962. doi: 10.1158/1078-0432.CCR-11-1595.
39. Waterhouse P, Penninger JM, Timms E, et al. Lymphoproliferative disorders with early lethality in mice deficient in Ctla-4. Science. 1995;270:985–988. doi: 10.1126/science.270.5238.985.
40. Leach DR, Krummel MF, Allison JP. Enhancement of antitumor

immunity by CTLA-4 blockade. Science. 1996;271:1734–1736. doi: 10.1126/science.271.5256.1734.

41. van Elsas A, Hurwitz AA, Allison JP. Combination immunotherapy of B16 melanoma using anti-cytotoxic T lymphocyte-associated antigen 4 (CTLA-4) and granulocyte/macrophage colony-stimulating factor (GM-CSF)-producing vaccines induces rejection of subcutaneous and metastatic tumors accompanied by autoimmune depigmentation. J Exp Med. 1999;190:355–366. doi: 10.1084/jem.190.3.355.
42. van Elsas A, Sutmuller RP, Hurwitz AA, et al. Elucidating the autoimmune and antitumor effector mechanisms of a treatment based on cytotoxic T lymphocyte antigen-4 blockade in combination with a B16 melanoma vaccine: comparison of prophylaxis and therapy. J Exp Med. 2001;194:481–489. doi: 10.1084/jem.194.4.481.
43. Lute KD, May KF, Jr, Lu P, et al. Human CTLA4 knock-in mice unravel the quantitative link between tumor immunity and autoimmunity induced by anti-CTLA-4 antibodies. Blood. 2005;106:3127–3133. doi: 10.1182/blood-2005-06-2298.
44. Keler T, Halk E, Vitale L, et al. Activity and safety of CTLA-4 blockade combined with vaccines in cynomolgus macaques. J Immunol. 2003;171:6251–6259. doi: 10.4049/jimmunol.171.11.6251.
45. Ghiotto M, Gauthier L, Serriari N, et al. PD-L1 and PD-L2 differ in their molecular mechanisms of interaction with PD-1. Int Immunol. 2010;22:651–660. doi: 10.1093/intimm/dxq049.
46. Latchman Y, Wood CR, Chernova T, et al. PD-L2 is a second ligand for PD-1 and inhibits T cell activation. Nat Immunol. 2001;2:261–268. doi: 10.1038/85330.
47. Dong H, Zhu G, Tamada K, et al. B7-H1, a third member of the B7 family, co-stimulates T-cell proliferation and interleukin-10 secretion. Nat Med. 1999;5:1365–1369. doi: 10.1038/70932.
48. Ishida Y, Agata Y, Shibahara K, et al. Induced expression of PD-1, a novel member of the immunoglobulin gene superfamily, upon programmed cell death. EMBO J. 1992;11:3887–3895.
49. Agata Y, Kawasaki A, Nishimura H, et al. Expression of the PD-1 antigen on the surface of stimulated mouse T and B lymphocytes. Int Immunol. 1996;8:765–772. doi: 10.1093/intimm/8.5.765.
50. Francisco LM, Salinas VH, Brown KE, et al. PD-L1 regulates the development, maintenance, and function of induced regulatory T

cells. J Exp Med. 2009;206:3015–3029. doi: 10.1084/jem.20090847.

51. Jin HT, Ahmed R, Okazaki T. Role of PD-1 in regulating T-cell immunity. Curr Top Microbiol Immunol. 2011;350:17–37.
52. Sheppard KA, Fitz LJ, Lee JM, et al. PD-1 inhibits T-cell receptor induced phosphorylation of the ZAP70/CD3zeta signalosome and downstream signaling to PKCtheta. FEBS Lett. 2004;574:37–41. doi: 10.1016/j.febslet.2004.07.083.
53. Barber DL, Wherry EJ, Masopust D, et al. Restoring function in exhausted CD8 T cells during chronic viral infection. Nature. 2006;439:682–687. doi: 10.1038/nature04444.
54. Dong H, Strome SE, Salomao DR, et al. Tumor-associated B7-H1 promotes T-cell apoptosis: a potential mechanism of immune evasion. Nat Med. 2002;8:793–800. doi: 10.1038/nm0902-1039c.
55. Niezgoda A, Niezgoda P, Czajkowski R. Novel approaches to treatment of advanced melanoma: a review on targeted therapy and immunotherapy. Biomed Res Int. 2015;2015:851387. doi: 10.1155/2015/851387.
56. Topalian SL, Sznol M, McDermott DF, et al. Survival, durable tumor remission, and long-term safety in patients with advanced melanoma receiving nivolumab. J Clin Oncol. 2014;32:1020–1030. doi: 10.1200/JCO.2013.53.0105.
57. Keating GM. Nivolumab: a review in advanced nonsquamous non-small cell lung cancer. Drugs. 2016;76:969–978. doi: 10.1007/s40265-016-0589-9.
58. Larkin J, Chiarion-Sileni V, Gonzalez R, et al. Combined nivolumab and ipilimumab or monotherapy in untreated melanoma. N Engl J Med. 2015;373:23–34. doi: 10.1056/NEJMoa1504030.
59. Kantoff PW, Higano CS, Shore ND, et al. Sipuleucel-T immunotherapy for castration-resistant prostate cancer. N Engl J Med. 2010;363:411–422. doi: 10.1056/NEJMoa1001294.
60. Lutz ER, Wu AA, Bigelow E, et al. Immunotherapy converts nonimmunogenic pancreatic tumors into immunogenic foci of immune regulation. Cancer Immunol Res. 2014;2:616–631. doi: 10.1158/2326-6066.CIR-14-0027.
61. Nummer D, Suri-Payer E, Schmitz-Winnenthal H, et al. Role of tumor endothelium in CD4+ CD25+ regulatory T cell infiltration of human pancreatic carcinoma. J Natl Cancer Inst. 2007;99:1188–1199. doi:

10.1093/jnci/djm064.

62. Sainz B, Jr, Carron E, Vallespinos M, et al. Cancer stem cells and macrophages: implications in tumor biology and therapeutic strategies. Mediat Inflamm. 2016;2016:9012369. doi: 10.1155/2016/9012369.

63. Bui JD, Uppaluri R, Hsieh CS, et al. Comparative analysis of regulatory and effector T cells in progressively growing versus rejecting tumors of similar origins. Cancer Res. 2006;66:7301–7309. doi: 10.1158/0008-5472.CAN-06-0556.

64. Diller ML, Kudchadkar RR, Delman KA, et al. Balancing inflammation: the link between Th17 and regulatory T cells. Mediat Inflamm. 2016;2016:6309219. doi: 10.1155/2016/6309219.

65. Shaw TJ, Martin P. Wound repair: a showcase for cell plasticity and migration. Curr Opin Cell Biol. 2016;42:29–37. doi: 10.1016/j.ceb.2016.04.001.

66. DuPage M, Bluestone JA. Harnessing the plasticity of CD4(+) T cells to treat immune-mediated disease. Nat Rev Immunol. 2016;16:149–163. doi: 10.1038/nri.2015.18.

67. Zhu J, Yamane H, Paul WE. Differentiation of effector CD4 T cell populations. Annu Rev Immunol. 2010;28:445–489. doi: 10.1146/annurev-immunol-030409-101212.

68. Palucka K, Banchereau J. Cancer immunotherapy via dendritic cells. Nat Rev Cancer. 2012;12:265–277. doi: 10.1038/nrc3258.

69. Martinez FO, Gordon S. The M1 and M2 paradigm of macrophage activation: time for reassessment. F1000Prime Rep. 2014;6:13. doi: 10.12703/P6-13.

70. Mantovani A, Cassatella MA, Costantini C, et al. Neutrophils in the activation and regulation of innate and adaptive immunity. Nat Rev Immunol. 2011;11:519–531. doi: 10.1038/nri3024.

71. Steinman RM. Dendritic cells: understanding immunogenicity. Eur J Immunol. 2007;37(Suppl 1):S53–S60. doi: 10.1002/eji.200737400.

72. Fujii S, Liu K, Smith C, et al. The linkage of innate to adaptive immunity via maturing dendritic cells in vivo requires CD40 ligation in addition to antigen presentation and CD80/86 costimulation. J Exp Med. 2004;199:1607–1618. doi: 10.1084/jem.20040317.

73. Maldonado RA, von Andrian UH. How tolerogenic dendritic cells induce regulatory T cells. Adv Immunol. 2010;108:111–165. doi:

10.1016/B978-0-12-380995-7.00004-5.

74. Lee JI, Ganster RW, Geller DA, et al. Cyclosporine A inhibits the expression of costimulatory molecules on in vitro-generated dendritic cells: association with reduced nuclear translocation of nuclear factor kappa B. Transplantation. 1999;68:1255–1263. doi: 10.1097/00007890-199911150-00007.
75. Penna G, Adorini L. 1 alpha, 25-dihydroxyvitamin D3 inhibits differentiation, maturation, activation, and survival of dendritic cells leading to impaired alloreactive T cell activation. J Immunol. 2000;164:2405–2411. doi: 10.4049/jimmunol.164.5.2405.
76. Gordon S, Martinez FO. Alternative activation of macrophages: mechanism and functions. Immunity. 2010;32:593–604. doi: 10.1016/j.immuni.2010.05.007.
77. Kidd JF, Pilkington MF, Schell MJ, et al. Paclitaxel affects cytosolic calcium signals by opening the mitochondrial permeability transition pore. J Biol Chem. 2002;277:6504–6510. doi: 10.1074/jbc.M106802200.
78. Sica A, Schioppa T, Mantovani A, et al. Tumour-associated macrophages are a distinct M2 polarised population promoting tumour progression: potential targets of anti-cancer therapy. Eur J Cancer. 2006;42:717–727. doi: 10.1016/j.ejca.2006.01.003.
79. Rath M, Muller I, Kropf P, et al. Metabolism via arginase or nitric oxide synthase: two competing arginine pathways in macrophages. Front Immunol. 2014;5:532. doi: 10.3389/fimmu.2014.00532.
80. Golpon HA, Fadok VA, Taraseviciene-Stewart L, et al. Life after corpse engulfment: phagocytosis of apoptotic cells leads to VEGF secretion and cell growth. FASEB J. 2004;18:1716–1718.
81. Zhao Y, Zhou FL, Li WP, et al. Slit2Robo1 signaling promotes the adhesion, invasion and migration of tongue carcinoma cells via upregulating matrix metalloproteinases 2 and 9, and downregulating Ecadherin. Mol Med Rep. 2016;14:1901–1906.
82. Jakubowska K, Pryczynicz A, Januszewska J, et al. Expressions of matrix metalloproteinases 2, 7, and 9 in carcinogenesis of pancreatic ductal adenocarcinoma. Dis Markers. 2016;2016:9895721. doi: 10.1155/2016/9895721.
83. Huber S, Hoffmann R, Muskens F, et al. Alternatively activated macrophages inhibit T-cell proliferation by Stat6-dependent

expression of PD-L2. Blood. 2010;116:3311–3320. doi: 10.1182/blood-2010-02-271981.

84. Zhang L, Conejo-Garcia JR, Katsaros D, et al. Intratumoral T cells, recurrence, and survival in epithelial ovarian cancer. N Engl J Med. 2003;348:203–213. doi: 10.1056/NEJMoa020177.
85. Geissler K, Fornara P, Lautenschlager C, et al. Immune signature of tumor infiltrating immune cells in renal cancer. Oncoimmunology. 2015;4:e985082. doi: 10.4161/2162402X.2014.985082.
86. Sato E, Olson SH, Ahn J, et al. Intraepithelial CD8+ tumor-infiltrating lymphocytes and a high CD8+/regulatory T cell ratio are associated with favorable prognosis in ovarian cancer. Proc Natl Acad Sci USA. 2005;102:18538–18543. doi: 10.1073/pnas.0509182102.
87. Li X, Kostareli E, Suffner J, et al. Efficient Treg depletion induces T-cell infiltration and rejection of large tumors. Eur J Immunol. 2010;40:3325–3335. doi: 10.1002/eji.201041093.
88. Bonavita E, Galdiero MR, Jaillon S, et al. Phagocytes as corrupted policemen in cancer-related inflammation. Adv Cancer Res. 2015;128:141–171. doi: 10.1016/bs.acr.2015.04.013.
89. Nathan C. Points of control in inflammation. Nature. 2002;420:846–852. doi: 10.1038/nature01320.
90. Beaman KD, Jaiswal MK, Katara GK, et al. Pregnancy is a model for tumors, not transplantation. Am J Reprod Immunol. 2016;76:3–7. doi: 10.1111/aji.12524.
91. Vakkila J, Lotze MT. Inflammation and necrosis promote tumour growth. Nat Rev Immunol. 2004;4:641–648. doi: 10.1038/nri1415.
92. Harris CC. The carcinogenicity of anticancer drugs: a hazard in man. Cancer. 1976;37:1014–1023. doi: 10.1002/1097-0142(197602)37:2+<1014::AID-CNCR2820370805>3.0.CO;2-Z.
93. Iyer SS, Pulskens WP, Sadler JJ, et al. Necrotic cells trigger a sterile inflammatory response through the Nlrp3 inflammasome. Proc Natl Acad Sci USA. 2009;106:20388–20393. doi: 10.1073/pnas.0908698106.
94. Thompson CB. Apoptosis in the pathogenesis and treatment of disease. Science. 1995;267:1456–1462. doi: 10.1126/science.7878464.
95. Casares N, Pequignot MO, Tesniere A, et al. Caspase-dependent immunogenicity of doxorubicin-induced tumor cell death. J Exp

Med. 2005;202:1691–1701. doi: 10.1084/jem.20050915.

96. Okada M, Adachi S, Imai T, et al. A novel mechanism for imatinib mesylate-induced cell death of BCR-ABL-positive human leukemic cells: caspase-independent, necrosis-like programmed cell death mediated by serine protease activity. Blood. 2004;103:2299–2307. doi: 10.1182/blood-2003-05-1605.
97. Campisi L, Cummings RJ, Blander JM. Death-defining immune responses after apoptosis. Am J Transplant. 2014;14:1488–1498. doi: 10.1111/ajt.12736.
98. Ferguson TA, Herndon J, Elzey B, et al. Uptake of apoptotic antigen-coupled cells by lymphoid dendritic cells and cross-priming of CD8(+) T cells produce active immune unresponsiveness. J Immunol. 2002;168:5589–5595. doi: 10.4049/jimmunol.168.11.5589.
99. Griffith TS, Kazama H, VanOosten RL, et al. Apoptotic cells induce tolerance by generating helpless CD8+ T cells that produce TRAIL. J Immunol. 2007;178:2679–2687. doi: 10.4049/jimmunol.178.5.2679.
100. Tomimori Y, Ikawa Y, Oyaizu N. Ultraviolet-irradiated apoptotic lymphocytes produce interleukin-10 by themselves. Immunol Lett. 2000;71:49–54. doi: 10.1016/S0165-2478(99)00163-7.
101. Weigert A, Tzieply N, von Knethen A, et al. Tumor cell apoptosis polarizes macrophages role of sphingosine-1-phosphate. Mol Biol Cell. 2007;18:3810–3819. doi: 10.1091/mbc.E06-12-1096.
102. Williams CA, Harry RA, McLeod JD. Apoptotic cells induce dendritic cell-mediated suppression via interferon-gamma-induced IDO. Immunology. 2008;124:89–101. doi: 10.1111/j.1365-2567.2007.02743.x.
103. Nakahashi-Oda C, Udayanga KG, Nakamura Y, et al. Apoptotic epithelial cells control the abundance of Treg cells at barrier surfaces. Nat Immunol. 2016;17:441–450. doi: 10.1038/ni.3345.
104. Kleinclauss F, Perruche S, Masson E, et al. Intravenous apoptotic spleen cell infusion induces a TGF-beta-dependent regulatory T-cell expansion. Cell Death Differ. 2006;13:41–52. doi: 10.1038/sj.cdd.4401699.
105. Wu C, Zhang Y, Jiang Y, et al. Apoptotic cell administration enhances pancreatic islet engraftment by induction of regulatory T cells and tolerogenic dendritic cells. Cell Mol Immunol. 2013;10:393–402. doi: 10.1038/cmi.2013.16.

14

Quantitative Mechanistic Modeling in Support of Pharmacological Therapeutics Development in Immuno-Oncology

Kirill Peskov [1,2], Ivan Azarov [1], Lulu Chu[3], Veronika Voronova[1], Yuri Kosinsky [1] and Gabriel Helmlinger [3]

[1] M&S Decisions, Moscow, Russia
[2] Computational Oncology Group, I.M. Sechenov First Moscow State Medical University of the Russian Ministry of Health, Moscow, Russia
[3] Quantitative Clinical Pharmacology, Early Clinical Development, IMED Biotech Unit, AstraZeneca Pharmaceuticals, Boston, MA, United States

ABSTRACT

Following the approval, in recent years, of the first immune checkpoint inhibitor, there has been an explosion in the development of immunomodulating pharmacological modalities for the treatment of various cancers. From the discovery phase to late-stage clinical testing and regulatory approval,

Citation: Peskov K, Azarov I, Chu L, Voronova V, Kosinsky Y, Helmlinger G. Quantitative Mechanistic Modeling in Support of Pharmacological Therapeutics Development in Immuno-Oncology. Front Immunol. 2019; 10:924. Published 2019 Apr 30. doi:10.3389/fimmu.2019.00924

challenges in the development of immuno-oncology (IO) drugs are multi-fold and complex. In the preclinical setting, the multiplicity of potential drug targets around immune checkpoints, the growing list of immuno-modulatory molecular and cellular forces in the tumor microenvironment—with additional opportunities for IO drug targets, the emergence of exploratory biomarkers, and the unleashed potential of modality combinations all have necessitated the development of quantitative, mechanistically-oriented systems models which incorporate key biology and patho-physiology aspects of immuno-oncology and the pharmacokinetics of IO-modulating agents. In the clinical setting, the qualification of surrogate biomarkers predictive of IO treatment efficacy or outcome, and the corresponding optimization of IO trial design have become major challenges. This mini-review focuses on the evolution and state-of-the-art of quantitative systems models describing the tumor vs. immune system interplay, and their merging with quantitative pharmacology models of IO-modulating agents, as companion tools to support the addressing of these challenges.

Keywords: immuno-oncology, mechanistic models, tumor vs. immune system, systems pharmacology, pharmacokinetics, pharmacodynamics, molecular and cellular biomarkers

INTRODUCTION

Immunotherapy of cancer has had a long history of development, starting from pioneering efforts in using coley toxins to treat patients—a therapeutic approach named after Dr. William Coley (1). Even though these earlier efforts never turned into a standard treatment, further investigations on the relationships between tumor cells and the immune system led to discoveries which unveiled fundamental principles underlying cancer progression, such as immune surveillance (2, 3), cancer dormancy (4), cancer immuno-editing (5), and the cancer immunity cycle (6). These discoveries were foundational for clinical successes and corresponding regulatory approvals in recent years, of therapies targeting the CTLA-4, PD-1, and PD-L1 immune checkpoints. In the wake of these successes, there has been an explosion in the development of immuno-modulating, anti-cancer pharmacological modalities, leading to the initiation of, literally, thousands of clinical trials (7, 8). However, from the discovery phase to late-stage clinical testing and regulatory approval, challenges in the development of immuno-oncology (IO) drugs are multi-fold and complex (9), with related complexities in the design of clinical trials; if unaddressed, these may lead to a decreased probability of success (10). Some of these challenges can be mapped to an incomplete mechanistic understanding of immune response dynamics and the interplay of such immune responses with tumor infiltration processes and tumor cell growth (11). These quantitative knowledge gaps hinder: (i) effective translation of novel promising therapeutic approaches into the clinic, (ii) identification of predictive response biomarkers, and (iii) search of therapeutic drug combinations which may overcome intrinsic or acquired resistance to existing standards of care (12). This mini-review focuses on quantitative, mechanistically-oriented modeling approaches which have been sought in IO, to address, at least partially, the abovementioned challenges and knowledge gaps.

EVOLUTION OF QUANTITATIVE, MECHANISTICALLY-ORIENTED IO SYSTEMS MODELING

Application of mathematical modeling in support of preclinical and clinical research, as well as decision-making in Oncology, has a long-standing history covering multiple problems and addressing a variety of research questions—today often referred to as computational oncology (13–15). Historical milestones include adaptations of the Gompertz model for treatment outcomes in breast cancer (16). These earlier efforts started from models with a simplistic empirical structure, based on an ordinary differential equation (ODE) describing tumor size growth using an exponential or sigmoidal function (17). Such a model, however, would not adequately describe the interplay between tumor cells and tissue vs. the immune system, since it entirely ignores the immune component (18). It is nevertheless valuable to mathematically describe treatment response effects following various chemotherapies, which are adequately captured by generalized Gompertzian kinetics (19). In fact, such modeling results provided a basis for the use of specific "dose-dense" chemotherapeutic regimens, which subsequently showed favorable outcomes in the treatment of breast cancer (20). Additionally, such empirical considerations allowed for a gradual evolution of modeling concepts, which today can be grounded in mechanistically-oriented principles, including for tumor *vs.* immune system interactions (Figure 1).

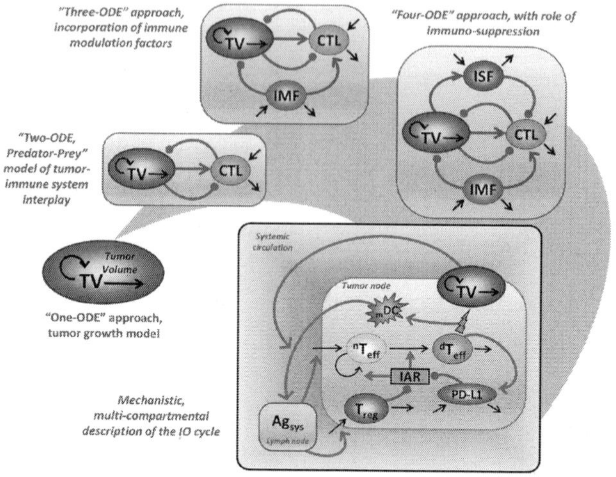

Figure 1: Evaluation of mathematical models which describe tumor vs. immune system interactions. "One-ODE" approach, simplistic description of tumor growth kinetics; "Two-ODE" approach, a typical "Predator-Prey" model, incorporating a basic description of tumor vs. immune system interactions; "Three-ODE" approach, incorporating additional immuno-modulating factor(s); "Four-ODE" approach, including considerations for immuno-suppression; mechanistic multi-compartmental model, taking into account essential biological principles underlying the IO cycle concept (21); TV, tumor volume; CTL, cytotoxic T lymphocytes; IMF, immuno-modulating factor; ISF, immuno-suppressive factor; mDC, level of mature dendritic cells; $_{n}T_{eff}$, non-differentiated T effectors cells; $_{d}T_{eff}$, differentiated T effectors cells; Treg, regulatory T cells; PD-L1, level of PD-L1 expression; Ag_{sys}, level of systemic antigen; IAR, immuno-activation rate function; green line, positive regulation, red line negative, regulation; back line, variable turnover.

Earlier efforts to describe tumor vs. immune system relationships via a general mathematical description appeared in the 1980's, following the pioneering IO work that introduced the concept of immune surveillance (2, 3). These mathematical models considered the addition of a second variable describing the dynamics of cytotoxic immune cells, which are able to attack tumor cells (22–24). The resultant "two-ODE" model actually follows a typical "predator-prey" model introduced by Alfred Lotka and Vito Volterra, in much earlier days, at the turn of the 20th century. In such a model, tumor cells may be interpreted as the "prey," whereas cytotoxic immune cells may be viewed as the "predator": their dynamic interplay may result in one possible system behavior reflective of cancer dormancy (4). Given the relative simplicity of such a "two-ODE" model and since the behavior of such a model could be assessed analytically, it gained immense popularity within the oncology modeling community and led to several theoretical hypotheses underlying fundamental principles of cancer progression. For example, it was shown, through modeling, that key parameters controlling tumor re-growth under steady-state conditions of cancer dormancy were those relating to activities of the immune system (25). A corollary result was that it is a reduction in the probability of achieving tumor cell kill, rather than a reduction in the probability of tumor cells being recognized by cytotoxic cells, which best explained immune evasion by tumor cells (26). Interestingly, this key result, derived theoretically at the time, has recently been supported by elegant modeling work linking high-level immunological and epidemiological data, which suggests that age-related decline in T cell output correlates better with risk of cancer diagnosis vs. age-related accumulation of somatic mutations in tumor cells (27).

With the explosive growth of experimental data surrounding the complexity of tumor vs. immune system interplay, "two-ODE" models experienced a further evolution with additional biological entities and mechanisms being taken into mathematical consideration. At this point and looking forward, many biological candidates were tested as the "third modeling variable," representing either specific immune cells or cytokines that modulate cytotoxic T lymphocyte (CTL) function (28). Such models were initially focused on including IL-2 function and effects, reflective of the potential importance of this cytokine and its associated dynamics in long-term tumor relapse (29). In further work, de Pillis et al. used a "three-ODE" model to reveal a difference between the dynamics of $CD8^+$ CTLs *vs.* natural killer cells, which supported the importance of considering multiple cell types in the overall anti-tumor immune activity (30). More recently, $CD4^+$ T helper cells were considered as the third component, in a quantitative, model-based investigation of adoptive cellular immunotherapy (31).

"Three-ODE" models, however, exhibit one significant structural limitation, namely they completely lack (an) immuno-suppressive component(s), which would be crucial when considering immune evasion mechanisms (32). Therefore, embedding a fourth variable into such models, to describe immuno-suppression, would seem rather natural; however, choices for the most appropriate candidate in this role are multi-fold. Several types of immuno-suppressive cells or molecules could be suitable candidates, including regulatory T cells (Tregs), myeloid-derived suppressor cells (MDSCs), or Type 2 tumor-associated macrophages, as well as cytokines such as TGFβ or IL-10. Thus, Arciero et al. chose TGFβ as the fourth model variable (33), while de Pillis et al. used Tregs as the principal immuno-suppressive component in their model (34). While these two modeling examples focused on immuno-suppressive effectors, other "four-ODE" models abound, declining a vast variety of immune players or tumor cell clones (35–41). Such a variety in potential key immuno-modulating factors made the generalization of any "three-ODE" or "four-ODE" model an overly difficult process, since any one of the models cited above can be challenged with newly generated experimental data featuring the importance of one vs. another immune factor. This may also explain, at least partially, the relatively minimal recognition, to date, of quantitative modeling approaches by immuno-oncologists (28, 42, 43).

On one hand, some of the biological complexities which compose the IO cycle, as summarized in recent reviews (6, 44, 45) clearly indicate the limitations of oversimplified models such as "prey-predator" models, which appear to be too remote from experimental reality and would not be applicable or of use for the majority of research relevant questions. On the other hand, increasing model complexity with additional mechanistic insights always comes with challenges of model calibration, as depicted in this famous quote by John von Neumann, "*with four parameters, I can fit an elephant and with five, I can make him wiggle his trunk*" [see in Dyson (46)]—pointing to the necessity of avoiding overparameterized "metastatic" models with unreliable extensions and loss of predictive power. Achieving such a balance in capturing necessary (not over-simplified) yet sufficient (not over-developed) features, and as constrained by the available data, is arguably one of the most difficult challenges in fit-for-purpose, parsimonious mechanistic model building and calibration. Overparameterization can easily negate all benefits brought forward by the incorporation of exquisite biological details of the system under consideration (47); models which attempt to explain everything may in fact not be useful, their predictive power remaining a question mark (48).

To address this challenge, part of the solution may reside in the combining of modeling methodologies developed previously and in other disciplines (49). This would result in a repository of prior information and knowledge validated elsewhere, to build mechanistic models in immuno-oncology which, on one hand, incorporate increasing system complexity and, on the other hand, avoid overparameterization based on newly generated data—thereby resulting, using terminology of a Bayesian mindset, in a posterior model based on existing, established quantitative priors (50).

If so, the question then becomes, "where to find such established prior models?" One obvious domain is quantitative immunology (51, 52), where the use of various modeling techniques by experimentalists has already gotten significantly more traction, arguably, than in other fundamental biological disciplines (53). For example, modeling has provided quantitative "inference frameworks" for immunology basics and fundamentals such as T cell activation, homeostasis or self / non-self recognition (54–57), immune receptor signaling (58), and understanding of T cell immunological memory (59). Prior models from quantitative immunology may then be combined with prior models from quantitative pharmacology (60–62), another field where modeling has provided quantitative "inference frameworks" (63) In the next section, we will discuss selected works which considered the combining of modeling methodologies, in attempts to develop pharmacologically-modulated posterior models, which were then used to prospectively address questions in the development of IO therapies (Table 1).

Table 1: Mechanistic models in support of IO therapy development

Description[a]	Application[a]	Limitations[a]	References
Cergutuzumab amunaleukin (CEA mAb-IL2v fusion protein) PK/PD described using a population NLME modeling approach	Model was used to identify optimal dosing regimen and support design of the clinical dose escalation study	Since mechanisms of tumor vs. immune system interactions have not been considered, the model cannot be generalized to other MoAs nor their combinations	(64)
Pembrolisumab (αPD-1 mAb) PK/PD described using a population NLME modeling approach	Model was used to estimate MABEL dose and was applied, accordingly, for FIH dose selection	1. Since mechanisms of tumor vs. immune system interactions have not been considered, the model cannot be generalized to other MoAs nor their combinations 2. Model is based on preclinical data only	(65)
DART against CD3 and P-cadherin PK/PD was described using a simple ODE modeling framework	1. Model was used to estimate MABEL dose and was applied, accordingly, for FIH dose selection 2. Model was further applied for identification/better characterization of PK/PD relationship and MoA	1. Since mechanisms of tumor vs. immune system interactions have not been considered, the model cannot be generalized to other MoAs nor their combinations 2. Model is based on preclinical data only and does not take into account variability	(66)

Multiple MoAs including vaccination (CyaA-E7), TLR9 agonist (CpG), chemotherapy (cyclophosphamide), and IL-12 administration were incorporated using a NLME modeling approach	Model was applied for a better understanding of synergistic effects in combination treatment	1. Due to the simple description of tumor vs. immune system interactions, the model cannot be generalized to other MoAs 2. Model is based on preclinical data only	(21, 67)
Multiple MoAs including αPD-1 and αPD-L1, αCTLA4 mAb, OX40 agonists, CXCR2 inhibitors, and RT were incorporated using a population NLME modeling approach	1. Model was applied for a better understanding of synergistic effects in combination treatment and identification of predictive biomarkers 2. Based on the model simulations, an optimal sequencing schedule was proposed for the combination treatment	1. Model is based on preclinical data only	(68, 69)
RT and αCTLA4 mAb were described using a simple ODE modeling framework	Model was applied to guide optimal combination treatment doses and schedules	1. Due to the simple description of tumor vs. immune system interactions, pharmacological interventions and limited validation with experimental data, the model cannot be generalized to other MoAs nor used for clinically relevant simulations 2. Model is based on preclinical data only and does not take into account variability	(70)
Mechanistic physiologically-based description of clinically-relevant immune cell fluxes and RT	1. Model was applied for a better understanding of ICD systemic effects 2. Optimal RT administration sites for metastatic solid tumors were identified	1. Limited validation with clinical data was performed during model development stage 2. Model does not take into account variability	(71)

Multiple MoAs including αPD-L1, BRAF and MEK inhibitors and vaccination (GVAX) were incorporated using a simple PDE modeling framework, to account for spatial immune species distribution within the tumor compartment	Model was applied for a better understanding of synergistic effects	1. Due to the simple description of tumor vs. immune system interactions, pharmacological interventions and limited validation with experimental data, the model cannot be generalized to other MoAs nor used for clinically relevant simulations 2. Model is based on preclinical data only and does not take into account variability	(72, 73)
Multiple MoAs including vaccination (UV-8101-RE), IL-2 neutralization, Treg cell depletion, androgen deprivation therapy and castration were incorporated using a simple ODE modeling framework	Model was applied to guide optimal combination treatment schemes	1. Due to the simple description of tumor vs. immune system interactions, pharmacological interventions and limited validation with experimental data, the model cannot be generalized to other MoAs nor used for clinically relevant simulations; 2. Model is based on the preclinical data only and does not take into account variability	(74)
Alloreactive cytotoxic-T-lymphocytes transfer was described using a simple ODE modeling framework	Model was applied for the identification of predictive biomarkers	1. Model does not take into account variability 2. Limited validation with clinical data was performed	(75)
IL-21 administration was described using a simple ODE modeling framework	Model was applied for a better MoA understanding and the identification of predictive biomarkers	1. Model is based on preclinical data only and does not take into account variability 2. Limited validation with clinical data was performed	(76)
Prostate cancer vaccination effects were described using a simple ODE modeling framework	Model was applied for an evaluation of personalized treatment strategies	1. Limited validation with clinical data was performed	(77)
Multiple MoAs including αPD-L1 mAb, BTK inhibitor (ibrutinib), and vaccination were incorporated using a simple ODE modeling framework	Model was applied for guiding optimal combination treatment schemes	1. Due to the simple description of tumor vs. immune system interactions, pharmacological interventions and limited validation with experimental data model cannot be generalized to other MoAs nor used for clinically relevant simulations; 2. Model is based on preclinical data only and does not take into account variability	(78)

αPD-L1 mAb clinical effects were described using a 3D ABM framework	Model was applied for an evaluation of personalized treatment strategies	1. Limited validation with clinical was performed 2. Systemic treatment effects were not considered	(79)
Generalized effects of adaptive immunity stimulation and stromal cell depletion were described using a 2D and 3D ABM framework	Model was applied for guiding optimal combination treatment schemes	1. Generic representation of treatment effects	(80, 81)

[a]*MoA, Mechanism of action; CEA, carcinoembryonic antigen; mAb, monoclonal antibody; NLME, nonlinear mixed effects; IO, immuno-oncology; PK, pharmacokinetics; PD, pharmacodynamics; MABEL, minimally anticipated biological effect level; FIH, first-in-human; RT, radiotherapy; ICD, immunologic cell death; ODE, ordinary differential equations; PDE, partial differential equation; ABM, agent-based modeling.*

MECHANISTIC MODELING IN SUPPORT OF IO THERAPY DEVELOPMENT

Applications of mechanistic modeling in support of preclinical and clinical research, commonly referred to as pharmacokinetic (PK)/pharmacodynamic (PD) modeling, are traditionally centered around the optimization of treatment dosing and scheduling—the "dose" representing a critical component of any drug development program (82). Such modeling approaches have thus been used in the development of IO agents such as PD-1 and PD-L1 inhibitors (83–85). In particular, mechanistic PKPD modeling has been applied in support of first-in-human dose selection of pembrolizumab, an anti PD-1 agent (65); this resulted in a seamless clinical trial design with a model-informed dose justification, which the US FDA accepted in the process of an accelerated regulatory review (86). Label updates with flat dosing schedules were subsequently granted, for both nivolumab and pembrolizumab, strongly supported by model-based simulations (87, 88). PKPD modeling has also been used in the translation of preclinical data for a conjugated IL-2 therapy, in particular to gain a better understanding of such a therapy's downstream effects (89). PKPD modeling has been further used in the development of bispecific biologics. Chen et al. used it for the estimation of the minimally anticipated biological effect level (MABEL) of a bispecific antibody targeting CD3 and p-cadherin (66), while Ribba et al. used it for guided dose escalation study design of cergutuzumab amunaleukin, a fusion protein consisting of IL-2 and a carcinoembryonic antigen (CEA) human monoclonal antibody (64). Such models are great examples of a "fit-for-purpose" quantitative approach, focused on addressing a specific pharmacological question. However, they do not take into account details of the tumor vs. immune system interactions, which would be critical to gain a better understanding of mechanisms of action (MoA) of immunotherapies.

Progressively adding components of tumor vs. immune system interactions into such PKPD models may well support the addressing of questions around pharmacologically-modulated IO biology, a topic of paramount importance in, for example, the search for therapeutic IO drug combinations (90). Such a systems approach may become an indispensable quantitative tool supporting "go/no-go" decisions in development programs, especially if sufficient biological knowledge for viable generalization is considered in the model (91). This prior knowledge is generally derived from two sources: (i) connectivity information to determine the system structure, e.g., molecular & cellular interactions, and their integration into patho-physiological processes; and (ii) quantitative data, for the calibration of model parameters. As discussed in the previous section, an imbalance in structural *vs*. quantitative information will in one way or another complicate integration into, and practical use of a mathematical model. For example, Lai and Friedman developed an elegant, yet complex model which includes a high number of biological elements, and considered their dynamics in space and time using partial differential equations (PDEs), to better understand the potential synergy between PD-(L)1 antagonists and either a GVAX vaccination or BRAFi/MEKi targeted therapies (72, 73). However, assessing the predictive power of such a model is impractical, given insufficient experimental data for model validation. Serre et al. provided another example of an elegant, yet insufficiently validated mathematical model describing the potential synergy between radiotherapy (RT) and immune checkpoint blockade (70).

One obvious way to improve model validation and hence model predictive power is to use rich experimental data, to rigorously constrain model parameters. This, however, requires the use of adequate statistical methods to properly quantify uncertainty and variability, which are inherent to any experimental biomedical and life sciences dataset (49, 92). In oncology drug development, quantitative data supporting MoA elucidation are typically generated at the preclinical stage. Parra-Guillen et al. for example, used a nonlinear mixed-effects (NLME) model and experimental data from syngeneic tumor models, to reveal the most influential immuno-adjuvant capable of boosting anti-tumor vaccination effects (21, 67). Such a modeling approach, which combines mechanistic features and mixed effects, allows one to incorporate individual-level data into the model, which may then describe not only mean trends, but also the full range of individual biomarker dynamics (93). A similar, combined mechanistic and mixed-effects approach was used to develop a model describing synergistic effects between RT and PD-(L)1 blockade in mice (68). This model, in fact, synthesizes a fit-for-purpose, yet sufficiently detailed mathematical description of the IO cycle, together with adequate model validation based on data from multiple experiments. As a result, this model can be used as a simulation tool for experimental study design, and is also adequate for determining optimal schedule and sequencing of RT + IO, and IO + IO treatment combinations (68, 69). Interestingly, despite the well-known challenges in translating oncology preclinical results into the clinic, simulation results from this preclinical modeling exercise were recently supported, in a qualitative sense, with clinical data and a corresponding meta-analysis (94, 95). For a quantitative translation, the Kosinsky et al. model would require adjustments for multiple quantitative differences that exist between mouse *vs*. human immune systems, e.g., appropriate expressions of immune checkpoints and turnover of specific T cells (96). Another modeling approach aimed at supporting the development of such an RT + IO combination therapy was proposed by Poleszczuk et al. who developed a physiologically-based model which considered a detailed incorporation of T cell trafficking and was used for the identification of an optimal site for RT administration, to maximally increase the probability of incremental anti-tumor immune effects (71). Predictions from such a comprehensive modeling effort were also recently supported by clinical results, which showed that RT administered to liver metastases triggered a higher immunological response (97). A mechanistic model has also been proposed by Peng et al. in the search of an optimal combination strategy against castration-resistant prostate cancer (74).

The modeling applications discussed to this point emphasize the importance of addressing multi-pronged questions, e.g., not only around dose finding, but also on the identification of an adequate time window for maximizing therapeutic benefits (98). This problem is particularly challenging in the development of combination therapies, where multiple options around which cancer indication, which combination agents, which scheduling per agent, and which sequencing of the agents make trial design enormously complex (99, 100). In recent years, platform design of clinical studies, driven by one master protocol, has gained momentum (101, 102)—a format which, in fact, benefits even further from a supportive quantitative mechanistic modeling approach (103).

MECHANISTIC MODELING IN SUPPORT OF IO BIOMARKER IDENTIFICATION

A third problem which is highly relevant in the development of IO therapies is the identification of predictive biomarkers. Indeed, there still is a lot of room for improving numbers of responder patients in pivotal IO trials, even in immunologically-active indications (104). Several computational models focusing on the identification of predictive biomarkers, with applications to personalized treatment against glioblastoma and prostate cancer have been developed (75, 77). These approaches have yet to find a general use in clinical practice. Part of the challenge arises from the biological complexity in the IO field, although there also are significant limitations from an experimental standpoint, such as differences in fresh vs. archived samples, difficulties in obtaining multiple biopsies per patient, with related risk and cost issues (105). One approach to alleviate some of these problems is the development of novel combinatorial biomarkers ("signatures") which may relate multiple, routinely measured markers with clinically meaningful biological phenotypes (106). In fact, such a consensus approach, "Immunoscore," has recently been validated in a large international study of colon cancer (107).

Another complicating factor in the development and interpretation of mechanistic modeling of IO data is the tremendous heterogeneity in tumor cell clones and elements of the surrounding immune microenvironment (108). A rapid development of novel experimental techniques may overcome this challenge, at least partially. Thus, the identification of specific gene expression signatures may help in further validating existing immunoscores and related biomarkers, even increasing their discriminatory ability (109), as recently shown with a PD-L1 expression signature which outperformed a standard PD-L1 immunohistochemistry (IHC) assay (110). Multiple immune signatures have now been identified, which allow for a better characterization of various aspects of anti-tumor immunity (111–116). Recent technological breakthroughs such as cytometry by time-of-flight (CyTOF) and single-cell mRNA sequencing (scRNA-seq) may further advance the utility and robustness of these immune signatures (117, 118); these techniques may allow for a deeper, more granular profiling of tumor and immune cell phenotypes involved in response or resistance to immunotherapies, in multiple indications (119–123). The importance in using quantitative models toward the selection and qualification (within the chain of events, from dosing to patient response) of IO biomarker signatures cannot be over-emphasized (108): immune biomarkers involve a high number of molecular and cellular species, and often exhibit complex temporal and spatial dynamics; these need to be properly framed in the context of a quantitative model, especially if the purpose is to relate multi-variate biomarker signatures to IO treatment effects and clinical endpoints (124). Quantitative modeling may also support the development of biomarkers in context, by integrating different data types, and following a model-based qualification of biomarkers as surrogate measures of efficacy and response. Such an approach has been proposed, recently, in the evaluation of neoantigen fitness as a surrogate measure of immunogenic quality of the existing neoantigen pool (125, 126). The progressive integration of such consensus, multi-variate combinatorial biomarkers into a unified, quantitative and mechanistic modeling framework will help overcome some of the limitations in the clinical use of IO biomarkers (127, 128).

OTHER MECHANISTIC MODELING APPROACHES WITH RELEVANCE TO IO

The above sections focused on traditional deterministic models, which make use of ODEs and PDEs for the description of IO systems dynamics. Other modeling techniques can be used to describe tumor *vs.* immune interactions. For example, cellular automata and agent-based models (ABMs) (129), as well as various hybrid models which link continuous and discrete modeling elements have been developed (130). Such models may be useful in raising new hypotheses, which may arise from emergent properties of the system based on existing data, rather than generating *bona fide* forward predictions. For example, a lattice gas automata technique has been used to gain a better understanding of a vaccination treatment mechanism and its corresponding anti-tumor immune response dynamics (131, 132). ABMs also represent a popular modeling technique, since they are well-suited to describe stochastic processes which do occur at various stages of the IO cycle. For example, Gong et al. developed an ABM to reveal spatio-temporal characteristics of PD-L1 blockade (79). In another publication, Kather et al. presented an elegant 2D ABM framework for an improved understanding of the role of stromal cells in colorectal cancer (CRC) (80). These authors determined that malignant cells hiding in the stroma cannot be eradicated completely, while stromal cells, at the same time, would not allow for rapid tumor progression. Consequently, simulations of an immuno-therapy illustrated how stroma permeabilization, concomitantly with immune activation, were able to markedly increase response to therapy *in silico*. Additionally, it was shown that a stroma-targeted therapy with insufficient activation of tumor-specific CTLs can lead to rapid tumor escape and hyper-progression (80). More recently, this model has been extended and generalized to a 3D spatial description, incorporating macrophage effects; it accurately reproduced the tissue architecture typically observed in CRC and can be used, similarly to ODE systems models, for the identification of effective IO therapeutic combinations (81).

CONCLUDING REMARKS

Following the approval, in recent years, of the first immune checkpoint inhibitors, the landscape of cancer treatment has changed dramatically and has shifted to a deep reconsideration of the role of the immune system in cancer progression and treatment. This led to an unprecedented number of clinical trials and generation of clinical data in the IO field. Clinical success rates, however, while improving significantly, are still relatively low. The observed imbalance, between the amount of biological and clinical data being generated *vs*. probability of trial success is not uncommon in biomedical disciplines, and calls for the development and updating of a companion, integrative, quantitative modeling framework with predictive value for MoAs and simulation value for study design purposes. As described by Sidney Brenner in his "*Sequences and Consequences*" landmark paper: "*We should welcome with open arms everything that modern technology has to offer us but we must learn to use it in new ways. Biology urgently needs a theoretical basis to unify it and it is only theory that will allow us to convert data to knowledge*" (133). We propose that quantitative, mechanistically-oriented modeling represents a means toward the establishment of such a "theoretical basis," pending proper integration of prior knowledge gained from biology and clinical research. One of the main factors limiting a wider application of quantitative systems modeling is its demand for rich experimental data necessary for precise parameter estimation. Historically, generation of such datasets in oncology research has been challenging, due to translational limitations of experimental preclinical models and sparse collection of tissue samples in clinical settings. Also, in the IO field, another challenge is the lack of predictive power for univariate biomarkers (e.g., PD-L1 IHC status or tumor mutational burden taken in isolation), which may unequivocally link immunologically-driven therapeutic effects to clinical response; a multivariate approach is clearly needed (128). Recent developments in multi-modality biomarkers and associated molecular signatures, together with innovative pharmacologies and clinical design under platform trials (134) will help in the progressive build-out and qualification of such a unified quantitative modeling framework, which in turn may help in predicting patient responses based on a given pharmacological intervention choice and multi-variate biomarker signatures.

AUTHOR CONTRIBUTIONS

KP, VV, and YK generated Figure 1. The Table was generated by IA, VV, and LC. All authors contributed to the writing of the manuscript.

ACKNOWLEDGMENTS

The authors would like to acknowledge that the modeling work presented here benefited from numerous discussions with associates from AstraZeneca and MedImmune, including members from Translational Sciences, Preclinical Biology & Pharmacology, Oncology Clinicians, and Quantitative Clinical Pharmacology.

REFERENCES

1. Coley WB. The treatment of inoperable sarcoma by bacterial toxins (the mixed toxins of the Streptococcus erysipelas and the Bacillus prodigiosus). Proc R Soc Med. (1910) 3:1–48.
2. Ehrlich P. Collected Papers of Paul Ehrlich: in Four Volumes Including a Complete Bibliography. London: : Pergamon Press; (1956).
3. Burnet FM. Immunological surveillance in neoplasia. Transplant Rev. (1971) 7:3–25.
4. Uhr JW, Scheuermann RH, Street NE, Vitetta ES. Cancer dormancy: opportunities for new therapeutic approaches. Nat Med. (1997) 3:505–9. 10.1038/nm0597-505
5. Dunn GP, Bruce AT, Ikeda H, Old LJ, Schreiber RD. Cancer immunoediting: from immunosurveillance to tumor escape. Nat Immunol. (2002) 3:991–8. 10.1038/ni1102-991
6. Chen DS, Mellman I. Oncology meets immunology: the cancer-immunity cycle. Immunity. (2013) 39:1–10. 10.1016/j.immuni.2013.07.012
7. Tang J, Shalabi A, Hubbard-Lucey VM. Comprehensive analysis of the clinical immuno-oncology landscape. Ann Oncol. (2018) 29:84–91. 10.1093/annonc/mdx755
8. Tang J, Yu JX, Hubbard-Lucey VM, Neftelinov ST, Hodge JP, Lin Y. Trial watch: the clinical trial landscape for PD1/PDL1 immune checkpoint inhibitors. Nat Rev Drug Discov. (2018) 17:854–5. 10.1038/nrd.2018.210
9. Fox BA, Schendel DJ, Butterfield LH, Aamdal S, Allison JP, Ascierto P, et al. . Defining the critical hurdles in cancer immunotherapy. J Transl Med. (2011) 9:214. 10.1186/1479-5876-9-214
10. Baik CS, Rubin EH, Forde PM, Mehnert JM, Collyar D, Butler MO, et al.. Immuno-oncology clinical trial design: limitations, challenges, and opportunities. Clin Cancer Res. (2017) 23:4992–5002. 10.1158/1078-0432.CCR-16-3066
11. Klinke DJ. Enhancing the discovery and development of immunotherapies for cancer using quantitative and systems pharmacology: interleukin-12 as a case study. J Immunother Cancer. (2015) 3:27. 10.1186/s40425-015-0069-x
12. Sharma P, Hu-Lieskovan S, Wargo JA, Ribas A. Primary, adaptive, and acquired resistance to cancer immunotherapy. Cell. (2017) 168:707–23. 10.1016/j.cell.2017.01.017

13. Gallasch R, Efremova M, Charoentong P, Hackl H, Trajanoski Z. Mathematical models for translational and clinical oncology. J Clin Bioinforma. (2013) 3:23. 10.1186/2043-9113-3-23
14. Altrock PM, Liu LL, Michor F. The mathematics of cancer: integrating quantitative models. Nat Rev Cancer. (2015) 15:730–45. 10.1038/nrc4029
15. Barbolosi D, Ciccolini J, Lacarelle B, Barlési F, André N. Computational oncology–mathematical modelling of drug regimens for precision medicine. Nat Rev Clin Oncol. (2016) 13:242–54. 10.1038/nrclinonc.2015.204
16. Norton L, Simon R, Brereton HD, Bogden AE. Predicting the course of Gompertzian growth. Nature. (1976) 264:542–5.
17. Araujo R. A history of the study of solid tumour growth: the contribution of mathematical modelling. Bull Math Biol. (2004) 66:1039–91. 10.1016/j.bulm.2003.11.002
18. Agur Z. From the evolution of toxin resistance to virtual clinical trials: the role of mathematical models in oncology. Future Oncol. (2010) 6:917–27. 10.2217/fon.10.61
19. Norton L. A Gompertzian model of human breast cancer growth. Cancer Res. (1988) 48:7067–71.
20. Norton L. Conceptual and practical implications of breast tissue geometry: toward a more effective, less toxic therapy. Oncologist. (2005) 10:370–81. 10.1634/theoncologist.10-6-370
21. Parra-Guillen ZP, Berraondo P, Ribba B, Troconiz IF. Modeling tumor response after combined administration of different immune-stimulatory agents. J Pharmacol Exp Ther. (2013) 346:432–42. 10.1124/jpet.113.206961
22. Stepanova NV. Course of the immune reaction during the development of a malignant tumor. Biophysics. (1980) 24:917–23.
23. Kuznetsov VA. A mathematical model for the interaction between cytotoxic T lymphocytes and tumour cells. Analysis of the growth, stabilization, and regression of a B-cell lymphoma in mice chimeric with respect to the major histocompatibility complex. Biomed Sci. (1991) 2:465–76.
24. Kuznetsov V, Makalkin I, Taylor M, Perelson A. Nonlinear dynamics of immunogenic tumors: parameter estimation and global bifurcation analysis. Bull Math Biol. (1994) 56:295–321. 10.1016/S0092-

8240(05)80260-5

25. Kuznetsov VA, Knott GD. Modeling tumor regrowth and immunotherapy. Math Comput Model. (2001) 33:1275–87. 10.1016/S0895-7177(00)00314-9
26. d'Onofrio A, Ciancio A. Simple biophysical model of tumor evasion from immune system control. Phys Rev E. (2011) 84:031910. 10.1103/PhysRevE.84.031910
27. Palmer S, Albergante L, Blackburn CC, Newman TJ. Thymic involution and rising disease incidence with age. Proc Natl Acad Sci USA. (2018) 115:1883–8. 10.1073/pnas.1714478115
28. Eftimie R, Bramson JL, Earn DJD. Interactions between the immune system and cancer: a brief review of non-spatial mathematical models. Bull Math Biol. (2011) 73:2–32. 10.1007/s11538-010-9526-3
29. Kirschner D, Panetta JC. Modeling immunotherapy of the tumor-immune interaction. J Math Biol. (1998) 37:235–52.
30. de Pillis LG, Radunskaya AE, Wiseman CL. A validated mathematical model of cell-mediated immune response to tumor growth. Cancer Res. (2005) 65:7950–8. 10.1158/0008-5472.CAN-05-0564
31. Dong Y, Miyazaki R, Takeuchi Y. Mathematical modeling on helper T cells in a tumor immune system. Discrete Contin Dyn Syst - Ser B. (2013) 19:55–72. 10.3934/dcdsb.2014.19.55
32. Zindl CL, Chaplin DD. Tumor immune evasion. Science. (2010) 328:697–8. 10.1126/science.1190310
33. Arciero JC, Jackson TL, Kirschner DE. A mathematical model of tumor-immune evasion and siRNA treatment. Discrete Contin Dyn Syst - Ser B. (2003) 4:39–58. 10.3934/dcdsb.2004.4.39
34. dePillis L, Caldwell T, Sarapata E, Williams H. Mathematical modeling of regulatory T cell effects on renal cell carcinoma treatment. Discrete Contin Dyn Syst - Ser B. (2013) 18:915–943. 10.3934/dcdsb.2013.18.915
35. Nani F, Freedman HI. A mathematical model of cancer treatment by immunotherapy. Math Biosci. (2000) 163:159–99. 10.1016/S0025-5564(99)00058-9
36. Szymanska Z. Analysis of immunotherapy models in the context of cancer dynamics. Int J Appl Math Comput Sci. (2008) 13:407–18.
37. Villasana M, Radunskaya A. A delay differential equation model for tumor growth. J Math Biol. (2003) 47:270–94. 10.1007/s00285-003-

0211-0

38. Byrne HM, Cox SM, Kelly CE. Macrophage-tumour interactions: *in vivo* dynamics. Discrete Contin Dyn Syst - Ser B. (2003) 4:81–98. 10.3934/dcdsb.2004.4.81
39. Bunimovich-Mendrazitsky S, Shochat E, Stone L. Mathematical model of BCG immunotherapy in superficial bladder cancer. Bull Math Biol. (2007) 69:1847–70. 10.1007/s11538-007-9195-z
40. Wilson S, Levy D. A mathematical model of the enhancement of tumor vaccine efficacy by immunotherapy. Bull Math Biol. (2012) 74:1485–500. 10.1007/s11538-012-9722-4
41. Nanda S, dePillis L, Radunskaya A. B cell chronic lymphocytic leukemia - a model with immune response. Discrete Contin Dyn Syst - Ser B. (2013) 18:1053–76. 10.3934/dcdsb.2013.18.1053
42. Walker R, Enderling H. From concept to clinic: mathematically informed immunotherapy. Curr Probl Cancer. (2016) 40:68–83. 10.1016/j.currproblcancer.2015.10.004
43. dePillis LG, Eladdadi A, Radunskaya AE. Modeling cancer-immune responses to therapy. J Pharmacokinet Pharmacodyn. (2014) 41:461–78. 10.1007/s10928-014-9386-9
44. Mellman I, Coukos G, Dranoff G. Cancer immunotherapy comes of age. Nature. (2011) 480:480–9. 10.1038/nature10673
45. Chen DS, Mellman I. Elements of cancer immunity and the cancer–immune set point. Nature. (2017) 541:321–30. 10.1038/nature21349
46. Dyson F. A meeting with enrico fermi. Nature. (2004) 427:297. 10.1038/427297a
47. Azeloglu EU, Iyengar R. Good practices for building dynamical models in systems biology. Sci Signal. (2015) 8:fs8. 10.1126/scisignal.aab0880
48. Gunawardena J. Models in biology: "accurate descriptions of our pathetic thinking." BMC Biol. (2014) 12:29. 10.1186/1741-7007-12-29
49. Tsigkinopoulou A, Baker SM, Breitling R. Respectful modeling: addressing uncertainty in dynamic system models for molecular biology. Trends Biotechnol. (2017) 35:518–29. 10.1016/j.tibtech.2016.12.008
50. Tarantola A. Popper, Bayes, and the inverse problem. Nat Phys. (2006) 2:492–4. 10.1038/nphys375

51. Andrew SM, Baker CTH, Bocharov GA. Rival approaches to mathematical modelling in immunology. J Comput Appl Math. (2007) 205:669–86. 10.1016/j.cam.2006.03.035
52. Ludewig B, Stein JV, Sharpe J, Cervantes-Barragan L, Thiel V, Bocharov G. A global "imaging" view on systems approaches in immunology: HIGHLIGHTS. Eur J Immunol. (2012) 42:3116–25. 10.1002/eji.201242508
53. Louzoun Y. The evolution of mathematical immunology. Immunol Rev. (2007) 216:9–20. 10.1111/j.1600-065X.2006.00495.x
54. Grossman Z, Paul WE. Autoreactivity, dynamic tuning and selectivity. Curr Opin Immunol. (2001) 13:687–98. 10.1016/S0952-7915(01)00280-1
55. Grossman Z. Mathematical modeling of thymopoiesis in HIV infection: real data, virtual data, and data interpretation. Clin Immunol. (2003) 107:137–9. 10.1016/S1521-6616(03)00122-0
56. Grossman Z, Min B, Meier-Schellersheim M, Paul WE. Concomitant regulation of T-cell activation and homeostasis. Nat Rev Immunol. (2004) 4:387–95. 10.1038/nri1355
57. Khailaie S, Bahrami F, Janahmadi M, Milanez-Almeida P, Huehn J, Meyer-Hermann M. A mathematical model of immune activation with a unified self-nonself concept. Front Immunol. (2013) 4:474. 10.3389/fimmu.2013.00474
58. Goldstein B, Faeder JR, Hlavacek WS. Mathematical and computational models of immune-receptor signalling. Nat Rev Immunol. (2004) 4:445–56. 10.1038/nri1374
59. Antia R, Ganusov VV, Ahmed R. The role of models in understanding CD8+ T-cell memory. Nat Rev Immunol. (2005) 5:101–11. 10.1038/nri1550
60. Danhof M. Systems pharmacology – towards the modeling of network interactions. Eur J Pharm Sci. (2016) 94:4–14. 10.1016/j.ejps.2016.04.027
61. Helmlinger G, Al-Huniti N, Aksenov S, Peskov K, Hallow KM, Chu L, et al. Drug-disease modeling in the pharmaceutical industry - where mechanistic systems pharmacology and statistical pharmacometrics meet. Eur J Pharm Sci. (2017) 109:S39–46. 10.1016/j.ejps.2017.05.028
62. Musante C, Ramanujan S, Schmidt B, Ghobrial O, Lu J, Heatherington A. Quantitative systems pharmacology: a case for disease models. Clin

Pharmacol Ther. (2017) 101:24–7. 10.1002/cpt.528

63. Black J. A personal view of pharmacology. Annu Rev Pharmacol Toxicol. (1996) 36:1–33. 10.1146/annurev.pa.36.040196.000245
64. Ribba B, Boetsch C, Nayak T, Grimm HP, Charo J, Evers S, et al. . Prediction of the optimal dosing regimen using a mathematical model of tumor uptake for immunocytokine-based cancer immunotherapy. Clin Cancer Res. (2018) 24:3325–33. 10.1158/1078-0432.CCR-17-2953
65. Lindauer A, Valiathan C, Mehta K, Sriram V, de Greef R, Elassaiss-Schaap J, et al. . Translational pharmacokinetic/pharmacodynamic modeling of tumor growth inhibition supports dose-range selection of the anti-PD-1 antibody pembrolizumab: translational pharmacokinetic/pharmacodynamic modeling. CPT Pharmacomet Syst Pharmacol. (2017) 6:11–20. 10.1002/psp4.12130
66. Chen X, Haddish-Berhane N, Moore P, Clark T, Yang Y, Li H, et al. . Mechanistic projection of first-in-human dose for bispecific immunomodulatory P-Cadherin LP-DART: an integrated PK/PD modeling approach. Clin Pharmacol Ther. (2016) 100:232–41. 10.1002/cpt.393
67. Parra-Guillen ZP, Berraondo P, Grenier E, Ribba B, Troconiz IF. Mathematical model approach to describe tumour response in mice after vaccine administration and its applicability to immune-stimulatory cytokine-based strategies. AAPS J. (2013) 15:797–807. 10.1208/s12248-013-9483-5
68. Kosinsky Y, Dovedi SJ, Peskov K, Voronova V, Chu L, Tomkinson H, et al. . Radiation and PD-(L)1 treatment combinations: immune response and dose optimization via a predictive systems model. J Immunother Cancer. (2018) 6:17. 10.1186/s40425-018-0327-9
69. Kosinsky Y, Chu L, Peskov K, Voronova V, Borodovsky A, Woessner R, et al. Abstract 2098: quantitative modeling as a systematic approach for drug combination evaluation in immuno-oncology. (IO). Cancer Res. (2018) 78:2098 10.1158/1538-7445.AM2018-2098
70. Serre R, Benzekry S, Padovani L, Meille C, Andre N, Ciccolini J, et al. . Mathematical modeling of cancer immunotherapy and its synergy with radiotherapy. Cancer Res. (2016) 76:4931–40. 10.1158/0008-5472.CAN-15-3567
71. Poleszczuk JT, Luddy KA, Prokopiou S, Robertson-Tessi M, Moros EG, Fishman M, et al. . Abscopal benefits of localized radiotherapy

depend on activated T-cell trafficking and distribution between metastatic lesions. Cancer Res. (2016) 76:1009–18. 10.1158/0008-5472.CAN-15-1423

72. Lai X, Friedman A. Combination therapy of cancer with cancer vaccine and immune checkpoint inhibitors: a mathematical model. PLoS ONE. (2017) 12:e0178479. 10.1371/journal.pone.0178479
73. Lai X, Friedman A. Combination therapy for melanoma with BRAF/MEK inhibitor and immune checkpoint inhibitor: a mathematical model. BMC Syst Biol. (2017) 11:70. 10.1186/s12918-017-0446-9
74. Peng H, Zhao W, Tan H, Ji Z, Li J, Li K, et al. . Prediction of treatment efficacy for prostate cancer using a mathematical model. Sci Rep. (2016) 6:21599. 10.1038/srep21599
75. Kronik N, Kogan Y, Vainstein V, Agur Z. Improving alloreactive CTL immunotherapy for malignant gliomas using a simulation model of their interactive dynamics. Cancer Immunol Immunother. (2008) 57:425–39. 10.1007/s00262-007-0387-z
76. Cappuccio A, Elishmereni M, Agur Z. Cancer immunotherapy by interleukin-21: potential treatment strategies evaluated in a mathematical model. Cancer Res. (2006) 66:7293–300. 10.1158/0008-5472.CAN-06-0241
77. Kronik N, Kogan Y, Elishmereni M, Halevi-Tobias K, Vuk-Pavlović S, Agur Z. Predicting outcomes of prostate cancer immunotherapy by personalized mathematical models. PLoS ONE. (2010) 5:e15482. 10.1371/journal.pone.0015482
78. Kim R, Woods T II, Radunskaya A. Mathematical modeling of tumor immune interactions: a closer look at the role of a PD-L1 inhibitor in cancer immunotherapy. SPORA J Biomath. (2018) 4:25–41. 10.30707/SPORA4.1Radunskaya
79. Gong C, Milberg O, Wang B, Vicini P, Narwal R, Roskos L, et al. . A computational multiscale agent-based model for simulating spatio-temporal tumour immune response to PD1 and PDL1 inhibition. J R Soc Interface. (2017) 14:20170320. 10.1098/rsif.2017.0320
80. Kather JN, Poleszczuk J, Suarez-Carmona M, Krisam J, Charoentong P, Valous NA, et al. . *In silico* modeling of immunotherapy and stroma-targeting therapies in human colorectal cancer. Cancer Res. (2017) 77:6442–52. 10.1158/0008-5472.CAN-17-2006
81. Kather JN, Charoentong P, Suarez-Carmona M, Herpel E, Klupp

F, Ulrich A, et al. . High-throughput screening of combinatorial immunotherapies with patient-specific *in silico* models of metastatic colorectal cancer. Cancer Res. (2018) 78:5155–63. 10.1158/0008-5472.CAN-18-1126

82. Venkatakrishnan K, Ecsedy J. Enhancing value of clinical pharmacodynamics in oncology drug development: an alliance between quantitative pharmacology and translational science. Clin Pharmacol Ther. (2017) 101:99–113. 10.1002/cpt.544
83. Zhao X, Wang X, Feng Y, Agrawal S, Shah D. Development of Antibody-Based Therapeutics: Application of PK-PD Modeling and Simulation Approaches for Immuno-Oncology Drugs. (2018). Available online at: 10.1007/978-981-13-0496-5 (accessed December 11, 2018).
84. Stroh M, Carlile D, Li C-C, Wagg J, Ribba B, Ramanujan S, et al. . Challenges and opportunities for quantitative clinical pharmacology in cancer immunotherapy: something old, something new, something borrowed, and something blue: something old, something new, something borrowed, and something blue. CPT Pharmacomet Syst Pharmacol. (2015) 4:495–7. 10.1002/psp4.12014
85. de Greef R, Elassaiss-Schaap J, Chatterjee M, Turner D, Ahamadi M, Forman M, et al. Pembrolizumab: role of modeling and simulation in bringing a novel immunotherapy to patients with melanoma: modeling and simulation of Pembrolizumab. CPT Pharmacomet Syst Pharmacol. (2017) 6:5–7. 10.1002/psp4.12131
86. Nayak S, Sander O, Al-Huniti N, de Alwis D, Chain A, Chenel M, et al. . Getting innovative therapies faster to patients at the right dose: impact of quantitative pharmacology towards first registration and expanding therapeutic use. Clin Pharmacol Ther. (2018) 103:378–83. 10.1002/cpt.978
87. Zhao X, Suryawanshi S, Hruska M, Feng Y, Wang X, Shen J, et al. . Assessment of nivolumab benefit–risk profile of a 240-mg flat dose relative to a 3-mg/kg dosing regimen in patients with advanced tumors. Ann Oncol. (2017) 28:2002–8. 10.1093/annonc/mdx235
88. Freshwater T, Kondic A, Ahamadi M, Li CH, de Greef R, de Alwis D, et al. . Evaluation of dosing strategy for pembrolizumab for oncology indications. J Immunother Cancer. (2017) 5:43. 10.1186/s40425-017-0242-5
89. Charych D, Khalili S, Dixit V, Kirk P, Chang T, Langowski J, et al. . Modeling the receptor pharmacology, pharmacokinetics,

and pharmacodynamics of NKTR-214, a kinetically-controlled interleukin-2. (IL2) receptor agonist for cancer immunotherapy. PLoS ONE. (2017) 12:e0179431. 10.1371/journal.pone.0179431

90. Zappasodi R, Merghoub T, Wolchok JD. Emerging concepts for immune checkpoint blockade-based combination therapies. Cancer Cell. (2018) 33:581–98. 10.1016/j.ccell.2018.03.005
91. Robertson-Tessi M, El-Kareh A, Goriely A. A mathematical model of tumor–immune interactions. J Theor Biol. (2012) 294:56–73. 10.1016/j.jtbi.2011.10.027
92. Kirk PDW, Babtie AC, Stumpf MPH. Systems biology. (un) certainties. Science. (2015) 350:386–8. 10.1126/science.aac9505
93. Davidian M, Giltinan DM. Nonlinear models for repeated measurement data: an overview and update. J Agric Biol Environ Stat. (2003) 8:387–419. 10.1198/1085711032697
94. Kotecha R, Kim JM, Miller JA, Chao ST, Mohammadi AM, Peereboom D, et al. Stereotactic radiosurgery (SRS) with immune checkpoint inhibitor therapy. (ICI) for patients with brain metastasis (BM): the impact of timing and sequencing. Int J Radiat Oncol. (2018) 102:e345–6. 10.1016/j.ijrobp.2018.07.1051
95. Lehrer EJ, Peterson JL, Zaorsky NG, Brown PD, Sahgal A, Chiang VL, et al. . Single versus multifraction stereotactic radiosurgery for large brain metastases: an international meta-analysis of 24 trials. Int J Radiat Oncol. (2018) 103:618–30. 10.1016/j.ijrobp.2018.10.038
96. De Boer RJ, Perelson AS. Quantifying T lymphocyte turnover. J Theor Biol. (2013) 327:45–87. 10.1016/j.jtbi.2012.12.025
97. Tang C, Welsh JW, de Groot P, Massarelli E, Chang JY, Hess KR, et al. . Ipilimumab with stereotactic ablative radiation therapy: phase I results and immunologic correlates from peripheral T cells. Clin Cancer Res. (2017) 23:1388–96. 10.1158/1078-0432.CCR-16-1432
98. Rothschilds AM, Wittrup KD. What, why, where, and when: bringing timing to immuno-oncology. Trends Immunol. (2019) 40:12–21. 10.1016/j.it.2018.11.003
99. Day D, Siu LL. Approaches to modernize the combination drug development paradigm. Genome Med. (2016) 8:115. 10.1186/s13073-016-0369-x
100. Bui NQ, Kummar S. Evolution of early phase clinical trials in oncology. J Mol Med. (2018) 96:31–8. 10.1007/s00109-017-1612-7

101. Woodcock J, LaVange LM. Master protocols to study multiple therapies, multiple diseases, or both. N Engl J Med. (2017) 377:62–70. 10.1056/NEJMra1510062
102. Simonsen KL, Fracasso PM, Bernstein SH, Wind-Rotolo M, Gupta M, Comprelli A, et al. . The fast real-time assessment of combination therapies in immuno-oncology (FRACTION) program: innovative, high-throughput clinical screening of immunotherapies. Eur J Cancer. (2018) 103:259–66. 10.1016/j.ejca.2018.07.127
103. Wages NA, Chiuzan C, Panageas KS. Design considerations for early-phase clinical trials of immune-oncology agents. J Immunother Cancer. (2018) 6:81. 10.1186/s40425-018-0389-8
104. Gibney GT, Weiner LM, Atkins MB. Predictive biomarkers for checkpoint inhibitor-based immunotherapy. Lancet Oncol. (2016) 17:e542–51. 10.1016/S1470-2045(16)30406-5
105. Wargo JA, Reddy SM, Reuben A, Sharma P. Monitoring immune responses in the tumor microenvironment. Curr Opin Immunol. (2016) 41:23–31. 10.1016/j.coi.2016.05.006
106. Galon J, Bruni D. Approaches to treat immune hot, altered and cold tumours with combination immunotherapies. Nat Rev Drug Discov. (2019) 18:197–218. 10.1038/s41573-018-0007-y
107. Pagès F, Mlecnik B, Marliot F, Bindea G, Ou F-S, Bifulco C, et al. . International validation of the consensus Immunoscore for the classification of colon cancer: a prognostic and accuracy study. Lancet. (2018) 391:2128–39. 10.1016/S0140-6736(18)30789-X
108. Finotello F, Eduati F. Multi-omics profiling of the tumor microenvironment: paving the way to precision immuno-oncology. Front Oncol. (2018) 8:430. 10.3389/fonc.2018.00430
109. Galon J, Angell HK, Bedognetti D, Marincola FM. The continuum of cancer immunosurveillance: prognostic, predictive, and mechanistic signatures. Immunity. (2013) 39:11–26. 10.1016/j.immuni.2013.07.008
110. Paré L, Pascual T, Seguí E, Teixidó C, Gonzalez-Cao M, Galván P, Rodríguez A, et al. . Association between PD1 mRNA and response to anti-PD1 monotherapy across multiple cancer types. Ann Oncol. (2018) 29:2121–8. 10.1093/annonc/mdy335
111. Ayers M, Lunceford J, Nebozhyn M, Murphy E, Loboda A, Kaufman DR, Albright A, et al. . IFN-γ-related mRNA profile predicts clinical response to PD-1 blockade. J Clin Invest. (2017) 127:2930–40.

10.1172/JCI91190

112. Li B, Cui Y, Diehn M, Li R. Development and validation of an individualized immune prognostic signature in early-stage nonsquamous non–small cell lung cancer. JAMA Oncol. (2017) 3:1529–37. 10.1001/jamaoncol.2017.1609
113. Morrison C, Pabla S, Conroy JM, Nesline MK, Glenn ST, Dressman D, et al.. Predicting response to checkpoint inhibitors in melanoma beyond PD-L1 and mutational burden. J Immunother Cancer. (2018) 6:32. 10.1186/s40425-018-0344-8
114. Duruisseaux M, Martínez-Cardús A, Calleja-Cervantes ME, Moran S, Castro de Moura M, Davalos V, et al.. Epigenetic prediction of response to anti-PD-1 treatment in non-small-cell lung cancer: a multicentre, retrospective analysis. Lancet Respir Med. (2018) 6:771–81. 10.1016/S2213-2600(18)30284-4
115. Auslander N, Zhang G, Lee JS, Frederick DT, Miao B, Moll T, et al. Robust prediction of response to immune checkpoint blockade therapy in metastatic melanoma. Nat Med. (2018) 24:1545–9. 10.1038/s41591-018-0157-9
116. Jiang P, Gu S, Pan D, Fu J, Sahu A, Hu X, et al.. Signatures of T cell dysfunction and exclusion predict cancer immunotherapy response. Nat Med. (2018) 24:1550–8. 10.1038/s41591-018-0136-1
117. Giladi A, Amit I. Single-cell genomics: a stepping stone for future immunology discoveries. Cell. (2018) 172:14–21. 10.1016/j.cell.2017.11.011
118. Papalexi E, Satija R. Single-cell RNA sequencing to explore immune cell heterogeneity. Nat Rev Immunol. (2017) 18:35–45. 10.1038/nri.2017.76
119. Tirosh I, Izar B, Prakadan SM, Wadsworth MH, Treacy D, Trombetta JJ, et al.. Dissecting the multicellular ecosystem of metastatic melanoma by single-cell RNA-seq. Science. (2016) 352:189–96. 10.1126/science.aad0501
120. Lavin Y, Kobayashi S, Leader A, Amir ED, Elefant N, Bigenwald C, et al.. Innate immune landscape in early lung adenocarcinoma by paired single-cell analyses. Cell. (2017) 169:750–65. 10.1016/j.cell.2017.04.014
121. Zheng C, Zheng L, Yoo J-K, Guo H, Zhang Y, Guo X, et al.. Landscape of infiltrating T cells in liver cancer revealed by single-

cell sequencing. Cell. (2017) 169:1342–56. 10.1016/j.cell.2017.05.035

122. Jerby-Arnon L, Shah P, Cuoco MS, Rodman C, Su M-J, Melms JC, et al. . A cancer cell program promotes T cell exclusion and resistance to checkpoint blockade. Cell. (2018) 175:984–97. 10.1016/j.cell.2018.09.006
123. Sade-Feldman M, Yizhak K, Bjorgaard SL, Ray JP, de Boer CG, Jenkins RW, et al. . Defining T cell states associated with response to checkpoint immunotherapy in melanoma. Cell. (2018) 175:998–1013. 10.1016/j.cell.2018.10.038
124. Lesterhuis WJ, Bosco A, Millward MJ, Small M, Nowak AK, Lake RA. Dynamic versus static biomarkers in cancer immune checkpoint blockade: unravelling complexity. Nat Rev Drug Discov. (2017) 16:264–72. 10.1038/nrd.2016.233
125. Łuksza M, Riaz N, Makarov V, Balachandran VP, Hellmann MD, Solovyov A, et al. . A neoantigen fitness model predicts tumour response to checkpoint blockade immunotherapy. Nature. (2017) 551:517–20. 10.1038/nature24473
126. Balachandran VP, Łuksza M, Zhao JN, Makarov V, Moral JA, Remark R, et al. . Identification of unique neoantigen qualities in long-term survivors of pancreatic cancer. Nature. (2017) 551:512–6. 10.1038/nature24462
127. Roh W, Chen P-L, Reuben A, Spencer CN, Prieto PA, Miller JP, et al. . Integrated molecular analysis of tumor biopsies on sequential CTLA-4 and PD-1 blockade reveals markers of response and resistance. Sci Transl Med. (2017) 9:eaah3560. 10.1126/scitranslmed.aah3560
128. Havel JJ, Chowell D, Chan TA. The evolving landscape of biomarkers for checkpoint inhibitor immunotherapy. Nat Rev Cancer. (2019) 19:133–50. 10.1038/s41568-019-0116-x
129. Chiacchio F, Pennisi M, Russo G, Motta S, Pappalardo F. Agent-based modeling of the immune system: netlogo, a promising framework. BioMed Res Int. (2014) 2014:1–6. 10.1155/2014/907171
130. Stéphanou A, Volpert V. Hybrid modelling in biology: a classification review. Math Model Nat Phenom. (2016) 11:37–48. 10.1051/mmnp/201611103
131. Pappalardo F, Lollini P-L, Castiglione F, Motta S. Modeling and simulation of cancer immunoprevention vaccine. Bioinformatics. (2005) 21:2891–7. 10.1093/bioinformatics/bti426

132. Pappalardo F, Motta S, Lollini P-L, Mastriani E. Analysis of vaccine's schedules using models. Cell Immunol. (2006) 244:137–40. 10.1016/j.cellimm.2007.03.002

133. Brenner S. Sequences and consequences. Philos Trans R Soc B Biol Sci. (2010) 365:207–12. 10.1098/rstb.2009.0221

134. Emens LA, Butterfield LH, Hodi FS, Marincola FM, Kaufman HL. Cancer immunotherapy trials: leading a paradigm shift in drug development. J Immunother Cancer. (2016) 4:42. 10.1186/s40425-016-0146-9

15

Sexual Dimorphism of Immune Responses: A New Perspective in Cancer Immunotherapy

Imerio Capone[1] , Paolo Marchetti[2] , Paolo Antonio Ascierto[3] , Walter Malorni[4] and Lucia Gabriele[1]

[1]Department of Oncology and Molecular Medicine, Istituto Superiore di Sanità, Rome, Italy
[2]Department of Oncology, Sapienza University of Rome, Sant'Andrea Hospital, Rome, Italy
[3]Unit of Melanoma, Cancer Immunotherapy and Innovative Therapy, Istituto Nazionale Tumori Fondazione G. Pascale (IRCCS), Naples, Italy
[4]Center for Gender-Specific Medicine, Istituto Superiore di Sanità, Rome, Italy

ABSTRACT

Nowadays, several types of tumors can benefit from the new frontier of immunotherapy, due to the recent increasing knowledge of the role of the immune system in cancer control. Among the new therapeutic strategies,

Citation: Capone I, Marchetti P, Ascierto PA, Malorni W, Gabriele L. Sexual Dimorphism of Immune Responses: A New Perspective in Cancer Immunotherapy. Front Immunol. 2018;9:552. Published 2018 Mar 21. doi:10.3389/fimmu.2018.00552.

there is the immune checkpoint blockade (ICB), able to restore an efficacious antitumor immunity and significantly prolong the overall survival (OS) of patients with advanced tumors such as melanoma and non-small cell lung cancer (NSCLC). Despite the impressive efficacy of these agents in some patients, treatment failure and resistance are frequently observed. In this regard, the signaling governed by IFN type I (IFN-I) has emerged as pivotal in orchestrating host defense. This pathway displays different activation between sexes, thus potentially contributing to sexual dimorphic differences in the immune responses to immunotherapy. This perspective article aims to critically consider the immune signals, with particular attention to IFN-I, that may differently affect female and male antitumor responses upon immunotherapy.

Keywords: cancer, immunotherapy, sexual dimorphism, interferon, immune response, immune checkpoint inhibitors, therapeutic vaccines

In the last decade the increasing knowledge of the role of the immune system in cancer control has led to the development of the immune checkpoint inhibitors (ICIs), that targeting proteins acting as negative regulators of T-cell activation, reverse the tumor-induced immune tolerance.

The main actors of this scene are two key molecules called cytotoxic T-lymphocyte-associated protein 4 (CTLA-4) and programmed cell death receptor 1 (PD-1) (1). CTLA-4 binds B7 receptors on antigen-presenting cells (APCs), thus blocking T-cell activation. Likewise, PD-1 expressed on activated T lymphocytes, upon interaction with its ligands PD-L1 and PD-L2 on APCs or tumor cells, limits their activity delivering negative signals. PD-1 is also expressed by regulatory T cells (Tregs), enhancing their function. The clinical use of ICIs has extraordinarily increased overall survival (OS) in patients with cancer, suggesting that targeting the immune checkpoint blockade (ICB) is a privileged strategy for fighting cancer (2, 3). A crucial part of this game is played by tumor-infiltrating T cells (4) and in many cases, the effectiveness of ICIs is limited by the lack of adequate antitumor immunity in the tumor microenvironment (TME) (5). To overcome this limitation, new combination therapies are being investigated (6). In this landscape, a new perspective for improving the efficacy of immunotherapy is to take into account sexually dimorphic differences of immune responses (7).

COMPONENTS OF ANTITUMOR IMMUNITY CRITICAL FOR IMMUNOTHERAPY

In the era of ICIs, T-cell immunity remains central for tumor regression (8). This is far from being simple and occurs only when diverse elements coexist either during cancer immunosurveillance or immunotherapeutic treatments (9). Cancer cells carry tumor antigens, and in particular neoantigens, that are the most capable of inducing an effective T-cell immunity. These antigens, mainly present in tumors with high mutational rate, generate T-cell responses upon processing and presention by DCs. Accordingly, CD8+ T cells infiltrates at high frequency these tumors and this associates with a patient survival advantage (10). Nevertheless, in specific conditions tumor-infiltrating DCs lack their immunostimulatory function and acquire immunosuppressive activity (11). This relies on different phenotypes and functions of DC subsets that provide additional variability to the onset of antitumor responses. Within human tumors, rare BDCA3+ DCs were found extremely competent in processing and cross-presenting antigens, driving the expansion of tumor-specific cytotoxic T cells (CTLs) (12). These DCs are the equivalent of the mouse CD103+ DCs, depending uniquely on DC-lineage committing transcription factors such as interferon regulatory factor 8 (IRF-8) (13). Tumor-infiltrating CD103+/BDCA3+ DCs represent privileged players for responses upon ICB (14, 15). However, in TME the most abundant DCs show immunosuppressive activity, including plasmacytoid DC (pDCs). Activation by toll-like receptor 7 (TLR7) ligand reverses pDC immunosuppressive function to such an extent that their administration to melanoma patients induces tumor-specific CD4+ and CD8+ T-cell responses (16, 17). PDCs are the main producers of IFN-α (18), a well-known cytokine linking innate and adaptive immunity and endowed with potent direct and immunomediated antitumor effects (19). Endogenous IFN-I is required to initiate antitumor response in the elimination phase of cancer immunoediting (20), and IFN-I production is also essential for tumor rejection by DC-stimulated activated T cells (21). IFN-I specifically improves the ability of CD8a+ CD103+ DCs to cross-prime tumor specific CD8+ T cells (22–24). However, although IFN-I is crucial for the outcome of ICB, its activation may produce an opposite role in a time-dependent manner (25). In melanoma, while early IFN-I activation correlates with an effective PD-1 blockade (5, 26), prolonged IFN-I signaling seems to favor resistance (27). This apparent discrepancy might not be surprising since, during persistent viral infections, IFN-I displays protective effects in early stages and becomes detrimental upon continued signaling activation in the chronic phase (28). Interestingly, CML patients in remission who stopped IFN-α treatment because of good clinical response developed higher protective T-cell memory response than patients who continued therapy (29). In TME, while IFN-I signaling is critical for survival and full activation of CD8+T cells, the continuous generation of terminally differentiated CD8+T cells might determine the failure of ICB by reducing progenitors able to respond (30). Thus, the generation of exhausted CD8+T cells may be the results of complex immunosuppressive interactions including prolonged IFN-I signaling (31). Indeed, IFN-I is critical for the induction of immune checkpoints or coinhibitory receptors, such as PD-1 and T-cell immunoglobulin- and mucin-containing molecule-3 (Tim-3) (32, 33), whose persistent overexpression characterizes exhausted T cells within TME (34). Moreover, acute and chronic IFN-I expression impacts Tregs differently. While IFN-α abrogates the suppressive activity of CD4+CD25+Foxp3+Tregs, thus hampering tumor evasion (35, 36), IFN-I signal blockade boosts the ICB-induced antitumor response by favoring the effector T cells to Tregs ratio (37). Of interest, Treg-mediated immune tolerance may occur through the control of IFN-I on indoleamine 2,3-dioxygenase (IDO) (38, 39), being melanoma peritumoral IDO expression by pDCs an early marker of resistance (40). Noteworthy, both Tregs and IFN-I also suppress the function of natural killer (NK) cells whose antigen-independent cytotoxicity is the second effector mechanism responsible for an efficacious antitumor response (41, 42). Altogether, these immune components build a functional framework for innate and acquired resistance to PD-1 blockade(43), where the balance between inflammation and suppression determined by the fine-tune regulation of IFN-I is crucial. This central achievement occurs through the modulation of different transcriptional programs involving the Janus kinase (JAK)-signal transducer and activator of transcription (STAT) and IFN regulatory factors (IRFs) families, as well as proteins of the phosphatidyl inositide3-kinase (PI3K) and mitogen-activated protein kinases (MAPK) pathways (44). Moreover, epigenetic signals, such as histone modification, DNA methylation, and microRNAs (miRs), represent a second key layer of regulation(45). Indeed, exhausted CD8+T cells within tumors display a distinct epigenetic profile and the limited remodeling may represent a pivotal component of resistance to PD-L1 blockade (46). Likewise, miRs modulate the expression of PD-L1 potentially accounting for innate resistance (47). Finally, it is worthy to mention drug toxicity as one of the major cause of reduced dosage, delayed drug administration and therapy discontinuation (48). Immune-related adverse events (irAEs), such as the loss of the protective function of intestinal barriers and changes occurring in the microbiota composition, represent the most frequent ICI-associated toxicities to which dysregulated activation of the IFN-I signaling may contribute (49, 50).

SEXUAL DIMORPHISM OF THE IMMUNE COMPONENTS OF HOST RESPONSE TO IMMUNOTHERAPY

Sexual dimorphism of the immune functions is a crucial element that has so far been largely ignored in the field of immunotherapy (51). These differences affect both innate and adaptive immune responses, leading to a considerable functional diversity between females and males (7) (Figure (Figure1).1). Sex variations include the number and activity of cells as well as intracellular and extracellular signals orchestrating the two branches of immunity. In the innate context, females own APCs that perform antigen presentation more vigorously, have neutrophils and macrophages endowed with higher phagocytic activity, and show a higher frequency of both progenitors and mature group 2 innate lymphoid cells (ILCs), key regulators of type-2 inflammatory responses (52). On the contrary, males exhibit enhanced numbers of NK cells (7). In the adaptive context, females exhibit higher CD4+T cell counts associated with an increased CD4+/CD8+T-cell ratio, along with Th2 prevalence, and greater proliferation and cytotoxicity of T cells. In contrast, males have higher CD8+and Treg cell counts associated with Th1 dominance, lower B cell numbers and basal immunoglobulin levels along with weaker antibody responses (53). Both hormonal and genetic differences concur in the sexual dimorphism of the immune system. The 17β-estradiol (E2) –estrogen receptor α (ER) axis is a key regulator of innate immune populations. E2 reduces mobility and inflammatory activity of neutrophils (54), and female mMDSCs seem to be more suppressive than the male counterpart (55). Importantly, the E2–ERα axis exerts a tight control on the functional responses of diverse DC subsets, also by modulating IFN-I production. Moreover, high levels of E2 promote epigenetic changes in DC precursors of females driving DC differentiation and robust IFN-I production (56) (Figure (Figure2).2). PDCs are the DC with major differences between the sexes. Their activity is driven by TLR7, whose gene located on X chromosome and under the E2–ERα signaling (57, 58) promotes high IFN-α production through enhanced expression of IRF-5 (59) (Figure (Figure2).2). Nevertheless, the development and function of other DC subsets are also affected by the E2–ERα axis (60). In female mice, CD103+conventional DCs (cDCs) are represented at very high levels in the cutaneous lymph nodes and CD103− cDCs show high expression of ERα in the lung (61, 62). Furthermore, E2 treatment has been reported to enhance IRF-4+DC capability to stimulate Th17 in a murine model of HSV-2 infection (63), while it induces IFN-α and IL-6 production as well as CD40, CD86, and MHCII expression in cDCs. In contrast, tolerogenic FOXO3-expressing DCs display lower frequency and reduced function in tumors from females compared with their male counterpart (64). Altogether, these evidences confirm a positive regulatory feedback loop between the E2–ERα and IFN-I signals in regulating the phenotype and function of DC populations (65). DC differentiation and function are also affected by other hormones, such as prolactin, progesterone and glucocorticoids, driving either a pro-inflammatory or a tolerogenic phenotype (66). In this light, the sex-specific components of the innate immunity become important in improving cancer immunotherapy. Many elements of the adaptive immunity are regulated by the E2–ERα axis as well. Low levels of estrogen, as during the luteal phase, favor Th1 polarization of CD4+T cells associated with increased production of IFN-γ, responsiveness to IL-12 through STAT-4 activation, and T proliferation. Conversely, high levels of estrogens, found in the follicular phase and during pregnancy, sustain Th2-mediated immunity characterized by IRF-1-mediated reduction of IFN-γ, IL-4 induction, and PD-1-overexpressing Tregs associated with reduced Th17 response (67). Nevertheless, estrogen together with other factors may directly stimulate Th17cells inducing IL-17 production (68). Thus, females exhibit activated and proliferating CD4+and CD8+T cells, characterized by preferential production of IFNγ and high-cytotoxicity activity, whereas males exhibit IL-17-producing T cells (69). This underlines how sex hormones become crucial in determining the efficacy of some therapies. B16 melanoma-bearing female mice, more than males, benefit from ICB, partially due to a greater PD-L1 blockade-mediated reduction of Treg function (70).

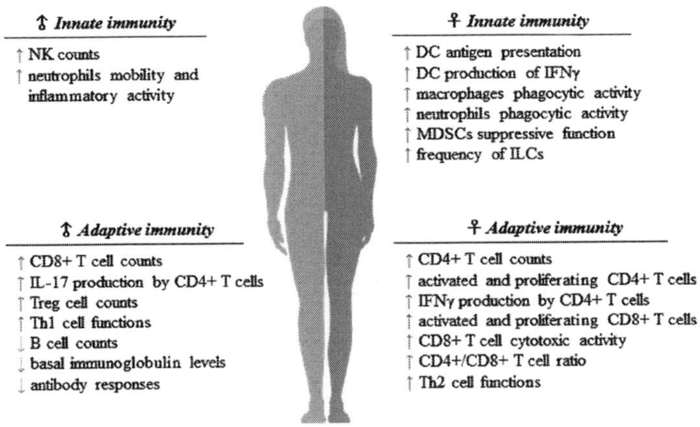

Figure 1: Sexual dimorphism of the immune responses. Immune components of both innate and adaptive immunity are differently regulated in females and males. Apparently, females display higher capability of mounting type-2 versus type-1 immune responses, whereas males seem to prefer type-1 immune responses, of which many traits are still unclear. As a matter of fact, the difference of the strength of type-1 immunity between sexes is smaller than that of type 2, preserving the onset of female inflammatory cell-mediated immune responses.

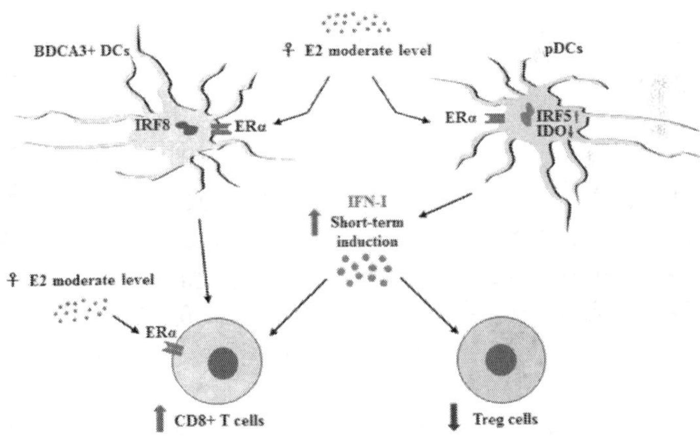

Figure 2: Sex-biased regulation of T-cell response through E2-induced IFN type I (IFN-I) production. The E2–ERα axis controls the functional responses of diverse DC subsets along with IFN-I production. In females, upon 17β-estradiol (E2) stimulation plasmacytoid DC (pDCs) express enhanced interferon regulatory factor 5 (IRF-5) and reduced indoleamine 2,3-dioxygenase (IDO) leading to transient production of high levels of IFN-α that, in turn, stimulate CD8+T-

cell activity and downregulate regulatory T (Treg) cells. CD8+T cell can also be stimulated by activation of BDCA3+DCs, whose activation is interferon regulatory factor 8 (IRF-8) dependent. Moreover, the activity of CD8+cells is directly modulated by the E2–ERα axis in a hormone dosage-dependent manner. In this context, IFN-I maybe the signal which, in some conditions such as the onset of the antitumor response restored by immune checkpoint inhibitor, drives a more powerful inflammatory cell-mediated immune response in female.

The molecular point of view of the sexual immune dimorphism is also extremely intriguing. Many key immune-related genes, such as FOXP3 and CD40L, are located on the X chromosome (71, 72), and numerous X-linked genes of T cells carry the estrogen response elements (EREs) in their promoter. Hence, T cells from women may display a sex-biased signature characterized by inflammatory/cytotoxic effector genes such as IFN-γ, granulysin (GNLY), granzyme A (GZMA), RIGI, LTβ, IL12Rβ2, OAS1, IF16, X3CL1, CX3CL2, and the cytokines IL-15 and IL-16 (73). Therefore, the X chromosome may be responsible, at least in part, for the immunological advantage of females, whose signals are potentially more suitable to be activated by immunotherapies. This assumption is confirmed by the finding that the X chromosome is highly enriched in miRs as about 118 are located in this position, whereas only two miRs have been identified on the Y chromosome (74). MiRs sex-specific expression may have an enormous regulatory power on immune responses, through the control of signals and function of specific immune cell populations. For instance, CD4+T cells from female lupus patients have been found to overexpress 18 X chromosome-linked miRs with respect to the male counterpart (75). The PD-L1 expression itself, directly or through the control on trascription factors, appears to be modulated by several X-linked miRs, including miR-221, miR-222, miR-106b, miR-20b, and miR-513 (47); of interest, the X-linked miR-424 targets both PD-L1 and CD80 regulating concurrently the PD-L1/PD-1 and CD80/CTLA-4 pathways (76). Notably, a close crosstalk has been reported between the E2–ERα axis estrogen activity and epigenetic regulation by X-linked miRs, since molecules such as miR-221 and miR222 bind and regulate the ERα transcript (74).

The impact of sex differences on antitumor immunity is critical. In melanoma patients, partially exhausted cytotoxic CD8+T lymphocytes (peCTLs), upon increased expression and engagement of CTLA-4 and PD-1, drive the dysregulation of the host response. Therefore, while both sexes with high peCTLs show a similar objective response rate (ORR) following anti-PD-1 monotherapy as well as combination therapy, females with low peCTLs exhibit higher ORR after combination therapy (77). In the sex-biased immune landscape the complex interplay between microbiome, host immune system, and tumor is pivotal (49). The composition of gut microbiome affects responsiveness to ICIs, as melanoma patients with a greater variety of gut bacteria have high frequence of tumor-infiltrating CTLs predicting a better prognosis (78). In a murine melanoma model, the commensal *Bifdobacterium* was shown to improve ICB through DC activation associated with higher tumor-infiltrating CD8+T cells (79). Also, irAEs, such as diarrhea and colitis due to gut bacteria composition changes and intestinal barrier dysfunction (49), may have a sex-biased occurrence as gut microbiome is under a hormone-dependent control (80).

THE IMPLEMENTATION OF IMMUNOTHERAPY ON A SEX-BASED PERSPECTIVE

Vaccines was one of the first cancer immunotherapy approach. Despite most of them elicit antigen-specific immune responses in several clinical settings, they have largely failed in achieving a survival benefit (81), due to both tumor-cell intrinsic and extrinsic factors circumventing immune recognition and creating a suppressive TME (82).

Among therapeutic cancer vaccines, DCs have been regarded as a promising approach (83) and have been tested on wide range of tumor types (84). A comprehensive meta-analysis has demonstrated tumor-specific T-cell response in 77% of prostate and 61% of Renal Cell Carcinoma (RCC) patients (85). Although a correlation between DC-induced antigen-specific immune responses and prolonged patient survival has been confirmed in many studies (86, 87), the ORR does not exceed 15–20% (88). Moreover, most of these studies are phase-I/-II trials, involving few patients, whose immune and clinical responses were not evaluated taking into account patient characteristics, including sex. Future promising areas of investigation will be the use of personalized vaccines targeting neoantigens and their combination with immunomodulatory agents limiting the inhibitory signals in the TME (82, 89). In this context, due also to the hormonal modulation of many immune populations, including DCs (66), it would be extremely beneficial to consider how to take advantage of the sex-specific immune components.

A major paradigm shift in cancer immunotherapy was the use of antibodies targeting the immune-inhibitory pathways to unleash anticancer T-cell responses (90). Ipilimumab, targeting CTLA-4, was the first antibody clinically approved for treating patients with advanced unresectable melanoma (91). Soon after, nivolumab and pembrolizumab were developed to target PD-1. Today, ICIs are in the front line of immunotherapy of various advanced cancers; partial or complete objective responses have been obtained in patients with melanoma (31–44%) (92), NSCLC (20%) (93), and RCC (22–25%) (94). However, while such therapies exhibit clinical efficacy in many patients, the lack of response in a significant fraction of them remains the major concern. In this regard, intratumoral-infiltrating T cells correlate with a favorable outcome in melanoma patients, to the extent that analyses of type, functional orientation, density, and spatial location of these cells have been developed into a predictive immune scoring system (95). Accordingly, ICI efficacy seems to correlate with an ongoing cellular immune response and patients who do not respond to therapy often present metastatic lesions poorly infiltrated with immune cells (96).

Nevetheless, other crucial components concur to ICI treatment failure, including DC breakdown in proper Ag presentation to effector T cells, persistence of exhausted T cells, and lack of T-cell memory in TME (4, 9). Upon therapy, this dysfunctional state can be reversed or not. In this light, IFN-I signaling has been identified as an important factor for both response and resistance to ICI therapy (25). In fact, IFN-I may have immuno-suppressive or stimulatory activities depending on the magnitude, the timing and the duration of the activation of the signaling. Therefore, the activation of the IFN system may be crucial for the initial response to anti-PD-1, potentially favored in females who have an enhanced aptitude to activate this signal. It may, however, be detrimental if prolonged, promoting escape and resistance in patients who first responded (26, 27). In this light, the implementation of ICI therapies should take into account also the timing of the activation of the sex-specific components of the antitumor response.

Recently, a meta-analysis evaluating sex-related differences in the response to ICIs was carried out (97). The study assessed the progression free survival (PFS) and the OS in selected 36 phase II/III clinical trials on patients with solid tumors, including melanoma, NSCLC, RCC, head and neck, and urotheial carcinoma, and treated in the first or second line with ipilimumab and/or anti-PD-1 antibodies. Overall, 3,274 patients, of which 2,007 males (61.3%) and 1,267 females (38.7%), were analyzed. The results showed a better OS associated with ipilimumab in males compared with females. Not statistically significant results were observed with anti-PD-1 neither for OS nor for PFS. It should be emphasized that this analysis presents some weaknesses and limitations depending on the heterogeneity of the trials, the different cancer types considered, the variability of treatment regimens including patients treated or not with previous therapies, and the absence of information about hormonal status and on PD-L1 expression according to sex.

CONCLUSION

Research in cancer therapeutics has largely focused on two distinct approaches. One is based on the characterization of mutations causing tumorigenesis pivotal for the use of drugs targeting altered proteins in cancer cells; the other treats cancer "indirectly," exploiting the activation of the immune system to generate an antitumor immunity. While the first approach elicits impressive, but often not durable, tumor responses, the second accomplishes durable clinical responses but only in a subset of patients, a fraction of which experiences relapse after initial encouraging responses. Mechanisms of primary or acquired resistance are the major obstacles in both cases. Sex-associated factors correlate with cancer incidence, outcome, and response to therapy, underlining that sex differences are critical in tumor–host interaction. In this framework, sex disparity in immunity have been recently "re-discovered" and IFN-I signal could play a pivotal role in this scenario. Thus, the sexual dimorphism of the immune signals, including the IFN-I ones, may be a new attractive perspective for optimizing immunotherapy. Moreover, this critical challenge could represent a future opportunity to better integrate immunotherapies with other conventional as well as targeted therapies.

ACKNOWLEDGMENTS

The authors are very grateful to Dr Maria Buoncervello for assistance with figures and to Dr Tonino Sofia for the English editing and proofreading of the manuscript.

REFERENCES

1. Dyck L, Mills KHG. Immune checkpoints and their inhibition in cancer and infectious diseases. Eur J Immunol (2017) 47(5):765–79.10.1002/eji.201646875
2. Ugurel S, Rohmel J, Ascierto PA, Flaherty KT, Grob JJ, Hauschild A, et al. Survival of patients with advanced metastatic melanoma: the impact of novel therapies. Eur J Cancer (2016) 53:125–34.10.1016/j.ejca.2015.09.013
3. Fusi A, Dalgleish A. The importance for immunoregulation for long-term cancer control. Future Oncol (2017) 13:1619–32.10.2217/fon-2017-0085
4. Zarour HM. Reversing T-cell dysfunction and exhaustion in cancer. Clin Cancer Res (2016) 22(8):1856–64.10.1158/1078-0432.CCR-15-1849
5. Wang Q, Wu X. Primary and acquired resistance to PD-1/PD-L1 blockade in cancer treatment. Int Immunopharmacol (2017) 46:210–9.10.1016/j.intimp.2017.03.015
6. Kyi C, Postow MA. Immune checkpoint inhibitor combinations in solid tumors: opportunities and challenges. Immunotherapy (2016) 8(7):821–37.10.2217/imt-2016-0002
7. Klein SL, Flanagan KL. Sex differences in immune responses. Nat Rev Immunol (2016) 16(10):626–38.10.1038/nri.2016.90
8. Hegde PS, Karanikas V, Evers S. The where, the when, and the how of immune monitoring for cancer immunotherapies in the era of checkpoint inhibition. Clin Cancer Res (2016) 22(8):1865–74.10.1158/1078-0432.CCR-15-1507
9. Vasaturo A, Verdoes M, de Vries J, Torensma R, Figdor CG. Restoring immunosurveillance by dendritic cell vaccines and manipulation of the tumor microenvironment. Immunobiology (2015) 220(2):243–8.10.1016/j.imbio.2014.11.007
10. Brown SD, Warren RL, Gibb EA, Martin SD, Spinelli JJ, Nelson BH, et al. Neo-antigens predicted by tumor genome meta-analysis correlate with increased patient survival. Genome Res (2014) 24(5):743–50.10.1101/gr.165985.113
11. Veglia F, Gabrilovich DI. Dendritic cells in cancer: the role revisited. Curr Opin Immunol (2017) 45:43–51.10.1016/j.coi.2017.01.002
12. Broz ML, Krummel MF. The emerging understanding of myeloid cells as partners and targets in tumor rejection. Cancer Immunol

Res (2015) 3(4):313–9.10.1158/2326-6066.CIR-15-0041

13. Collin M, Bigley V. Human dendritic cell subsets: an update. Immunology (2018).10.1111/imm.12888
14. Salmon H, Idoyaga J, Rahman A, Leboeuf M, Remark R, Jordan S, et al. Expansion and activation of CD103(+) dendritic cell progenitors at the tumor site enhances tumor responses to therapeutic PD-L1 and BRAF inhibition. Immunity (2016) 44(4):924–38.10.1016/j.immuni.2016.03.012
15. Roberts EW, Broz ML, Binnewies M, Headley MB, Nelson AE, Wolf DM, et al. Critical role for CD103(+)/CD141(+) dendritic cells bearing CCR7 for tumor antigen trafficking and priming of T cell immunity in melanoma. Cancer Cell (2016) 30(2):324–36.10.1016/j.ccell.2016.06.003
16. Le Mercier I, Poujol D, Sanlaville A, Sisirak V, Gobert M, Durand I, et al. Tumor promotion by intratumoral plasmacytoid dendritic cells is reversed by TLR7 ligand treatment. Cancer Res (2013) 73(15):4629–40.10.1158/0008-5472.CAN-12-3058
17. Tel J, Aarntzen EH, Baba T, Schreibelt G, Schulte BM, Benitez-Ribas D, et al. Natural human plasmacytoid dendritic cells induce antigen-specific T-cell responses in melanoma patients. Cancer Res (2013) 73(3):1063–75.10.1158/0008-5472.CAN-12-2583
18. Swiecki M, Colonna M. The multifaceted biology of plasmacytoid dendritic cells. Nat Rev Immunol (2015) 15(8):471–85.10.1038/nri3865
19. Ivashkiv LB, Donlin LT. Regulation of type I interferon responses. Nat Rev Immunol (2014) 14(1):36–49.10.1038/nri3581
20. Parker BS, Rautela J, Hertzog PJ. Antitumour actions of interferons: implications for cancer therapy. Nat Rev Cancer (2016) 16(3):131–44.10.1038/nrc.2016.14
21. Burnette BC, Liang H, Lee Y, Chlewicki L, Khodarev NN, Weichselbaum RR, et al. The efficacy of radiotherapy relies upon induction of type i interferon-dependent innate and adaptive immunity. Cancer Res (2011) 71(7):2488–96.10.1158/0008-5472.CAN-10-2820
22. Fuertes MB, Kacha AK, Kline J, Woo SR, Kranz DM, Murphy KM, et al. Host type I IFN signals are required for antitumor CD8+ T cell responses through CD8{alpha}+ dendritic cells. J Exp Med (2011) 208(10):2005–16.10.1084/jem.20101159

23. Schiavoni G, Mattei F, Gabriele L. Type I interferons as stimulators of DC-mediated cross-priming: impact on anti-tumor response. Front Immunol (2013) 4:483.10.3389/fimmu.2013.00483
24. Corrales L, Matson V, Flood B, Spranger S, Gajewski TF. Innate immune signaling and regulation in cancer immunotherapy. Cell Res (2017) 27(1):96–108.10.1038/cr.2016.149
25. Minn AJ, Wherry EJ. Combination cancer therapies with immune checkpoint blockade: convergence on interferon signaling. Cell (2016) 165(2):272–5.10.1016/j.cell.2016.03.031
26. Zaretsky JM, Garcia-Diaz A, Shin DS, Escuin-Ordinas H, Hugo W, Hu-Lieskovan S, et al. Mutations associated with acquired resistance to PD-1 blockade in melanoma. N Engl J Med (2016) 375(9):819–29.10.1056/NEJMoa1604958
27. Benci JL, Xu B, Qiu Y, Wu TJ, Dada H, Twyman-Saint Victor C, et al. Tumor interferon signaling regulates a multigenic resistance program to immune checkpoint blockade. Cell (2016) 167(6):1540–54.e12.10.1016/j.cell.2016.11.022
28. Teijaro JR. Pleiotropic roles of type 1 interferons in antiviral immune responses. Adv Immunol (2016) 132:135–58.10.1016/bs.ai.2016.08.001
29. Ilander M, Kreutzman A, Rohon P, Melo T, Faber E, Porkka K, et al. Enlarged memory T-cell pool and enhanced Th1-type responses in chronic myeloid leukemia patients who have successfully discontinued IFN-alpha monotherapy. PLoS One (2014) 9(1):e87794.10.1371/journal.pone.0087794
30. Katlinski KV, Gui J, Katlinskaya YV, Ortiz A, Chakraborty R, Bhattacharya S, et al. Inactivation of interferon receptor promotes the establishment of immune privileged tumor microenvironment. Cancer Cell (2017) 31(2):194–207.10.1016/j.ccell.2017.01.004
31. Snell LM, McGaha TL, Brooks DG. Type I interferon in chronic virus infection and cancer. Trends Immunol (2017) 38(8):542–57.10.1016/j.it.2017.05.005
32. Terawaki S, Chikuma S, Shibayama S, Hayashi T, Yoshida T, Okazaki T, et al. IFN-alpha directly promotes programmed cell death-1 transcription and limits the duration of T cell-mediated immunity. J Immunol (2011) 186(5):2772–9.10.4049/jimmunol.1003208
33. Schwartz JA, Clayton KL, Mujib S, Zhang H, Rahman AK, Liu J, et

al. Tim-3 is a marker of plasmacytoid dendritic cell dysfunction during HIV infection and is associated with the recruitment of IRF7 and p85 into lysosomes and with the submembrane displacement of TLR9. J Immunol (2017) 198(8):3181–94.10.4049/jimmunol.1601298

34. Tsai HF, Hsu PN. Cancer immunotherapy by targeting immune checkpoints: mechanism of T cell dysfunction in cancer immunity and new therapeutic targets. J Biomed Sci (2017) 24(1):35.10.1186/s12929-017-0341-0
35. Yu Y, Huang R, Zong X, He X, Mo W. INFalpha-2b inhibitory effects on CD4(+)CD25(+)FOXP3(+) regulatory T cells in the tumor microenvironment of C57BL/6 J mice with melanoma xenografts. BMC Cancer (2016) 16:397.10.1186/s12885-016-2473-0
36. Bacher N, Raker V, Hofmann C, Graulich E, Schwenk M, Baumgrass R, et al. Interferon-alpha suppresses cAMP to disarm human regulatory T cells. Cancer Res (2013) 73(18):5647–56.10.1158/0008-5472.CAN-12-3788
37. Fend L, Yamazaki T, Remy C, Fahrner C, Gantzer M, Nourtier V, et al. Immune checkpoint blockade, immunogenic chemotherapy or IFN-alpha blockade boost the local and abscopal effects of oncolytic virotherapy. Cancer Res (2017) 77(15):4146–57.10.1158/0008-5472.CAN-16-2165
38. Munn DH, Sharma MD, Johnson TS, Rodriguez P. IDO, PTEN-expressing Tregs and control of antigen-presentation in the murine tumor microenvironment. Cancer Immunol Immunother (2017) 66(8):1049–58.10.1007/s00262-017-2010-2
39. Lemos H, Mohamed E, Huang L, Ou R, Pacholczyk G, Arbab AS, et al. STING promotes the growth of tumors characterized by low antigenicity via IDO activation. Cancer Res (2016) 76(8):2076–81.10.1158/0008-5472.CAN-15-1456
40. Chevolet I, Speeckaert R, Haspeslagh M, Neyns B, Kruse V, Schreuer M, et al. Peritumoral indoleamine 2,3-dioxygenase expression in melanoma: an early marker of resistance to immune control? Br J Dermatol (2014) 171(5):987–95.10.1111/bjd.13100
41. Rautela J, Huntington ND. IL-15 signaling in NK cell cancer immunotherapy. Curr Opin Immunol (2017) 44:1–6.10.1016/j.coi.2016.10.004

42. Paolini R, Bernardini G, Molfetta R, Santoni A. NK cells and interferons. Cytokine Growth Factor Rev (2015) 26(2):113–20.10.1016/j.cytogfr.2014.11.003
43. O'Donnell JS, Long GV, Scolyer RA, Teng MW, Smyth MJ. Resistance to PD1/PDL1 checkpoint inhibition. Cancer Treat Rev (2017) 52:71–81.10.1016/j.ctrv.2016.11.007
44. Schreiber G. The molecular basis for differential type I interferon signaling. J Biol Chem (2017) 292(18):7285–94.10.1074/jbc.R116.774562
45. Chen K, Liu J, Cao X. Regulation of type I interferon signaling in immunity and inflammation: a comprehensive review. J Autoimmun (2017) 83:1–11.10.1016/j.jaut.2017.03.008
46. Pauken KE, Sammons MA, Odorizzi PM, Manne S, Godec J, Khan O, et al. Epigenetic stability of exhausted T cells limits durability of reinvigoration by PD-1 blockade. Science (2016) 354(6316):1160–5.10.1126/science.aaf2807
47. Grenda A, Krawczyk P. New dancing couple: PD-L1 and microRNA. Scand J Immunol (2017) 86:130–4.10.1111/sji.12577
48. Linardou H, Gogas H. Toxicity management of immunotherapy for patients with metastatic melanoma. Ann Transl Med (2016) 4(14):272.10.21037/atm.2016.07.10
49. Botticelli A, Zizzari I, Mazzuca F, Ascierto PA, Putignani L, Marchetti L, et al. Cross-talk between microbiota and immune fitness to steer and control response to anti PD-1/PDL-1 treatment. Oncotarget (2017) 8(5):8890–9.10.18632/oncotarget.12985
50. Kotredes KP, Thomas B, Gamero AM. The protective role of type I interferons in the gastrointestinal tract. Front Immunol (2017) 8:410.10.3389/fimmu.2017.00410
51. Mirandola L, Wade R, Verma R, Pena C, Hosiriluck N, Figueroa JA, et al. Sex-driven differences in immunological responses: challenges and opportunities for the immunotherapies of the third millennium. Int Rev Immunol (2015) 34(2):134–42.10.3109/08830185.2015.1018417
52. Laffont S, Blanquart E, Savignac M, Cenac C, Laverny G, Metzger D, et al. Androgen signaling negatively controls group 2 innate lymphoid cells. J Exp Med (2017) 214(6):1581–92.10.1084/jem.20161807
53. Pennell LM, Galligan CL, Fish EN. Sex affects immunity. J Autoimmun (2012) 38(2–3):J282–91.10.1016/j.jaut.2011.11.013

54. Zilio S, Serafini P. Neutrophils and granulocytic MDSC: the Janus god of cancer immunotherapy. Vaccines (Basel) (2016) 4(3):E31.10.3390/vaccines4030031
55. Lu C, Redd PS, Lee JR, Savage N, Liu K. The expression profiles and regulation of PD-L1 in tumor-induced myeloid-derived suppressor cells. Oncoimmunology (2016) 5(12):e1247135.10.1080/2162402X.2016.1247135
56. Kovats S. Estrogen receptors regulate innate immune cells and signaling pathways. Cell Immunol (2015) 294(2):63–9.10.1016/j.cellimm.2015.01.018
57. Seillet C, Laffont S, Tremollieres F, Rouquie N, Ribot C, Arnal JF, et al. The TLR-mediated response of plasmacytoid dendritic cells is positively regulated by estradiol in vivo through cell-intrinsic estrogen receptor alpha signaling. Blood (2012) 119(2):454–64.10.1182/blood-2011-08-371831
58. Laffont S, Rouquie N, Azar P, Seillet C, Plumas J, Aspord C, et al. X-Chromosome complement and estrogen receptor signaling independently contribute to the enhanced TLR7-mediated IFN-alpha production of plasmacytoid dendritic cells from women. J Immunol (2014) 193(11):5444–52.10.4049/jimmunol.1303400
59. Griesbeck M, Ziegler S, Laffont S, Smith N, Chauveau L, Tomezsko P, et al. Sex differences in plasmacytoid dendritic cell levels of IRF5 drive higher IFN-alpha production in women. J Immunol (2015) 195(11):5327–36.10.4049/jimmunol.1501684
60. Laffont S, Seillet C, Guery JC. Estrogen receptor-dependent regulation of dendritic cell development and function. Front Immunol (2017) 8:108.10.3389/fimmu.2017.00108
61. Bajana S, Roach K, Turner S, Paul J, Kovats S. IRF4 promotes cutaneous dendritic cell migration to lymph nodes during homeostasis and inflammation. J Immunol (2012) 189(7):3368–77.10.4049/jimmunol.1102613
62. Heng TS, Painter MW, Immunological Genome Project Consortium . The Immunological Genome Project: networks of gene expression in immune cells. Nat Immunol (2008) 9(10):1091–4.10.1038/ni1008-1091
63. Anipindi VC, Bagri P, Roth K, Dizzell SE, Nguyen PV, Shaler CR, et al. Estradiol enhances CD4+ T-cell anti-viral immunity by priming

vaginal DCs to induce Th17 responses via an IL-1-dependent pathway. PLoS Pathog (2016) 12(5):e1005589.10.1371/journal.ppat.1005589

64. Thompson MG, Peiffer DS, Larson M, Navarro F, Watkins SK. FOXO3, estrogen receptor alpha, and androgen receptor impact tumor growth rate and infiltration of dendritic cell subsets differentially between male and female mice. Cancer Immunol Immunother (2017) 66(5):615–25.10.1007/s00262-017-1972-4
65. Panchanathan R, Shen H, Zhang X, Ho SM, Choubey D. Mutually positive regulatory feedback loop between interferons and estrogen receptor-alpha in mice: implications for sex bias in autoimmunity. PLoS One (2010) 5(5):e10868.10.1371/journal.pone.0010868
66. Mackern-Oberti JP, Jara EL, Riedel CA, Kalergis AM. Hormonal modulation of dendritic cells differentiation, maturation and function: implications for the initiation and progress of systemic autoimmunity. Arch Immunol Ther Exp (2017) 65(2):123–36.10.1007/s00005-016-0418-6
67. Fish EN. The X-files in immunity: sex-based differences predispose immune responses. Nat Rev Immunol (2008) 8(9):737–44.10.1038/nri2394
68. Arsenovic-Ranin N, Kosec D, Pilipovic I, Nacka-Aleksic M, Bufan B, Stojic-Vukanic Z, et al. Sex and age as determinants of rat T-cell phenotypic characteristics: influence of peripubertal gonadectomy. Mol Cell Biochem (2017) 431(1–2):169–85.10.1007/s11010-017-2989-x
69. Zhang MA, Rego D, Moshkova M, Kebir H, Chruscinski A, Nguyen H, et al. Peroxisome proliferator-activated receptor (PPAR)alpha and -gamma regulate IFNgamma and IL-17A production by human T cells in a sex-specific way. Proc Natl Acad Sci U S A (2012) 109(24):9505–10.10.1073/pnas.1118458109
70. Lin PY, Sun L, Thibodeaux SR, Ludwig SM, Vadlamudi RK, Hurez VJ, et al. B7-H1-dependent sex-related differences in tumor immunity and immunotherapy responses. J Immunol (2010) 185(5):2747–53.10.4049/jimmunol.1000496
71. Markle JG, Fish EN. SeXX matters in immunity. Trends Immunol (2014) 35(3):97–104.10.1016/j.it.2013.10.006
72. Xu A, Wang R, Freywald A, Stewart K, Tikoo S, Xu J, et al. CD40 agonist converting CTL exhaustion via the activation of the mTORC1

pathway enhances PD-1 antagonist action in rescuing exhausted CTLs in chronic infection. Biochem Biophys Res Commun (2017) 484(3):662–7.10.1016/j.bbrc.2017.01.172

73. Hewagama A, Patel D, Yarlagadda S, Strickland FM, Richardson BC. Stronger inflammatory/cytotoxic T-cell response in women identified by microarray analysis. Genes Immun (2009) 10(5):509–16.10.1038/gene.2009.12

74. Dai R, Ahmed SA. Sexual dimorphism of miRNA expression: a new perspective in understanding the sex bias of autoimmune diseases. Ther Clin Risk Manag (2014) 10:151–63.10.2147/TCRM.S33517

75. Hewagama A, Gorelik G, Patel D, Liyanarachchi P, McCune WJ, Somers E, et al. Overexpression of X-linked genes in T cells from women with lupus. J Autoimmun (2013) 41:60–71.10.1016/j.jaut.2012.12.006

76. Smolle MA, Calin HN, Pichler M, Calin GA. Noncoding RNAs and immune checkpoints-clinical implications as cancer therapeutics. FEBS J (2017) 284(13):1952–66.10.1111/febs.14030

77. Loo K, Tsai KK, Mahuron K, Liu J, Pauli ML, Sandoval PM, et al. Partially exhausted tumor-infiltrating lymphocytes predict response to combination immunotherapy. JCI Insight (2017) 2(14):93433.10.1172/jci.insight.93433

78. Gut microbes may up PD-1 inhibitor response. Cancer Discov (2017) 7(5):448.10.1158/2159-8290.CD-NB2017-039

79. Sivan A, Corrales L, Hubert N, Williams JB, Aquino-Michaels K, Earley ZM, et al. Commensal *Bifidobacterium* promotes antitumor immunity and facilitates anti-PD-L1 efficacy. Science (2015) 350(6264):1084–9.10.1126/science.aac4255

80. Markle JG, Frank DN, Adeli K, von Bergen M, Danska JS. Microbiome manipulation modifies sex-specific risk for autoimmunity. Gut Microbes (2014) 5(4):485–93.10.4161/gmic.29795

81. Melero I, Gaudernack G, Gerritsen W, Huber C, Parmiani G, Scholl S, et al. Therapeutic vaccines for cancer: an overview of clinical trials. Nat Rev Clin Oncol (2014) 11(9):509–24.10.1038/nrclinonc.2014.111

82. van der Burg SH, Arens R, Ossendorp F, van Hall T, Melief CJ. Vaccines for established cancer: overcoming the challenges posed by immune evasion. Nat Rev Cancer (2016) 16(4):219–33.10.1038/nrc.2016.16

83. Steinman RM. Decisions about dendritic cells: past, present, and future. Annu Rev Immunol (2012) 30:1–22.10.1146/annurev-

immunol-100311-102839

84. Bracci L, Capone I, Moschella F, Proietti E, Belardelli F. Exploiting dendritic cells in the development of cancer vaccines. Expert Rev Vaccines (2013) 12(10):1195–210.10.1586/14760584.2013.836905
85. Draube A, Klein-Gonzalez N, Mattheus S, Brillant C, Hellmich M, Engert A, et al. Dendritic cell based tumor vaccination in prostate and renal cell cancer: a systematic review and meta-analysis. PLoS One (2011) 6(4):e18801.10.1371/journal.pone.0018801
86. Sheikh NA, Petrylak D, Kantoff PW, Dela Rosa C, Stewart FP, Kuan LY, et al. Sipuleucel-T immune parameters correlate with survival: an analysis of the randomized phase 3 clinical trials in men with castration-resistant prostate cancer. Cancer Immunol Immunother (2013) 62(1):137–47.10.1007/s00262-012-1317-2
87. GuhaThakurta D, Sheikh NA, Fan LQ, Kandadi H, Meagher TC, Hall SJ, et al. Humoral immune response against nontargeted tumor antigens after treatment with Sipuleucel-T and its association with improved clinical outcome. Clin Cancer Res (2015) 21(16):3619–30.10.1158/1078-0432.CCR-14-2334
88. Anguille S, Smits EL, Lion E, van Tendeloo VF, Berneman ZN. Clinical use of dendritic cells for cancer therapy. Lancet Oncol (2014) 15(7):e257–67.10.1016/S1470-2045(13)70585-0
89. Yarchoan M, Johnson BA, III, Lutz ER, Laheru DA, Jaffee EM. Targeting neoantigens to augment antitumour immunity. Nat Rev Cancer (2017) 17(4):209–22.10.1038/nrc.2016.154
90. Littman DR. Releasing the brakes on cancer immunotherapy. Cell (2015) 162(6):1186–90.10.1016/j.cell.2015.08.038
91. Wolchok JD, Hodi FS, Weber JS, Allison JP, Urba WJ, Robert C, et al. Development of ipilimumab: a novel immunotherapeutic approach for the treatment of advanced melanoma. Ann N Y Acad Sci (2013) 1291:1–13.10.1111/nyas.12180
92. Robert C, Schachter J, Long GV, Arance A, Grob JJ, Mortier L, et al. Pembrolizumab versus ipilimumab in advanced melanoma. N Engl J Med (2015) 372(26):2521–32.10.1056/NEJMoa1503093
93. Garon EB, Rizvi NA, Hui R, Leighl N, Balmanoukian AS, Eder JP, et al. Pembrolizumab for the treatment of non-small-cell lung cancer. N Engl J Med (2015) 372(21):2018–28.10.1056/NEJMoa1501824

94. Motzer RJ, Rini BI, McDermott DF, Redman BG, Kuzel TM, Harrison MR, et al. Nivolumab for metastatic renal cell carcinoma: results of a randomized phase II trial. J Clin Oncol (2015) 33(13):1430–7.10.1200/JCO.2014.59.0703
95. Erdag G, Schaefer JT, Smolkin ME, Deacon DH, Shea SM, Dengel LT, et al. Immunotype and immunohistologic characteristics of tumor-infiltrating immune cells are associated with clinical outcome in metastatic melanoma. Cancer Res (2012) 72(5):1070–80.10.1158/0008-5472.CAN-11-3218
96. Galon J, Fox BA, Bifulco CB, Masucci G, Rau T, Botti G, et al. Immunoscore and immunoprofiling in cancer: an update from the melanoma and immunotherapy bridge 2015. J Transl Med (2016) 14:273.10.1186/s12967-016-1029-z
97. Botticelli A, Onesti CE, Zizzari I, Cerbelli B, Sciattella P, Occhipinti M, et al. The sexist behaviour of immune checkpoint inhibitors in cancer therapy? Oncotarget (2017) 8(59):99336–46.10.18632/oncotarget.22242

16

Recent Advances in Targeting the EGFR Signaling Pathway for the Treatment of Metastatic Colorectal Cancer

Yuji Miyamoto [1], Koichi Suyama [2] and Hideo Baba [1]

[1]Department of Gastroenterological Surgery, Graduate School of Medical Sciences, Kumamoto University, 1-1-1 Honjo, Kumamoto 860-8556, Japan
[2]Cancer Center, Kumamoto University Hospital, Kumamoto 860-8556, Japan

ABSTRACT

Outcomes for metastatic colorectal cancer (mCRC) patients have been improved by treatment with anti-epidermal growth factor receptor (anti-EGFR) antibodies, particularly when combined with predictive biomarkers to select patients lacking *RAS* mutations. New technologies such as liquid biopsy and next-generation sequencing have revealed that potential mechanisms of resistance to anti-EGFR therapies act through acquired mutations of *KRAS* and the *EGFR* ectodomain. Mutations in cross-talking molecular effectors that participate in downstream EGFR signaling are also negative predictors for anti-EGFR therapy. In the current review, we

Citation: Miyamoto, Y.; Suyama, K.; Baba, H. Recent Advances in Targeting the EGFR Signaling Pathway for the Treatment of Metastatic Colorectal Cancer. Int. J. Mol. Sci. 2017, 18, 752. https://doi.org/10.3390/ijms18040752.

describe recent advances in anti-EGFR therapy and discuss new treatment strategies to target downstream RAS-MAPK signaling in mCRC.

Keywords: metastatic colorectal cancer, epidermal growth-factor receptor signaling (EGFR), chemotherapy, secondary resistance

INTRODUCTION

In recent decades, significant improvements have been seen in the survival of patients with metastatic colorectal cancer (mCRC). This has been driven to a large extent by the approval of new drugs, including irinotecan, oxaliplatin, capecitabine, several humanized monoclonal antibodies that target either vascular endothelial growth factor (VEGF; bevacizumab, aflibercept, and ramucirumab) or the epidermal growth factor receptor (EGFR; cetuximab and panitumumab), and most recently, regorafenib and trifluridine/tipiracil (TAS-102) [1,2,3,4,5,6,7,8,9,10,11,12,13]. Clinical benefit from these drugs is now well established for patients with mCRC, with median overall survival (OS) of over 30 months (M) [14,15,16]. In addition, a number of studies have focused on the identification of more reliable and specific predictive biomarkers for choosing the most appropriate treatment for an individual patient. The only established biomarker for the treatment of mCRC patients is tumor *RAS* mutational status, which is a negative predictive marker for anti-EGFR therapy [17]. Currently, *BRAF* mutational testing is also recommended by the National Comprehensive Cancer Network (NCCN) [18] and the European Society for Medical Oncology (ESMO) [19].

Unlike anti-VEGF therapy, the mechanisms of resistance in anti-EGFR therapy are well-studied, as are drugs inhibiting downstream RAS-MAPK signaling. Indeed, several recent clinical trials targeting RAS signaling have shown promising activity in chemorefractory mCRC. In this review, we focus on recent clinical and preclinical studies of EGFR inhibitors, their resistance mechanisms, and new downstream inhibitors of the EGFR pathway.

CLINICAL ADVANCES IN ANTI-EGFR ANTIBODIES

Cetuximab and panitumumab are both monoclonal antibodies directed against the extracellular domain of the EGFR, which block ligand binding and lead to inhibition of the downstream RAS-RAF-MEK-ERK signaling pathway. Several randomized clinical trials have established the effectiveness of both drugs in combination with fluorouracil (5-FU) plus irinotecan (FOLFIRI) [8,20] and 5-FU plus oxaliplatin (FOLFOX) [10,21] for patients with wild-type *RAS* mCRC. The differences between cetuximab and panitumumab might be derived from their different protein class characteristics or species (cetuximab is 13% mouse and 87% human, panitumumab is 100% human) [22]. As an IgG1 antibody, cetuximab exerts additional antitumor effects by mediating antibody-dependent cellular cytotoxicity [23]. However, the significance of this effect is not fully understood. The ASPECCT study [24], a phase III randomized controlled trial, indicated that panitumumab was non-inferior to cetuximab and that these agents provided a similar OS benefit in patients with chemotherapy-refractory wild-type *KRAS* exon 2 mCRC. The primary endpoint, median OS was 10.4 M in the panitumumab group and 10.0 M in the cetuximab group (hazard ratio, HR = 0.97, 95% confidence interval, CI = 0.84–1.11, $p < 0.0007$ for non-inferiority). These results also showed that the incidence of grade 3 or 4 hypomagnesemia was greater in patients receiving panitumumab (7%) than in those receiving cetuximab (3%), although the incidence of severe skin toxicities was similar in the two groups.

THE EFFECT OF RAS STATUS ON ANTI-EGFR THERAPIES

The search for a predictive biomarker for anti-EGFR therapies was initially directed toward EGFR expression, which has been reported to be increased in 49% to 82% of mCRC [25,26]. Anti-EGFR therapies inhibit downstream signaling pathways, but *EGFR* expression status, as assessed using immunohistochemistry, does not predict treatment efficacy [27,28]. Data from the BOND study indicated that the intensity of immunohistochemical EGFR staining in colorectal tumor cells did not correlate with the objective response rate (ORR) to cetuximab [29]. Subsequently, mutations conferring resistance to anti-EGFR therapies were identified in codons 12 and 13 of exon 2 of the *KRAS* gene, which result in constitutive activation of the RAS-RAF-MEK-ERK pathway [8,21,30,31]. Activating mutations in *KRAS* are detected in approximately 40% of mCRC [31], with good concordance between the primary tumors and matched distant metastases [32,33].

More recent studies have found that resistance to anti-EGFR therapy can also be mediated by lower-frequency mutations in *KRAS* exon 3 or 4, or in *NRAS* exon 2, 3, or 4 [11,32,34]. Exclusion of patients with any *RAS* mutation identifies a population that is more likely to benefit from anti-EGFR therapies [35]. In the PRIME trial [33], 17% of patients without mutations in *KRAS* exon 2 did have mutations in *KRAS* exon 3 or 4, or in *NRAS* exon 2, 3, or 4. All of these *RAS* mutations predicted a lack of response to panitumumab, and in fact, their presence was associated with inferior progression-free survival (PFS) and OS in patients receiving panitumumab plus FOLFOX compared with FOLFOX alone. Median OS was 25.8 M versus 20.2 M (HR = 0.77, 95% CI = 0.64–0.94, $p = 0.009$) in wild-type *RAS* populations, in favor of the combination of panitumumab and FOLFOX. Similar results were presented for all *RAS* genotypes in the CRYSTAL [36] and OPUS [37] trials, in which randomized patients received first-line cetuximab in combination with FOLFIRI or FOLFOX respectively [38]. In addition, a meta-analysis of nine randomized controlled trials of anti-EGFR antibodies for mCRC demonstrated the predictive value of *RAS* mutational profiles for both PFS and OS [35]. These results indicate that the anti-EGFR antibodies should be restricted to mCRC patients whose tumors lack all *KRAS* and *NRAS* mutations.

KRAS^{G13D} MUTATION

Preclinical studies and retrospective data from phase III trials suggest that patients with *KRAS*^{G13D} mutant tumors might benefit from cetuximab. De Roock et al. reported that patients carrying the *KRAS*^{G13D} mutation and treated with cetuximab had prolonged OS (HR = 0.50, 95% CI = 0.31–0.81, $p = 0.005$) and PFS (HR = 0.51, 95% CI = 0.32–0.81, $p = 0.004$) compared with patients whose tumors harbored other *KRAS* mutations [39]. Another retrospective analysis was performed on a pooled data set from the CRYSTAL and OPUS studies [40]. Within *KRAS* mutation subgroups, cetuximab plus chemotherapy versus chemotherapy alone significantly improved PFS (median, 7.4 versus 6.0 M, HR = 0.47, $p = 0.039$) and tumor response rate (41% versus 22%, odds ratio, OR = 3.38, $p = 0.042$) in patients with *KRAS*^{G13D} mutant tumors. Based on these results, two prospective phase II trials were performed; however, both trials showed no activity for cetuximab monotherapy in patients with *KRAS*^{G13D} mutant mCRC. In the phase II ICECREAM trial [41], there was no statistically significant improvement in disease control at 6 M for patients with *KRAS*^{G13D} mutant tumors treated with either cetuximab monotherapy or cetuximab plus irinotecan. Currently, anti-EGFR therapies are not routinely recommended for patients with *KRAS*^{G13D} mutant tumors.

ANTI-EGFR THERAPIES VERSUS BEVACIZUMAB IN FIRST LINE CHEMOTHERAPY

A highly important unresolved issue is whether the anti-VEGF bevacizumab or an anti-EGFR antibody is the optimal first-line biologic agent to be added to chemotherapy for patients with wild-type *RAS* mCRC. Two phase III trials, FIRE-3 [14] and CALGB/SWOG 80405 [42], and the phase II PEAK trial [43] have directly compared the addition of bevacizumab versus cetuximab or panitumumab to FOLFOX/FOLFIRI in terms of efficacy outcomes. The FIRE-3 study [14] was conducted to evaluate the superiority of FOLFIRI plus cetuximab to FOLFIRI plus bevacizumab as a first-line treatment for mCRC patients with wild-type *KRAS*. In patients with only wild-type *RAS* tumors, the median OS was significantly longer with cetuximab (33.1 M versus 25.6 M, HR = 0.70, 95% CI = 0.53–0.92, $p = 0.011$), but the median PFS was almost identical (10.4 M versus 10.2 M, HR = 0.93, 95% CI = 0.74–1.17, $p = 0.54$). A similar pattern of outcomes was observed in the PEAK trial [43], primarily comparing FOLFOX plus either panitumumab or bevacizumab in patients with previously untreated, wild-type *KRAS* mCRC. In patients with all wild-type *RAS*, PFS was 13.0 M in the panitumumab arm and 9.5 M in the bevacizumab arm (HR = 0.65, 95% CI = 0.44–0.96, $p = 0.029$), and median OS was 41.3 M in the panitumumab arm and 28.9 M in the bevacizumab arm (HR = 0.63, 95% CI = 0.39–1.02, $p = 0.058$).

The phase III CALGB/SWOG 80405 trial [42] could not demonstrate a benefit for initial cetuximab versus bevacizumab in combination with either FOLFIRI or FOLFOX, even in patients with only wild-type *RAS* tumors. In patients with wild-type *RAS* tumors, there was no significant difference in OS and PFS between the cetuximab and bevacizumab in combination with chemotherapy. However, there was a higher response rate achieved in the cetuximab arm (68.6% versus 53.6%, $p < 0.01$). A meta-analysis of these randomized controlled trials supports a potential benefit of first-line EGFR inhibitors plus chemotherapy versus bevacizumab plus chemotherapy with respect to OS (HR = 0.8, 95% CI = 0.68–0.93, $p = 0.004$) and ORR (OR = 0.57, 95% CI = 0.42–0.76) in patients with wild-type *RAS* mCRC [44]. Another phase III trial currently in progress, PARADIGM [45], is comparing panitumumab versus bevacizumab in combination with FOLFOX.

PRIMARY TUMOR LOCATION AS A PROGNOSTIC AND PREDICTIVE BIOMARKER IN MCRC

Another topic of interest is the prognostic and predictive significance of primary tumor location in mCRC patients with wild-type *RAS* tumors. Genetic and phenotypic differences may be derived from pre-existing disparities between the left and right colon with respect to embryonic origin, blood supply, innervation, lymphatic drainage, and lumen environment. Associations between a particular marker and prognosis or response to chemotherapy have been found to be site-specific [46,47,48,49].

Recently, retrospective analyses were conducted in patients with wild-type *RAS* mCRC from two large phase III trials, CRYSTAL and FIRE-3, in which mCRC were classified as left-sided or right-sided [50]. In the CRYSTAL trial, patients with right-sided tumors in the FOLFIRI plus cetuximab treatment group had a significantly shorter PFS and OS compared with equivalent patients with left-sided tumors (PFS: 8.1 M versus 12.0 M, HR = 1.77, 95% CI = 1.08–2.91, $p = 0.02$; OS: 28.7 M versus 18.5 M, HR = 1.93, 95% CI = 1.24–2.99, $p = 0.003$). Furthermore, the addition of cetuximab to FOLFIRI in patients with left-sided tumors significantly improved PFS (12.0 M versus 8.9 M, HR = 0.50, 95% CI = 0.34–0.72, $p < 0.001$), OS (28.7 M versus 21.7 M, HR = 0.65, 95% CI = 0.50–0.86, $p = 0.002$), and ORR (72.5% versus 40.6%, OR = 3.99, 95% CI = 2.40–6.62, $p < 0.001$), as expected based on the overall population. In patients with right-sided tumors, however, the addition of cetuximab conferred more limited benefits on PFS (8.1 M versus 7.1 M, HR = 0.87, 95% CI = 0.47–1.62, $p = 0.66$), OS (18.5 M versus 15.0 M, HR = 1.8, 95% CI = 0.65–1.81, $p = 0.76$), and ORR (42.4% versus 33.3%, OR = 1.45, 95% CI = 0.58–3.64, $p = 0.43$).

Results consistent with these were obtained from the FIRE-3 trial. Patients with left-sided tumors who received FOLFIRI plus cetuximab had significantly longer OS than those receiving FOLFIRI plus bevacizumab (38.3 M versus 28.0 M, HR = 0.63, 95% CI = 0.48–0.85, $p < 0.002$), but with no significant differences in PFS or ORR. In contrast, among patients with right-sided tumors, there were no significant differences in OS, PFS, or ORR for FOLFIRI plus cetuximab versus FOLFIRI plus bevacizumab. Moreover, multivariable analyses revealed that, in both studies, there was a significant interaction between primary tumor location and treatment modality for OS of patients with wild-type *RAS* tumors.

These results confirm the prognostic value of primary tumor location in wild-type *RAS* populations, and further suggest an interaction between tumor location and response to treatment. Right-sided tumors do not significantly benefit from the addition of cetuximab, in contrast to a marked benefit for left-sided tumors [48,51]. This difference may be derived from the molecular/genetic background, and/or from embryological or epidemiological factors. For example, right-sided tumors are associated with *BRAF* mutation, DNA hypermethylation (the CpG island methylator phenotype, CIMP), and hypermutated consensus molecular subtypes [52]. Comprehensive molecular and genetic analysis of specimens derived from phase III trials is needed to identify biological differences between right- and left-sided colons.

RESISTANCE MECHANISM TO ANTI-EGFR THERAPIES

Even among patients with wild-type *RAS* mCRC that initially responds to anti-EGFR therapies, the majority will eventually experience disease progression. This implies acquired resistance to anti-EGFR therapy. Therefore, multiple studies have focused on exploring resistance mechanisms, and it seems that several biomarkers and pathways are involved in the development of resistance to anti-EGFR therapy. The two principal downstream effectors of EGFR activation are the RAS-MAPK pathway and the PI3K-AKT-mTOR pathway, both of which control cell growth and proliferation (Figure 1) [53]. Several investigators have assessed circulating tumor DNA (ctDNA) in the blood of mCRC patients during anti-EGFR therapy and found that undetectable low-frequency *KRAS*-mutant clones may be selected by anti-EGFR treatment [54,55,56,57].

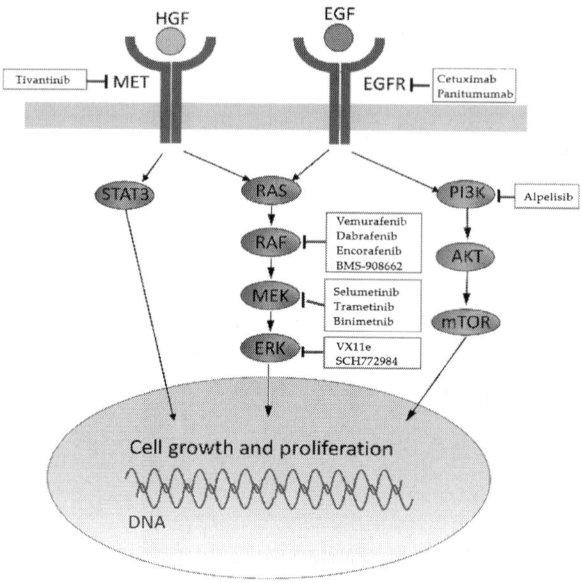

Figure 1: Current strategies for targeting the epidermal growth factor receptor (EGFR) pathway. EGF receptors (Blue) activate survival signaling pathways including RAS/MAPK and PI3K/AKT signaling pathway leading to cell growth and proliferation. HGF/MET (green) signaling also promotes cell growth and proliferation through RAS/MAPK and STAT3 phosphorylation. Regular arrow: activates, Arrow ending with a straight line: inhibits.

Using BEAMing technology, which can detect somatic mutations in small amounts of ctDNA, Morelli et al. found that acquired mutations in *KRAS* were seen in 44% of patients initially characterized as having wild-type *KRAS* tumors, following the development of resistance to anti-EGFR therapy [54]. Interestingly, atypical *KRAS* codon 61 and 146 mutations were more frequent among these newly diagnosed mutations. In addition, the mutations were detectable as low-allele-frequency clones in 35% of patients with plasma mutations, suggesting that they were already present in the treatment-naïve primary tumors. The presence of low-frequency *KRAS* mutations in these patients was correlated with a poorer response to anti-EGFR treatment. These data suggest that a high-sensitivity mutation detection technique that provides quantitative information about the presence of mutant alleles in the gene of interest would allow for the selection of patients most likely to benefit from anti-EGFR treatment.

A second mechanism of resistance entails mutations in the extracellular domain (ECD) of *EGFR* that prevent binding of the drug to the receptor. Although secondary *EGFR* mutation is rarely detected after resistance to anti-EGFR therapies, Montagut and colleagues discovered an acquired *EGFR* mutation, S492R, and proved its association with acquired resistance to cetuximab in mCRC [58]. The S492R mutation is within the EGFR ectodomain, and a bulky side chain at this position could interfere with cetuximab binding. The $EGFR^{S492R}$ mutation was detected in two of ten patients who experienced disease progression after a prior response to cetuximab with chemotherapy. Interestingly, one patient with cetuximab resistance harboring the $EGFR^{S492R}$ mutation responded to treatment with panitumumab. This result was also confirmed in vitro by using NIH3T3 fibroblasts expressing wild-type or S492R-mutant *EGFR*. In cells carrying the S492R mutation, only panitumumab blocked the activation of EGFR. Overall, these findings identified the S492R mutation as a mechanism of acquired but not primary resistance to cetuximab.

Some new EGFR inhibitors appear to overcome resistance to cetuximab or panitumumab that is due to the emergence of mutations in the *EGFR* ECD. Sym004 is a novel 1:1 mixture of two non-overlapping anti-EGFR monoclonal antibodies that target different epitopes of the *EGFR* ECD. A unique feature of Sym004 is its ability to mediate rapid EGFR internalization and subsequent degradation via EGFR cross-linking [59,60,61]. Preclinical studies with Sym004 showed superior antitumor activity as compared with cetuximab or panitumumab, as well as activity in models of acquired cetuximab resistance [60,62]. In a phase I study of mCRC patients whose tumors were resistant to cetuximab or panitumumab, Sym004 demonstrated a 44% tumor shrinkage rate and a 13% response rate [59]. It is currently being tested in a dose-optimization, phase II study in refractory mCRC patients (NCT02083653), as well as in combination with FOLFIRI (NCT02568046) [63].

MM-151 is a combination of three fully human IgG1 monoclonal antibodies that can simultaneously engage distinct, non-overlapping epitopes on the EGFR [64,65]. MM-151 was designed using a systems biology approach to substantially improve on the mechanisms that enable potent EGFR antagonism, *EGFR* downregulation, and immune effector function. In a phase I study examining MM-151 as a monotherapy or in combination with irinotecan, three mCRC patients achieved partial responses, and eight had stable disease for more than 4 M on single-agent MM-151 [66]. The most common adverse events were toxicities related to the EGFR pathway, including rashes (70%), hypomagnesemia (24%), fatigue, dry skin, and diarrhea (21%). After MM-151 treatment, liquid biopsies of patients with *EGFR* ECD mutations at baseline showed decreased or stabilized *EGFR* ECD mutant DNA concentrations that paralleled response assessments using radiological methods.

Activation of the PI3K/AKT/mTOR signaling pathway has also been implicated as an important mechanism in the resistance to EGFR inhibitors. *PIK3CA* mutations occur at a prevalence of 10%–20%, are mainly located at hot spots in exons 20 and 9, and constitutively activate downstream signaling [67]. Retrospective studies have investigated the potential role of *PIK3CA* mutations as predictors of resistance to anti-EGFR treatment in mCRC patients, with conflicting results. A systematic review was conducted to evaluate the association between *PIK3CA* mutations and resistance to anti-EGFR therapy in mCRC [68]. The results showed a trend of worse PFS and a significantly shorter OS (HR = 1.43, 95% CI = 1.02–2.00) compared with patients with wild-type tumors. However, a large prospective study is needed to confirm this result.

Additional possible mechanisms of resistance to anti-EGFR therapies have been proposed in patients with *RAS* and *BRAF* wild-type mCRC [69]. Genetic aberrations of the receptor tyrosine kinases *c-MET* and *ERBB2* are identified as the bypass mechanisms for acquired resistance to anti-EGFR therapies. c-MET and its ligand HGF have been previously implicated in both primary and acquired resistance to anti-EGFR therapies. For example, among 103 patients with *KRAS* wild-type mCRC who received anti-EGFR therapies, those with high serum levels of HGF had shorter PFS (4.4 M versus 6.4 M, $p = 0.001$) and OS (8.0 M versus 15.3 months; $p = 0.001$) compared with those with low levels of HGF [70]. Another study demonstrated that c-MET overexpression significantly correlated with shorter PFS (3 M versus 5 M, $p = 0.018$) and OS (11 M versus 10 M, $p = 0.037$) compared with low/normal expression [71]. Whole-exome analysis revealed that the c-MET copy number was increased in CRC tissues from patients who developed resistance to anti-EGFR therapies [72]. Moreover, the c-MET amplification locus could be detected in circulating tumor DNA, suggesting that the early initiation of c-MET inhibitors in those patients who respond to anti-EGFR therapies and do not display emergence of *KRAS* mutations in blood tests during anti-EGFR therapies [72].

ERBB2 gene amplifications were also described as drivers of acquired resistance to anti-EGFR therapies in *RAS* and *BRAF* wild-type mCRC. Yonesaka et al. showed that activation of HER2 signaling in cancer cells, through either *HER2* gene amplification or upregulation of the HER2/HER3 ligand heregulin, is enabled to acquire the resistance to anti-EGFR therapies [73]. Large xenograft cohorts from patient-derived demonstrated that ErbB-2 amplification was shown in cetuximab-resistant patients and in *RAS/BRAF/PI3KCA* wild-type patients [74]. Moreover, HER2 amplification was detected in ctDNA from patients with mCRC and acquired resistance to anti-EGFR antibody therapy [75,76]. These data suggest that combination therapies against EGFR and ERBB2 in *RAS* wild-type tumors might be useful.

BRAF MUTATION IN COLORECTAL CANCER

BRAF is a serine/threonine protein kinase that plays an important role in the *EGFR*-mediated *MAPK* pathway [77], where it is activated by the small GTPase, *RAS*. The *BRAF* gene is somatically mutated in approximately 5% to 10% of mCRC [78,79]. The most common *BRAF* mutation in CRC is the V600E alteration, which causes a constitutive activation of the MAPK pathway, independent of RAS activity [80]. In addition, *BRAF* mutations are mutually exclusive with *RAS* mutations [81]. The $BRAF^{V600E}$ mutation is associated with proximal location, greater age, female gender, microsatellite instability (MSI), CIMP, high grade, and mucinous histology in colorectal cancer [81]. *BRAF* mutations have also been shown to be associated with poor prognosis in mCRC. Moreover, *BRAF* mutations also appear to have predictive value, with increasing evidence that *BRAF* mutations predict a lack of benefit from anti-EGFR therapy, even in wild-type *RAS* mCRC. Currently, consensus-based guidelines from NCCN [18] and ESMO [19] both recommend avoiding the use of cetuximab or panitumumab for patients with *BRAF*-mutated cancers.

COMBINATION THERAPIES WITH BRAF INHIBITORS

BRAF mutations are commonly identified in melanoma, in which the *BRAF* mutation rate is more than 60%, offering hope that inhibition of BRAF kinase activity could benefit melanoma patients. The BRAF inhibitors, vemurafenib and dabrafenib, have shown response rates of up to approximately 50% in metastatic melanoma patients with the *BRAF*^{V600E} mutation [82,83]. However, BRAF inhibitor monotherapies appear to be ineffective in *BRAF*^{V600E} mutant mCRC, with a response rate of approximately 5% [79]. Accumulating evidence suggests that these BRAF inhibitors induce negative feedback reactivation of EGFR, which in turn activates MAPK via CRAF and RAS [84,85,86].

Preclinical studies indicated that combinations of BRAF inhibitors with inhibitors of upstream receptors show an improved efficacy in *BRAF*-mutated mCRC [87]. Based on these data, BRAF inhibitor combinations have been evaluated in clinical trials for patients with *BRAF*-mutated mCRC in recent years, and are showing signs of improved efficacy compared with RAF inhibition alone (Table 1) [85,88,89,90]. Kopetz and colleagues presented the results of a phase II/III trial in which they randomly assigned 49 patients with *BRAF*-mutated mCRC to receive irinotecan, cetuximab, and vemurafenib and 50 to receive only irinotecan and cetuximab [91]. Patients receiving the three-drug regimen showed significantly longer PFS compared with those receiving the doublet regimen (4.4 M versus 2.0 M, HR = 0.42, 95% CI = 0.26–0.66, $p = 0.0002$). Notably, Grade 3 or 4 toxicities in patients in the three-drug group included neutropenia (28%), diarrhea (22%), anemia (13%), and nausea (15%). However, these are similar to a prior second line study of cetuximab plus irinotecan. These results for combination therapy including a BRAF inhibitor are consistent with earlier preclinical studies.

Table 1: Clinical trials of drugs targeting EGFR downstream pathway

Strategy	Treatment	Genomic Profile	Phase	N	ORR (%)	PFS (M)	OS (M)	Clinical Development	Reference
BRAF monotherapy	Vemurafenib	BRAFmut	I/II	21	5	2.1	7.7	Complete	NCT00405587 [93]
BRAF monotherapy	Vemurafenib	BRAFmut	II	10	0	4.5	9.3	Complete	NCT01524978 [94]
BRAF + MEK	Dabrafenib + Trametinib	BRAFmut	I/II	43	12	3.5	(–)	Complete	NCT01726738 [88]
BRAF + EGFR	Vemurafenib + Cetuximab	BRAFmut	II	27	4	3.7	7.1	Complete	NCT01524978 [94]
BRAF + EGFR	Vemurafenib + Panitumumab	BRAFmut	Pilot	15	13	3.2	7.6	Complete	NCT01791309 [89]
BRAF + EGFR	Encorafenib + Cetuximab	BRAFmut	II	50	22	4.2	12.4	Ongoing	NCT01719380 [90]
BRAF + EGFR	Dabrafenib + Panitumumab	BRAFmut	I/II	20	10	3.4	(–)	Ongoing	NCT01750918 [85]

BRAF + EGFR	BMS-908662 + Cetuximab	Kras mut or BRAF mut	I/II	17 *	(−)	(−)	(−)	Ongoing	NCT01086267
BRAF + EGFR + CT	Vemurafenib + Cetuximab + irinotecan	BRAFmut	I	12	50	(−)	(−)	Ongoing	NCT01787500
BRAF + EGFR + CT	Cetuximab + irinotecan ± Vemurafenib	BRAFmut	II (RCT)	78 *	(−)	(−)	(−)	Ongoing	NCT02164916
BRAF + EGFR + MEK	Dabrafenib + Panitumumab + Trametinib	BRAFmut	I/II	83	18	(−)	(−)	Ongoing	NCT01750918 [85]
BRAF + EGFR ± MEK	Encorafenib + Cetuximab ± Binimetinib	None	III	645 *	(−)	(−)	(−)	Ongoing	NCT02928224
BRAF + EGFR + PI3K	Encorafenib + Cetuximab + Alpelisib	BRAFmut	II	52	27	5.4	13.1	Ongoing	NCT01719380 [90]
MEK + CT	Selumetinib + irinotecan	Kras mut or BRAF mut	II	32	9	(−)	(−)	Complete	NCT01116271 [95]
MEK + EGFR	Binimetinib + Panitumumab	RAS wt or RAS mut	I/II	90 *	(−)	(−)	(−)	Ongoing	NCT01927341
MEK + EGFR	Panitumumab + Trametinib	BRAFmut	I/II	31	0	2.8		Ongoing	NCT01750918 [85]
MEK + EGFR	Trametinib + Panitumumab	KRAS BRAF wt	II	26 *	(−)	(−)	(−)	Ongoing	NCT02399943
MET + EGFR + CT	Cetuximab + irinotecan ± Tivantinib	KRAS wt	I/II (RCT)	60	45	8.3	19.8	Complete	NCT01075048 [96]
MEK + AKT	Selumetinib + MK-2206	KRAS wt or KRAS mut	II	21 *	(−)	(−)	(−)	Ongoing	NCT01333475

ORR: objective response rate; PFS: progression free survival; OS: overall survival; M: months; CT: chemotherapy; RCT: randomized control trial. * Planed sample size, (-): not reported.

Despite combination therapy initially preventing the development of acquired resistance in patients with *BRAF*-mutated mCRC, these patients ultimately develop secondary resistance to treatment and experience disease progression. Alterations in MAPK pathway genes, including *KRAS* amplification, *BRAF* amplification, and a *MEK1* mutation, were found in resistant tumors and were not present in matched pretreatment tumors [92]. Importantly, an ERK inhibitor retained the ability to suppress the MAPK pathway and could overcome each of the identified acquired resistance mechanisms in *BRAF*-mutated mCRC [92]. These results suggest that ERK inhibitors could be important components of future therapeutic strategies for *BRAF*-mutated mCRC, either alone or in combination with BRAF and EGFR inhibitors.

CONCLUSIONS

EGFR-targeting drugs directly affect cancer cells and may be affected by accumulating drug resistance mutations in the RAS-RAF-MEK-ERK pathway. Several clinical trials are currently underway to elucidate which treatment has the best outcome. We believe that liquid biopsy and next-generation sequencing technology will help to identify and monitor predictive biomarkers for acquired resistance to anti-EGFR therapies. One can hope that a comprehensive understanding of resistance mechanisms will contribute to developing more effective strategies to overcome acquired resistance in the future.

ACKNOWLEDGMENTS

The authors would like to thank Edanz Editing, Minato-Ku Fukuoka City, Fukuoka, Japan for reviewing the article for language and grammar errors.

AUTHOR CONTRIBUTIONS

Yuji Miyamoto designed and drafted the manuscript; Koichi Suyama and Hideo Baba discussed and revised the manuscript.

REFERENCES

1. Saltz, L.B.; Cox, J.V.; Blanke, C.; Rosen, L.S.; Fehrenbacher, L.; Moore, M.J.; Maroun, J.A.; Ackland, S.P.; Locker, P.K.; Pirotta, N.; et al. Irinotecan plus fluorouracil and leucovorin for metastatic colorectal cancer. Irinotecan Study Group. *N. Engl. J. Med.* 2000, *343*, 905–914.
2. De Gramont, A.; Figer, A.; Seymour, M.; Homerin, M.; Hmissi, A.; Cassidy, J.; Boni, C.; Cortes-Funes, H.; Cervantes, A.; Freyer, G.; et al. Leucovorin and fluorouracil with or without oxaliplatin as first-line treatment in advanced colorectal cancer. *J. Clin. Oncol.* 2000, *18*, 2938–2947.
3. Cassidy, J.; Clarke, S.; Diaz-Rubio, E.; Scheithauer, W.; Figer, A.; Wong, R.; Koski, S.; Lichinitser, M.; Yang, T.-S.; Rivera, F.; et al. Randomized phase III study of capecitabine plus oxaliplatin compared with fluorouracil/folinic acid plus oxaliplatin as first-line therapy for metastatic colorectal cancer. *J. Clin. Oncol.* 2008, *26*, 2006–2012.
4. Hurwitz, H.; Fehrenbacher, L.; Novotny, W.; Cartwright, T.; Hainsworth, J.; Heim, W.; Berlin, J.; Baron, A.; Griffing, S.; Holmgren, E.; et al. Bevacizumab plus irinotecan, fluorouracil, and leucovorin for metastatic colorectal cancer. *N. Engl. J. Med.* 2004, *350*, 2335–2342.
5. Saltz, L.B.; Clarke, S.; Díaz-Rubio, E.; Scheithauer, W.; Figer, A.; Wong, R.; Koski, S.; Lichinitser, M.; Yang, T.-S.S.; Rivera, F.; et al. Bevacizumab in combination with oxaliplatin-based chemotherapy as first-line therapy in metastatic colorectal cancer: A randomized phase III study. *J. Clin. Oncol.* 2008, *26*, 2013–2019.
6. Van Cutsem, E.; Tabernero, J.; Lakomy, R.; Prenen, H.; Prausova, J.; Macarulla, T.; Ruff, P.; van Hazel, G.A.; Moiseyenko, V.; Ferry, D.; et al. Addition of aflibercept to fluorouracil, leucovorin, and irinotecan improves survival in a phase III randomized trial in patients with metastatic colorectal cancer previously treated with an oxaliplatin-based regimen. *J. Clin. Oncol.* 2012, *30*, 3499–3506.
7. Tabernero, J.; Yoshino, T.; Cohn, A.L.; Obermannova, R.; Bodoky, G.; Garcia-Carbonero, R.; Ciuleanu, T.-E.E.; Portnoy, D.C.; van Cutsem, E.; Grothey, A.; et al. Ramucirumab versus placebo in combination with second-line FOLFIRI in patients with metastatic colorectal carcinoma that progressed during or after first-line therapy with bevacizumab, oxaliplatin, and a fluoropyrimidine (RAISE): A randomised, double-

blin. *Lancet Oncol.* 2015, *16*, 499–508.

8. Van Cutsem, E.; Kohne, C.H.; Hitre, E.; Zaluski, J.; Chang Chien, C.-R.; Makhson, A.; D'Haens, G.; Pinter, T.; Lim, R.; Bodoky, G.; et al. Cetuximab and chemotherapy as initial treatment for metastatic colorectal cancer. *N. Engl. J. Med.* 2009, *360*, 1408–1417.
9. Van Cutsem, E.; Lenz, H.J.; Kohne, C.H.; Heinemann, V.; Tejpar, S.; Melezinek, I.; Beier, F.; Stroh, C.; Rougier, P.; van Krieken, J.H.; et al. Fluorouracil, leucovorin, and irinotecan plus cetuximab treatment and RAS mutations in colorectal cancer. *J. Clin. Oncol.* 2015, *33*, 692–700.
10. Douillard, J.Y.; Siena, S.; Cassidy, J.; Tabernero, J.; Burkes, R.; Barugel, M.; Humblet, Y.; Bodoky, G.; Cunningham, D.; Jassem, J.; et al. Randomized, phase III trial of panitumumab with infusional fluorouracil, leucovorin, and oxaliplatin (FOLFOX4) versus FOLFOX4 alone as first-line treatment in patients with previously untreated metastatic colorectal cancer: The PRIME study. *J. Clin. Oncol.* 2010, *28*, 4697–4705.
11. Douillard, J.Y.; Oliner, K.S.; Siena, S.; Tabernero, J.; Burkes, R.; Barugel, M.; Humblet, Y.; Bodoky, G.; Cunningham, D.; Jassem, J.; et al. Panitumumab-FOLFOX4 treatment and RAS mutations in colorectal cancer. *N. Engl. J. Med.* 2013, *369*, 1023–1034.
12. Grothey, A.; van Cutsem, E.; Sobrero, A.; Siena, S.; Falcone, A.; Ychou, M.; Humblet, Y.; Bouché, O.; Mineur, L.; Barone, C.; et al. Regorafenib monotherapy for previously treated metastatic colorectal cancer (CORRECT): An international, multicentre, randomised, placebo-controlled, phase 3 trial. *Lancet* 2013, *381*, 303–312.
13. Mayer, R.J.; van Cutsem, E.; Falcone, A.; Yoshino, T.; Garcia-Carbonero, R.; Mizunuma, N.; Yamazaki, K.; Shimada, Y.; Tabernero, J.; Komatsu, Y.; et al. Randomized trial of TAS-102 for refractory metastatic colorectal cancer. *N. Engl. J. Med.* 2015, *372*, 1909–1919.
14. Heinemann, V.; von Weikersthal, L.F.; Decker, T.; Kiani, A.; Vehling-kaiser, U.; Scholz, M.; Müller, S.; Link, H.; Niederle, N.; Rost, A.; et al. FOLFIRI plus cetuximab versus FOLFIRI plus bevacizumab as first-line treatment for patients with metastatic colorectal cancer (FIRE-3): A randomised, open-label, phase 3 trial. *Lancet Oncol.* 2014, *15*, 1065–1075.
15. Yamada, Y.; Takahari, D.; Matsumoto, H.; Baba, H.; Nakamura, M.; Yoshida, K.; Yoshida, M.; Iwamoto, S.; Shimada, K.; Komatsu, Y.; et al. Leucovorin, fluorouracil, and oxaliplatin plus bevacizumab versus

S-1 and oxaliplatin plus bevacizumab in patients with metastatic colorectal cancer (SOFT): An open-label, non-inferiority, randomised phase 3 trial. *Lancet Oncol.* 2013, *14*, 1278–1286.

16. Loupakis, F.; Cremolini, C.; Masi, G.; Lonardi, S.; Zagonel, V.; Salvatore, L.; Cortesi, E.; Tomasello, G.; Ronzoni, M.; Spadi, R.; et al. Initial therapy with FOLFOXIRI and bevacizumab for metastatic colorectal cancer. *N. Engl. J. Med.* 2014, *371*, 1609–1618.
17. Waring, P.; Tie, J.; Maru, D.; Karapetis, C.S. RAS mutations as predictive biomarkers in clinical management of metastatic colorectal cancer. *Clin. Colorectal Cancer* 2016, *15*, 95–103.
18. National Comprehensive Cancer Network (NCCN). *Guidelines Version 1. 2016 Colon Cancer*; National Comprehensive Cancer Network: Washington, PA, USA, 2016.
19. Van Cutsem, E.; Cervantes, A.; Adam, R.; Sobrero, A.; van Krieken, J.H.; Aderka, D.; Aranda Aguilar, E.; Bardelli, A.; Benson, A.; Bodoky, G.; et al. ESMO consensus guidelines for the management of patients with metastatic colorectal cancer. *Ann. Oncol.* 2016, *27*, 1386–1422.
20. Peeters, M.; Price, T.J.; Cervantes, A.; Sobrero, A.F.; Ducreux, M.; Hotko, Y.; Andre, T.; Chan, E.; Lordick, F.; Punt, C.J.A.; et al. Randomized phase III study of panitumumab with fluorouracil, leucovorin, and irinotecan (FOLFIRI) compared with FOLFIRI alone as second-line treatment in patients with metastatic colorectal cancer. *J. Clin. Oncol.* 2010, *28*, 4706–4713.
21. Bokemeyer, C.; Bondarenko, I.; Makhson, A.; Hartmann, J.T.; Aparicio, J.; De Braud, F.; Donea, S.; Ludwig, H.; Schuch, G.; Stroh, C.; et al. Fluorouracil, leucovorin, and oxaliplatin with and without cetuximab in the first-line treatment of metastatic colorectal cancer. *J. Clin. Oncol.* 2009, *27*, 663–671.
22. Snyder, L.C.; Astsaturov, I.; Weiner, L.M. Overview of monoclonal antibodies and small molecules targeting the epidermal growth factor receptor pathway in colorectal cancer. *Clin. Colorectal Cancer* 2005, *5*, S71–S80.
23. Seo, Y.; Ishii, Y.; Ochiai, H.; Fukuda, K.; Akimoto, S.; Hayashida, T.; Okabayashi, K.; Tsuruta, M.; Hasegawa, H.; Kitagawa, Y. Cetuximab-mediated ADCC activity is correlated with the cell surface expression level of EGFR but not with the KRAS/BRAF mutational status in colorectal cancer. *Oncol. Rep.* 2014, *31*, 2115–2122.

24. Price, T.J.; Peeters, M.; Kim, T.W.; Li, J.; Cascinu, S.; Ruff, P.; Suresh, A.S.; Thomas, A.; Tjulandin, S.; Zhang, K.; et al. Panitumumab versus cetuximab in patients with chemotherapy-refractory wild-type KRAS exon 2 metastatic colorectal cancer (ASPECCT): A randomised, multicentre, open-label, non-inferiority phase 3 study. *Lancet Oncol.* 2014, *15*, 569–579.
25. Antonacopoulou, A.G.; Tsamandas, A.C.; Petsas, T.; Liava, A.; Scopa, C.D.; Papavassiliou, A.G.; Kalofonos, H.P. EGFR, HER-2 and COX-2 levels in colorectal cancer. *Histopathology* 2008, *53*, 698–706.
26. McKay, J.A.; Murray, L.J.; Curran, S.; Ross, V.G.; Clark, C.; Murray, G.I.; Cassidy, J.; McLeod, H.L. Evaluation of the epidermal growth factor receptor (EGFR) in colorectal tumours and lymph node metastases. *Eur. J. Cancer* 2002, *38*, 2258–2264.
27. Saltz, L.B.; Meropol, N.J.; Loehrer, P.J.S.; Needle, M.N.; Kopit, J.; Mayer, R.J. Phase II trial of cetuximab in patients with refractory colorectal cancer that expresses the epidermal growth factor receptor. *J. Clin. Oncol.* 2004, *22*, 1201–1208.
28. Hecht, J.R.; Mitchell, E.; Neubauer, M.A.; Burris, H.A., 3rd; Swanson, P.; Lopez, T.; Buchanan, G.; Reiner, M.; Gansert, J.; Berlin, J. Lack of correlation between epidermal growth factor receptor status and response to Panitumumab monotherapy in metastatic colorectal cancer. *Clin. Cancer Res.* 2010, *16*, 2205–2213.
29. Cunningham, D.; Humblet, Y.; Siena, S.; Khayat, D.; Bleiberg, H.; Santoro, A.; Bets, D.; Mueser, M.; Harstrick, A.; Verslype, C.; et al. Cetuximab monotherapy and cetuximab plus irinotecan in irinotecan-refractory metastatic colorectal cancer. *N. Engl. J. Med.* 2004, *351*, 337–345.
30. Graham, D.M.; Coyle, V.M.; Kennedy, R.D.; Wilson, R.H. Molecular subtypes and personalized therapy in metastatic colorectal cancer. *Curr. Colorectal Cancer Rep.* 2016, *12*, 141–150.
31. Amado, R.G.; Wolf, M.; Peeters, M.; Van Cutsem, E.; Siena, S.; Freeman, D.J.; Juan, T.; Sikorski, R.; Suggs, S.; Radinsky, R.; et al. Wild-type KRAS is required for panitumumab efficacy in patients with metastatic colorectal cancer. *J. Clin. Oncol.* 2008, *26*, 1626–1634.
32. Peeters, M.; Kafatos, G.; Taylor, A.; Gastanaga, V.M.; Oliner, K.S.; Hechmati, G.; Terwey, J.-H.; van Krieken, J.H. Prevalence of RAS mutations and individual variation patterns among patients with metastatic colorectal cancer: A pooled analysis of randomised

controlled trials. *Eur. J. Cancer* 2015, *51*, 1704–1713.

33. Han, C.-B.; Li, F.; Ma, J.-T.; Zou, H.-W. Concordant KRAS mutations in primary and metastatic colorectal cancer tissue specimens: A meta-analysis and systematic review. *Cancer Investig.* 2012, *30*, 741–747.

34. Loupakis, F.; Ruzzo, A.; Cremolini, C.; Vincenzi, B.; Salvatore, L.; Santini, D.; Masi, G.; Stasi, I.; Canestrari, E.; Rulli, E.; et al. KRAS codon 61, 146 and BRAF mutations predict resistance to cetuximab plus irinotecan in KRAS codon 12 and 13 wild-type metastatic colorectal cancer. *Br. J. Cancer* 2009, *101*, 715–721.

35. Sorich, M.J.; Wiese, M.D.; Rowland, A.; Kichenadasse, G.; McKinnon, R.A.; Karapetis, C.S. Extended RAS mutations and anti-EGFR monoclonal antibody survival benefit in metastatic colorectal cancer: A meta-analysis of randomized, controlled trials. *Ann. Oncol.* 2015, *26*, 13–21.

36. Van Cutsem, E.; Kohne, C.H.; Lang, I.; Folprecht, G.; Nowacki, M.P.; Cascinu, S.; Shchepotin, I.; Maurel, J.; Cunningham, D.; Tejpar, S.; et al. Cetuximab plus irinotecan, fluorouracil, and leucovorin as first-line treatment for metastatic colorectal cancer: Updated analysis of overall survival according to tumor KRAS and BRAF mutation status. *J. Clin. Oncol.* 2011, *29*, 2011–2019.

37. Bokemeyer, C.; Bondarenko, I.; Hartmann, J.T.; de Braud, F.; Schuch, G.; Zubel, A.; Celik, I.; Schlichting, M.; Koralewski, P. Efficacy according to biomarker status of cetuximab plus FOLFOX-4 as first-line treatment for metastatic colorectal cancer: The OPUS study. *Ann. Oncol. Off. J. Eur. Soc. Med. Oncol.* 2011, *22*, 1535–1546.

38. Bokemeyer, C.; van Cutsem, E.; Rougier, P.; Ciardiello, F.; Heeger, S.; Schlichting, M.; Celik, I.; Köhne, C.H. Addition of cetuximab to chemotherapy as first-line treatment for KRAS wild-type metastatic colorectal cancer: Pooled analysis of the CRYSTAL and OPUS randomised clinical trials. *Eur. J. Cancer* 2012, *48*, 1466–1475.

39. De Roock, W.; Jonker, D.J.; di Nicolantonio, F.; Sartore-Bianchi, A.; Tu, D.; Siena, S.; Lamba, S.; Arena, S.; Frattini, M.; Piessevaux, H.; et al. Association of KRAS p.G13D mutation with outcome in patients with chemotherapy-refractory metastatic colorectal cancer treated with cetuximab. *JAMA* 2010, *304*, 1812–1820.

40. Tejpar, S.; Celik, I.; Schlichting, M.; Sartorius, U.; Bokemeyer, C.; van Cutsem, E. Association of KRAS G13D tumor mutations with outcome in patients with metastatic colorectal cancer treated with first-

line chemotherapy with or without cetuximab. *J. Clin. Oncol.* 2012, *30*, 3570–3577.

41. Segelov, E.; Thavaneswaran, S.; Waring, P.M.; Desai, J.; Robledo, K.P.; Gebski, V.J.; Elez, E.; Nott, L.M.; Karapetis, C.S.; Lunke, S.; et al. Response to cetuximab with or without irinotecan in patients with refractory metastatic colorectal cancer harboring the KRAS G13D mutation: Australasian gastro-intestinal trials group ICECREAM study. *J. Clin. Oncol.* 2016, *34*, 2258–2264.
42. Lenz, H.J.; Niedzwiecki, D.; Innocenti, F.; Blanke, C.; Mahony, M.R.; O'Neil, B.H.; Shaw, J.E.; Polite, B.; Hochster, H.; Atkins, J.; et al. CALGB/SWOG 80405: Phase III trial of FOLFIRI or mFOLFOX6 with bevacizumab or cetuximab for patients with expanded RAS analyses in untreated metastatic adenocarcinoma of the colon or rectum. *Ann. Oncol.* 2014, *32*, 5.
43. Schwartzberg, L.S.; Rivera, F.; Karthaus, M.; Fasola, G.; Canon, J.L.; Hecht, J.R.; Yu, H.; Oliner, K.S.; Go, W.Y. PEAK: A randomized, multicenter phase II study of panitumumab plus modified fluorouracil, leucovorin, and oxaliplatin (mFOLFOX6) or bevacizumab plus mFOLFOX6 in patients with previously untreated, unresectable, wild-type KRAS exon 2 metastatic colorectal. *J. Clin. Oncol.* 2014, *32*, 2240–2247.
44. Heinemann, V.; Rivera, F.; O'Neil, B.H.; Stintzing, S.; Koukakis, R.; Terwey, J.H.H.; Douillard, J.Y.Y. A study-level meta-analysis of efficacy data from head-to-head first-line trials of epidermal growth factor receptor inhibitors versus bevacizumab in patients with RAS wild-type metastatic colorectal cancer. *Eur. J. Cancer* 2016, *67*, 11–20.
45. Yoshino, T.; Uetake, H.; Tsuchihara, K.; Shitara, K.; Yamazaki, K.; Oki, E.; Sato, T.; Naitoh, T.; Komatsu, Y.; Takeshi, K.; et al. PARADIGM study: A multicenter, randomized, phase III study of 5-fluorouracil, leucovorin, and oxaliplatin (mFOLFOX6) plus panitumumab or bevacizumab as first-line treatment in patients with RAS (KRAS/NRAS) wild-type metastatic colorectal cancer. *J. Clin. Oncol.* 2016, *34*, TPS776.
46. Papagiorgis, P.C. Segmental distribution of some common molecular markers for colorectal cancer (CRC): Influencing factors and potential implications. *Tumor Biol.* 2016, *37*, 5727–5734.
47. Shen, H. Different treatment strategies and molecular features between right-sided and left-sided colon cancers. *World J.*

Gastroenterol. 2015, *21*, 6470.

48. Ciombor, K.K.; Goldberg, R.M. Primary tumor sidedness as prognostic and predictive biomarker in metastatic colorectal cancer. *JAMA Oncol.* 2016.
49. Hamada, H.; Meno, C.; Watanabe, D.; Saijoh, Y. Establishment of vertebrate left-right asymmetry. *Nat. Rev. Genet.* 2002, *3*, 103–113.
50. Tejpar, S.; Stintzing, S.; Ciardiello, F.; Tabernero, J.; Van Cutsem, E.; Beier, F.; Esser, R.; Lenz, H.-J.; Heinemann, V. Prognostic and predictive relevance of primary tumor location in patients with ras wild-type metastatic colorectal cancer retrospective analyses of the CRYSTAL and FIRE-3 Trials. *JAMA Oncol.* 2016.
51. Petrelli, F.; Tomasello, G.; Borgonovo, K.; Ghidini, M.; Turati, L.; Dallera, P.; Passalacqua, R.; Sgroi, G.; Barni, S. Prognostic survival associated with left-sided vs. right-sided colon cancer. *JAMA Oncol.* 2016.
52. Lee, M.; Advani, S.; Morris, J.; Jiang, Z.; Manyam, G.; Menter, D.; Broom, B.; Eng, C.; Overman, M.; Maru, D.; et al. Association of primary (1°) site and molecular features with progression-free survival (PFS) and overall survival (OS) of metastatic colorectal cancer (mCRC) after anti-epidermal growth factor receptor (αEGFR) therapy. *J. Clin. Oncol.* 2016, *34*, 3506.
53. Ciardiello, F.; Tortora, G.; Magrassi, S.F.; Lanzara, A. EGFR antagonists in cancer treatment. *N. Engl. J. Med.* 2008, *358*, 1160–1174.
54. Morelli, M.P.; Overman, M.J.; Dasari, A.; Kazmi, S.M.A.; Mazard, T.; Vilar, E.; Morris, V.K.; Lee, M.S.; Herron, D.; Eng, C.; et al. Characterizing the patterns of clonal selection in circulating tumor DNA from patients with colorectal cancer refractory to anti-EGFR treatment. *Ann. Oncol.* 2015, *26*, 731–736.
55. Diaz, L.A., Jr.; Williams, R.T.; Wu, J.; Kinde, I.; Hecht, J.R.; Berlin, J.; Allen, B.; Bozic, I.; Reiter, J.G.; Nowak, M.A.; et al. The molecular evolution of acquired resistance to targeted EGFR blockade in colorectal cancers. *Nature* 2012, *486*, 4–7.
56. Taly, V.; Pekin, D.; Benhaim, L.; Kotsopoulos, S.K.; Le Corre, D.; Li, X.; Atochin, I.; Link, D.R.; Griffiths, A.D.; Pallier, K.; et al. Multiplex picodroplet digital PCR to detect KRAS mutations in circulating DNA from the plasma of colorectal cancer patients. *Clin. Chem.* 2013, *59*, 1722–1731.

57. Bettegowda, C.; Sausen, M.; Leary, R.J.; Kinde, I.; Wang, Y.; Agrawal, N.; Bartlett, B.R.; Wang, H.; Luber, B.; Alani, R.M.; et al. Detection of circulating tumor DNA in early- and late-stage human malignancies. *Sci. Transl. Med.* 2014, *6*, 224ra24.

58. Montagut, C.; Dalmases, A.; Bellosillo, B.; Crespo, M.; Pairet, S.; Iglesias, M.; Salido, M.; Gallen, M.; Marsters, S.; Tsai, S.P.; et al. Identification of a mutation in the extracellular domain of the epidermal growth factor receptor conferring cetuximab resistance in colorectal cancer. *Nat. Med.* 2012, *18*, 221–223.

59. Dienstmann, R.; Patnaik, A.; Garcia-Carbonero, R.; Cervantes, A.; Benavent, M.; Rosello, S.; Tops, B.B.J.; van der Post, R.S.; Argiles, G.; Skartved, N.J.; et al. Safety and activity of the first-in-class Sym004 anti-EGFR antibody mixture in patients with refractory colorectal cancer. *Cancer Discov.* 2015, *5*, 598–609.

60. Pedersen, M.W.; Jacobsen, H.J.; Koefoed, K.; Hey, A.; Pyke, C.; Haurum, J.S.; Kragh, M. Sym004: A novel synergistic anti-epidermal growth factor receptor antibody mixture with superior anticancer efficacy. *Cancer Res.* 2010, *70*, 588–597.

61. Koefoed, K.; Steinaa, L.; Soderberg, J.N.; Kjaer, I.; Jacobsen, H.J.; Meijer, P.-J.; Haurum, J.S.; Jensen, A.; Kragh, M.; Andersen, P.S.; et al. Rational identification of an optimal antibody mixture for targeting the epidermal growth factor receptor. *MAbs* 2011, *3*, 584–595.

62. Iida, M.; Brand, T.M.; Starr, M.M.; Li, C.; Huppert, E.J.; Luthar, N.; Pedersen, M.W.; Horak, I.D.; Kragh, M.; Wheeler, D.L. Sym004, a novel EGFR antibody mixture, can overcome acquired resistance to cetuximab. *Neoplasia* 2013, *15*, 1196–1206.

63. Hanna, D.L.; Lenz, H.-J. Novel therapeutics in metastatic colorectal cancer: Molecular insights and pharmacogenomic implications. *Expert Rev. Clin. Pharmacol.* 2016, *2433*, 1–18.

64. Kearns, J.D.; Bukhalid, R.; Sevecka, M.; Tan, G.; Gerami-Moayed, N.; Werner, S.L.; Kohli, N.; Burenkova, O.; Sloss, C.M.; King, A.M.; et al. Enhanced targeting of the EGFR network with MM-151, an oligoclonal anti-EGFR antibody therapeutic. *Mol. Cancer Ther.* 2015, *14*, 1625–1636.

65. Arena, S.; Siravegna, G.; Mussolin, B.; Kearns, J.D.; Wolf, B.B.; Misale, S.; Lazzari, L.; Bertotti, A.; Trusolino, L.; Adjei, A.A.; et al. MM-151 overcomes acquired resistance to cetuximab and panitumumab in colorectal cancers harboring EGFR extracellular

domain mutations. *Sci. Transl. Med.* 2016, *8*, 324ra14.

66. Lieu, C.; Harb, W.; Beeram, M.; Power, P.; Kearns, J.; Nering, R.; Moyo, V.; Wolf, B.; Adjei, A. Phase 1 trial of MM-151, a novel oligoclonal anti-EGFR antibody combination in patients with refractory solid tumors. *J. Clin. Oncol.* 2014, *32*, 2518.
67. Martincorena, I.; Campbell, P.J. Somatic mutation in cancer and normal cells. *Science* 2015, *349*, 1483–1489.
68. Therkildsen, C.; Bergmann, T.K.; Henrichsen-Schnack, T.; Ladelund, S.; Nilbert, M. The predictive value of KRAS, NRAS, BRAF, PIK3CA and PTEN for anti-EGFR treatment in metastatic colorectal cancer: A systematic review and meta-analysis. *Acta Oncol.* 2014, *53*, 852–864.
69. Misale, S.; di Nicolantonio, F.; Sartore-Bianchi, A.; Siena, S.; Bardelli, A. Resistance to anti-EGFR therapy in colorectal cancer: From heterogeneity to convergent evolution. *Cancer Discov.* 2014, *4*, 1269–1280.
70. Takahashi, N.; Yamada, Y.; Furuta, K.; Honma, Y.; Iwasa, S.; Takashima, A.; Kato, K.; Hamaguchi, T.; Shimada, Y. Serum levels of hepatocyte growth factor and epiregulin are associated with the prognosis on anti-EGFR antibody treatment in KRAS wild-type metastatic colorectal cancer. *Br. J. Cancer* 2014, *110*, 2716–2727.
71. Inno, A.; di Salvatore, M.; Cenci, T.; Martini, M.; Orlandi, A.; Strippoli, A.; Ferrara, A.M.; Bagal, C.; Cassano, A.; Larocca, L.M.; et al. Is there a role for IGF1R and c-MET pathways in resistance to cetuximab in metastatic colorectal cancer? *Clin. Colorectal Cancer* 2011, *10*, 325–332.
72. Bardelli, A.; Corso, S.; Bertotti, A.; Hobor, S.; Valtorta, E.; Siravegna, G.; Sartore-Bianchi, A.; Scala, E.; Cassingena, A.; Zecchin, D.; et al. Amplification of the MET receptor drives resistance to anti-EGFR therapies in colorectal cancer. *Cancer Discov.* 2013, *3*, 658–673.
73. Yonesaka, K.; Zejnullahu, K.; Okamoto, I.; Satoh, T.; Cappuzzo, F.; Souglakos, J.; Ercan, D.; Rogers, A.; Roncalli, M.; Takeda, M.; et al. Activation of ERBB2 signaling causes resistance to the EGFR-directed therapeutic antibody cetuximab. *Sci. Transl. Med.* 2011, *3*, 99ra86.
74. Bertotti, A.; Migliardi, G.; Galimi, F.; Sassi, F.; Torti, D.; Isella, C.; Corà, D.; di Nicolantonio, F.; Buscarino, M.; Petti, C.; et al. A molecularly annotated platform of patient- derived xenografts ("xenopatients") identifies HER2 as an effective therapeutic target in cetuximab-resistant

colorectal cancer. *Cancer Discov.* 2011, *1*, 508–523.

75. Takegawa, N.; Yonesaka, K.; Sakai, K.; Ueda, H.; Watanabe, S.; Nonagase, Y.; Okuno, T.; Takeda, M.; Maenishi, O.; Tsurutani, J.; et al. HER2 genomic amplification in circulating tumor DNA from patients with cetuximab-resistant colorectal cancer. *Oncotarget* 2016, *7*, 3453–3460.
76. Siravegna, G.; Mussolin, B.; Buscarino, M.; Corti, G.; Cassingena, A.; Crisafulli, G.; Ponzetti, A.; Cremolini, C.; Amatu, A.; Lauricella, C.; et al. Clonal evolution and resistance to EGFR blockade in the blood of colorectal cancer patients. *Nat. Med.* 2015, *21*, 795–801.
77. Davies, H.; Bignell, G.R.; Cox, C.; Stephens, P.; Edkins, S.; Clegg, S.; Teague, J.; Woffendin, H.; Garnett, M.J.; Bottomley, W.; et al. Mutations of the BRAF gene in human cancer. *Nature* 2002, *417*, 949–954.
78. Yokota, T.; Ura, T.; Shibata, N.; Takahari, D.; Shitara, K.; Nomura, M.; Kondo, C.; Mizota, A.; Utsunomiya, S.; Muro, K.; et al. BRAF mutation is a powerful prognostic factor in advanced and recurrent colorectal cancer. *Br. J. Cancer* 2011, *104*, 856–862.
79. Di Nicolantonio, F.; Martini, M.; Molinari, F.; Sartore-Bianchi, A.; Arena, S.; Saletti, P.; de Dosso, S.; Mazzucchelli, L.; Frattini, M.; Siena, S.; et al. Wild-type BRAF is required for response to panitumumab or cetuximab in metastatic colorectal cancer. *J. Clin. Oncol.* 2008, *26*, 5705–5712.
80. Samatar, A.A.; Poulikakos, P.I. Targeting RAS–ERK signalling in cancer: Promises and challenges. *Nat. Rev. Drug Discov.* 2014, *13*, 928–942.
81. Rajagopalan, H.; Bardelli, A.; Lengauer, C.; Kinzler, K.W.; Vogelstein, B.; Velculescu, V.E. Tumorigenesis: RAF/RAS oncogenes and mismatch-repair status. *Nature* 2002, *418*, 934.
82. Chapman, P.B.; Hauschild, A.; Robert, C.; Haanen, J.B.; Ascierto, P.; Larkin, J.; Dummer, R.; Garbe, C.; Testori, A.; Maio, M.; et al. Improved survival with vemurafenib in melanoma with $BRAF^{V600E}$ mutation. *N. Engl. J. Med.* 2011, *364*, 2507–2516.
83. Hauschild, A.; Grob, J.-J.; Demidov, L.V.; Jouary, T.; Gutzmer, R.; Millward, M.; Rutkowski, P.; Blank, C.U.; Miller, W.H.J.; Kaempgen, E.; et al. Dabrafenib in BRAF-mutated metastatic melanoma: A multicentre, open-label, phase 3 randomised controlled trial. *Lancet*

(London, England) 2012, *380*, 358–365.

84. Lito, P.; Pratilas, C.A.; Joseph, E.W.; Tadi, M.; Halilovic, E.; Zubrowski, M.; Huang, A.; Wong, W.L.; Callahan, M.K.; Merghoub, T.; et al. Relief of profound feedback inhibition of mitogenic signaling by RAF inhibitors attenuates their activity in $BRAF^{V600E}$ melanomas. *Cancer Cell* 2012, *22*, 668–682.
85. Corcoran, R.; André, T.; Yoshino, T.; Bendell, J.; Atreya, C.; Schellens, J.; Ducreux, M.; McRee, A.; Siena, S.; Middleton, G.; et al. Efficacy and circulating tumor DNA (ctDNA) analysis of the BRAF inhibitor dabrafenib (D), MEK inhibitor trametinib (T), and anti-EGFR antibody panitumumab (P) in patients (pts) with BRAF V600E–mutated (BRAFm) metastatic colorectal cancer (mCRC). *Ann Oncol.* 2016, *27*, 4550.
86. Corcoran, R.B.; Ebi, H.; Turke, A.B.; Coffee, E.M.; Nishino, M.; Cogdill, A.P.; Brown, R.D.; Della Pelle, P.; Dias-Santagata, D.; Hung, K.E.; et al. EGFR-mediated re-activation of MAPK signaling contributes to insensitivity of BRAF mutant colorectal cancers to RAF inhibition with vemurafenib. *Cancer Discov.* 2012, *2*, 227–235.
87. Yang, H.; Higgins, B.; Kolinsky, K.; Packman, K.; Bradley, W.D.; Lee, R.J.; Schostack, K.; Simcox, M.E.; Kopetz, S.; Heimbrook, D.; et al. Antitumor activity of BRAF inhibitor vemurafenib in preclinical models of BRAF-mutant colorectal cancer. *Cancer Res.* 2012, *72*, 779–789.
88. Corcoran, R.B.; Atreya, C.E.; Falchook, G.S.; Kwak, E.L.; Ryan, D.P.; Bendell, J.C.; Hamid, O.; Messersmith, W.A.; Daud, A.; Kurzrock, R.; et al. Combined BRAF and MEK inhibition with dabrafenib and trametinib in BRAF V600-mutant colorectal cancer. *J. Clin. Oncol.* 2015, *33*, 4023–4031.
89. Yaeger, R.; Cercek, A.; O'Reilly, E.; Reidy, D.; Kemeny, N.; Wolinsky, T.; Gollub, M.; Lacouture, M.; Rosen, N.; Vakiani, E.; et al. Pilot study of vemurafenib and panitumumab combination therapy in patients with $BRAF^{V600E}$ mutated metastatic colorectal cancer. *J. Clin. Oncol.* 2015, *33*, 611.
90. Tabernero, J.; Geel, R.; Guren, T.; Yaeger, R.; Spreafico, A.; Faris, J.; Yoshino, T.; Yamada, Y.; Kim, T.; Bendell, J.; et al. Phase 2 results: Encorafenib (ENCO) and cetuximab (CETUX) with or without alpelisib (ALP) in patients with advanced BRAF-mutant colorectal cancer (BRAFm CRC). *J. Clin. Oncol.* 2016, *34*, 3544.

91. Kopetz, S.; McDonough, S.; Morris, V.; Lenz, H.; Magliocco, A.; Atreya, C.; Diaz, L.; Allegra, C.; Wang, S.; Lieu, C.; et al. Randomized trial of irinotecan and cetuximab with or without vemurafenib in BRAF-mutant metastatic colorectal cancer (SWOG 1406). *J. Clin. Oncol.* 2017, *35* (Suppl. S4), 520.
92. Ahronian, L.G.; Sennott, E.M.; van Allen, E.M.; Wagle, N.; Kwak, E.L.; Faris, J.E.; Godfrey, J.T.; Nishimura, K.; Lynch, K.D.; Mermel, C.H.; et al. Clinical acquired resistance to RAF inhibitor combinations in BRAF-mutant colorectal cancer through MAPK pathway alterations. *Cancer Discov.* 2015, *5*, 358–367.
93. Kopetz, S.; Desai, J.; Chan, E.; Hecht, J.R.; O'Dwyer, P.J.; Maru, D.; Morris, V.; Janku, F.; Dasari, A.; Chung, W.; et al. Phase II pilot study of vemurafenib in patients with metastatic braf-mutated colorectal cancer. *J. Clin. Oncol.* 2015, *33*, 4032–4038.
94. Hyman, D.M.; Puzanov, I.; Subbiah, V.; Faris, J.E.; Chau, I.; Blay, J.-Y.; Wolf, J.; Raje, N.S.; Diamond, E.L.; Hollebecque, A.; et al. Vemurafenib in multiple nonmelanoma cancers with $BRAF^{V600}$ mutations. *N. Engl. J. Med.* 2015, *373*, 726–736.
95. Hochster, H.; Messersmith, W.; O'Neil, B.; Groshen, S.; Cohen, D.; Denlinger, C.; Gold, P.; Eckhardt, S.; Locker, G.; Ames, P.; et al. Second-line therapy of KRAS-mutated (KRASm) metastatic colorectal cancer (CRC) with the MEK inihibitor selumetinib ([SEL], AZ6244, ARRY-142886) in combination with irinotecan (IRI): An AGICC study. *J. Clin. Oncol.* 2013, *31*, 380.
96. Eng, C.; Hart, L.; Severtsev, A.; Gladkov, O.; Mueller, L.; Kopp, M.; Vladimirov, V.; Langdon, R.; Kotiv, B.; Barni, S.; et al. A randomized, placebo-controlled, phase I/II study of tivantinib (ARQ 197) in combination with cetuximab and irinotecan in patients (pts) with KRAS wild-type (WT) metastatic colorectal cancer (CRC) who had received previous front-line systemic therapy. *J. Clin. Oncol.* 2013, *31*, 3508.

INDEX

D

E

F

G

H

I

L

M